Pyridine Nucleotide-Dependent Dehydrogenases

Pyridine Nucleotide-Dependent Dehydrogenases

Proceedings of the 2nd International Symposium held at the University of Konstanz, West Germany, March 28 – April 1, 1977
FEBS Symposium No. 49

Edited by Horst Sund

W DE G

Walter de Gruyter · Berlin · New York 1977

Editor
Horst Sund, Dr. rer. nat.
Professor of Biochemistry, Fachbereich Biologie
Universität Konstanz, Konstanz, Federal Republic of Germany

CIP-Kurztitelaufnahme der Deutschen Bibliothek

Pyridine nucleotide-dependent dehydrogenases:
proceedings of the . . . internat. symposium . . . –
Berlin, New York: de Gruyter.
2. 1977. Held at the University of Konstanz, West Germany,
March 28-April 1, 1977: FEBS-Symposium No. 49. – 1. Aufl. – 1977.
ISBN 3-11-007091-X
NE: Universität <Konstanz>; Federation of European
Biochemical societies

Printing: Mercedes Druck, Berlin. – Binding: Lüderitz & Bauer, Buchgewerbe GmbH, Berlin
Printed in Germany

Opening Remarks and Preface

In September 1969, seven-and-a-half years ago, we discussed during the "First International Symposium" here in Konstanz the problems concerning the pyridine nucleotide-dependent dehydrogenases. At a meeting in 1975, some of the participants of the Konstanz symposium had the feeling that we should meet again to discuss the progress and the problems in this field.

However, I had the impression that nobody, unfortunately, was willing to do the job of the organization of another meeting. Since I had experience with our first meeting, I commenced with the organization of the second meeting. The response by many colleagues was very positive, and now we are here to see what has happened in the field in which we are engaged. I hope that somebody else will invite us for the third meeting.

Two types of symposia are common, those with precirculated papers and those without. We have chosen the former because of the advantage that most of the time may be devoted to discussion, provided of course that all participants have read the papers prior to the symposium. You have received the preprints of the lectures. Unfortunately, there was some delay in preparing the preprints, and therefore some of them reached you a little late. I hope the majority of us will be familiar with the material and that we will have fruitful discussions after an extended summary of the results given by each speaker.

In order not to restrict the discussion I suggest not to adhere to a strict timetable. It is intended to have four lectures in the morning and three after lunch, except Thursday afternoon. In addition, based on suggestions by Freddie Gutfreund, the program contains a round-table discussion on "Functional Evidence for or against Symmetrie", the problem of "Half-of-the-Sites Reactivity" which is controversial. If you wish to introduce changes you are free to do so.

It is intended to publish the lectures together with the discussion remarks, but it is not intended to record the discussions directly. The participants are requested to provide a written report of what they considered worth including in the publication.

This week's symposium is generously sponsored by the Deutsche Forschungsgemeinschaft under its program 'Internationale wissenschaftliche Fachkonferenz der Deutschen Forschungsgemeinschaft' and FEBS under its 'FEBS Symposia Program'. The support given by the University of Konstanz, where the meeting takes place, by the Gesellschaft der Freunde und Förderer der Universität Konstanz, by the Byk Gulden Lomberg Chemische Fabrik GmbH, Konstanz, and by Boehringer Mannheim, is gratefully acknowledged.

With this I conclude my opening remarks. I hope you will have pleasant and stimulating days in Konstanz.

Konstanz, March 28, 1977. H. Sund

The present volume contains the proceedings of the FEBS Symposium No. 49: "Pyridine Nucleotide-Dependent Dehydrogenases" held at the University of Konstanz, West Germany, on March 28-April 1, 1977.

I would like to acknowledge the cooperation of the staff of the Verlag Walter de Gruyter, Berlin. For her brillant assistance in all phases of organization of the Symposium, I also wish to express my gratitude to my secretary, Mrs. Silvia Lecointre (formerly Mrs. Silvia Lau).

Konstanz, July 15, 1977. H. Sund

Contents

SECTION III. CHEMICAL MECHANISM AND COENZYME BINDING
(Chairmen: J.I.Harris and M.Rossmann)

SECTION IV. STRUCTURE FUNCTION RELATIONSHIP
(Chairman: A.C.T.North)

SECTION V. KINETICS AND REGULATION
(Chairmen: K.Mosbach and C.Veeger)

List of Contributors

M.J. Adams, Laboratory of Molecular Biophysics, University of Oxford, Oxford, England

L. Andersson, Chemical Center, Biochemistry 2, The Lund Institute of Technology, Lund, Sweden

I.G. Archibald, Laboratory of Molecular Biophysics, Department of Zoology, Oxford, England

A.M.-J. Au, Department of Biochemistry, University of California, Riverside, California, USA

A. Baici, Technisch-Chemisches Laboratorium der ETH Zürich, Zürich, Switzerland

L.J. Banaszak, Department of Biological Chemistry, Washington University, St. Louis, USA

D. Becker, Institut für Biophysik und Physikalische Biochemie, Universität Regensburg, Regensburg, Germany

S.A. Bernhard, Institute of Molecular Biology, University of Oregon, Eugene, USA

J.F. Biellmann, Institut de Chimie, Université Louis Pasteur, Strasbourg, France

J.J. Birktoft, Dept. of Bioligcal Chemistry, Division of Biology and Biomedical Sciences, Washington University School of Medicine, St. Louis, USA

G. Biesecker, Scripps Clinic and Research Foundation, La Jolla, USA

G. Blankenhorn, Fachbereich Biologie, Universität Konstanz, Konstanz, Germany

A. Bohn, Fachbereich Biologie, Universität Konstanz, Konstanz, Germany

W. Boos, Fachbereich Biologie, Universität Konstanz, Konstanz, Germany

C.-I. Bränden, Department of Chemistry, Agricultural College, Uppsala, Sweden

G. Branlant, Institut de Chimie, Université Louis Pasteur, Strasbourg, France

J.H. Coates, Molecular Enzymology Laboratory, Department of Biochemistry, University of Bristol, Bristol, England

H. Dieter, Institut für Toxikologie, Universität Düsseldorf, Düsseldorf, Germany

A. Dubied, Medizinisch-chemisches Institut der Universität Bern, Bern, Switzerland

M.F. Dunn, Institute of Molecular Biology, University of Oregon, Eugene, USA

H. Dutler, Laboratorium für Organische Chemie, ETH Zürich, Zürich, Switzerland

H. Eisenberg, Polymer Department, The Weizmann Institute of Science, Rehovot, Israel

P.C. Engel, Department of Biochemistry, University of Sheffield, Sheffield, England

W. Eventoff, Department of Biological Sciences, Purdue University, West Lafayette, USA

E. Fontan, Département de Biochimie et Génétique Microbienne, Institut Pasteur, Paris, France

B.Y. Foucaud, Institut de Chimie, Université Louis Pasteur, Strasbourg, France

A. Gafni, Chemical Physics Department, The Weizmann Institute of Science, Rehovot, Israel

R.M. Garavito, Department of Biological Sciences, Purdue University, West Lafayette, USA

H. Grisebach, Biologisches Institut II, Universität Freiburg, Freiburg, Germany

H. Gutfreund, Department of Biochemistry, University of Bristol, Bristol, England

H.R. Halvorson, Edsel B. Ford Institute, Henry Ford Hospital, Detroit, USA

M.J. Hardmann, Department of Biochemistry, University of Bristol, Bristol, England

J.I. Harris, Medical Research Council Laboratory of Molecular Biology, Cambridge, England

H. Heilmann, Institut für Physiologische Chemie, Ruhr-Universität, Bochum, Germany

J.R. Helliwell, Laboratory of Molecular Biophysics, Department of Zoology, Oxford, England

P. Hemmerich, Fachbereich Biologie, Universität Konstanz, Konstanz, Germany

H.-J. Hinz, Fachbereich Biologie, Universität Regensburg, Regensburg, Germany

J.J. Holbrook, Department of Biochemistry, University of Bristol, Bristol, England

M. Hollaway, Department of Biochemistry, University College London, London, England

R. Jaenicke, Institut für Biophysik und Physikalische Biochemie, Universität Regensburg, Regensburg, Germany

R. Jeck, Gustav-Embden Zentrum der Biologischen Chemie, Universität Frankfurt, Frankfurt, Germany

R. Joppich-Kuhn, Technisch-Chemisches Institut der ETH Zürich, Zürich, Switzerland

H. Jörnvall, Department of Biological Chemistry, UCLA, Los Angeles, USA

N. Kellershohn, Laboratoire E.P.C.M., Université de Paris-Sud, Orsay,France

M. Kempfle, Physiologisch-chemisches Institut, Universität Bonn, Bonn, Germany

H.H. Kiltz, Abteilung Chemie, Ruhr-Universität, Bochum, Germany

W. Knappe, Fachbereich Biologie, Universität Konstanz, Konstanz, Germany

R. Koberstein, Fachbereich Biologie, Universität Konstanz, Konstanz, Germany

S.C. Koerber, Department of Biochemistry, University of California, Riverside, California, USA

J. Krul, Department of Biochemistry, Agricultural University, Wageningen, The Netherlands

W. Leicht, Polymer Department, The Weizmann Institute of Science, Rehovot, Israel

K.D. Lucast, Edsel B. Ford Institute, Henry Ford Hospital, Detroit, USA

P.L. Luisi, Technisch-chemisches Laboratorium der ETH, Zürich, Switzerland

O.P. Malhotra, Department of Chemistry, Banares Hindu University, India

B. Mannervik, Department of Biochemistry, University of Stockholm, Stockholm, Sweden

K. Markau, Fachbereich Biologie, Universität Konstanz, Konstanz, Germany

M. Mevarech, Polymer Department, The Weizmann Institute of Science, Rehovot, Israel

R.G. Morris, Department of Biochemistry, University of California, Riverside, California, USA

K. Mosbach, Chemical Center, Biochemistry 2, The Lund Institute of Technology, Lund, Sweden

K. Mühlegger, Boehringer-Mannheim, Biochemica-Werk Tutzing, Tutzing,Germany

A.C.T. North, Astbury Department of Biophysics, The University of Leeds, Leeds, England

A. Olomucki, Collège de France, Biochimie Cellulaire, Paris, France

H.H. Osborne, Department of Biochemistry, University College London, London, England

J. Otto, Institut für Physiologische Chemie, Physikalische Biochemie und Zellbiologie, Universität München, München, Germany

D.M. Parker, Department of Biochemistry, University of Bristol, Bristol, England

D. Peetz, Fachbereich Biologie, Universität Konstanz, Konstanz, Germany

R.N. Perham, Department of Biochemistry, University of Cambridge, Cambridge, England

M. Perucho, Max-Planck-Institut für Genetik, Berlin, Germany

O. Pfenninger, Institute of Molecular Biology, University of Oregon, Eugene, USA

G.W.E. Plaut, Department of Biochemistry, Temple University, Philadelphia, USA

I.R. Rasched, Fachbereich Biologie, Universität Konstanz, Konstanz, Germany

B.S. Reddy, Max-Planck-Institut für Experimentelle Medizin, Göttingen, Germany

M.G. Rossmann, Department of Biological Sciences, Purdue University, West Lafayette, USA

R. Rudolph, Institut für Biophysik und Physikalische Biochemie, Universität Regensburg, Regensburg, Germany

G. Saliman, Department of Biochemistry, University of California, Riverside, California, USA

W. Saenger, Max-Planck-Institut für Experimentelle Medizin, Göttingen, Germany

P. Schack, Fysisk-Kemisk Institut, The Technical University of Denmark, Lyngby, Denmark

R. Scheek, Biochemical Laboratory, B.C.P. Jansen-Institute, Amsterdam, The Netherlands

F. Schmid, Fachbereich Biologie, Universität Regensburg, Regensburg, Germany

J. Schwarzwälder, Institut für Biophysik und Physikalische Biochemie, Universität Regensburg, Regensburg, Germany

B. Schwendimann, Department of Biochemistry, University of Geneva, Geneva, Switzerland

F. Seydoux, Laboratoire d'Enzymologie Physicochimique et Moléculaire, Université Paris-Sud, Orsay, France

J.D. Shore, Edsel B. Ford Institute for Medical Research, Henry Ford Hospital, Detroit, USA

C.M. Smith, Temple University School of Medicine, Philadelphia, USA

G.M.L. Spotorno, Department of Biochemistry, University College London, London, England

J. Südi, Biochemisches Institut, Universität Kiel, Kiel, Germany

H. Sund, Fachbereich Biologie, Universität Konstanz, Konstanz, Germany

H. Theorell, Medicinska Nobelinstitutet, Biokemiska Avdelningen, Stockholm, Sweden

F. Thomé-Beau, Technisch-Chemisches Institut der ETH, Zürich, Switzerland

H.-J. Torff, Fachbereich Biologie, Universität Regensburg, Regensburg, Germany

W. Trommer, Institut für Organische Chemie, Universität Stuttgart, Stuttgart, Germany

P. Truffa-Bachi, Département de Biochimie et Génétique Microbienne, Institut Pasteur, Paris, France

C. Veeger, Department of Biochemistry, Agricultural University, Wageningen, The Netherlands

R.-A. Walk, Biochemie der Morphogenese, Ruhr-Universität, Bochum, Germany

J. Walker, MRC Laboratory of Molecular Biology, Cambridge, England

L. Wallén, Institut de Chimie, Université Louis Pasteur, Strasbourg, France

K. Weber, Fachbereich Biologie, Universität Konstanz, Konstanz, Germany

G. Weimann, Boehringer-Mannheim, Biochemica-Werk Tutzing, Tutzing, Germany

M. Weininger, Department of Biological Chemistry, Washington University, St. Louis, USA

H. Wengenmayer, Biologisches Institut II, Universität Freiburg, Freiburg, Germany

M.M. Werber, Polymer Department, The Weizmann Institute of Science, Rehovot, Israel

Ch. Woenckhaus, Gustav-Embden Zentrum der Biologischen Chemie, Universität Frankfurt, Frankfurt, Germany

A.J. Wonacott, MRC Laboratory of Molecular Biology, Cambridge, England

D. Wyrambik, Biologisches Institut II, Universität Freiburg, Freiburg, Germany

M. Zeppezauer, Fachbereich Biochemie, Universität des Saarlandes, Saarbrücken, Germany

Section I. Primary and Conformation

Chairman: H. Eisenberg

CONFORMATIONAL ADAPTATIONS AMONG DEHYDROGENASES

Michael G. Rossmann, R. Michael Garavito and William Eventoff
Department of Biological Sciences, Purdue University, West Lafayette, Indiana 47907, U.S.A.

ABSTRACT

A comparison of known dehydrogenase structures shows a divergent evolutionary relationship in their common dinucleotide binding domains and a convergence of catalytic residues at their active centers. Features common to the active centers of lactate dehydrogenase (LDH) and glyceraldehyde-3-phosphate dehydrogenase (GAPDH) are the nicotinamide ring, an imidazole ring, as well as a guanidinium group. It is shown that the L specificity of the substrate in LDH demands an A-side specific reaction and, similarly, the B-side specific reaction of GAPDH is a consequence of a D tetrahedral intermediate.

GAPDH differs from LDH in forming a stable acylated intermediate and in the cooperativity between its subunits. To explore these, more subtle, properties of GAPDH, trifluoroacetonyl bromide (TFA) was used as an alkylating agent. Four sites were substituted per molecule, but the conformation of TFA was different in subunits on opposite sides of the R-axis. These results were consistent with the ^{19}F nuclear magnetic resonance work of Bode *et al.* (1) who find two distinct environments for the CF_3 group, as well as the suggestion of Buehner *et al.* (2) that the spatially close active centers (related by the R-axis) provide the mechanism for cooperativity and half-of-the-sites reactivity.

The arrangement of catalytic groups of GAPDH is found to be similar to those in papain (3). Thus, the processes of acylation and de-acylation of the sulfhydryl groups may have some common features in these two diverse enzymes.

INTRODUCTION

Eight years ago, at the time of the first Konstanz meeting on "Pyridine Nucleotide-Dependent Dehydrogenases", structural information on dehydrogenases was almost non-existent. Carl Brändén and Michael Rossmann presented low resolution balsa wood models of liver alcohol dehydrogenase and lactate dehydrogenase, respectively. They became aware, during the course of the meeting, that there appeared to be a remarkable similarity of the subunit structures of these two enzymes, and made the essentially correct deduction that tertiary structure is conserved to a greater extent than quaternary structure (4). While theirs was a thrilling discovery, the next few years gave even more exciting results. This paper will review some of these results and then consider the fundamental principles involved in hydride transfer on the basis of current knowledge. Although a comparison of the nucleotide binding domains in various dehydrogenases has been extensively discussed (5-7), a brief recapitulation is justified here in view of the events at the first Konstanz conference. Surprisingly little, if anything, has been written on the convergence of the spatial arrangement of the catalytic groups within the active center. This paper will attempt to alter that situation.

THE COMMON NUCLEOTIDE BINDING DOMAIN

A common structural domain, whose function is to bind NAD^+, has been found in lactate dehydrogenase (LDH; 8-11), in soluble malate dehydrogenase (sMDH; 12,13), in liver alcohol dehydrogenase (LADH; 14-17), and in glyceraldehyde-3-phosphate dehydrogenase (GAPDH; 18,19). Its structure consists of six parallel strands within a β-pleated sheet connected by helices in a sequential manner as shown in Figure 1. The bound NAD has an extended and open conformation as in Figure 2. The domain can be divided into two similar mononucleotide binding units related by a pseudo twofold axis. The first three strands of the sheet (βA, βB, βC) are

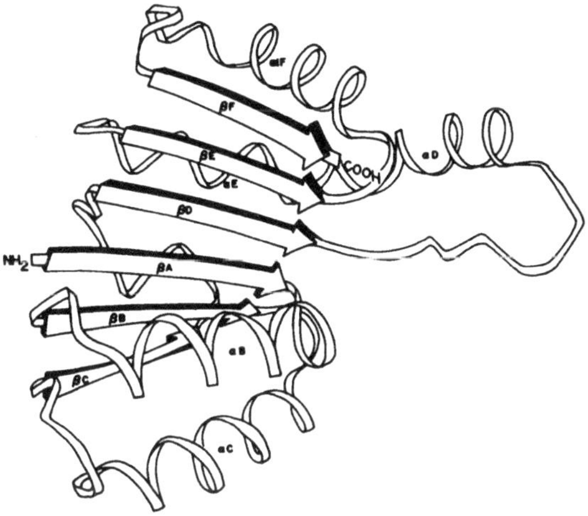

Figure 1 Schematic drawing of the NAD^+ binding region in dehydrogenases. The various secondary structural elements have been labeled according to the usually accepted nomenclature. The coenzyme binds to the carboxy end of the sheet.

associated with binding AMP, while the last three strands (βD, βE, βF) relate to the binding of NMN. The position of this domain within the complete polypeptide chain is variable (Fig. 3).

The first few structures of $NADP^+$ linked enzymes are now emerging. 6-Phosphogluconate dehydrogenase is to be described, at least to low resolution, at this meeting (20). Dihydrofolate reductase has just been solved in J. Kraut's laboratory in La Jolla, California. According to early accounts (M. J. Adams & J. Kraut, private communication), it would seem probable that neither structure has exactly the same domain as is found in NAD^+ linked enzymes. Although the basic motif of a parallel β-pleated sheet with helices on either side is likely in these $NADP^+$ dependent enzymes, the topology of their connectivity could be different. In view of other results (vide infra), it is probable that the coenzyme will bind to the carboxy end of the sheet.

The similarity of structure of the different NAD^+ binding domains permits their spatial superposition (6,7,21,22) to determine equivalent amino acid residues. These procedures replace the search for homology by inspection of linear amino acid sequences, and may be used either where the tertiary

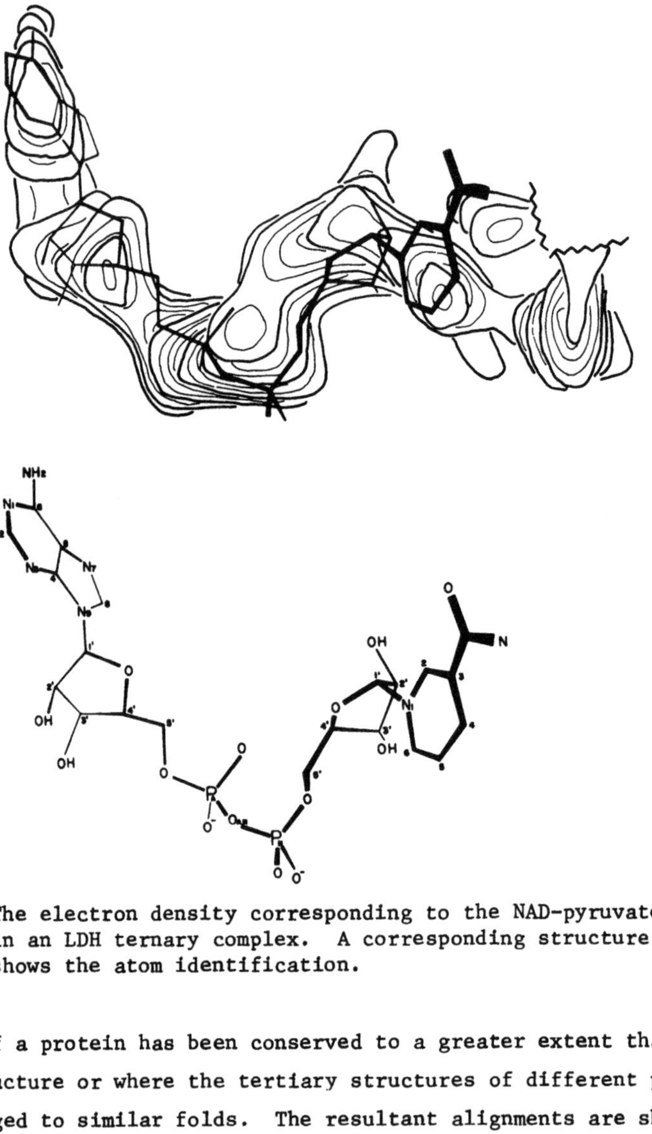

Figure 2 The electron density corresponding to the NAD-pyruvate adduct in an LDH ternary complex. A corresponding structure of NAD^+ shows the atom identification.

structure of a protein has been conserved to a greater extent than the primary structure or where the tertiary structures of different proteins have converged to similar folds. The resultant alignments are shown in Table 1. Inspection of this table shows a variety of features which permit some degree of generalization (see the "Function" row in Table 1). The repeated alternation of β-sheet and α-helix is marked by the hydrophobic residues (marked H and S in Table 1) within the enclosed cavity formed by these secondary structural elements. Such residues are likely to be guides for the folding process of the domain. The conservation of glycines 30 and 33 (LDH numbering), at the corner between βA and αB,

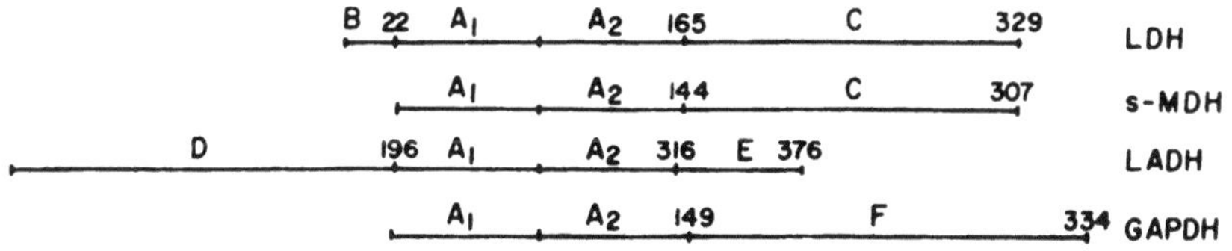

Figure 3 The position of the NAD binding domain within the complete polypeptide chain is variable. Each domain consists of two mononucleotide binding units A_1 and A_2. The catalytic domains, C, of LDH and sMDH are similar in structure, but different to the catalytic domains D and E in LADH and F in GAPDH. The primary function of the amino terminal arm B in LDH is to stabilize the quaternary structure.

suggests that these also are essential for the domain folding. The tendency for lysines or arginines in the second position along the polypeptide chain within the β-pleated sheet and the (albeit less striking) occurrence of negatively charged residues at the carboxy end of the sheet may also be guides for the folding process. Functional residues within this domain are also conserved. Glycines 28 and 99 are related to the positioning of the adenine and nicotinamide ribose moieties, respectively, and similarly, aspartate 53 binds to the 02' atom of the adenine ribose in the NAD^+ linked enzymes.

In Table 1 is shown the alignment of fragments of the beef and Neurospora crassa glutamate dehydrogenase (GluDH). This correspondence has been suggested by Rossmann et al. (6) and Wootton (29). Indeed, Wootton predicts two complete nucleotide binding folds by searching for the correct folding residues and patterns within these sequences. Wootton's complete alignments have been omitted, however, in view of the uncertainty of their precise correspondence with the other dehydrogenase structures. Wootton (29) associates the two domains with the two NADP and NAD binding sites expected for beef GluDH. The specific GluDH for which the sequence has been determined is $NADP^+$ dependent. While it is conceivable that the change from aspartic to glutamic acid residue at LDH position 53 might be sufficient to alter the function of this domain from NAD^+ to $NADP^+$ binding, yet a substitution of a basic residue might have been expected. Clearly, a structure determination of GluDH is essential for a more complete understanding of this enzyme.

Table 1

Alignment of Amino Acid Sequences in NAD Binding Domain of Dehydrogenases

	βA		αB	βB			βC	
	2	2 3	3 4	4	5	7	7 8 8	8 9
LDH[a]	2	9 3	4 0	7	4	5	6 1 3	4 0
1 Dogfish M_4	NKITVVG	VGAVG	MACAISILMKDLA	DEVALVD	VMEDK	T	AKIVSGK	DYSVSAG
2 Pig M_4	NKITVVG	VGAVG	MACAISILMKELA	KEIALVD	VMEDK	T	PKIVSGK	DYNVTAN
3 Chicken M_4	NKXXXXX	XXXXX	MACAISILMKBLA	BZLALVB	VVZBK	T	PKITSGK	DYSVTAH
4 Pig H_4	NKITVVG	VGQVG	MACAISILGKSLT	DELALVD	VLEDK	T	PKIVANK	DYSVTAN
5 Chicken H_4	NKXXXXX	XXQXX	MACAISILGKGLC	DELTLVD	VLEDK	T	HGIVABK	BYAVTAN
	1	2 2	2 2	2	2		2	2 2 2
	9	0 0	0 1	1	2		3	4 5 5
LADH[b]	3	0 4	5 1	7	4		6	3 1 6
6 Horse liver	STÇAVFG	LGGVG	LSVIMGCKAAG-A	ARIIGVD	INKDK		GATECVN	PQDYKKP
7 Rat liver	STCAVFG	LGGVG	LSVVIGCKTAG-A	AKIIAVD	INKDK		GATDCIN	PQDYTKP
		1	1 1	2	3		6	7
GAPDH	1	8 2	3 9	6	3		9	6
8 Pig[c]	VKVGVNG	FGRIG	RLVTRAAFNSGKV	DIVAIND	PFID		KAITIFQ	E
9 Lobster[c]	SKIGING	FGRIG	RLVLRAALSCG-A	QVVAVND	PFIA		KKITVFN	E
10 Yeast[c]	VRVAING	FGRIG	RLVMRISLSRPBV	ZVVASBB	PFIB		KKIATYQ	E
11 B. stearothermophilus[d]	VKVGING	FGRIG	RNVFRAALKNPDI	EVVAVND	LT-N		KEIIVKA	E
	2	2 2	2 2	2	2			
	4	5 5	5 6	6	7			
GluDH[e]	5	2 6	7 3	9	6			
12 Bovine & chicken	KTFAVQG	FGNVG	LHSMRYLHRFG-A	KCVAVGE	SDGSI			
	2	2 2	2 2	2	2			
	1	2 3	3 3	4	5			
	9	6 0	1 7	3	0			
13 *Neurospora*	KRVALSG	SGNVA	QYAALKLIELG-A	TVVSLSD	SKGAL			
Function	+H H F	C C	SS SS C	H H F̲			+H H	- S

	βD	loop	αD	αE	βE	α1F	βF	
	92	98 104 105	108 115	121 125 129 AB	133	140 148	158 164	
1	S KLVVIT	AGARQQEGE	SRLNLVQRNVNIF	KFIIPNIVKHSPN	CIILVVS	NPVDVLTYV	M HRIIGSG	-C
2	S RLVVIT	AGARQQEGE	SRLNLVQRNVNIF	KFIIPNIVKYT/SPN	CKLLVVS	NPVDILTYV	K NRVIGSG	-C
3	S KLVIVT	AGARQQEGE	SRLNLVQRNVNIF	KFIIPNVVKYSPD	CKXXXXX	XXXXXXXXV	G HRVIGAG	-C
4	S KIVVVT	AGVRQQEGE	SRLNLVQRNVNVF	KFIIPEIVKYSPN	CIIIVVS	NPVDILTYV	K HRVIGSG	-C
5	S KIVVVT	AGVRQQEGE	SRLNLVQRNVNVF	KFIIPZIVXXXXX	X VBS	NPVDILTYV	K HRVIGSG	-C
Lobster tail[g]		RQKEGE	SRLSLVQR				RVIGSG	-T
	263	269		272 283 286	287	294 302	312	319
6	V DFSFEV	IG		R LDTMVTALSCCQY	GVSVIVG	VPPDSQNLS	RTWKGAI	FG
7	V DFXFEV	IG		R LDTMAXXLLSCHC	GVSVIVG	VPPVAQSLS	RTWKGAI	FG
	90	96		106	115	121 125	141	148
8	T AYVVES	TGV		AGAHLKGGAK	-RVIISA	PSADA	D LKIVSNA	SC
9	A EYIVES	TGV		ASAHFKGGAK	-KVVISA	PSADA	D MTVVSNA	SC
10	S VIAIDS	TGV		AQKHIDAGAK	-KVVITA	PSSTA	S LKIVSNA	SC
11	V DIVVES	TGR		AAKHLEAGAK	-KVIISA	PAVEN	A HHVISNA	SC
Function	H H	F		SS SS	+H H H	S S	+H H H	

a. Sequences 1 - 5 of LDH are as in Eventoff et al. (23).
b. Sequences 6 and 7 of horse and rat liver ADH are as in Eklund et al. (24).
c. Sequences 8, 9 and 10 of GAPDH are as in Jones and Harris (25).
d. Sequence 11 of B. stearothermophilus GAPDH is by private communication from J. I. Harris.
e. Sequences 12 and 13 are as in Wootton et al. (26).
f. Symbols in Function row: H, hydrophobic residues in sheet opposite helix;
S, hydrophobic residues in helix opposite sheet;
+, positively charged residues in second position of sheet;
-, negatively charged residues at termination of sheet;
C, conserved residues or deletions;
F, residues involved in binding function.
g. Partial sequence of lobster tail LDH as reported by Taylor and Oxley (27).
Structural alignments taken from Eventoff and Rossmann (28).
Sequence alignments for bovine and Neurospora GluDH taken from Rossmann et al. (6) and Wootton (29).

Table 2 Mean distance (Å) between atoms in superimposed β-pleated sheets (top half) and resultant mean distance of equivalent atoms in coenzyme (lower half).

	LDH	LADH	GAPDH	
LDH		1.1	1.5	C_α Atoms
LADH	3.3		1.7	
NAD Atoms GAPDH	1.8	2.8		

(a)

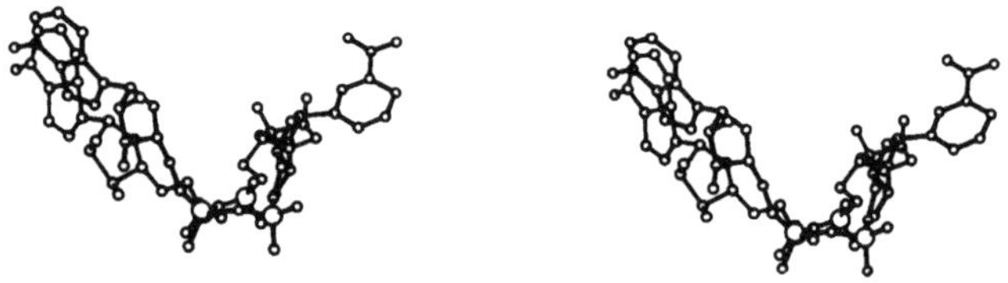

(b)

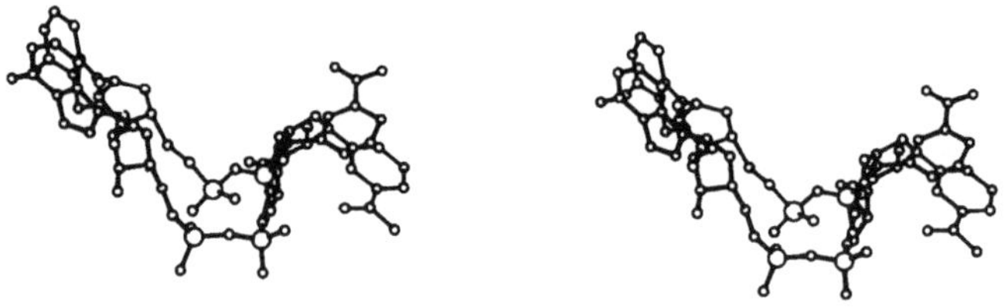

(c)

Figure 4 Comparison of the NAD^+ conformation and position between (a) sMDH and LDH, (b) LADH and LDH (only ADPR is shown for LADH; see ref. 30), and (c) GAPDH and LDH. The structure for LDH is in double outline in every case.

Table 3 Matrix showing the average minimum base change per codon among various dehydrogenases aligned as shown in Table 1.

		1	2	3	4	5	6
Dogfish LDH	1		1.17	1.17	1.08	1.21	1.19
Pig GAPDH	2			0.32	0.51	1.24	1.14
Lobster GAPDH	3				0.52	1.21	1.03
Yeast GAPDH	4					1.26	1.03
Horse LADH	5						1.14
Bovine GluDH	6						

Superposition of the protein domain results in the accurate alignment of the atoms of the various bound NAD^+ molecules (Table 2), demonstrating the equivalence of the coenzyme binding modes. These results are shown graphically in Figure 4 with respect to the NAD^+ conformation in ternary complexes of LDH. A comparison of the residues involved in the binding of NAD^+ to the various dehydrogenases has been discussed by Rossmann et al. (5) and Ohlsson et al. (7).

The high conservation of the structure and function of the NAD^+ binding domains, as well as evidence for remnants of a common ancestral amino acid sequence (Table 3), gives strong support for a divergent evolutionary process proceeding from a common primordial NAD^+ binding structure. The need for glycolysis in even the earliest prokaryotes, requiring the presence of various dehydrogenases, and the very slow rate of accepting point mutations, indicates that the NAD^+ binding structure has remained largely unchanged during the three billion years of biological evolution on Earth.

SUPER-SECONDARY STRUCTURES IN NUCLEOTIDE BINDING FOLDS

The structures of a substantial number of enzymes have been determined, since the previous Konstanz meeting, which have folding topologies remarkably similar to that observed in the NAD^+ linked dehydrogenases. Furthermore, many of these have the function of binding nucleotides with the cofactor attachment site almost invariably at the carboxy end of the β-pleated sheet. Rao and Rossmann (21) were first to remark upon this

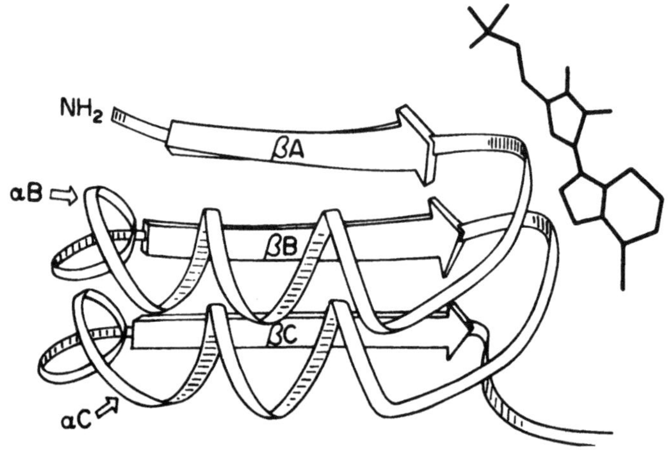

Figure 5 A possible primordial mononucleotide binding fold.

situation in a comparison of the flavodoxin structure with LDH. Subsequently, the same structure was found in adenylate kinase (31) and phosphoglycerate kinase (32,33). The rough duplication of a mononucleotide binding fold within the complete NAD^+ binding domain has, therefore, led to the suggestion of an ancient primordial mononucleotide binding fold (Fig. 5).

Such concepts have had to be modified in light of the variety of structures (34-36) which contain "β-α-β" folds with internal β-pleated sheets surrounded by roughly parallel α-helices. The topology of connectivity is not always the same and anti-parallel strands occur in some cases. Hence, the idea of super-secondary structures was introduced (21) to indicate stable folding units. Jane Richardson (36) and Sternberg and Thornton (35) have shown that the hand of "crossover structures" (Fig. 6) is almost invariably right-handed, just as α-helices are invariably right-handed. A mononucleotide binding fold thus consists of only two β-α-β structural units in sequence. Sternberg and Thornton (35) follow the earlier arguments of Schulz and Schirmer (37) by relating the number of possible structures that can be constructed from a succession of β-α-β units with the probability that any given structure has occurred by chance. They differ from Schulz in that they consider left-handed crossover structures as impossible and thus find greater probability that any given structure

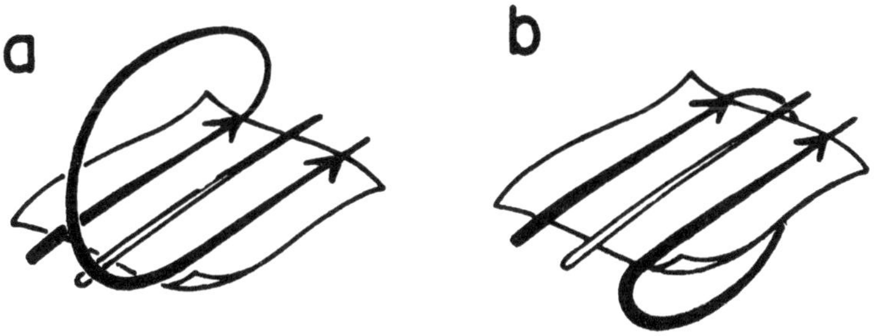

Figure 6 (a) A right-handed ± 2x crossover connection. (b) A left-handed ±2x crossover connection. Direction is not indicated for the skipped strand, since it may be either parallel or anti-parallel to the others.

can occur by convergence. Nevertheless, the probability that each NAD^+ binding structure in the various known dehydrogenases originated independently is still very low.

A COMPARISON OF THE ACTIVE CENTERS IN LDH AND GAPDH

A substantial body of information is now available on the structural properties of LDH (9-11) and GAPDH (18,19)(Table 4). Diagrammatic representations of the active centers of these two enzymes are shown in Figure 7. The following common features have been recognized at the active sites of these enzymes.

1. The hydride acceptor is the C_4 atom of nicotinamide.
2. The proton acceptor is an imidazole group (His 195 in LDH and His 176 in GAPDH).
3. The proton is transferred from the substrate to the ε nitrogen of a histidine for both LDH and GAPDH.
4. The substrate is situated between the nicotinamide and imidazole moiety at a site strongly attractive to anions (40,46).
5. The carboxyamide group is oriented by Ser 163 or the main chain carbonyl of residue 139 in LDH and by Asn 313 in GAPDH.
6. The histidine residues are oriented and polarized by hydrogen bonding to possibly Asp 168 or the main chain carbonyl of residue 164 in LDH and by Tyr 311 or Ser 238 in GAPDH.

Table 4

Structural Studies on LDH and GAPDH

Enzyme	Compound	Resolution	References
I. Lactate Dehydrogenase			
A.	Dogfish M_4 Apo LDH	2.0 Å	8-11,38
	1. Complexed with coenzyme fragments	5.0 Å	39
	2. Binding of citrate ions	2.8 Å	40
B.	Dogfish M_4 LDH Ternary Complexes		
	1. LDH:NAD-pyruvate	3.0 Å	9-11,38
	2. LDH:NAD:oxalate	3.0 Å	11
	3. LDH:NADH:oxamate	3.0 Å	11
C.	Pig M_4 LDH Ternary Complexes		
	1. LDH:NAD-pyruvate	very low	41
	2. LDH:NADH:oxamate	very low	41
	3. LDH:NAD:oxalate	very low	41
D.	Pig H_4 LDH Ternary Complexes		
	1. LDH:NAD:oxalate	very low	41
	2. LDH:NADH:oxamate	2.5 Å	23,42
E.	Mouse Apo LDH-X	2.9 Å	43
F.	*B. stearothermophilus* LDH	low	J. N. Jansonius, private communication
II. Glyceraldehyde-3-Phosphate Dehydrogenase			
A.	Lobster Holo GAPDH	2.9 Å	18,19,44
	1. Binding of 8-BrNAD	2.9 Å	44
	2. Binding of citrate ions	2.9 Å	44
	3. Trifluoroacetonyl NAD ternary complex	2.9 Å	45
B.	Lobster Apo GAPDH	very low	46
C.	*B. stearothermophilus* Holo and Apo GAPDH	2.5 Å	A. J. Wonacott, private communication
D.	Human Holo GAPDH	5.0 Å	47
E.	Human Holo and Apo GAPDH	very low	48

Note: Compounds grouped within a subheading A, B,.... are isomorphous.

Features 1, 2 and 3 were superimposed by a least squares procedure (49) to determine the relative spatial positions of the catalytic groups. It was immediately apparent (Fig. 8) that a similar orientation of the active center residues superimposed only the nicotinamide end of the coenzyme, and thus did not align the structures of the NAD^+ binding domain. Furthermore, although the plane of the histidine rings 195 in LDH and 176 in GAPDH did not superimpose, nevertheless, the catalytic N_δ-N_ε axis of the imidazole group was similarly oriented and positioned.

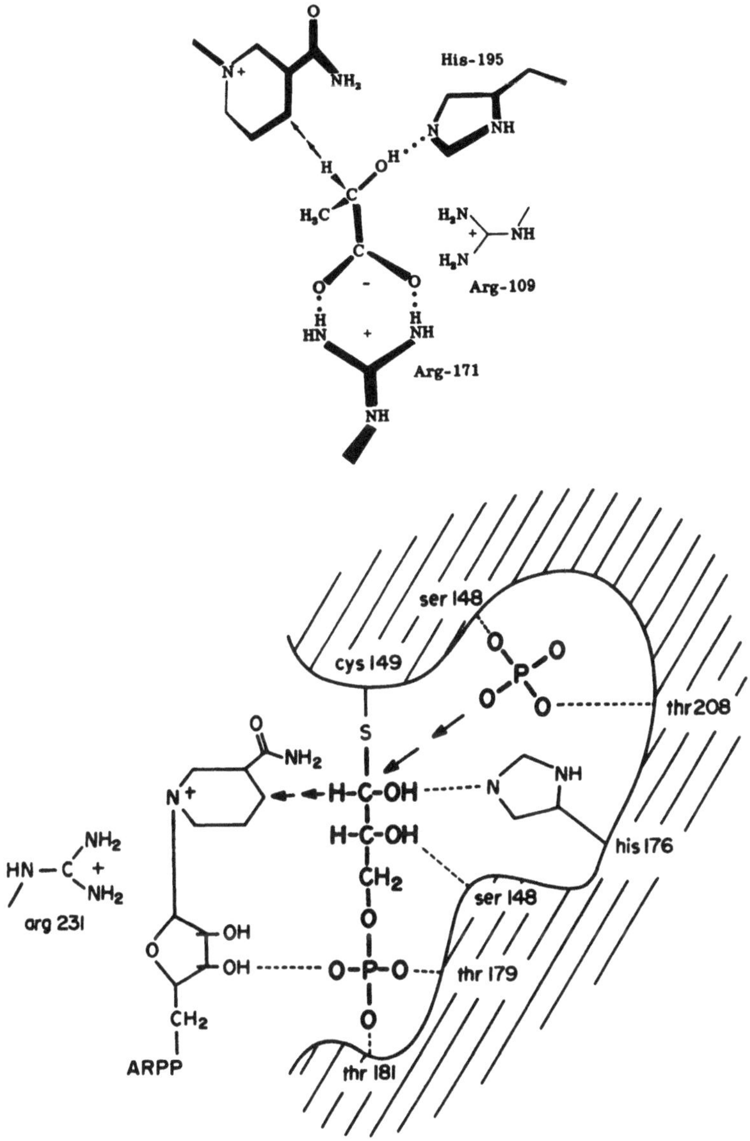

(a)

(b)

Figure 7 Diagrammatic representation of the active centers of (a) LDH and (b) GAPDH. The position of the arginines is distorted in an endeavor to simplify the diagrams.

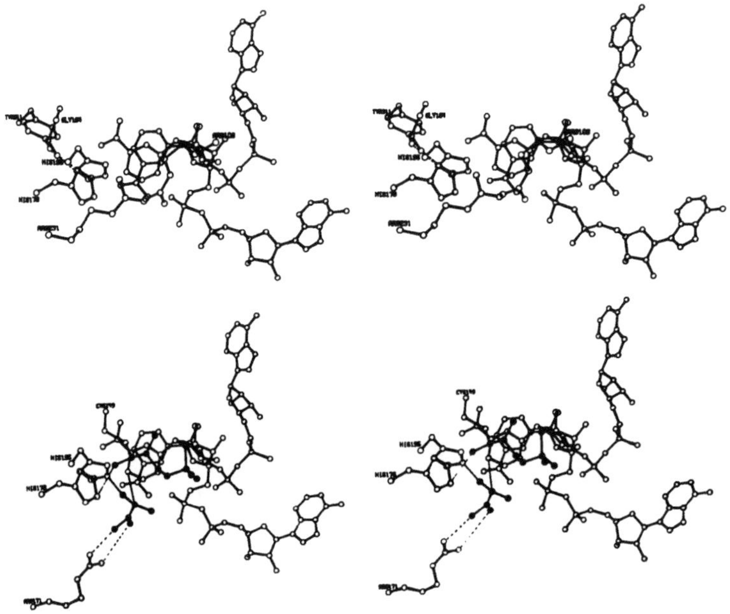

(a)

(b)

Figure 8 Superposition of (a) the catalytic residues in LDH and GAPDH in the red subunit and (b) the corresponding substrates showing the tetrahedral hemiacetal intermediate for GAPDH and lactate for LDH.

Most interestingly of all, the guanidinium group of arginine 109 in LDH (when the loop is down) superimposed on the guanidinium group of arginine 231 in GAPDH. As the major function of the LDH loop movement is to introduce the positive charge on the guanidinium group close to the substrate and nicotinamide ring during catalysis, it can be reasonably assumed that arginine 231 in GAPDH and arginine 109 in LDH must have similar functions. While no conformational changes of arginine 231 have yet been found, it will be of some interest to observe its orientation for the apo-enzyme. In view of the similarity of the function of these arginines, they were used to help orient the catalytic sites of GAPDH and LDH (Table 5). The importance of arginine 109 in LDH had long been recognized, although the significance of its function has probably been underestimated. Both Adams *et al.* (38) and Holbrook *et al.* (10) suggest that the insertion of the positive charge into the active center pocket helps to neutralize the negative charge on the substrate (Fig. 7a). However, possibly an even more important function of the arginine may be

Table 5 Least squares distances between the catalytically equivalent atoms of LDH and GAPDH (red subunit).

LDH Residue	Atom	GAPDH Residue	Atom	Distance (Å)
His 195	N_δ	His 176	N_δ	1.5
His 195	N_ε	His 176	N_ε	1.2
Nicotinamide	C_3	Nicotinamide	C_5	2.1
Nicotinamide	C_4	Nicotinamide	C_4	1.9
Nicotinamide Ribose	$C_{1'}$	Nicotinamide Ribose	$C_{1'}$	1.6
Nicotinamide Ribose	$O_{1'}$	Nicotinamide Ribose	$O_{1'}$	1.6
Nicotinamide Ribose	$O_{2'}$	Nicotinamide Ribose	$O_{2'}$	0.5
Nicotinamide Ribose	$O_{3'}$	Nicotinamide Ribose	$O_{3'}$	2.8
Nicotinamide Phosphorous	P	Nicotinamide Phosphorous	P	2.4
Arg 109	C_ζ	Arg 231	C_ζ	1.5

Note: The selected ribose atoms were used to provide maximum information on the orientation of this group.

to regulate the movement of the positive charge between the nicotinamide and histidine.

Figure 9 shows diagrammatically the relationship of the substrate to the coenzyme in LDH and in GAPDH. The histidines and nicotinamide rings are viewed similarly. However, as is clear from Figure 8, the substrate-orienting residues arginine 171 of LDH and cysteine 149 of GAPDH do not superimpose in space. Thus, while the bond to the "reactive" carbon is made from "below" in LDH, it is made from "above" in GAPDH (atoms C_2 in Figs. 9a and 9b). This implies that the chirality of lactate in LDH and the hemiacetal intermediate of glyceraldehyde in GAPDH are opposite. Furthermore, the methyl group of lactate points "down", forcing the carboxyamide group to point "up" to avoid steric hindrance in LDH. Correspondingly, the H-C-OH function points "up" and the carboxyamide group "down" in GAPDH. In other words, the specificity of the substrate is determined by the position of the orienting ligands (Arg 171 in LDH

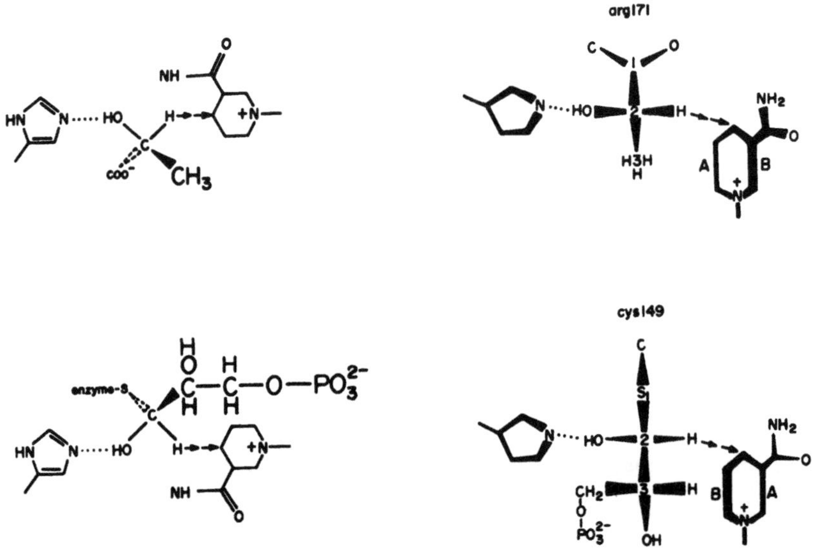

Figure 9 Comparison of the hand of the reactive carbon atom in (a) LDH and (b) GAPDH. The methyl group of the L-lactate substrate avoids steric interaction with the carboxyamide group when the A side of the nicotinamide ring faces the active center. Similarly, a D configuration of the tetrahedral intermediate for GAPDH requires that the B side of the nicotinamide ring faces the substrate. Furthermore, the hydroxyl group at the 3 position of the substrate (numbering S = 1) is kept away from the coenzyme only for D-glyceraldehyde.

and Cys 149 in GAPDH) relative to the nicotinamide and histidine rings. The positioning of the glyceraldehyde molecule close to the nicotinamide ring then determines the D-specificity of the C_3 atom (Fig. 9b), thus situating the hydroxyl group away from the nicotinamide ring.

Another feature is the apparently closer approach of the histidine and nicotinamide moieties to each other in LDH than in GAPDH. However, Figure 8 is a comparison of the LDH:NAD-pyruvate ternary complex with the lobster holo-GAPDH structure as observed in the red or yellow subunits (19). All the ternary complexes of LDH which have been studied so far (Table 4) are exceedingly similar and probably differ by less than 0.5 Å in their atomic positions (11). In GAPDH the R- or P-axis related subunits differ by less than 1.5 Å (19). The greater proximity of the catalytic groups in ternary complexes of LDH as compared to the holo-GAPDH

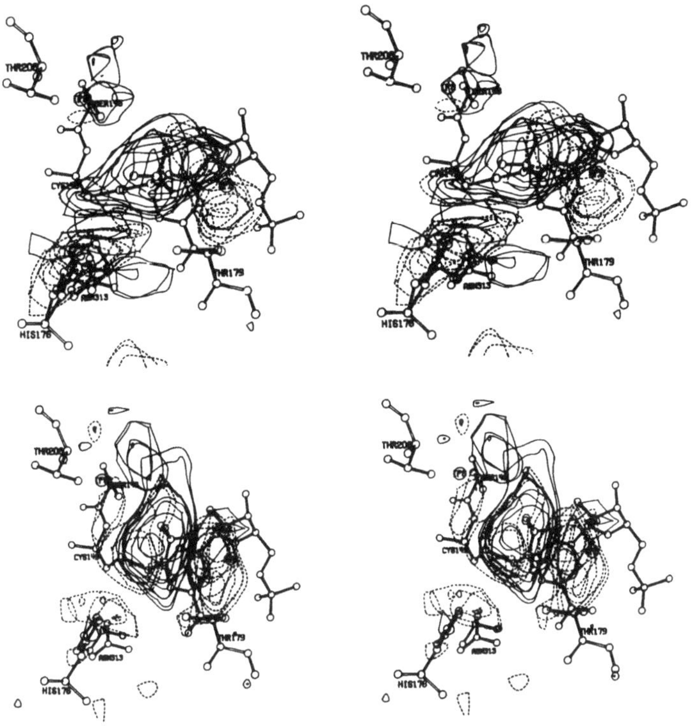

(a)

(b)

Figure 10 Stereo pairs showing the difference electron density ($F_{TFA}-F_{holo}$) with superimposed TFA and protein structure for (a) the red and yellow, and (b) the blue and green subunits of GAPDH.

structure is thus likely to be a significant difference with respect to model building errors. Thus, it is possible that the LDH structure shown in Figure 8 is "tighter" than the GAPDH structure, not because of differences between the enzymes but rather as a consequence of comparing a ternary with a binary complex. This hypothesis is given credence by the movement of His 176 toward the nicotinamide ring in the formation of the trifluoroacetonyl NAD complex in GAPDH (see below and ref. 45). Similarly, the nicotinamide moves away from the histidine in a binary complex of LDH relative to the ternary complex conformation (39).

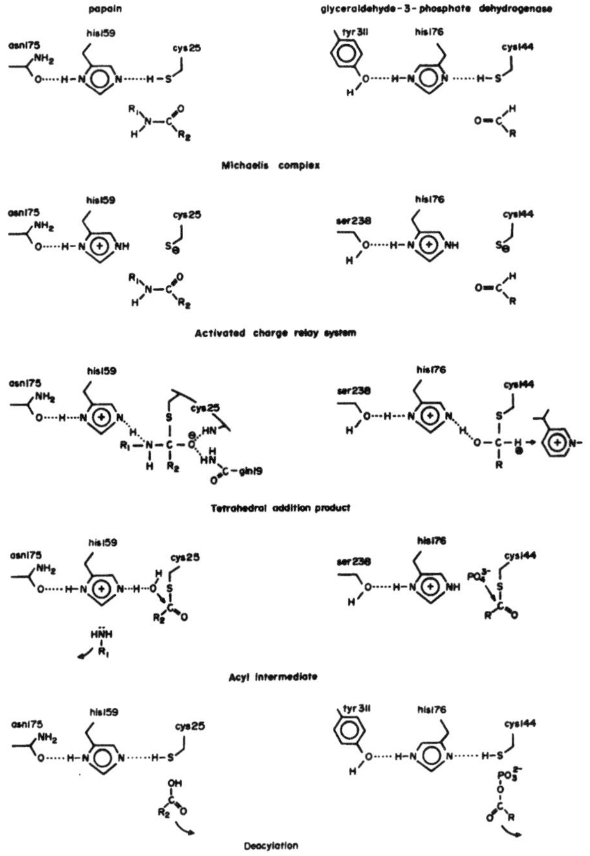

Figure 11 Comparison of probable catalytic steps of papain and GAPDH.

ACYLATED INTERMEDIATES

A study (45) of lobster holo-GAPDH alkylated with trifluoroacetonyl (TFA) has recently been completed in an attempt to study a possible analogue of the natural acylated intermediate:

$$\mathrm{E{-}S{-}\overset{\overset{\displaystyle O}{\|}}{C}{-}\underset{\underset{\displaystyle H}{|}}{\overset{\overset{\displaystyle OH}{|}}{C}}{-}CH_2{-}O{-}PO_3}$$ acylated intermediate

$$\mathrm{E{-}S{-}CH_2{-}\overset{\overset{\displaystyle O}{\|}}{C}{-}CF_3}$$ alkylated analogue.

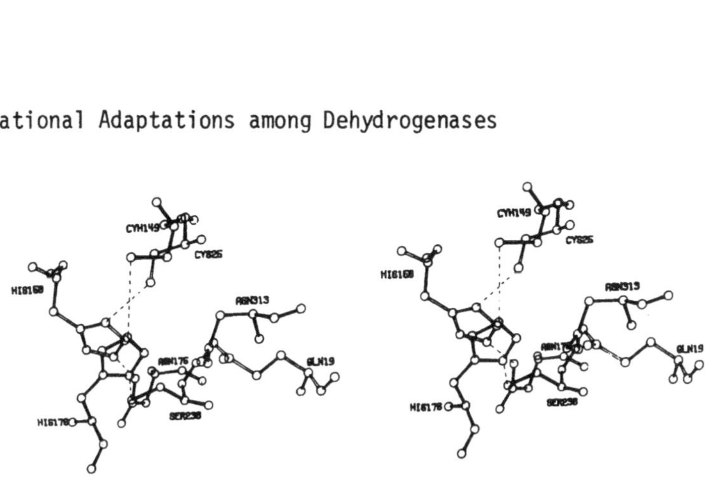

Figure 12 Superposition of the active centers of papain and GAPDH in the green subunit.

The alkylation was achieved according to the general scheme

$$E\text{-}SH \quad + \quad Br\text{-}CH_2\text{-}\overset{\overset{\displaystyle O}{\|}}{C}\text{-}CF_3 \quad \longrightarrow \quad E\text{-}S\text{-}CH_2\text{-}\overset{\overset{\displaystyle O}{\|}}{C}\text{-}CF_3 \quad + \quad HBr \quad .$$

A series of chloromethyl ketones have been used in a similar manner to alkylate the thiol protease papain (3) and the serine proteases subtilisin BPN' (50) and γ-chymotrypsin (51).

The alkylated GAPDH could be crystallized isomorphously to holo-GAPDH, and thus gave the opportunity of studying the alkylated product by difference maps. Bode et al. (1) had already shown by ^{19}F nuclear magnetic resonance spectroscopy that the $C^{19}F_3$ group had two distinct environments. It was, therefore, satisfying that the TFA moiety within the red and yellow subunits clearly showed a different conformation than in the blue and green subunits (Fig. 10). This result was also consistent with the observations of Buehner et al. (2,18) in speculating that the cooperative pair of subunits would be red - green and yellow - blue as these had spatially close active centers related by the R-axis. Furthermore, Moras et al. (19) had tentatively identified some structural difference between subunits related by the R-axis. Thus, the GAPDH-TFA results start to provide a structural basis for half-of-the-sites phenomenon in GAPDH. However, the precise mechanism by which information is transferred between the red and blue (or between the yellow and green) subunits is not yet clear.

Table 6 Least squares distances between the catalytically equivalent atoms of GAPDH (green subunit) and papain.

GAPDH		Papain		
Residue	Atom	Residue	Atom	Distance (Å)
His 176	N_{ϵ}	His 159	N_{δ}	1.0
His 176	N_{δ}	His 159	N_{ϵ}	0.8
Cys 149	C_{β}	Cys 25	C_{β}	0.9
Cys 149	S_{γ}	Cys 25	S_{γ}	1.3
Asn 313	N/O_{δ_1}	Gln 19	N/O_{ϵ_1}	1.5
Asn 313	N/O_{δ_1}	Gln 19	N/O_{ϵ_2}	1.6
Asn 313	C_{γ}	Gln 19	C_{δ}	1.4
Ser 238	O_{γ}	Asn 175	N/O	1.9

In Figure 10 is shown the position of the TFA moiety as found from the 2.9 Å resolution difference maps. The $SO_4^{=}$ ion present at the $PO_4^{=}$ binding site of the natural substrate (19,46) has been displaced in all four subunits. However, the position of the CF_3 group is quite different and the histidine and nicotinamide have moved closer to the substrate in the red/yellow subunits. That is, under certain conditions, the catalytic groups move closer to the substrate in ternary complexes of GAPDH as in LDH.

A comparison of the probable catalytic steps for papain and GAPDH is given in Figure 11. A tetrahedral intermediate is formed in both cases by the nucleophilic attack of a negatively charged cysteine before the formation of the acylated enzyme. Although the role of the histidine is not identical, yet it provides the initial condition for creating the nucleophile. Subsequently, the deacylation step is brought about by the introduction of water or inorganic phosphate. The active centers of papain and GAPDH were, therefore, compared in three-dimensions (Fig. 12). The distances between spatially equivalent atoms is shown in Table 6. It is apparent that the distribution of the catalytically important groups is exceedingly similar. The superposition of Gln 19 on Asn 313 suggests that the role of orienting the substrate carbonyl played by Gln 19 in papain may be duplicated by Asn 313 in GAPDH. However, Asn 313 also orients the carboxyamide group of the coenzyme in the latter case.

The similarity of arrangements of catalytic groups between chymotrypsin and subtilisin is frequently taken as an example of convergent evolution (52). Drenth et al. (3) have commented upon the resemblance of the catalytic mechanisms of the serine and cysteine proteases. The discussion in this paper shows that a degree of convergence exists among NAD^+ linked dehydrogenases (in particular GAPDH), cysteine proteases and serine proteases. The greater structural correspondence of the green subunit of GAPDH with papain on the one hand and between the red subunit of GAPDH and LDH on the other is consistent with the observation of Moras et al. (19) that GAPDH subunits modify their conformation according to whether they are catalyzing dehydrogenation or de-esterification.

CONCLUSIONS

Comparison of polypeptide folds and active center geometries among NAD^+ linked dehydrogenases has shown the essential features important to this group of enzymes. These comparisons can be extended to other, apparently unrelated, enzymes. Thus, the work on dehydrogenases has initiated a search for, and hopefully an understanding of, some basic principles of enzyme architecture and function.

ACKNOWLEDGMENTS

We are grateful to Dr. Patrick Argos for stimulating discussions and to Sharon Wilder for help in preparation of the manuscript. The work was supported by the National Science Foundation (grant no. BMS74-23537) and the National Institutes of Health (grant no. GM 10704). R.M.G. is the recipient of a National Institutes of Health Molecular and Cellular Biology Training Grant.

REFERENCES

1. Bode, J., Blumenstein, M. and Raftery, M. A.: ^{19}F nuclear magnetic resonance studies of structure and function relationships in trifluoroacetonylated rabbit muscle glyceraldehyde-3-phosphate dehydrogenase, Biochemistry 14, 1153-1160 (1975).
2. Buehner, M., Ford, G. C., Moras, D., Olsen, K. W. and Rossmann, M. G.: D-glyceraldehyde-3-phosphate dehydrogenase: three-dimensional structure and evolutionary significance, Proc. Natl. Acad. Sci. U.S. 70, 3052-3054 (1973).
3. Drenth, J., Kalk, K. H. and Swen, H. M.: Binding of chloromethyl ketone substrate analogues to crystalline papain, Biochemistry 15, 3731-3738 (1976).
4. Brändén, C. I. and Rossmann, M. G.: Discussion remarks, in Pyridine Nucleotide-Dependent Dehydrogenases (Sund, H., ed.), Springer-Verlag, Heidelberg, pp. 133-134 (1970).
5. Rossmann, M. G., Liljas, A., Brändén, C. I. and Banaszak, L. J.: Evolutionary and structural relationships among dehydrogenases, in The Enzymes (Boyer, P. D., ed.), 3rd edn., Vol. XI, Academic Press, New York, pp. 61-102 (1975).
6. Rossmann, M. G., Moras, D. and Olsen, K. W.: Chemical and biological evolution of a nucleotide-binding protein, Nature (London) 250, 194-199 (1974).
7. Ohlsson, I., Nordström, B. and Brändén, C. I.: Structural and functional similarities within the coenzyme binding domains of dehydrogenases, J. Mol. Biol. 89, 339-354 (1974).
8. Adams, M. J., Ford, G. C., Koekoek, R., Lentz, P. J. Jr., McPherson, A. Jr., Rossmann, M. G., Smiley, I. E., Schevitz, R. W. and Wonacott, A. J.: Structure of lactate dehydrogenase at 2.8 Å resolution, Nature (London) 227, 1098-1103 (1970).
9. Rossmann, M. G., Adams, M. J., Buehner, M., Ford, G. C., Hackert, M. L., Lentz, P. J. Jr., McPherson, A. Jr., Schevitz, R. W. and Smiley, I. E.: Structural constraints on possible mechanisms of lactate dehydrogenase as shown by high resolution studies of the apoenzyme and a variety of enzyme complexes, Cold Spring Harbor Symp. Quant. Biol. 36, 179-191 (1971).

10. Holbrook, J. J., Liljas, A., Steindel, S. J. and Rossmann, M. G.: Lactate dehydrogenase, in The Enzymes (Boyer, P. D., ed.), 3rd edn., Vol. XI, Academic Press, New York, pp. 191-292 (1975).

11. White, J. L., Hackert, M. L., Buehner, M., Adams, M. J., Ford, G. C., Lentz, P. J. Jr., Smiley, I. E., Steindel, S. J. and Rossmann, M. G.: A comparison of the structures of apo dogfish M_4 lactate dehydrogenase and its ternary complexes, J. Mol. Biol. 102, 759-779 (1976).

12. Hill, E., Tsernoglou, D., Webb, L. and Banaszak, L. J.: Polypeptide conformation of cytoplasmic malate dehydrogenase from an electron density map at 3.0 Å resolution, J. Mol. Biol. 72, 577-591 (1972).

13. Banaszak, L. J. and Bradshaw, R. A.: Malate dehydrogenase, in The Enzymes (Boyer, P. D., ed.), 3rd edn., Vol. XI, Academic Press, New York, pp. 369-396 (1975).

14. Brändén, C. I., Eklund, H., Nordström, B., Boiwe, T., Söderlund, G., Zeppezauer, E., Ohlsson, I. and Åkeson, Å.: Structure of liver alcohol dehydrogenase at 2.9-Å resolution, Proc. Natl. Acad. Sci. U.S. 70, 2439-2442 (1973).

15. Brändén, C. I., Jörnvall, H., Eklund, H. and Furugren, B.: Alcohol dehydrogenases, in The Enzymes (Boyer, P. D., ed.), 3rd edn., Vol. XI, Academic Press, New York, pp. 104-190 (1975).

16. Eklund, H., Nordström, B., Zeppezauer, E., Söderlund, G., Ohlsson, I., Boiwe, T. and Brändén, C. I.: The structure of horse liver alcohol dehydrogenase, FEBS Lett. 44, 200-204 (1974).

17. Eklund, H., Nordström, B., Zeppezauer, E., Söderlund, G., Ohlsson, I., Boiwe, T., Söderberg, B. O., Tapia, O., Brändén, C. I. and Åkeson, Å.: Three-dimensional structure of horse liver alcohol dehydrogenase at 2.4 Å resolution, J. Mol. Biol. 102, 27-59 (1976).

18. Buehner, M., Ford, G. C., Moras, D., Olsen, K. W. and Rossmann, M. G.: Three-dimensional structure of D-glyceraldehyde-3-phosphate dehydrogenase, J. Mol. Biol. 90, 25-49 (1974).

19. Moras, D., Olsen, K. W., Sabesan, M. N., Buehner, M., Ford, G. C. and Rossmann, M. G.: Studies of asymmetry in the three-dimensional structure of lobster D-glyceraldehyde-3-phosphate dehydrogenase, J. Biol. Chem. 250, 9137-9162 (1975).

20. Adams, M. J., Archibald, I. G. and Helliwell, J. R.: X-ray diffraction studies on sheep liver 6-phosphogluconate dehydrogenase at 6 Å

resolution, in Second Symposium on Pyridine Nucleotide-Dependent Dehydrogenases (Sund, H., ed.), Walter de Gruyter and Co., Berlin, in press (1977), p. 72.

21. Rao, S. T. and Rossmann, M. G.: Comparison of super-secondary structures in proteins, J. Mol. Biol. 76, 241-256 (1973).

22. Rossmann, M. G. and Argos, P.: Exploring structural homology of proteins, J. Mol. Biol. 105, 75-95 (1976).

23. Eventoff, W., Rossmann, M. G., Taylor, S. S., Torff, H. J., Meyer, H., Keil, W. and Kiltz, H. H.: Structural adaptations of lactate dehydrogenase isozymes, Proc. Natl. Acad. Sci. U.S., in press (1977).

24. Eklund, H., Brändén, C. I. and Jörnvall, H.: Structural comparisons of mammalian, yeast and bacillar alcohol dehydrogenases, J. Mol. Biol. 102, 61-73 (1976).

25. Jones, G. M. T. and Harris, J. I.: Glyceraldehyde-3-phosphate dehydrogenase: amino acid sequence of enzyme from baker's yeast, FEBS Lett. 22, 185-189 (1972).

26. Wootton, J. C., Chambers, G. K., Holder, A. A., Baron, A. J., Taylor, J. G., Fincham, J. R. S., Blumenthal, K. M., Moon, K. and Smith, E. L.: Amino-acid sequence of NADP-specific glutamate dehydrogenase of Neurospora crassa, Proc. Natl. Acad. Sci. U.S. 71, 4361-4365 (1974).

27. Taylor, S. S. and Oxley, S. S.: Homologies in the active site regions of lactate dehydrogenases, Arch. Biochem. Biophys. 175, 373-383 (1976).

28. Eventoff, W. and Rossmann, M. G.: The evolution of dehydrogenases and kinases, in CRC Critical Reviews in Biochemistry (Fasman, G. D., ed.), CRC Press, Cleveland, pp. 111-140 (1975).

29. Wootton, J. C.: The coenzyme-binding domains of glutamate dehydrogenases, Nature (London) 252, 542-546 (1974).

30. Abdallah, M. A., Biellmann, J. F., Nordström, B. and Brändén, C. I.: The conformation of adenosine diphosphoribose and 8-bromoadenosine diphosphoribose when bound to liver alcohol dehydrogenase, Eur. J. Biochem. 50, 475-481 (1975).

31. Schulz, G. E., Elzinga, M., Marx, F. and Schirmer, R. H.: Three-dimensional structure of adenyl kinase, Nature (London) 250, 120-123 (1974).

32. Blake, C. C. F. and Evans, P. R.: Structure of horse muscle phos-

phoglycerate kinase. Some results on the chain conformation, substrate binding and evolution of the molecule from a 3 Å Fourier map, J. Mol. Biol. 84, 585-601 (1974).

33. Bryant, T. N., Watson, H. C. and Wendell, P. L.: Structure of yeast phosphoglycerate kinase, Nature (London) 247, 14-17 (1974).

34. Levitt, M. and Chothia, C.: Structural patterns in globular proteins, Nature (London) 261, 552-558 (1976).

35. Sternberg, M. J. E. and Thornton, J. M.: On the conformation of proteins: the handedness of the β-strand - α-helix - β-strand unit, J. Mol. Biol. 105, 367-382 (1976).

36. Richardson, J. S.: Handedness of crossover connections in β sheets, Proc. Natl. Acad. Sci. U.S. 73, 2619-2623 (1976).

37. Schulz, G. E. and Schirmer, R. H.: Topological comparison of adenyl kinase with other proteins, Nature (London) 250, 142-144 (1974).

38. Adams, M. J., Buehner, M., Chandrasekhar, K., Ford, G. C., Hackert, M. L., Liljas, A., Rossmann, M. G., Smiley, I. E., Allison, W. S., Everse, J., Kaplan, N. O. and Taylor, S. S.: Structure-function relationships in lactate dehydrogenase, Proc. Natl. Acad. Sci. U.S. 70, 1968-1972 (1973).

39. Chandrasekhar, K., McPherson, A. Jr., Adams, M. J. and Rossmann, M. G.: Conformation of coenzyme fragments when bound to lactate dehydrogenase, J. Mol. Biol. 76, 503-518 (1973).

40. Adams, M. J., Liljas, A. and Rossmann, M. G.: Functional anion binding sites in dogfish M_4 lactate dehydrogenase, J. Mol. Biol. 76, 519-531 (1973).

41. Hackert, M. L., Ford, G. C. and Rossmann, M. G.: Molecular orientation and position of the pig M_4 and H_4 isoenzymes of lactate dehydrogenase in their crystal cells, J. Mol. Biol. 78, 665-673 (1973).

42. Eventoff, W., Hackert, M. L. and Rossmann, M. G.: A low resolution crystallographic study of porcine heart lactate dehydrogenase, J. Mol. Biol. 98, 249-258 (1975).

43. Musick, W. D. L., Adams, A. D., Rossmann, M. G., Wheat, T. E. and Goldberg, E.: A low-resolution study of testicular lactate dehydrogenase using the molecular replacement technique, J. Mol. Biol. 104, 659-668 (1976).

44. Olsen, K. W., Moras, D., Rossmann, M. G. and Harris, J. I.: Sequence

variability and structure of D-glyceraldehyde-3-phosphate dehydrogenase, J. Biol. Chem. 250, 9313-9321 (1976).

45. Garavito, R. M., Berger, D. and Rossmann, M. G.: Molecular asymmetry in an abortive ternary complex of lobster glyceraldehyde-3-phosphate dehydrogenase, Biochemistry, submitted for publication (1977).

46. Olsen, K. W., Garavito, R. M., Sabesan, M. N. and Rossmann, M. G.: Studies on coenzyme binding to glyceraldehyde-3-phosphate dehydrogenase, J. Mol. Biol. 107, 577-584 (1976).

47. Watson, H. C., Duée, E. and Mercer, W. D.: Low resolution structure of glyceraldehyde 3-phosphate dehydrogenase, Nature New Biol. 240, 130-133 (1972).

48. Gorjunov, A. I., Andreeva, N. S., Baranowski, T. and Wolny, M.: Molecular symmetry of human glyceraldehyde 3-phosphate dehydrogenase and its transformation during coenzyme binding, J. Mol. Biol. 69, 421-426 (1972).

49. Rossmann, M. G. and Argos, P.: A comparison of the heme binding pocket in globins and cytochrome b_5, J. Biol. Chem. 250, 7525-7532 (1975).

50. Robertus, J. D., Alden, R. A., Birktoft, J. J., Kraut, J., Powers, J. C. and Wilcox, P. E.: An X-ray crystallographic study of the binding of peptide chloromethyl ketone inhibitors to subtilisin BPN', Biochemistry 11, 2439-2449 (1972).

51. Segal, D. M., Powers, J. C., Cohen, G. H., Davies, D. R. and Wilcox, P. E.: Substrate binding site in bovine chymotrypsin A_γ. A crystallographic study using peptide chloromethyl ketones as site-specific inhibitors, Biochemistry 10, 3728-3738 (1971).

52. Kraut, J., Robertus, J. D., Birktoft, J. J., Alden, R. A., Wilcox, P. E. and Powers, J. C.: The aromatic substrate binding site in subtilisin BPN' and its resemblance to chymotrypsin, Cold Spring Harbor Symp. Quant. Biol. 36, 117-123 (1971).

Received February 2, 1977

DISCUSSION

Werber: What is the mechanism by which Arg-171 in lactate dehydrogenase and Arg-231 in glyceraldehyde 3-phosphate dehydrogenase do destabilize the tetrahedral intermediates formed by the respective substrates of these enzymes?

Rossmann: Charge repulsion, but this is only a suggestion.

Bernhard: The acyl-S-CH_2-E bonds in papain and in holo-glyceraldehyde 3-phosphate dehydrogenase are electronically the same, and atypical of thiol ester bonds. Removal or replacement of NAD^+ from holo-acyl-glyceraldehyde 3-phosphate dehydrogenase restores the usual (stable) thiol-ester bond. Therefore, by superposition analogy, papain acyl-enzym is a rigid model for the holo-acyl enzyme of glyceraldehyde 3-phosphate dehydrogenase.
(1) Do these two structures superimpose to indicate the excited acyl-enzyme bond?
(2) Does the papain acyl-enzyme structure indiçate why holo-acyl-glyceraldehyde 3-phosphate dehydrogenase binds NAD^+ so poorly relative to holo-enzyme?

Rossmann: Superposition of papain and glyceraldehyde 3-phosphate dehydrogenase shows a good superposition of the thioester bond of alkyl adducts. Thus it can be concluded that the active tetrahedral intermediates would superimpose equally well.
The binding of NAD^+ or NADH to glyceraldehyde 3-phosphate dehydrogenase can best be compared with that in lactate dehydrogenase. Although the positive charge on Arg-231 will certainly affect the binding of NAD^+ (but not so much of NADH) it is difficult to determine binding energies directly from the inspection of structure.

Dunn: Isn't it really true that the superpositing of substrate structures is dictated by the lengths of chemical bonds and the arrangement of catalytic groups to optimize the chemical transformation rather than to convergent evolution? It seems to me that the structures must superimpose if catalysis is to occur, provided the handedness is the same and similar substrate structures are involved.

Rossmann: Convergent evolution means exactly what you say, namely the occurrance of a few similar characters (such as the catalytic residues) in the presence of a far greater number of differences (e.g. the protein fold).

Saenger: In all dehydrogenases, NAD^+ is bound to the NAD^+ binding domain in a similar conformation. This conformation, however, is unusual since the orientation about the C(4')-C(5') bond is gauche, trans rather than the preferred gauche, gauche. In fact in all the crystals structures of 5'-ribonucleotides and of polynucleotides, the gauche, trans conformation was never observed. The change of two gauche, gauche conformations to two gauche, trans corresponds to about 10 kcal/mole. Is anything known about the reason why NAD^+ adopts this unusual form?

Rossmann: The conformation of NAD when bound to enzymes is not so unusual, as it is always essentially the same. Apparently the conformation of NADP when bound to enzymes is also likely to be similar. Thus there would appear to be two types of dinucleotide structures: Those bound to enzyme and those in solution.

Hollaway: Is there any indication of changes in ion-pairing interactions associated with the movement of Arg-231 in glyceraldehyde 3-phosphate dehydrogenase (or Arg-109 in lactate dehydrogenase) on binding NAD?

Rossmann: We do not see any, but there do seem to be changes in the number of water molecules bound.

Veeger: Ernster et al. (Biochem. Biophys.Acta (1957) 26, 416-421) published that lactate dehydrogenase is rather active with $NADP^+$ at pH 6. How can you accommodate a bulky phosphate group at the 2'-OH-ribose vicinity, if it is, as you propose in close interaction with the peptide chain by Asp-53 and perhaps Gly-28. Furthermore the $NADP^+$ dependent activity delines to zero at pH 6.5, while that with NADH increases.

Rossmann: There is some conformational change of the protein structures in the vicinity of the adenine ribose even on binding NAD^+. Presumably this would have to be much higher on binding $NADP^+$. An increase of pH, altering the charge distribution may affect the ease with which such conformational changes take place.

Trommer: L-specific lactate dehydrogenase does not only bind D-lactate but even forms a productive complex. D-lactate is oxidized, however, at a very low rate. This is not simply caused by a contamination with L-lactate because you get a conversion to pyruvate of more than 10 %. We have found the same for a substrate-analogous reaction with a nicotinamide derivative to the 5-position of which D-lactate is covalently attached via a methylene spacer. Does this seem possible at all from your structural data?

Rossmann: It is difficult to understand how D-lactate can bind productively in the known structure of L-lactate specific lactate dehydrogenase.

Banaszak: Do L-isomers and D-isomers bind to A-side dehydrogenases?

Rossmann: I think Dr. Trommer's comment has already answered this point.

Luisi: Is it too early to compare the various dehydrogenases in terms of water content in the active site?

Rossmann: Yes!

THE PRIMARY STRUCTURES OF CHICKEN LACTATE DEHYDROGENASE M_4 AND H_4 ISOENZYMES

Hans-Joachim Torff, Dagobert Becker and Johanna Schwarzwälder
Institut für Biophysik und Physikalische Biochemie,
Universität Regensburg, 8400 Regensburg, West Germany

INTRODUCTION

Lactate dehydrogenase (LDH;EC 1.1.1.27) is a well-known enzyme of the glycolytic pathway catalysing the reversible reaction of lactate to pyruvate, with NAD as coenzyme. The tetrameric molecule has a molecular weight of 144 000 (1). In higher vertebrates five isoenzymes are found due to two types of subunits, M and H, leading to the homo-tetramers M_4 and H_4 and the three hybrids MH_3, M_2H_2 and M_3H (2,3). The M_4 isoenzyme predominates in sceletal muscle, the H_4 isoenzyme in heart tissue. The two polypeptide chains, M and H, differ in their amino acid composition yielding unique amino acid sequences. The properties of the isoenzymes therefore depend to their subunit composition. As to expect, the greatest differences are found between the M_4 and H_4 isoenzymes. Some of those properties are the specific activity, the isoelectric point, immunological behaviour, substrate affinity, substrate inhibition and coenzyme binding. With respect to these and other properties the M_4 isoenzymes of different, related species behave more similar than the M_4 to the H_4 isoenzyme of the same species; the same is true for the H_4 isoenzymes.
Most of the information characterizing the lactate dehydrogenases and studying the mechanism of action, has been obtained by kinetic investigations, physicochemical studies and chemical modification (4). For further understanding of mechanisms on the molecular level structural investigations have been performed. So the three-dimensional structures of dogfish M_4

and pig H_4 lactate dehydrogenase have been determined (5,6). Preliminary amino acid sequences of LDH-M_4 from dogfish have been published (7,8,9) and the complete amino acid sequence has now been elucidated (10). Very recently the primary structures of pig M_4 and H_4 isoenzymes have been reported (11). We want to present the primary structures of the chicken M_4 and H_4 isoenzymes.

THE AMINO ACID SEQUENCES OF THE M AND H POLYPEPTIDE CHAINS OF CHICKEN LACTATE DEHYDROGENASE

The amino acid compositions of lactate dehydrogenase from a number of vertebrates have been determined including the M_4 and H_4 isoenzymes from chicken (12,13). If one compares the amino acid content of these enzymes from different species an outstanding feature is, that the number of histidine residues in the M_4 enzymes of birds is about double the amount compared

Table 1 Amino acid compositions of the M and H polypeptide chains of chicken lactate dehydrogenase as derived from the amino acid sequences

Amino acid	H chain	M chain	Difference M to H chain
Asp	32	31	- 1
Thr	18	12	- 6
Ser	26	26	----
Glu	29	24	- 5
Pro	9	11	+ 2
Gly	25	26	+ 1
Ala	21	20	- 1
Val	41	37	- 4
Met	6	9	+ 3
Ile	19	24	+ 5
Leu	38	31	- 7
Tyr	8	5	- 3
Phe	5	7	+ 2
His	8	18	+ 10
Lys	25	30	+ 5
Arg	9	9	----
Trp	6	6	----
Cys	7	5	- 2
	332	331	- 1

to the H_4 enzymes. In Table 1 the amino acid residues are given as derived from the amino acid sequences. As can be seen the M chain compared to the H chain has, apart from the difference in the histidines, some more lysine and some less aspartic and glutamic acid residues being in agreement with the observation that in general the M_4 isoenzymes have the highest isoelectric point that means are the most basic ones.
The amino acid sequences of the M and H polypeptide chains of chicken lactate dehydrogenase are shown in Fig.1. Some peptides have been reported (14,15,16,17).

1 10
CH:Ala-Thr-Leu-Lys-Glu-Lys-Leu-Ile-Thr-Pro-Val-Ala-Ala-Gly-Ser-
CM:----Ser-Leu-Lys-Asp-His-Leu-Ile-His-Asn-Val-His-Lys-Glu-Glu-

20 30
CH:Thr-----Val-Pro-Ser-Asn-Lys-Ile-Thr-Val-Val-Gly-Val-Gly-Gln-
CM:His-Ala-His-Ala-His-Asn-Lys-Ile-Ser-Val-Val-Gly-Val-Gly-Ala-

40
CH:Val-Gly-Met-Ala-Cys-Ala-Ile-Ser-Ile-Leu-Gly-Lys-Gly-Leu-Cys-
CM:Val-Gly-Met-Ala-Cys-Ala-Ile-Ser-Ile-Leu-Met-Lys-Asp-Leu-Ala-

50 60
CH:Asp-Glu-Leu-Ala-Leu-Val-Asp-Val-Leu-Glu-Asp-Lys-Leu-Lys-Gly-
CM:Asp-Glu-Leu-Thr-Leu-Val-Asp-Val-Val-Glu-Asp-Lys-Leu-Lys-Gly-

70
CH:Glu-Met-Met-Asp-Leu-Gln-His-Gly-Ser-Leu-Phe-Leu-Gln-Thr-His-
CM:Glu-Met-Met-Asp-Leu-Gln-His-Gly-Ser-Leu-Phe-Leu-Lys-Thr-Pro-

80 90
CH:Lys-Ile-Val-Ala-Asp-Lys-Asp-Tyr-Ala-Val-Thr-Ala-Asn-Ser-Lys-
CM:Lys-Ile-Thr-Ser-Gly-Lys-Asp-Tyr-Ser-Val-Thr-Ala-His-Ser-Lys-

100
CH:Ile-Val-Val-Val-Thr-Ala-Gly-Val-Arg-Gln-Gln-Glu-Gly-Glu-Ser-
CM:Leu-Val-Ile-Val-Thr-Ala-Gly-Ala-Arg-Gln-Gln-Glu-Gly-Glu-Ser-

110 120
CH:Arg-Leu-Asn-Leu-Val-Gln-Arg-Asn-Val-Asn-Val-Phe-Lys-Phe-Ile-
CM:Arg-Leu-Asn-Leu-Val-Gln-Arg-Asn-Val-Asn-Ile-Phe-Lys-Phe-Ile-

130
CH: Ile-Pro-Gln-Ile-Val-Lys-Tyr-Ser-Pro-Asn-Cys-Val-Ile-Leu-Val-
CM: Ile-Pro-Asn-Val-Val-Lys-Tyr-Ser-Pro-Asp-Cys-Lys-Leu-Leu-Ile-

140 150
CH: Val-Ser-Asn-Pro-Val-Asp-Ile-Leu-Thr-Tyr-Val-Thr-Trp-Lys-Leu-
CM: Val-Ser-Asn-Pro-Val-Asp-Ile-Leu-Thr-Tyr-Val-Ala-Trp-Lys-Ile-

160
CH: Ser-Gly-Leu-Pro-Lys-His-Arg-Val-Ile-Gly-Ser-Gly-Cys-Asn-Leu-
CM: Ser-Gly-Phe-Pro-Lys-His-Arg-Val-Ile-Gly-Ser-Gly-Cys-Asn-Leu-

170 180
CH: Asp-Thr-Ala-Arg-Phe-Arg-Tyr-Leu-Met-Ala-Glu-Arg-Leu-Gly-Ile-
CM: Asp-Ser-Ala-Arg-Phe-Arg-His-Leu-Met-Gly-Glu-Arg-Leu-Gly-Ile-

190
CH: His-Pro-Thr-Ser-Cys-His-Gly-Trp-Ile-Leu-Gly-Glu-His-Gly-Asp-
CM: His-Pro-Leu-Ser-Cys-His-Gly-Trp-Ile-Val-Gly-Gln-His-Gly-Asp-

200 210
CH: Ser-Ser-Val-Ala-Val-Trp-Ser-Gly-Val-Asn-Val-Ala-Gly-Val-Ser-
CM: Ser-Ser-Val-Pro-Val-Trp-Ser-Gly-Val-Asn-Val-Ala-Gly-Val-Ser-

220
CH: Leu-Gln-Gln-Leu-Asp-Pro-Ala-Met-Gly-Thr-Asp-Lys-Asp-Ser-Glu-
CM: Leu-Lys-Ala-Leu-His-Pro-Asp-Met-Gly-Thr-Asp-Ala-Asp-Lys-Glu-

230 240
CH: Asn-Trp-Lys-Glu-Val-His-Lys-Gln-Val-Val-Glu-Ser-Ala-Tyr-Glu-
CM: His-Trp-Lys-Glu-Val-His-Lys-Gln-Val-Val-Asp-Ser-Ala-Tyr-Glu-

250
CH: Val-Ile-Arg-Leu-Lys-Gly-Tyr-Thr-Asn-Trp-Ala-Ile-Gly-Leu-Ser-
CM: Val-Ile-Lys-Leu-Lys-Gly-Tyr-Thr-Ser-Trp-Ala-Ile-Gly-Leu-Ser-

260 270
CH: Val-Ala-Glu-Leu-Cys-Glu-Thr-Met-Leu-Lys-Asn-Leu-Tyr-Arg-Val-
CM: Val-Ala-Asp-Leu-Ala-Glu-Thr-Ile-Met-Lys-Asn-Leu-Arg-Arg-Val-

280
CH: His-Ser-Val-Ser-Thr-Leu-Val-Lys-Gly-Thr-Tyr-Gly-Ile-Glu-Asn-
CM: His-Pro-Ile-Ser-Thr-Ala-Val-Lys-Gly-Met-His-Gly-Ile-Lys-Asp-

290 300
CH: Asp-Val-Phe-Leu-Ser-Leu-Pro-Cys-Val-Leu-Ser-Ala-Ser-Gly-Leu-
CM: Asp-Val-Phe-Leu-Ser-Val-Pro-Cys-Val-Leu-Gly-Ser-Ser-Gly-Ile-

```
                                             310
CH:Thr-Ser-Val-Ile-Asn-Gln-Lys-Leu-Lys-Asp-Asp-Glu-Val-Ala-Gln-
CM:Thr-Asp-Val-Val-Lys-Met-Ile-Leu-Lys-Pro-Asp-Glu-Glu-Glu-Lys-

                        320                                 330
CH:Leu-Lys-Lys-Ser-Ala-Asp-Thr-Leu-Trp-Ser-Ile-Gln-Lys-Asp-Leu-
CM:Ile-Lys-Lys-Ser-Ala-Asp-Thr-Leu-Trp-Gly-Ile-Gln-Lys-Glu-Leu-

CH:Lys-Asp-Leu
CM:Gln-----Phe
```

Fig. 1 The amino acid sequences of chicken heart (CH) and chicken muscle (CM) lactate dehydrogenase. Identical positions are indicated by boxed areas. Position 17 in the CH-sequence and positions 1 and 332 in the CM-sequence are deletions.

THE AMINO ACID SEQUENCES OF CHICKEN AND PIG H_4 AND M_4 LACTATE DEHYDROGENASE ISOENZYMES

Since the primary structures of pig H_4 and M_4 have been published recently (11), a comparison of the two pairs of isoenzymes shall be given in Fig. 2.

```
   1                                      20
CH:A-T-L-K-E-K-L-I-T-P-V-A-A-G-S-T---V-P-S-N-K-I-T-V-V-G-V-G-
PH:A-T-L-K-E-K-L-I-A-P-V-A-Q-Q-E-T-T-I-P-N-B-K-I-T-V-V-G-V-G-
PM:A-T-L-K-D-Q-L-I-H-N-L-L-K-E-E---H-V-P-H-N-K-I-T-V-V-G-V-G-
CM:--S-L-K-D-H-L-I-H-N-V-H-K-E-E-H-A-H-A-H-N-K-I-S-V-V-G-V-G-

                         40
CH:Q-V-G-M-A-C-A-I-S-I-L-G-K-G-L-C-D-E-L-A-L-V-D-V-L-E-D-K-L-
PH:Q-V-G-M-A-C-A-I-S-I-L-G-K-S-L-T-D-E-L-A-L-V-D-V-L-E-D-K-L-
PM:A-V-G-M-A-C-A-I-S-I-L-M-K-E-L-A-D-E-I-A-L-V-D-V-M-E-D-K-L-
CM:A-V-G-M-A-C-A-I-S-I-L-M-K-D-L-A-D-E-L-T-L-V-D-V-V-E-D-K-L-

     60                                   80
CH:K-G-E-M-M-D-L-Q-H-G-S-L-F-L-Q-T-H-K-I-V-A-D-K-D-Y-A-V-T-A-
PH:K-G-E-M-M-D-L-Q-H-G-S-L-F-L-Q-T-P-K-I-V-A-N-K-D-Y-S-V-T-A-
PM:K-G-E-M-M-D-L-Q-H-G-S-L-F-L-R-T-P-K-I-V-S-G-K-D-Y-N-V-T-A-
CM:K-G-E-M-M-D-L-Q-H-G-S-L-F-L-K-T-P-K-I-T-S-G-K-D-Y-S-V-T-A-
```

```
                           100
CH:N-S-K-I-V-V-V-T-A-G-V-R-Q-Q-E-G-E-S-R-L-N-L-V-Q-R-N-V-N-V-
PH:N-S-K-I-V-V-V-T-A-G-V-R-Q-Q-E-G-E-S-R-L-N-L-V-Q-R-N-V-N-V-
PM:N-S-R-L-V-V-I-T-A-G-A-R-Q-Q-E-G-E-S-R-L-N-L-V-Q-R-N-V-N-I-
CM:H-S-K-L-V-I-V-T-A-G-A-R-Q-Q-E-G-E-S-R-L-N-L-V-Q-R-N-V-N-I-
         120                                     140
CH:F-K-F-I-I-P-Q-I-V-K-Y-S-P-N-C-V-I-L-V-V-S-N-P-V-D-I-L-T-Y-
PH:F-K-F-I-I-P-Q-I-V-K-Y-S-P-N-C-I-I-I-V-V-S-N-P-V-D-I-L-T-Y-
PM:F-K-F-I-I-P-N-I-V-K-Y-S-P-N-C-K-L-L-V-V-S-N-P-V-D-I-L-T-Y-
CM:F-K-F-I-I-P-N-V-V-K-Y-S-P-D-C-K-L-L-I-V-S-N-P-V-D-I-L-T-Y-
                               160
CH:V-T-W-K-L-S-G-L-P-K-H-R-V-I-G-S-G-C-N-L-D-T-A-R-F-R-Y-L-M-
PH:V-A-W-K-L-S-G-L-P-K-H-R-V-I-G-S-G-C-N-L-D-S-A-R-F-R-Y-L-M-
PM:V-A-W-K-I-S-G-F-P-K-N-R-V-I-G-S-G-C-N-L-D-S-A-R-F-R-Y-L-M-
CM:V-A-W-K-I-S-G-F-P-K-H-R-V-I-G-S-G-C-N-L-D-S-A-R-F-R-H-L-M-
             180                                     200
CH:A-E-R-L-G-I-H-P-T-S-C-H-G-W-I-L-G-E-H-G-D-S-S-V-A-V-W-S-G-
PH:A-E-K-L-G-V-H-P-S-S-C-H-G-W-I-L-G-E-H-G-D-S-S-V-A-V-W-S-G-
PM:G-E-R-L-G-V-H-P-L-S-C-H-G-W-I-L-G-E-H-G-D-S-S-V-P-V-W-S-G-
CM:G-E-R-L-G-I-H-P-L-S-C-H-G-W-I-V-G-Q-H-G-D-S-S-V-P-V-W-S-G-
                                   220
CH:V-N-V-A-G-V-S-L-Q-Q-L-D-P-A-M-G-T-D-K-D-S-E-N-W-K-E-V-H-K-
PH:V-N-V-A-G-V-S-L-Q-Q-L-B-P-Z-M-G-T-D-N-D-S-E-N-W-K-E-V-H-K-
PM:V-N-V-A-G-V-S-L-K-N-L-H-P-E-L-G-T-D-A-D-K-E-H-W-K-A-V-H-K-
CM:V-N-V-A-G-V-S-L-K-A-L-H-P-D-M-G-T-D-A-D-K-E-H-W-K-E-V-H-K-
                 240                                      260
CH:Q-V-V-E-S-A-Y-E-V-I-R-L-K-G-Y-T-N-W-A-I-G-L-S-V-A-E-L-C-E-
PH:M-V-V-E-S-A-Y-E-V-I-K-L-K-G-Y-T-N-W-A-I-G-L-S-V-A-D-L-I-E-
PM:E-V-V-D-S-A-Y-E-V-I-K-L-K-G-Y-T-S-W-A-I-G-L-S-V-A-D-L-A-E-
CM:Q-V-V-D-S-A-Y-E-V-I-K-L-K-G-Y-T-S-W-A-I-G-L-S-V-A-D-L-A-E-
                                       280
CH:T-M-L-K-N-L-Y-R-V-H-S-V-S-T-L-V-K-G-T-Y-G-I-E-N-D-V-F-L-S-
PH:S-M-L-K-N-L-S-R-I-H-P-V-S-T-M-V-Q-G-M-Y-G-I-E-N-E-V-F-L-S-
PM:S-I-M-K-N-L-R-R-V-H-P-I-S-T-M-I-K-G-L-Y-G-I-K-E-N-V-F-L-S-
CM:T-I-M-K-N-L-R-R-V-H-P-I-S-T-A-V-K-G-M-H-G-I-K-D-D-V-F-L-S-
                     300
CH:L-P-C-V-L-S-A-S-G-L-T-S-V-I-N-Q-K-L-K-D-D-E-V-A-Q-L-K-K-S-
PH:L-P-C-V-L-N-A-R-G-L-T-S-V-I-N-Q-K-L-K-D-D-E-V-A-Q-L-K-N-S-
PM:V-P-C-I-L-G-Q-N-G-I-S-D-V-V-K-V-T-L-T-P-E-E-E-A-H-L-K-K-S-
CM:V-P-C-V-L-G-S-S-G-I-T-D-V-V-K-M-I-L-K-P-D-E-E-E-K-I-K-K-S-
   320                       333
CH:A-D-T-L-W-S-I-Q-K-D-L-K-D-L
PH:A-D-T-L-W-G-I-Q-K-D-L-K-D-L
PM:A-D-T-L-W-G-I-Q-K-E-L-Q---F
CM:A-D-T-L-W-G-I-Q-K-E-L-Q---F
```

Fig. 2 Comparison of the chicken H_4 (CH) and M_4 (CM) and pig H_4(PH) and M_4 (PM) lactate dehydrogenase isoenzymes. One-letter-nomenclature has been used; boxed areas indicate identical positions.

DISCUSSION

The sequences shall be discussed here from a more statistical point of view. The functional role of distinct amino acids, the meaning of certain substitutions in discrete positions and the correlation of the sequences to the tertiary structures of dogfish M_4 and pig H_4 will be given together with the dogfish M_4 and pig M_4 and H_4 sequences in a lecture by M.G. Rossmann and are discussed elsewhere (18).

A comparison of the chicken sequences show that 252 (76%) of the residues occur in identical sequence. The nonidentical residues are not spread statistically over all the sequence but accumulate at the N- and C-terminus. So only 5 (25%) of the first N-terminal 20 residues and 17 (56%) of the last 30 C-terminal ones are identical. On the other hand as to expect, sequences representing active center domains only show one or two substitutions in sequencies up to 20 and more amino acid residues. What the distribution of less frequent amino acids is concerned, all the 6 tryptophanes are in the same positions in both enzymes. The 5 phenylalanines in the H chain will be found in the same positions in the M chain what is true as well for 5 methionines out of 6. From the 9 arginine residues in each of the sequences 8 are in the same positions. A species-specific feature may be the accumulation of histidines in the N-terminal 20 residues of the muscle enzyme; 6 of the 18 histidine residues are found here.

Since now from two species, pig and chicken, the M_4 and H_4 isoenzyme sequences are available, and additional the M_4 from dogfish, more principal common features may be seen by comparison as well as random or species-specific ones and be correlated to known results of other experiments. From the comparison of the four sequences in Fig. 2 it shows, that 66% of the positions are identical. If only the H_4 isoenzymes are compared, about 90% of the sequences become identical and the C-terminal 34 amino acid residens even show only two substitutions although this C-terminal sequence has quite a low degree of homology if one compares the H_4 and M_4 isoenzymes of either pig or chicken. The longest sequence of identical residues in both types of enzymes is that including the loop region, which undergoes a conformational change on coenzyme binding. So, in both H_4 enzymes this sequence has been completely conserved over a range of 47 amino acids (pos. 85-131).

REFERENCES

1. Jaenicke, R. and Knof, S. : Molecular Weight and Quaternary Structur of Lactic Dehydrogenase. Europ. J. Biochem., 4, 157-163 (1968).
2. Wieland, T. and Pfleiderer, G. : Nachweis der Heterogenität von Milchsäure-dehydrogenase verschiedenen Ursprungs durch Trägerelektrophorese. Biochem. Z., 329, 112-116 (1957).
3. Appels, E. and Markert, C.L. : Dissociation of Lactate Dehydrogenase into subunits with Guanidine Hydrochloride. Biochem. Biophys. Res. Commun., 6, 171-176 (1961).
4. Everse, J. and Kaplan, N.O. : Lactate Dehydrogenases: Structure and Function. Adv. Enzymol., 37, 61-133 (1973).
5. White, J.L., Hackert, M.L., Buehner, M., Adams, M.J., Ford, G.C., Lentz, P.J.Jr., Smiley, I.E., Steindel, S.J. and Rossmann, M.G. : A Comparison of the Structures of Apo Dogfish M_4 Lactate Dehydrogenase and its Ternary Complexes. J. Mol. Biol., 102, 759-779 (1976).
6. Eventoff, W., Hackert, M.L. and Rossmann, M.G. : Low-resolution Crystallographic Study of Porcine Heart Lactate Dehydrogenase. J. Mol. Biol., 98, 249-258 (1975).
7. Taylor, S.S., Oxley, S.S., Allison, W.S. and Kaplan, N.O. : Amino Acid Sequence of Dogfish M_4 Lactate Dehydrogenase. Proc. Natl. Acad. Sci. USA, 70, 1790-1794 (1973).
8. Holbrook, J.J., Liljas, A., Steindel, S.J. and Rossmann, M.G., The Enzymes, (Boyer ed.), Vol. XI, 191-192 (1975).

9. Taylor, S.S., Allison, W.S. and Kaplan, N.O.: The Amino Acid Sequence of the Tryptic Peptides Isolated from Dogfish M_4 Lactate Dehydrogenase. J.Biol.Chem. 250,8740-8747(1975).
10. Taylor, S.S., in press, private communication.
11. Kiltz, H.-H., Keil, W., Griesbach, M. Petry, K. and Meyer, H.: The Primary Structure of Porcine Lactate Dehydrogenase: Isoenzymes M_4 and H_4. Z.Physiol.Chem. 358,122-127(1977).
12. Pesce, A., McKay, R.H., Stolzenbach, F., Cahn, R.D. and Kaplan, N.O. : The Comparative Enzymology of Lactic Dehydrogenases. J. Biol. Chem., 239,1753-1761(1964).
13. Pesce, A., Fondy, T.P., Stolzenbach, F., Castillo, F. and Kaplan, N.O. : The Comparative Enzymology of Lactic Dehydrogenases. J. Biol. Chem. 242,2151-2167(1967).
14. Fondy, T.P., Everse, J., Driscoll, G.A., Castillo, F., Stolzenbach, F.E. and Kaplan, N.O.: The Comparative Enzymology of Lactic Dehydrogenases. J.Biol.Chem., 240,4219-4234(1965)
15. Mella, K., Fölsche, E.Th.J., Torff, H.-J. and Pfleiderer, G. : Die Aminosäuresequenzen um den essentiellen Cysteinrest in Schweineherz-, Schweinemuskel- und Hühnermuskel-Lactatdehydrogenase. Z. Physiol. Chem. 349,891-895(1968).
16. Pfleiderer, G., Woenckhaus, C., Jeckel, D. and Mella, K. : Studies on Structure and Active Sites of Lactate Dehydrogenase from pig Heart and pig Muscle. In Pyridine Nucleotide-Dependant Dehydrogenases, H. Sund, ed., Springer-Verlag, pp. 145-155 (1970).
17. Brummel, M.C., Sanborn, B.M. and Stegink, L.D. : Chicken Heart H_4 Lactate Dehydrogenase: N-Terminal and C-Terminal Residues. Arch. Biochem. Biophys. 143,330-335(1971).
18. Eventoff, W., Rossmann, M.G., Taylor, S.S., Torff, H.-J., Meyer, H., Keil, W. and Kiltz, H.-H.: Structural Adaptations of Lactate Dehydrogenase Isozymes. Proc. Natl. Acad. Sci., USA, submitted, (1976).

Received February 28, 1977

DISCUSSION

Sund: Table 1 of your reprint shows 332 amino acid residues in both cases but your slide shows only 331 residues in the M enzyme and, therefore, a difference of one in the total number of amino acid residues.

Torff: That's right. As I mentioned in my lecture, in the reprint, I sent you, the glycine in position 1 of the M_4 isozyme, has been given erroneously; so, this slide of Table 1 is the correct one with 332 amino acid residues in the H_4 isozyme and 331 in the M_4 and a difference of 1 amino acid between both.

Harris: Have you been able to correlate sequence changes in the different form of LDH: (a) with features of the three-dimensional structure, such as coenzyme binding, subunit contacts, active site, buried and surface residues etc., and (b) with the different properties manifest by the different isozyme forms?

Torff: We have built a model of the LDH subunit according to the coordinates of the apoenzyme from dogfish muscle LDH given by Michael Rossmann and his coworkers to correlate sequence results and known experimental data to the tertiary structure. But since there is a common publication of the authors of the LDH sequences and the tertiary structure people in press, I think Michael Rossmann will be the competent man here to give a comment.

Rossmann: It has been a pleasure to see the emergence of LDH sequence results such as those reported by Joachim Torff just now. I would like to recall the history of the work and draw your attention to some correction to the literature.
The first tentative complete sequence of LDH was that of Susan Taylor's published, with her permission, in the article by Holbrook et al. in "The Enzymes", Vol. XI, 3rd ed., (page 204). Subsequently Hans-Hermann Kiltz, working on pig M_4 LDH and H_4 LDH, showed that the sequence needed some rearrangement between the residues around 133-146 with those preceding 251. This was simultaneously confirmed by Susan Taylor. Later, at the Hamburg conference, it was possible to arrange Joachim Torff's peptides according to the corrected sequences, giving us for the first time three completed M and two H primary structures. It appears that the most important M to H change is the substitution of a Gln (H) to an Ala (M) of position 31. This effects the NMN phosphate, altering slightly the position of the nicotinamide ring. The largest number of changes occur in the R-axis contact between the amino terminal arm (1-22) and residues around 300 in a different subunit. On the other hand the Q-axis contact area is highly conserved showing that the MDH dimer is an older evolutionary event than the formation of the LDH tetramer by virtue of the additional arm. Truely the external residues are less conserved than the internal amino acids.

Gutfreund: Have you done any kinetic experiments on the chicken enzymes?

Torff: No, we haven't. Those experiments concerning kinetic constants and other characteristics of the chicken enzymes have been done by N.O. Kaplan and his coworkers some fifteen years ago.

Kiltz: I am very impressed on the much higher degree of homology between pig and chicken H chains as compared with the M chains. A close inspection of the amino acid exchanges in respect to the three-dimensional structure (see the modell I have brought with me) shows, that most of them occur at the surface of the tetrameric molecule. This means that the surfaces of the H_4 isozymes are much more similar as the M_4 isozymes. I see here a connection to the observation, that in general the M_4 isozymes are strong immunogens whereas the H_4 isozymes are very weak immunogens.

Otto: I wish to add some amino acid sequences of a further dehydrogenase, the glycerophosphate dehydrogenase (GDH) from rabbit muscle. The glycerophosphate dehydrogenase has two identical polypeptide chains of about 360 amino acids. I am going to show you the last two thirds of the amino acid sequence (Fig. 1). We have sequenced an additional part of 37 amino acids. Comparing these results with other dehydrogenases there are two interesting alignments. The first is a pentapeptide (Fig. 2), which is homologous to peptides containg the histidine 195 (or 193) of the active center in the LDH and the reactive histidine in the mitochondrial malate dehydrogenase (MDH), respectively. It seems to be obvious that this pentapeptide contains the His of the active center of the glycerophosphate dehydrogenase. The other interesting alignment seems to be homologous to the NMN binding unit of LDH (Fig. 3). It is the longest homologous alignment that is evident in

Ac(Gly Ala) ______________________________

______________ $(X)_n$ ______________

____

________ Met Ser Val Leu Met Gly Ala Asn Ile Ala Ser Glu Val Ala

Asp Glu Lys Phe Cys Glu Thr Thr Ile Gly Cys Lys Asp Gln Ala Gln Gly Gln

Leu Leu Lys Gln Leu Met ______________________

______________ $(X)_m$ ______________

______________ Met Pro Arg Val Thr Met(140) Trp Val Phe Glu

Glu Asp Ile Gly Gly Lys(150) Lys Leu Thr Glu Ile Ile Asn Thr His Glu(160) Glu Asn

Val Lys Tyr Leu Pro Gly His Lys(170) Leu Pro Pro Asn Val Val Ala Val Pro Asp(180)

Val Val Lys Ala Ala Ala Asp Ala Asp Ile(190) Leu Ile Phe Val Val Pro His Gln

Phe Ile(200) Gly Lys Ile Cys Asp Glu Ile Lys Gly His(210) Leu Lys Ala Asn Ala Ile

Gly Ile Ser Leu(220) Ile Lys Gly Val Asn Glu Gly Pro Lys Gly(230) Leu Lys Leu Ile

Ser Glu Val Ile Gly Glu(240) His Leu Gly Ile Pro Met Ile Ala Phe Ala(250) Lys Leu

Phe Cys Ser Gly Pro Val Ser Pro(260) Ala Thr Phe Leu Glu Ser Cys Gly Val Ala(270)

Asp Leu Ile Thr Thr Cys Tyr Gly Gly Arg(280) Asn Arg Lys Val Ala Glx Ala Phe

Ala Arg(290) Thr Gly Lys Ser Ile Glu Gln Leu Glu Lys(300) Glu Met Leu Asn Gly Gln

Lys Leu Gln Gly(310) Pro Glu Thr Ala Arg Glu Leu His Ser Ile(320) Leu Gln His Lys

Gly Leu Val Asp Trp Phe(330) Phe Leu Phe Met Ala Val Tyr Lys Val Cys(340) Tyr Gln

Gly Gln Pro Val Gly Glu Phe Ile(350) Arg Cys Leu Gln Asn His Pro Glu His Met(360)

Amino acid sequence of glycerophosphate dehydrogenase from rabbit muscle

(n + m = 94)

Fig. 1

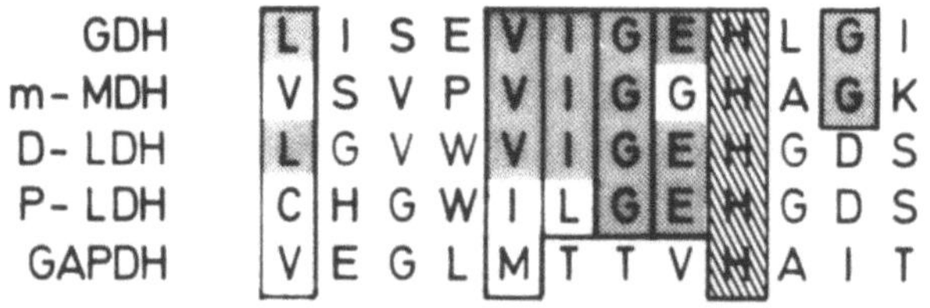

Fig. 2 Alignment of partial sequences containing the His residues in the active site of some dehydrogenases with the homologous part from the GDH sequence given in Fig. 1 (D-LDH and P-LDH, LDH from dogfish and pig, respectively; GAPDH from pig). Identical or conservatively exchanged residues are indicated by boxes.

any two different dehydrogenases. Calculations using the method of Chou and Fasman may indicate that there is an α helix between the positions 174 and 188. This would be in an agreement with the α E part of the LDH. A comparison of the two homologous alignments shows that an elongation may have occurred at the beginning of the chain, provided that there is no missing link of about 40 amino acids between the CNBr peptides No. 6 and 7 or No. 8 and 9, because the overlaps have not been established with absolute certainty. Nevertheless, our results suggest, the highest degree of family connection of all investigated dehydrogenases is between LDH and GDH. I agree with Dr. Rossmann that the method of X-ray crystallography is a powerful tool to demonstrate the similarities of the nucleotide binding enzymes and the extended group of related proteins. I am convinced, however, that the analysis of amino acid sequences is still necessary for understanding the principles of folding and to characterize the close relationship within the family.

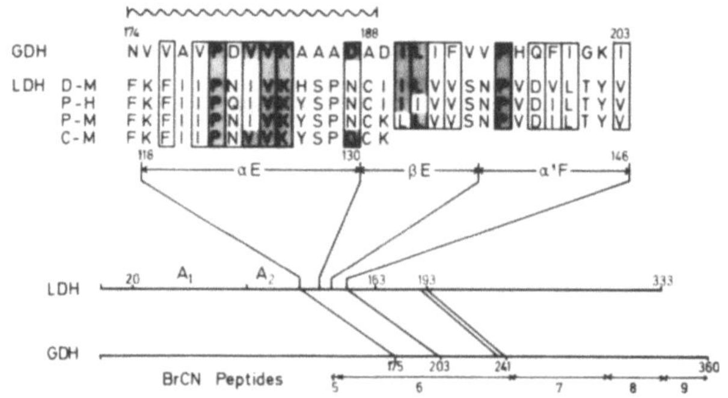

Fig. 3 Comparison of the GDH sequence underlined in Fig. 1 (broken line) with the sequence 117 to 146 of LDH subunits (D: dogfish; P: pig; C: chicken) and their positions on the polypeptide chains of both enzymes. LDH 193 and GDH 241 are the histidine residues shown in Fig. 2.

STRUCTURE AND PROPERTIES OF GLYCERALDEHYDE 3-PHOSPHATE DEHYDROGENASE FROM THERMOPHILIC MICROORGANISMS

J. Ieuan Harris and John E. Walker
Medical Research Council Laboratory of Molecular Biology
Hills Road, Cambridge CB2 2QH, England

Glyceraldehyde 3-phosphate dehydrogenase (GPDH) which catalyses the oxidative phosphorylation of glyceraldehyde 3-phosphate in the presence of the coenzyme NAD is a tetrameric enzyme comprising four chemically identical subunit polypeptide chains (for review see ref. 1). Amino acid sequence studies have shown that the subunit consists of about 330 amino acids and the three-dimensional structure of enzyme/NAD complex (holoenzyme) from lobster muscle has been determined (2,3). A knowledge of the structure of NAD-free (apo)enzyme is necessary to identify the structural changes that occur when NAD binds by a negatively cooperative mechanism to the tetrameric apoprotein. Enzyme from which NAD has been totally removed tends to be unstable and attempts to prepare crystals of apoenzyme suitable for high resolution X-ray diffraction analysis have not been successful. We therefore turned our attention to glyceraldehyde 3-phosphate dehydrogenase from thermophilic microorganisms (4,5,6) with the expectation that their greater stability could be utilised to advantage in studies of the structure and mechanism of action of this "evergreen" enzyme.

Bacillus stearothermophilus and Thermus aquaticus are thermophiles that achieve optimal growth at temperatures of 60-65°C and 70-75°C respectively. Glyceraldehyde 3-phosphate dehydrogenase has been isolated from the two thermophiles (5,6) and, as had been anticipated, the pure enzymes in both holo and apo forms were considerably more stable than their mesophile counterparts. Enzyme from B. stearothermophilus has been crystallised with and without bound NAD (5) as well as with a range of intermediate NAD stoichiometries between zero and four moles per tetramer (7). A study of this enzyme (7) involving parallel amino acid sequence and X-ray crystallographic analysis has been undertaken and the combined results are summarised in this and in an accompanying paper by A.J. Wonacott and G. Biesecker (8). In addition to its relevance to the structure and the mechanisms of coenzyme binding and subunit cooperativity in glyceraldehyde 3-phosphate dehydrogenase this study has also, for the first time, made it possible to compare the sequence and three-dimensional structure of the same multisubunit enzyme from a mesophile (lobster) and a thermophile (B. stearothermophilus).

MOLECULAR STRUCTURE

The complete amino acid sequence of the subunit polypeptide chain of B. stearothermophilus GPDH (Fig. 1) shows it to be highly homologous to the sequence of the lobster muscle enzyme. About a half (51%) of the residues are identical in the two proteins [as compared with a 70% identity between lobster and pig, and a 68% homology between lobster and yeast (1)]. This degree of sequence homology is virtually the same as that found previously (6) in a comparison of the sequence of another procaryotic thermophile (T. aquaticus) enzyme with that of the lobster enzyme. Moreover these experimentally determined values are in close agreement with the figure of 50% homology predicted using a figure of two accepted point mutations per 100 residues per 10^8 years and on the assumption that eukaryotes diverged from procaryotes 3×10^9 years ago (9). Comparison of the sequences of five GPDHs (Fig. 2), three from eukaryotic mesophiles, and two from procaryotic thermophiles, shows that 33% (110 out of the 330 residues in the subunit polypeptide chain) of the residues are conserved. This degree of sequence homology in proteins from such evolutionarily diverse sources shows that the sequence of glyceraldehyde 3-phosphate dehydrogenase has been conserved to a much greater extent than that of other glycolytic enzymes such as alcohol dehydrogenase (10) and triosephosphate isomerase (11).

Now that the three-dimensional structures of lobster muscle (2,3) and B. stearothermophilus enzymes (7,8) are known it is of interest to relate conserved residues to the three-dimensional structure [cf. (12)]. Such a comparison (Figs. 2 and 3) shows that residues involved in binding substrates (NAD and G-3P), and in subunit interfaces are conserved to a greater extent than others. Residues thought to be important for catalysis, such as Cys-149, His-176 (and possibly Tyr-311) are retained in stretches of

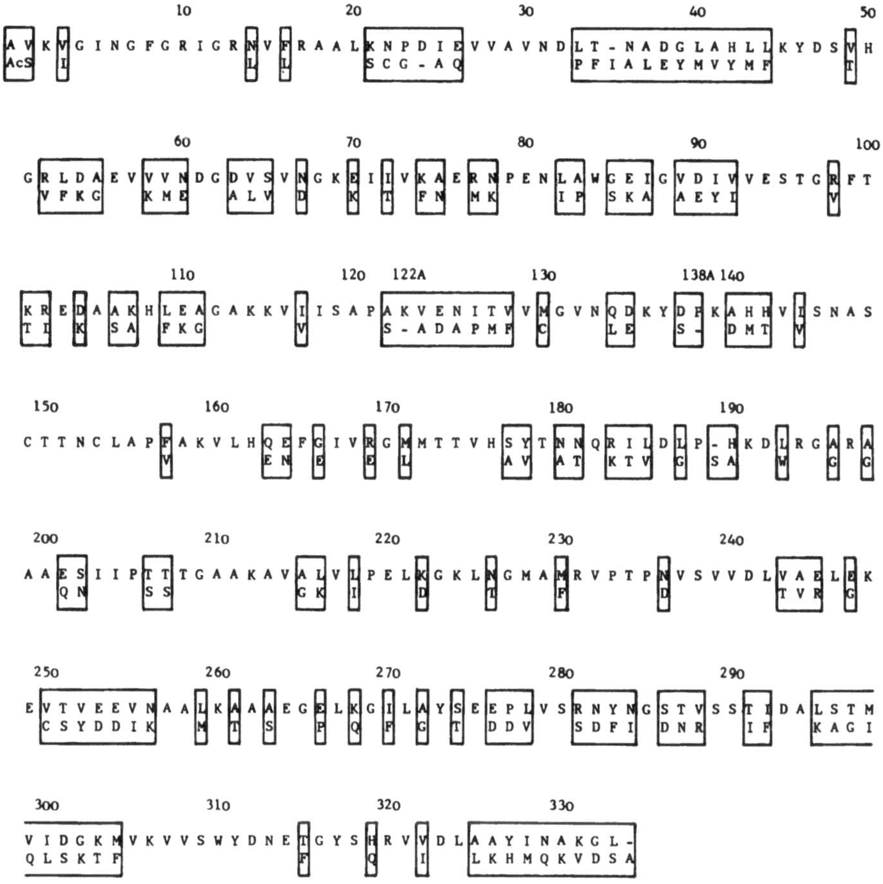

Figure 1 Comparison of the primary structures of glyceraldehyde 3-phosphate dehydrogenase from B. stearothermophilus (upper) and lobster (lower).

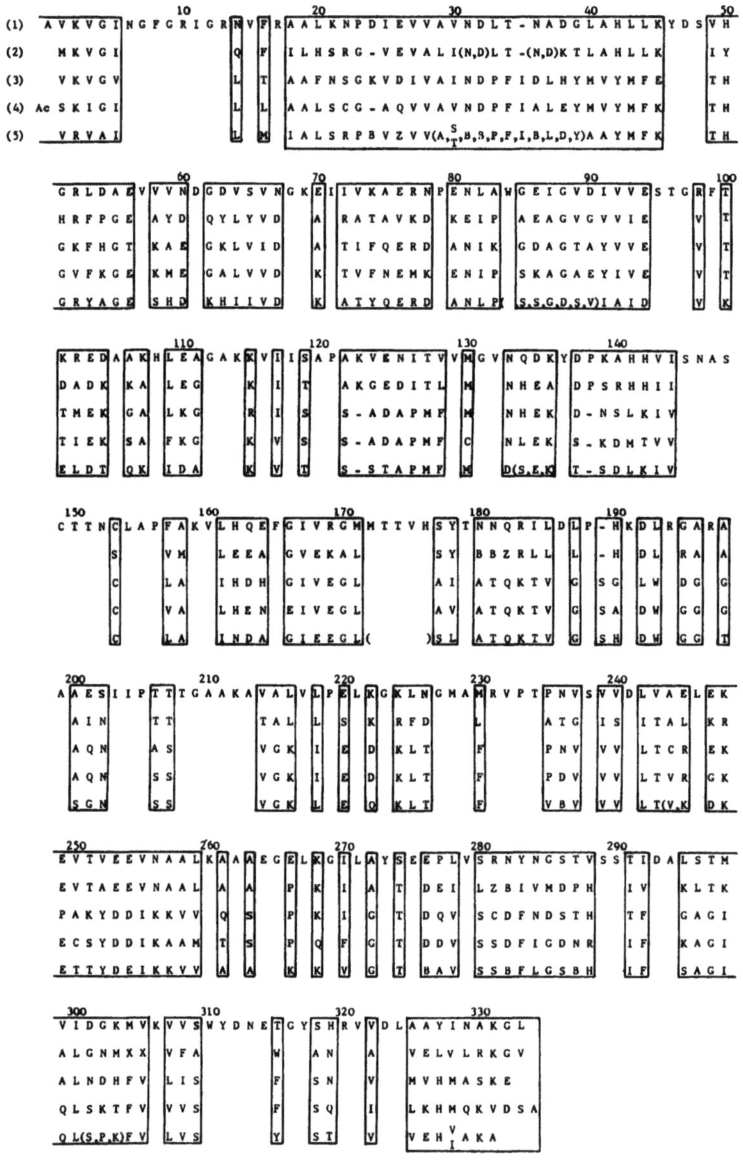

Figure 2 Primary structures of glyceraldehyde 3-phosphate dehydrogenase from B. stearothermophilus (1), T. aquaticus (2), pig (3), lobster (4) and yeast (5).

conserved residues but others, invariant in the mesophile (lobster, pig and yeast) enzymes, and thought to be important, are not conserved in the two procaryotic thermophile sequences. Among these are Cys-153 and Lys-183. Although Cys-153 is conserved in all other known GPDHs (including the B. stearothermophilus enzyme) the three-dimensional structure does not indicate any special role for it. Its replacement by serine in the T. aquaticus protein may therefore have occurred for reasons of stability since it is known that formation of an intrachain disulphide bond with Cys-149 can occur and lead to irreversible inactivation. Similarly Lys-183, conserved in all known mesophile sequences, had been thought to be important, because of its participation in a specific acyl-transfer reaction involving Cys-149 both in muscle and yeast enzymes, and possible involvement in binding NAD [cf. (1)]. Its replacement by arginine in the two thermophile enzymes shows that acyl-transfer reactions to Lys-183 are not important for enzyme function, but the significance (if any) of the change from lysine to arginine must await the further interpretation of the three-dimensional structure in the S-loop (7) part of the molecule (see below). The overall structure of the subunit of the thermophile holo-GPDH is closely similar to that described for lobster GPDH, consisting of the two folding domains shown in Fig. 3. The first domain, residues 0-148, in which the secondary and tertiary structural features of the lobster enzyme are reproduced quite accurately, contains most of the residues involved in NAD binding. Side chains implicated in catalysis are contained in the second domain, the region comprising residues 149-333. In this domain the interaction of the β-sheet regions, β_5 with β_6, and β_7 with β_8, is different from that of the lobster enzyme and gives rise to a modified hydrogen-bonding pattern. This change results in the burying of aspartic acid 293 which has significance for thermostability (see below). An important feature of this domain is an irregular S-shaped loop of polypeptide chain (residues 178-201) which is in contact with the NAD and with the S-loop and the α_c-helix of the subunit related by the dyad axis marked R [cf. (7,8)]. The S-loop also interacts with several amino acids across the P-axis.

The sequence of the S-loop has been highly conserved in the three mesophile sequences; the two thermophile sequences, although very similar to each

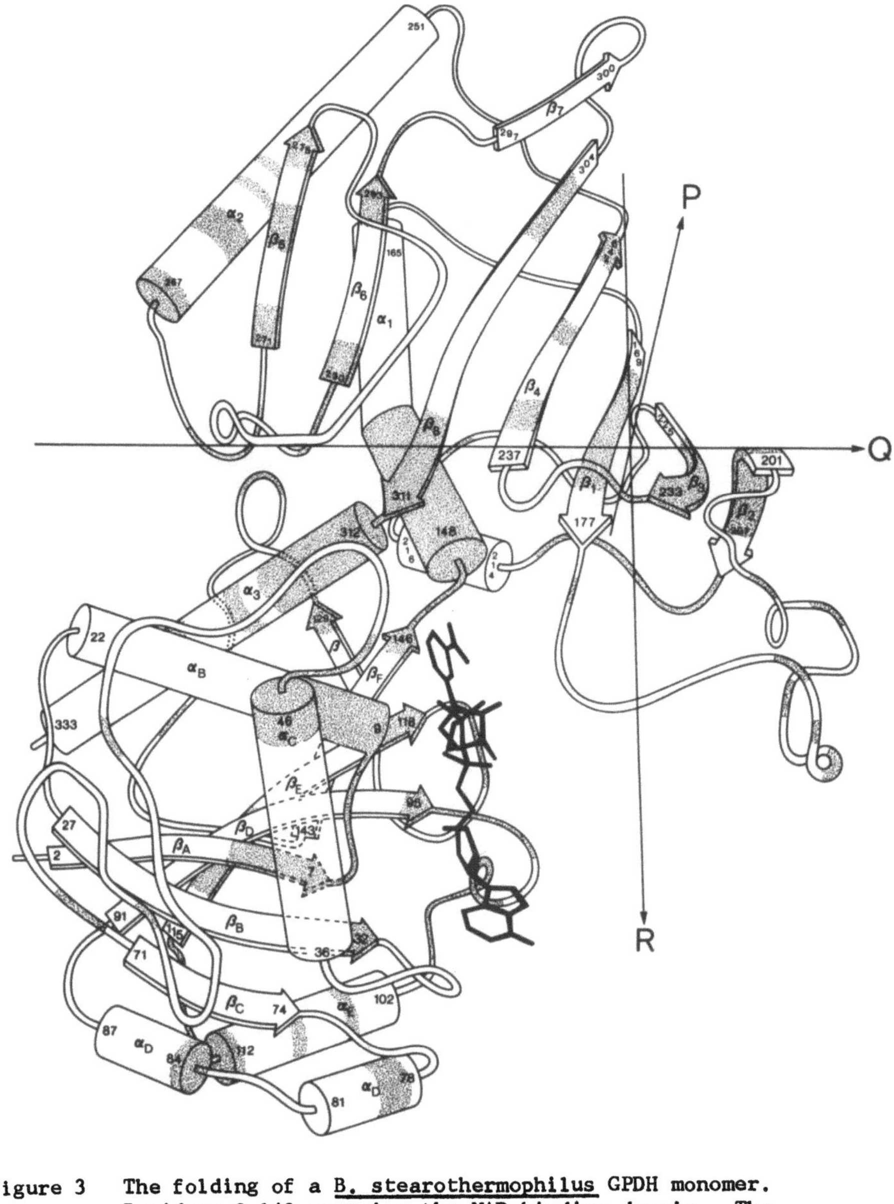

Figure 3 The folding of a B. stearothermophilus GPDH monomer. Residues 0-149 comprise the NAD binding domain. The catalytic (C-terminal) domain has an antiparallel β-sheet region which forms an interface with the equivalent region of the P-axis related subunit. The S-loop is residues 178-201. Shaded regions correspond to residues conserved in all five species of enzyme.

other, are however quite different (Fig. 4). This is surprising since

			180										190										200	
P:	I	T	A	T	Q	K	T	V	D	G	P	S	G	K	L	W	R	D	G	R	G	A	A	Q
L:	V												A		D			G			G		A	Q
Y:	L												H		D			G			T		S	G

<u>B. st</u>:	Y	T	N	N	Q	R	I	L	D	L	P	-	H	K	D	L	R	G	A	R	A	A	A	E
<u>T. aq</u>:							L					-						R						I
*		T							D		P			K			R			R		A		

Figure 4 Amino acid sequence of "S-loop" parts of GPDH subunits: P (pig), L (lobster), Y (yeast), <u>B. stearothermophilus</u> and <u>T. aquaticus</u>
* Residues conserved in five enzymes.

the four S-loops form the core of the tetramer. Most of their residues are internal and essential to the interactions across the R-axis (cf. Fig. 3). The conserved residues (Fig. 4) are, however, similarly oriented in both the mesophile and thermophile structures. Moreover although the conformation of the S-loop is flexible it is partly maintained by the packing of side chains and by interactions with neighbouring subunits through contacts across the P- and R-axes. The S-loop contacts are more hydrophobic in the two thermophile enzymes and this may be a factor in the thermal stabilisation of apoenzyme [see also ref. (8)].

Residues forming the interface between subunits related by the P-axis are highly conserved while residues in the Q-axis interface, although few in number, retain the same specific interactions as in the lobster enzyme, cf. (12). In the active site the side chains interacting with NAD are almost identical to those found in lobster GPDH. Similarly residues so far implicated in substrate binding and catalysis are also conserved, except for the replacement of threonine-181 by asparagine. Two strong electron dense features interpreted in the lobster structure (2,3) to be bound anions that occupy the sites of the phosphate groups of the substrates during the catalytic reaction are also present in the thermophile enzyme structure (7,8). Indeed, the close similarity between the lobster and B. stearothermophilus structures suggests that the catalytic reaction is also the same and it is likely therefore that the only structural differences that are of fundamental importance are those that enable the thermophile enzyme to be formed and to function effectively at elevated temperatures.

THERMAL STABILITY

In considering the phenomenon of thermophily it is important to recognise that the thermophile proteins need to be thermodynamically as well as kinetically stable. Moreover since the free energy of activation (ΔG) is logarithmically proportional to the rate constant for thermal inactivation large differences in thermal stability can be accounted for by relatively small differences in ΔG. Thus, for example, in the case of triosephosphate isomerase a difference in activation energy of only 2.2 kcal/mole accounts for a 30-fold difference in thermal stability between the chicken muscle and B. stearothermophilus enzymes (13). Glyceraldehyde 3-phosphate dehydrogenase from B. stearothermophilus and T. aquaticus are stable to 70°C and 90°C respectively (6) and the additional stabilisation energy needed to produce even these dramatically increased thermal stabilities does not amount to more than five to ten kcal/mole. It follows from considerations such as these that the structural changes in GPDH that are associated with its thermal stabilisation are unlikely to be extensive.

The availability of the three-dimensional structures of a thermophilic enzyme and its mesophilic counterpart enables us to search for the stereochemical basis of the extra thermal stability. Several significant features have been identified.

Inter Subunit Ionic Bonds

B. stearothermophilus GPDH contains more arginine residues [16] than the lobster enzyme [9] and two of these arginines are capable of forming additional ionic bonds across subunit interfaces.

A buried ion-pair is formed between Arg-194 in one subunit and Asp-293 from the P-axis related subunit. These residues are conserved in the five sequenced species of GPDH but the ionic interaction cannot occur in the lobster enzyme because the side chain carboxyl group of Asp-293 exists in a different conformation and is accessible to the surface of the molecule as shown in Fig. 5.

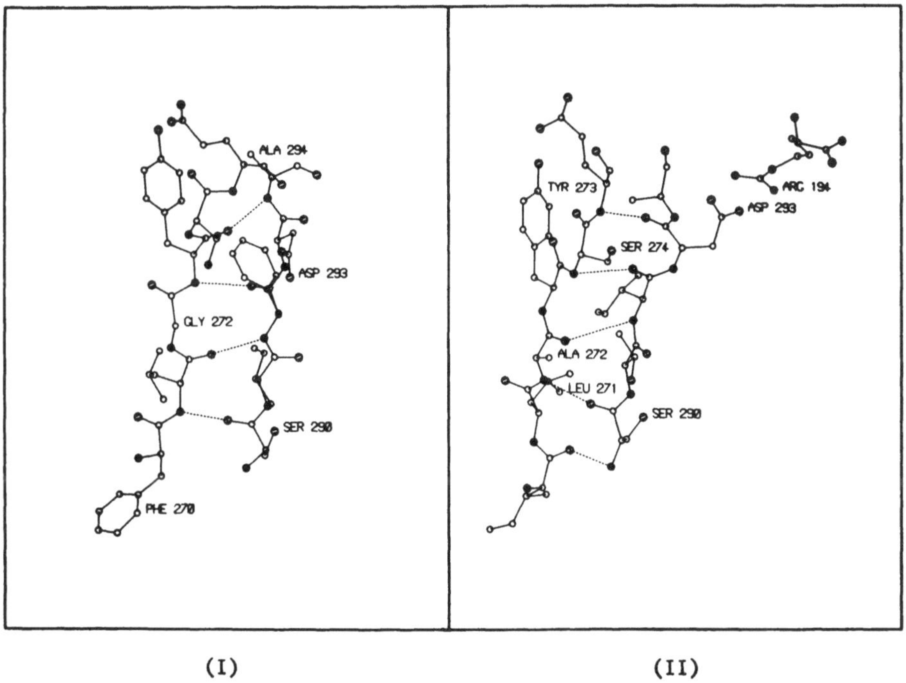

Figure 5 Conformation of β-pleated sheets β_5 and β_6 in the lobster (I) and B. stearothermophilus (II) enzymes. A buried ionic bond can form between aspartic acid 293 and arginine 194 from the P-axis related subunit in the thermophile enzyme only.

The different conformation of the side chain carboxyl group is part of a more extensive structural change involving the parallel β-sheets β_5 and β_6 (Fig. 3, residues 271-275 and 290-293) which, in the thermophile structure, have moved relative to each other, bringing the carboxyl of Asp-293 into the interior of the molecule close to the guanidino group of Arg-194 of the P-axis related subunit. The hydrogen bonding pattern between the two β-sheets is also more extensive in the thermophile enzyme. It should be noted that this particular ionic bond is not formed as the direct result of a sequence change. Moreover, its existence would not have been predicted merely by relating the sequence of the thermophile enzyme to the molecular model of the lobster enzyme.

The second ionic interaction involves the formation of a double ionic bond between Arg-281 from two Q-axis related subunits and Glu-201 from the R- and P-axis related subunits as shown in Fig. 6.

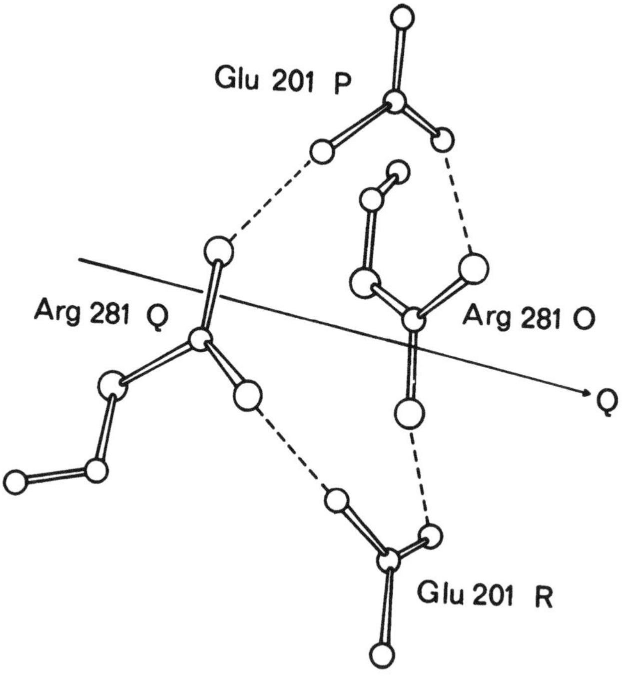

Figure 6 Conformation of the double ionic bond in B. stearothermophilus GPDH. The bond is formed between arginine 281 of the original (Arg-281 O) and Q-axis related subunits (Arg-281 Q) and glutamic acid 201 from the R-axis (Glu 201 R) and P-axis (Glu-201 P) related subunits. A second double ionic bond is present in the tetrameric enzyme and comprises arginine 281 from P- and R-axis related subunits and glutamic acid 201 from original and Q-axis related subunits.

These interactions are shielded from solvent and bring the S-loops from two subunits into contact with the other two subunits in the centre of the tetrameric molecule.

These additional intersubunit ionic bonds undoubtedly contribute towards the thermal stabilisation of the active tetrameric structure of glyceraldehyde 3-phosphate dehydrogenase from B. stearothermophilus. In this connection it is of interest to note that the residues forming S-loop contacts in the thermophile enzyme are in themselves more hydrophobic and these interactions could be expected to reinforce the stabilisation energy that results from the formation of the double ionic bond itself.

Intra Subunit Ionic Bonds

While inter subunit ionic bonds contribute to the stabilisation of the quaternary structure we would anticipate that the overall stability of a multisubunit enzyme such as GPDH requires in addition the stabilisation of the tertiary structure of individual subunits. In this connection it may be recalled that internal ionic bonds have also been identified in the structure of mesophile GPDHs. Thus, for example, Lys-306 forms an intrasubunit ionic bond with Asp-241 in the structure of the lobster muscle enzyme (2,3). This particular buried ionic bond is shielded from solvent by its location in the P-axis subunit interface and is thereby thought to exercise a stabilising influence on the structure of the lobster muscle enzyme (3,12). Moreover Lys-306 and Asp-241 are conserved in all five sequences (cf. Figs. 2 and 3) and the internal ionic bond that forms between them is also present in the structure of the B. stearothermophilus (and, by inference, T. aquaticus) enzyme (7,8).

The discussion has so far focussed upon the possible stabilising effects of buried ionic (and hydrophobic) bonds but that is not to say that surface ionic bonds may not also play a role in thermal stabilisation of the tertiary structure of enzyme subunits. The two thermophile GPDHs [both have isoelectric pHs of about 4.5 (6)] contain more charged groups than the three mesophile enzymes (cf. Fig. 2), and inspection of the molecular model of the B. stearothermophilus enzyme shows that there are numerous possibilities for the formation of surface hydrogen and ionic bonds (for example, the ϵ-NH_2 group of Lys-303 could interact with the γ-COOH group of Glu-247). Moreover in ferredoxin thermal stabilisation of thermophile forms has been correlated with the formation of external ionic bonds linking residues near the N-terminus to others near the C-terminus of this monomeric protein (14). Surface ionic bonds are also found in thermolysin (15) although in the case of this enzyme Ca^{++} ions are known to exert the major stabilising effect.

Glyceraldehyde 3-phosphate dehydrogenase from T. aquaticus is even more stable to heat and to protein denaturing agents than the B. stearothermophilus enzyme (6). There is no reason to suppose that the two thermophiles are closely related organisms and it was therefore of interest to seek to identify the source of thermal stability in T. aquaticus GPDH by relating its sequence (6) to the molecular model of the B. stearothermophilus enzyme. Such a comparison shows (Fig. 7) that 200 of the residues (61%) are identical and that most of the sequence differences are to be found on the surface of the nucleotide binding domain. Asp-293 and Arg-194 are conserved in the two proteins and, given the close similarity in the sequences of the S-loop regions (Fig. 4), it seems likely that an inter subunit ionic bond formed between these two residues also contributes to the thermal stabilisation of the T. aquaticus enzyme. The double ionic bond in the B. stearothermophilus protein involving Arg-281 and Glu-201, however, cannot occur, since these two residues in B. stearothermophilus GPDH are replaced by glutamine and isoleucine respectively. It follows that the extra stabilisation energy derived from this source in the B. stearothermophilus enzyme must be sought elsewhere in the T. aquaticus structure. Consequently, although ionic bonds at the surface as well as internally are likely to be important factors in the thermal stabilisation of thermophile proteins, contributions from other types of bonds would also appear to be necessary. A detailed analysis of all the factors involved becomes extremely complicated, since the sum of a large number of stabilising and destabilising effects must be considered and little information is available about the relative strengths of ionic, hydrogen, and hydrophobic bonds at elevated temperatures.

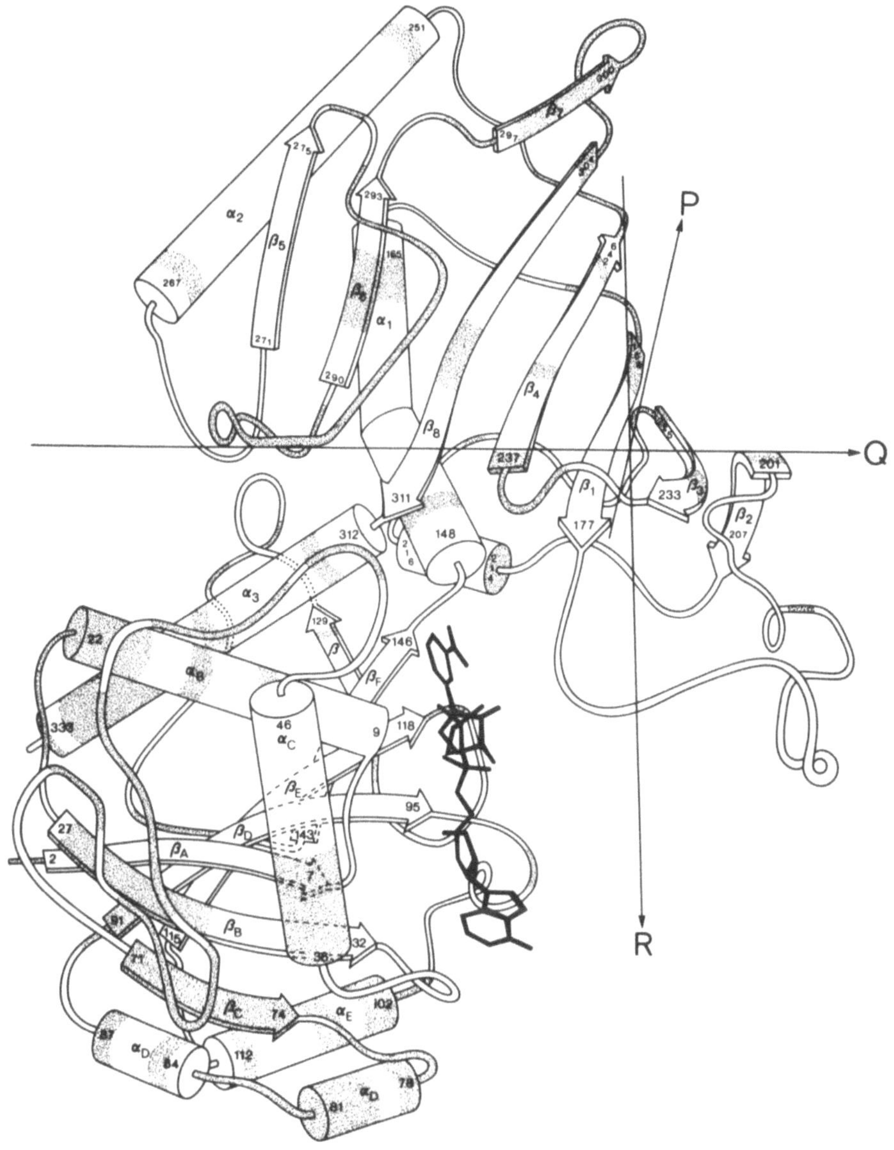

Figure 7 Folding diagram of one subunit of B. stearothermophilus GPDH showing the locations of amino acid residues that

CONCLUSION

Our results show that the structure and mode of action of a multi-subunit enzyme (glyceraldehyde 3-phosphate dehydrogenase) from a thermophile resembles that of its mesophile counterpart very closely. Comparison of the independently determined three-dimensional structures of the lobster muscle and B. stearothermophilus enzymes reveals relatively small differences in structure such as the formation of six additional inter sub-unit ionic bonds per tetramer. These exercise a considerable stabilising effect on the tetrameric enzyme. Thermophile GPDHs are not only stable at elevated temperatures but also retain their active tetrameric structures at lower temperatures in the presence of organic solvents, as well as in chaotropic agents such as 8 M urea and 1% dodecylsulphate. This extra stability confers many practical advantages in so far as it greatly extends the range of conditions that may be employed for enzyme purification and crystallisation, and permits the application of chemical and physicochemical methods under conditions that would denature less stable proteins. Hence this facilitates the determination of primary and three-dimensional structures and allows studies of relationships between structure and mechanism that could not otherwise be made. This view is reinforced by the outcome of recent studies with phosphofructokinase (16) and superoxide dismutase (17) and tyrosyl tRNA synthetase (18) from B. stearothermophilus.

ACKNOWLEDGEMENTS

We thank Alan Wonacott and Greg Biesecker for many stimulating and helpful model-side discussions. We thank Judy Smith for drawing Figs. 5 and 6.

REFERENCES

1. Harris, J.I., Waters, M.: Glyceraldehyde 3-phosphate dehydrogenase, in "The Enzymes" Vol. 13, 3rd Edition, Academic Press, pp. 1-50 (1976).

2. Buehner, M., Ford, G.C., Moras, D., Olsen, K.W.,Rossmann, M.G.: Three-dimensional structure of D-glyceraldehyde 3-phosphate dehydrogenase. J. Mol. Biol. 90, 25-49 (1974).

3. Moras, D., Olsen, K.W., Sabesan, M.N., Buehner, M., Ford, G.C., Rossmann, M.G.: Studies of asymmetry in the three-dimensional structure of lobster D-glyceraldehyde 3-phosphate dehydrogenase. J. Biol. Chem. 250, 9137-9162 (1975).

4. Singleton, R., Kimmel, J.R., Amelunxen, R.E.: The amino acid composition and other properties of thermostable glyceraldehyde 3-phosphate dehydrogenase from Bacillus stearothermophilus. J. Biol. Chem. 244, 1623-1630 (1969).

5. Suzuki, K., Harris, J.I.: Glyceraldehyde 3-phosphate dehydrogenase from Bacillus stearothermophilus. FEBS Letters 13, 217-220 (1971).

6. Hocking, J.D., Harris, J.I.: Purification by affinity chromatography of thermostable glyceraldehyde 3-phosphate dehydrogenase from Thermus aquaticus. FEBS Letters 34, 280-284 (1973).

idem. Glyceraldehyde 3-phosphate dehydrogenase from an extreme thermophile, Thermus aquaticus. "Enzymes and Proteins from Thermophilic Micro-organisms", H. Zuber, Ed., pp. 121-133, Birkhäuser Verlag, 1976.

7. Biesecker, G., Harris, J.I., Thierry, J.-C., Walker, J.E., Wonacott, A.J.: Sequence and structure of D-glyceraldehyde 3-phosphate dehydrogenase from Bacillus stearothermophilus. Nature 266, 328-333 (1977).

8. Wonacott, A.J., Biesecker, G.: Symmetry and NAD-dependent structural changes in D-glyceraldehyde 3-phosphate dehydrogenase. This volume.

9. Dayhoff, M.O., Park, C.M., McLaughlin, P.J.: Building a phylogenetic tree: Cytochrome C., in "Atlas of Protein Sequence and Structure" Vol. 5, M.O. Dayhoff, Ed., pp. 7-16 (1972).

10. Jornvall, H.: The primary structure of yeast alcohol dehydrogenase. Eur. J. Biochem. 72, 425-442 (1977).

11. Artavanis, S., Harris, J.I.: The primary structure of triosephosphate isomerase from Bacillus stearothermophilus. Eur. J. Biochem. In press (1977).

12. Olsen, K.W., Moras, D., Rossmann, M.G., Harris, J.I.: Sequence variability and structure of D-glyceraldehyde 3-phosphate dehydrogenase. J. Biol. Chem. 250, 9313-9321 (1976).

13. Fahey, R.C., Kolb, E., Harris, J.I.: Triose phosphate isomerase from Bacillus stearothermophilus. Biochem. J. 124, 77P (1971) and Fahey, R.C., unpublished work.

14. Perutz, M.F., Raidt, H.: Stereochemical basis of heat stability in bacterial ferredoxins and in haemoglobin A2. Nature 255, 256-259 (1975).

15. Coleman, P.M., Jansonius, J.N., Matthews, B.W.: The structure of thermolysin: an electron density map at 2.3 A resolution. J. Mol. Biol. 70, 701-724 (1972).

16. Hudson, P., Hengartner, H., Harris, J.I.: Structure of phosphofructokinase from B. stearothermophilus. Biochem. Soc. Trans. 5. In press (1977).

17. Brock, C.J., Harris, J.I., Sato, S.: Superoxide dismutase from B. stearothermophilus. Preparation of stable apoprotein and reconstistution of fully active Mn enzyme. J. Mol. Biol. 107, 175-178 (1976).

18. Irwin, M.J., Nyborg, J., Reid, B.R., Blow, D.M.: The crystal structure of tyrosyl-transfer RNA synthetase at 2.7 Å resolution. J. Mol. Biol. 105, 577-586 (1976).

Received March 26, 1977

DISCUSSION

Rossmann: Is it not unlikely that a conserved residue has different conformations in GPDH from different species? Is it not a more probable explanation that the Asp 293-Arg 194 salt bridge in B. stearothermophilus GPDH in fact represents an error in the structure of the lobster muscle enzyme?

Harris: In this particular connection it is not necessary to assume that the structure of the lobster muscle enzyme is incorrect. Your question does, however, illustrate an important point. In asking it you are assuming that one can compare the structure of thermophilic and mesophilic forms of a given enzyme simply by relating the amino acid sequence of the one to the three-dimensional structure of the other. Our results with GPDH show that it is not always valid to do so.
The two residues that form this particular intersubunit ionic bond (i.e. Asp 293-Arg 194) in B. stearothermophilus GPDH (but not in lobster) are nevertheless conserved residues that also occur in the same relative positions in the primary structure of the lobster enzyme. The existence of this potentially stabilising interaction would not therefore have been predicted simply by relating the sequence of the thermophile enzyme to the molecular model of the lobster enzyme. It also follows that structural changes involved in thermal stabilisation do not necessarily correlate directly with sequence changes that arise by point mutations at a given locus in the sequence. Perhaps Alan Wonacott would like to comment further on this point.

Wonacott: Although residues Asp 293 and Arg 194 are themselves conserved in all species, the structure of the β_5-β_6 sheet region has a modified hydrogen-bonding pattern and the S-loop conformation which carries Arg 194 is also different. It is the structural changes in parts close to these two residues which lead to the formation of the salt-bridge.

Walker: When we compare the sequences of the S-loops we must remember that we are not only comparing mesophile with thermophile but also eukaryote with prokaryote. At the moment we have no sequence of the S-loop of a GPDH from a mesophilic bacterium. We should therefore be cautious in interpreting the differences between the sequences of the prokaryote and eukaryote GPDH S-loops solely in terms of thermostabilisation.

Harris: Thank you for mentioning this point. It would be interesting to know if some of the changes that have occurred in the two thermophile sequences (such as the insertion of a His (residue 142), the replacement of Lys 183 by Arg, as well as the parallel changes in the S-loop sequences) have also occurred in mesophilic procaryotes. Perhaps some day someone will sequence the relevant parts of the E. coli enzyme.

Trommer: I should like to comment on the fact that you find 16 arginine residues in the thermophilic enzyme as compared to nine in the lobster enzyme. Recently Pfleiderer and Tüngler have modified LDH by acetimidination of 17 out of 24 lysines per polypeptide chain. This reaction in fact converts lysine residues to arginine like residues shifting the pK of this group from 10.5 to 12.5.
Besides an increased stability against alkaline denaturation this enzyme, which has still a residual activity of 75 %, shows a significantly enhanced thermal stability. At 55° C it is indefinitely stable whereas the native enzyme shows half-life times in the order of 150 min. Even at 70° C, which leads to immediate denaturation of the native form, the acetimidinated derivative has half-life times of 20 min. I think this proves the great importance of the surface charge in stabilising thermophilic enzymes.

Harris: That is a very interesting observation. There are probably many different ways by which the structure of a protein can be stabilised to heat. We are, after all, only looking for stabilisation energies of the order of 5 kcal/mole, and the surface charge may be one way in which this can be achieved. On the other hand, it does not necessarily follow that such a protein could function adequately in vivo where it might need to interact with other proteins in the cell.
Perhaps I should also make it clear that we do not attach particular significance to the fact that the two thermophile GPDHs contain more arginine residues than their mesophile counterparts. It is not the total number that is important, but their location in the three-dimensional structure. The extra arginines that I talked about are special in the sense that they are not on the surface, but buried in intersubunit interfaces, so that they do not contribute to the surface charge.

Kiltz: The drastic increase in heat stability of acetimidated H_4 LDH can not be explained by a change of the number of salt-bridges, as the number of basic groups is not changed.

Harris: I hope that I did not give the impression that salt-bridges are the only means of thermal stabilisation! Buried charges are unusual and consequently draw attention to themselves. Note that although T. aquaticus GPDH is considerably more thermostable than the B. stearothermophilus enzyme it contains fewer internal salt-bridges. Thus the double ionic bond involving Arg-281 and Glu 201 that is a feature of the B. stearothermophilus protein cannot occur in the T. aquaticus enzyme because of changes in sequence. The stabilisation energy provided by these bonds in the B. stearothermophilus enzyme must therefore be derived in some other way in T. aquaticus.

Werber: In relationship with the ionic interactions that you suggested as a possible cause of the extra stability of thermophilic enzymes, I would like to make a comment about halophilic enzymes. These enzymes are very stable at high salt concentrations (about 4 M) where probably ionic interactions will probably not be operative. Moreover, whereas the thermophilic dehydrogenases you have discussed have more arginine residues than the mesophilic ones, this is not the case in halophilic enzymes which, on the contrary, have an excess of acidic residues. Would that mean that there exist two completely different mechanisms of stabilisation of enzymes?

Jaenicke: From the increase in hydrophobicity and from the increased number of ion pairs one would expect high hydrostatic pressure to be a tool to investigate differences between mesophilic and thermophilic organisms. Does high pressure affect the quaternary structure or the activity of your enzymes? Are thermophilic bacteria particularly sensitive towards high hydrostatic pressure?

Harris: We have not ourselves studied the effect of high pressure on the structure and activity of thermophile enzymes. It is however an interesting suggestion and I hope that your question will stimulate further work in this field.

Werber: Did you check if the stability of thermophilic dehydrogenases depends on salt concentration?

Harris: GPDH from B. stearothermophilus is stable at high ionic strength. For example, the rate of inactivation at 65° C is not affected by salt concentrations of up to 3 M NaCl.
As I mentioned earlier, I do not attach particular significance to the total number of arginine residues. Although the two thermophile GPDHs contain more arginines, the total number of lysine plus arginine residues is about the same in the mesophile and thermophile enzymes. Moreover, the thermophile enzymes possess acidic isoelectric points (pH 4.2) indicating that they also possess an excess of acidic side chains. In these respects the thermophile GPDHs resemble your halophilic enzymes.

Veeger: Since at ambient temperature there is not a large difference between the catalytic center activities of thermophiles and mesophiles, is it fair to superimpose your sequence on the three-dimensional structure of the lobster enzyme? Since drastic conformational changes are induced by increase in temperature, is it fair to draw conclusions on high temperature stability with respect to an enzyme of which the structure is recorded at room temperature? One should do this at 50 - 70° C.

Harris: The three-dimensional structure of B. stearothermophilus GPDH was not determined by superimposing its sequence on the three-dimensional structure of the lobster muscle enzyme. It was determined quite independently by Alan Wonacott and his colleagues, by X-ray diffraction analysis of enzyme crystals. The structure that we propose was obtained by relating the electron density maps to the amino acid sequence, and was thus determined without reference to the structure of the lobster enzyme.
In answer to your second point it is, of course, true that the X-ray diffraction analysis of crystals was undertaken at ambient temperatures. As far as the organism itself is concerned, 60° C does represent "ambient temperature" and we have no reason to think that the structure of the enzyme changes in any fundamental way between 25° and 60° C. We have, for example, compared spectra at 25° and 60° C and we also know that an Arrhenius plot relating enzyme activity to temperature is linear in this range. Do you worry about results obtained with mesophile enzymes at, say, 4° C (and in Douzou's hands at temperatures well below 0° C)? In terms of the Absolute

scale of temperature the difference between 25° and 60° C is not very great and the only major difference that we have found between lobster and B. stearothermophilus GPDHs is that the mesophile protein denatures and precipitates at 60 - 65° C, whereas the thermophile enzyme does not.

Hollaway: It has been shown that pig GPDH denatures more rapidly as you dilute the enzyme, i.e. it is a dissociated species that is denaturing. Therefore the thermophilic enzyme could be stabilised by (a) decreasing the tendency to dissociate, (b) increasing the stability of the dimeric species. Have you any indication about the tendency of the thermophilic enzyme to dissociate at high temperature, such as hybridisation studies, or dependence of rate of heat denaturation on enzyme concentration? In any case it seems that one of the reasons for having oligomeric proteins is that they gain stability.

Harris: We have not studied GPDH from these points of view but we do have some relevant data on B. stearothermophilus triosephosphate isomerase (TIM). In the case of this dimeric enzyme a study of the effect of enzyme concentration on the rate of thermal inactivation showed that the initial rate is lower at high active protein concentration than at low (< 100 μg/ml) concentrations. Inactivation appears to involve dissociation to the monomer and the rate of inactivation increases as the concentration of active dimer decreases. At 60° C the thermophilic enzyme inactivates at a rate that is approximately 35 times slower than that of mesophile (chicken muscle) TIM.
We were able to prepare the hybrid dimer consisting of one thermophilic (B. stearothermophilus) chain and one mesophilic (chicken muscle) chain following exposure of the parental dimers to urea at acid pH. Thermal inactivation of the hybrid showed a distinctly biphasic inactivation profile identical to that found for a 1:1 mixture of the parental dimers. Examination of heated samples by gel-electrophoresis showed that the disappearance of the hybrid dimer is accompanied by dissociation to the respective monomers; thermophile monomers are stable at 60° C and are capable of reforming active dimers, whereas the mesophile monomers are unstable and denature at this temperature. B. stearothermophilus monomers are thus thermodynamically as well as kinetically stable at 60° C.
The experiments demonstrate that stabilisation of a thermophile protein involves the stabilisation of the tertiary structure of individual monomers as well as of the quaternary structure (e.g. of intersubunit contacts) of the active dimeric or tetrameric forms.

COMPARATIVE ASPECTS OF STRUCTURAL STUDIES OF ALCOHOL DEHYDROGENASES

Hans Jörnvall
Department of Chemistry, Karolinska Institutet, S-104 01 Stockholm 60, Sweden

INTRODUCTION

Alcohol dehydrogenase is a well-studied enzyme (1). Much information has been obtained by crystallographic, chemical and kinetic analyses and by comparisons of the enzyme structures from different species. Recent additions to the diversity of the structural knowledge include the primary structure of yeast alcohol dehydrogenase (2) and some amino acid sequence information on the Drosophila enzyme (3). Comparisons of these structures with previously available data show that the enzyme has been subject to extensive variations in many properties from which further structural, functional and evolutionary conclusions about alcohol dehydrogenases may be deduced. Some of these aspects have been considered in detail elsewhere (4-6) and emphasis will presently be placed on overall comparisons and on some of the properties where successive or multiple changes may be traced.

COMPARISONS BETWEEN THE YEAST AND MAMMALIAN ENZYMES

Amino Acid Exchanges

A schematic representation of the aligned subunits of the horse liver and yeast enzymes is shown in Fig. 1, from which the general properties of the comparison are clearly visible:
There is an overall positional identity (vertical bars in the figure) which is low but significant (25%),
it differs between various segments of the molecules (approximate range: 10-40%),
insertions/deletions occur (represented by holes in the horizontal lines in the figure and affecting 41 positions), and
even functionally highly important residues (arrows in the figure) are not in strictly conserved segments.

The overall identity and the nature of exchanged residues establish that the enzymes are functionally and evolutionarily related. The differences show that this relationship is distant and that even important segments have been subject to changes. The latter aspect is of particular interest in regard to possible gradual changes of the enzyme. Residues in many more positions than those specially shown by arrows in Fig. 1 are of functional importance but those indicated suffice to show the pattern and are recognized as three active site zinc ligands, two residues suggested in a proton release mechanism and five residues characterized in coenzyme interactions (7). It may be noticed from Fig. 1 that these

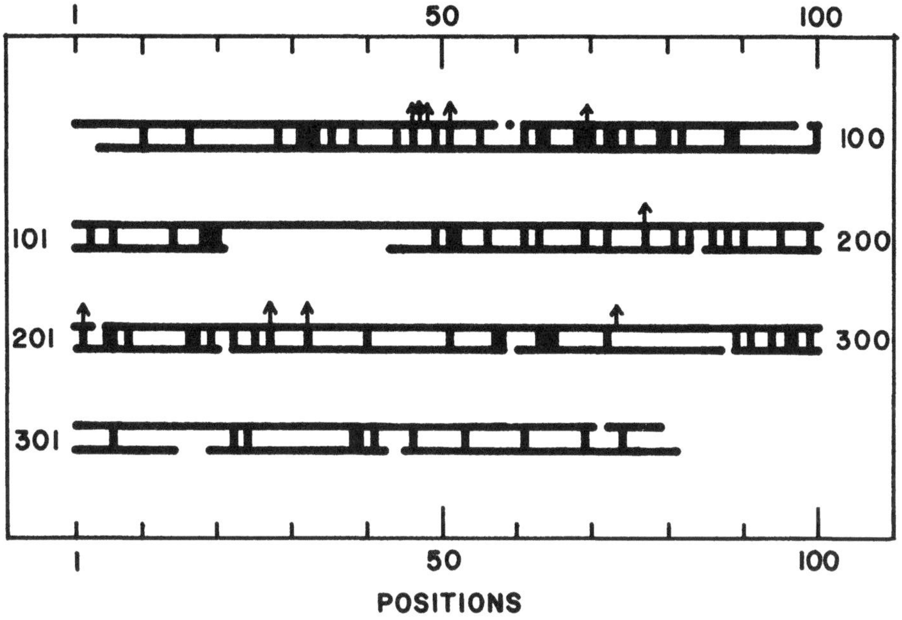

Fig. 1 Schematic comparison of aligned amino acid sequences of horse liver and yeast alcohol dehydrogenases.
In all pairs of horizontal lines the horse enzyme is represented by the top line and the yeast enzyme by the bottom line. Empty spaces in these lines indicate deletions (or insertions in the other sequence). Alignment is as in ref. 5, and gives a total of 381 positions when an extra residue (115 in ref. 5) present only in the rat enzyme is excluded.

residues tend to be in areas of higher similarity but still adjacent to variable structures. In contrast to the situation in several known proteins, alcohol dehydrogenases have no conserved continuous amino acid sequence which reflects the active site or other important structures. Furthermore, even some of the residues with presumably essential functions are exchanged, like (in the numbering system of the horse enzyme) Ser/Thr-48 in the possible proton release, Arg/His-47 in the coenzyme-phosphate binding site, and Ile/Ser-269 also affecting this site (1,7-9). It is true that the exchanged residues may largely replace each other's functions but the non-conserved nature of these positions is anyway noteworthy and may illustrate gradual changes in the enzymes, affecting even essential properties. In this context, however, it should be stressed that the whole structures are compatible with a mainly conserved enzymatic mechanism (6).

Table 1 Conserved residues in comparisons of yeast and horse liver alcohol dehydrogenases.
The first figure gives the number of positions in which each type of residue is conserved, the second figure gives the ratio (in percent) of this number in relation to the total occurrence of each amino acid.

Cys	6	(55%)	Glu	5	(24%)	Val	9	(24%)	Phe	2	(15%)
Asp	6	(38%)	Gln	1	(13%)	Met	0	(0%)	Trp	0	(0%)
Asn	1	(10%)	Pro	3	(18%)	Ile	3	(14%)	Lys	6	(22%)
Thr	2	(10%)	Gly	20	(49%)	Leu	5	(20%)	His	2	(24%)
Ser	2	(9%)	Ala	9	(29%)	Tyr	2	(22%)	Arg	2	(20%)

Tertiary Structures

Comparisons of the yeast and mammalian enzymes in relation to the crystallographically determined structure of horse liver alcohol dehydrogenase suggest that the tertiary structures are probably more closely related than the primary structures (4,6):
Many compensating exchanges in the hydrophobic cores are noticeable, where a pair of closely positioned residues are mutually replaced.
Glycine residues are by far the most common in positions with conserved residues, as shown in Table 1. In fact, although glycine is among the most abundant residues, almost half of them are conserved which should be compared to the overall degree of identity between the enzymes of only 25%. In relative ratios only cysteine is equally conserved as glycine (Table 1) but cysteine is less abundant and all those conserved are attributable to metal-liganding positions.
Substitutions involving charged residues do not affect positions where they would require altered polypeptide conformations.
The large deletion in the yeast enzyme affects a superficial structure (5) and all the smaller gaps in either enzyme (Fig. 1) can also be accommodated in the model of the horse liver enzyme (6).

The compensated exchanges and the general glycine conservancy are compatible with unaltered polypeptide conformations at those places where space requirements are most critical. Similarly, the charge distributions are consistent with stable structures without further rearrangements. It therefore seems highly probable that the subunit of the yeast enzyme has a largely conserved tertiary structure as compared to those of the mammalian enzymes in spite of the large amino acid sequence differences.

Quaternary Structures

The yeast enzyme is a tetramer and the horse enzyme a dimer (1) showing that at least some of the monomer associations are different. The areas of known subunit interactions in the mammalian enzyme are among those that are least conserved when the aligned yeast and horse liver enzymes are compared (5). This is in contrast to the situation observed for another dehydrogenase, where the quaternary structure is not changed between the enzyme species compared (10). It therefore seems possible that the amino acid replacements, affecting superficial areas and segments participating in subunit interactions, can be coupled with the differences in quaternary structures, although exact associations may not be judged from model studies (6).

It may be noticed that the variable region in the comparison between the yeast and horse enzymes corresponding to the segment participating in subunit interactions, is also non-conserved in comparisons of the horse and rat enzymes (11), which are both dimers. This might indicate small changes even between enzymes with identical quaternary structures.

MULTIPLE CHANGES

Gradual or multiple changes in proteins are of particular interest in regard to structural-functional relationships and to possible appearances of new features. In these respects, alcohol dehydrogenases offer several examples. Some of these are exemplified by the above comparisons or the following considerations, and the diversity is summarized in Table 2.

Functionally Important Areas

Certain exchanges affecting functionally important areas are given in the previous paragraphs. Another type of differences where multiple changes may be noticed is at the substrate binding site. Thus, most mammalian alcohol dehydrogenases have a broad substrate specificity including 3β-hydroxysteroids. However, one monomer (the E-subunit) of horse isozymes and the monomer of the yeast enzyme lack this activity. Interestingly, this reduction in the type of substrates accepted is apparently brought about by different mechanisms. In the horse E-subunit a few replacements (1) adjacent to the substrate binding region produce this difference, probably from an Asp/(Ser)-substitution at position 115 (12) or a Phe/Leu-substitution at position 110 (4). In contrast, the reduced space relationships in the yeast enzyme may be more directly produced in the substrate pocket region, presumably by two insertions and by replacements to bulky residues (4). These two different types of changes in the enzymes show that substrate restrictions can be produced in different ways and at separate times. They also show, as do the replacements affecting other functionally important residues, that even characteristic properties of the enzymes are subject to changes.

Metal Binding

The binding of the non-catalytic zinc atom, occurring in the horse enzyme (1), is one property for which different states may be noticed. Thus, on the one extreme, the zinc atom and its four cysteine ligands are present in the horse enzyme (7); in another case, the ligands are conserved in the primary structure of the yeast enzyme but in an altered loop in the tertiary structure (5) and the state of the zinc atom is then unclear (13,14); finally, in the Drosophila enzyme at least some of the cysteine

Table 2 Variabilities in alcohol dehydrogenases

Amino acid replacements	Subunit sizes	Number of different genetic loci characterized in one organism	Metal contents	Quaternary structures
75% of all residues, including some functionally important residues	from 24000 to 40000	1, 2 or 3	2 Zn atoms per subunit in mammalian enzymes; 1 or 2 in the yeast subunit; lower values or different ligands in the Drosophila protein	dimers or tetramers

ligands are not even present. Furthermore, changes in the region of the non-catalytic zinc atom are not confined to the comparison of these distantly related enzymes, but are also found between the otherwise more conserved mammalian enzymes (1), and then include an insertion (11) as well as possible chemical modifications associated with enzyme subforms (15). The region of the second zinc atom is therefore variable in comparisons of both distantly and closely related enzymes.

Monomer Sizes and Genetic Representations

In these two properties, there are also considerable differences in alcohol dehydrogenases. Regarding sizes, the known monomers vary from about 230 residues in the Drosophila enzyme to about 375 in the mammalian enzymes, again with the yeast enzyme in between, with about 350 residues.

Regarding genetic representation, multiple loci for alcohol dehydrogenase have been shown in yeast, man and horse but thus far not for the enzyme from other species (1). Therefore, separate and multiple duplicatory events of the whole gene must have occurred during evolution. This conclusion is furthermore supported by the isozyme variations, since they often are apparently species-specific in both functional properties and known amino acid sequences. Thus, at least two of the yeast enzymes are distinct and work physiologically in opposite directions, the human isozymes have a complex variability and the horse isozymes show the ethanol/steroid differences (1). Regarding amino acid sequences, isozyme variations are not completely determined, and may be too early to judge, but known exchanges in some isozymes from two mammalian species are at different positions (1). Presently, therefore, isozymes do not appear by any criterion to occur in uniformly found types similar to the major forms in the lactate dehydrogenases (16).

Conclusions

The sequence information confirms that different alcohol dehydrogenases are evolutionarily related and probably more conserved in tertiary structure than in primary structure. The divergencies in sequence, quaternary structure, the second metal atom, monomer size and gene duplications (Table 2) suggest that evolutionary changes have affected potentially important areas at the substrate pocket, at the active site and in the segments for subunit interactions. Finally, for several of these properties, similar events seem to have occurred at both distant and recent times, as evidenced by some parallel changes in comparisons involving either the yeast/horse enzyme pair or the inter-mammalian variations.

These properties of alcohol dehydrogenases may be considered in relation to mechanisms for evolutionary changes in general. Wider connections with other types of dehydrogenases or other enzymes are strongly suggested from known tertiary structures (17) and then include variations and evolutionary relationships among certain segments. Similar but even extended and modified relations may not be excluded from coincidences in amino acid sequences (5). The latter interpretations are difficult to judge and cannot be distinguished from random variations in the absence of known interconnecting forms. The presently discussed variations, however, definitely reflect evolutionary events. Of particular interest for tracing more distant connections, is the variation in monomer size and genetic representation for alcohol dehydrogenases, which establish the occurrence of changes in the gene architecture. Also of possible interest, is the noticeable variation in functional residues and metal binding in alcohol dehydrogenases, since other characterized dehydrogenases have different catalytic mechanisms and no metals (17).

ACKNOWLEDGEMENT

This work was supported by grants from the Swedish Medical Research Council (project 13X-3532).

REFERENCES

1. Brändén, C.-I., Jörnvall, H., Eklund, H. & Furugren, B.: Alcohol dehydrogenases. The Enzymes, 3rd ed., 11, 103-190 (1975).
2. Jörnvall, H.: The primary structure of yeast alcohol dehydrogenase. Eur. J. Biochem., 72, 425-442 (1977).
3. Schwartz, M. & Jörnvall, H.: Structural analyses of mutant and wild-type alcohol dehydrogenases from Drosophila melanogaster. Eur. J. Biochem. 68, 159-168 (1976).
4. Eklund, H., Brändén, C.-I. & Jörnvall, H.: Structural comparisons of mammalian, yeast and bacillar alcohol dehydrogenases. J. Mol. Biol. 102, 61-73 (1976).
5. Jörnvall, H.: Differences between alcohol dehydrogenases. Eur. J. Biochem., 72, 443-452 (1977).
6. Jörnvall, H., Eklund, H. & Brändén, C.-I.: Yeast alcohol dehydrogenase.

Positional comparisons in relation to the tertiary structure of the corresponding horse enzyme. Manuscript in preparation (1977).
7. Eklund, H., Nordström, B., Zeppezauer, E., Söderlund, G., Ohlsson, I., Boiwe, T., Söderberg, B.-O., Tapia, O., Brändén, C.-I. & Åkeson, Å.: Three-dimensional structure of horse liver alcohol dehydrogenase at 2.4 Å resolution. J. Mol. Biol. 102, 27-59 (1976).
8. Zeppezauer, E., Jörnvall, H. & Ohlsson, I.: Carboxymethylation of horse-liver alcohol dehydrogenase in the crystalline state. The active-site zinc region and general anion-binding site of the enzyme correlated in primary and tertiary structures. Eur. J. Biochem. 58, 95-104 (1975).
9. Lange, L.G., III, Riordan, J.F., Vallee, B.L. & Brändén, C.I.: The role of arginyl residues in directing carboxymethylation of horse liver alcohol dehydrogenase. Biochemistry 14, 3497-3502 (1975).
10. Olsen, K.W., Moras, D., Rossmann, M.G. & Harris, J.I.: Sequence variability and structure of D-glyceraldehyde-3-phosphate dehydrogenase. J. Biol. Chem. 250, 9313-9321 (1975).
11. Jörnvall, H.: Functional aspects of structural studies of alcohol dehydrogenases. In Alcohol and Aldehyde Metabolizing Systems (Thurman, R.G., Yonetani, T., Williamson, J.R. & Chance, B., eds.), Academic Press, 23-32 (1974).
12. Brändén, C.-I.: In Alcohol and Aldehyde Metabolizing Systems, 2nd Int. Symp., Academic Press, in press (1977).
13. Veillon, C. & Sytkowski, A.J.: The intrinsic zinc atoms of yeast alcohol dehydrogenase. Biochem. Biophys. Res. Commun. 67, 1494-1500 (1975).
14. Klinman, J.P. & Welsh, K.: The zinc content of yeast alcohol dehydrogenase. Biochem. Biophys. Res. Commun. 70, 878-884 (1976).
15. Jörnvall, H.: Differences in thiol groups and multiple forms of rat liver alcohol dehydrogenase. Biochem. Biophys. Res. Commun. 53, 1096-1101 (1973).
16. Holbrook, J.J., Liljas, A., Steindel, S.J. & Rossmann, M.G.: Lactate dehydrogenase. The Enzymes, 3rd ed., 11, 191-292 (1975).
17. Rossmann, M.G., Liljas, A., Brändén, C.-I. & Banaszak, L.J.: Evolutionary and structural relationships among dehydrogenases. The Enzymes, 3rd ed., 11, 61-102 (1975).

Received January 26, 1977

DISCUSSION

Walk: Do you know, whether the different isoenzymes in horse liver and in man are localized in the same subcellular compartment?

Jörnvall: Yes, they are localized in the same compartment. All the mammalian isozymes are cytoplasmic.

Banaszak: Can you say anything about the residues involved in contributing invariant "atoms" such as C=O?

Jörnvall: Residues contributing important polypeptide backbone structures have side chains, even if they are exchanged, that are still compatible with these functions.

Kiltz: The sequence around Ser 139, the carbonyl group of which hydrogen bonds to the carboxamide of the nicotinamide in H-type LDH, is highly conserved in all five known LDH sequences.

Luisi: A comment more than a question. Your statement, that the tertiary structures of liver and yeast ADH may be rather similar to each other despite the differences in primary structure, implies that polypeptide chains which differ considerably in amino acid composition may have the same (or quite similar) conformations. This aspect of protein structure is now well established on the basis of X-ray work on several systems - it is however not an easy one to rationalize in terms of conformational energy considerations. Perhaps one can rationalize more easily the opposite situation, namely when large conformational changes are brought about by subtle differences in structure.

Jörnvall: Yes, similarities between distantly related proteins appear in general to be larger in tertiary than in primary structures. However, an opposite statement in the case of differences need not be true, and since destabilizing exchanges can have quite extensive effects on the tertiary structure.

Kiltz: Did you ever consider the effect of the amino acid exchange at Ile 269/Arg 47 on the conformation of NAD, in special on the nicotinamide, as this exchange is at the phosphate binding site of NAD. In the case of the LDH isozymes the main difference is found exactly at this position, that is in the H isozyme there is a glutamine whereas in the M isozyme an alanine. This exchange produces a different position of the nicotinamide explaning the lower specificity of the H_4 isozyme.

Jörnvall: We did not discover evidence for any conformational changes of NAD.

Engel: You mentioned that in yeast ADH the four cysteine ligands of the second zinc in horse liver ADH are conserved. If these residues are not involved in binding zinc in the yeast enzyme, why are they conserved?

Jörnvall: Alternative functions may of course be visualized but zinc binding seems to be a good explanation. Importantly, however, the surrounding structural variability may weaken this binding in relation to the horse enzyme, and in this way contribute an explanation to the different zinc values reported in the literature for the yeast enzyme.

Hinz: Is it possible to draw the conclusion on the basis of your and the X-ray data, that the lower degree of amino acid conservation in ADH is associated with the lower substrate specificity of ADH in comparison to LDH?

Jörnvall: In a sense, yes. Generally, structures that are highly conserved during evolution represent constant and important functions, whereas structural variability reflects non-constant or less important functions. In the case of LDH, its functional role and importance are well established. For ADH, however, different functions have been suggested and species-specific enzymologies are known. Consequently, the lower degree of structural conservation in ADH is consistent, and it may further illustrate the low substrate specificity of the mammalian enzyme and other variable properties of ADH.

Dunn: Do you think any of the liver isozymes evolved for the purpose of metabolizing ethanol?

Jörnvall: I think ADH is still submitted to large evolutionary changes and that isozymes may have several functions, including metabolism of intestinally produced ethanol and other alcohols.

Dunn: Certainly none of the liver isozyme substrate site structures appear to be designed for binding ethanol.

Jörnvall: Yes, I agree. This does at least not appear to be the sole function in any liver isozyme.

Bernhard: Are the small changes in amino acid sequence among liver ADH isozymes unusual or are they found in other enzymes?

Jörnvall: Evolution of isozymes by gene duplications and subsequent point mutations is a gradual and time-dependent process after the duplicatory events. In comparison with the amino acid exchanges among liver ADH isozymes, both larger and smaller differences between related polypeptide chains are known (larger: LDH isozymes, smaller: some globin polypeptide chains where in certain cases even identical polypeptide chains appear to be produced from different loci). The small changes among liver ADH isozymes form an indication of recent duplications, a conclusion which is also compatible with the finding of different number of ADH-loci between closely related species.

Harris: Would you like to comment on possible sequence similarities among different types of dehydrogenases?

Jörnvall: Some sequence coincidences have been notices (as recently reported in Eur. J. Biochem. 72, (1977) 443-452). Their significance is unknown but the known divergence even within alcohol dehydrogenases show that sequence similarities from possibly still more remote connections may be small and discontinuous. They are therefore difficult to distinguish from random distributions.

Rossmann: The NAD-binding domain has been shown to be homologous in tertiary structure among known NAD-linked dehydrogenases. Thus, this part of the sequences must be homologous. The catalytic domains have different tertiary structures and no homology can be expected.

Jörnvall: The above-mentioned sequence coincidences support the conclusions from the similarities in tertiary structure but they also show that even for other regions, more distant homology may not be excluded.

Harris/Jörnvall: Knowledge of additional structures is apparently necessary for a complete understanding of all variations.

Perham: One proposal for the evolution of dimers from monomers and of tetramers from dimers is the generation of a hydrophobic "sticky" patch on the surface of the subunit. You have told us that you cannot yet explain why the yeast enzyme is a tetramer whereas the liver enzyme is a dimer, but can you see any evidence for the evolution of hydrophobic patches in subunit association?

Jörnvall: The known subunit interactions in the horse liver enzyme are hydrophobic and thus support the proposal. In the yeast enzyme subunit interactions are uncharacterized but regarding possible hydrophobic surfaces in the monomer, there are two areas where amino acid exchanges from the alternatives in the mammalian enzyme are extensive.

X-ray Diffraction Studies on Sheep Liver 6-Phosphogluconate Dehydrogenase at 6Å Resolution

M.J. Adams, I.G. Archibald, J.R. Helliwell.
Laboratory of Molecular Biophysics, Department of Zoology, South Parks Road, Oxford, U.K.

INTRODUCTION

The enzyme 6-phosphogluconate dehydrogenase (6-PGDH), is an $NADP^+$ dependent dimer of molecular weight 94,000 Daltons. (Silverberg and Dalziel, 1973). It is a member of the pentose phosphate pathway and acts as an oxidative decarboxylase yielding ribulose-5-phosphate and NADPH.

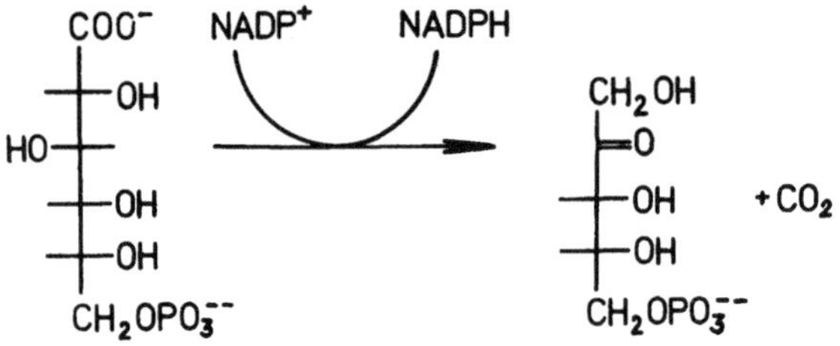

The two subunits of the dimer act independently and there is no specific requirement for Mg^{2+} ions. (Villet and Dalziel, 1972). The mechanism of the oxidation has been studied under various conditions with different results. A random order mechanism has been suggested in buffers such as tris-acetate (Dyson, D'Orazio and Hanson, 1973) while an ordered binding of co-enzyme followed by substrate has been indicated in phosphate buffer at pH 7. (Silverberg, 1973). Inhibition by various nucleotide phosphates, co-enzyme fragments and some other metabolic intermediates has been studied. (Dyson and D'Orazio, 1973). The enzyme has been crystallised from ammonium sulphate and crystals were first reported by Silverberg, Dalziel and Adams (1973).

THE STRUCTURE AT 6Å

The structure determination by the method of isomorphous replacement has been described (Adams, Helliwell and Bugg, 1977) and will only be reviewed briefly here. The enzyme crystallises in the spacegroup $C222_1$ with $\underline{a}$ = 72.7Å, b = 148.2Å, $\underline{c}$ = 102.9Å with a single subunit per asymmetric unit. Soaking crystals in solutions of salts containing heavy atoms yielded two suitable heavy atom derivatives. The most satisfactory conditions for substitution were 22μM potassium dicyano-aurate(I), $KAu(CN)_2$, for 22 hours and 2.7mM potassium tetracyanoplatinate (II), $K_2Pt(CN)_4$, for 24 hours. Heavy atom positions were found by use of both Patterson maps and direct methods. For each derivative in turn, measurements of anomalous dispersion were combined with isomorphous differences to resolve the phase ambiguity. The resultant phase sets, each calculated using only one derivative were combined with the iso-morphous differences of the second derivative and used to calculate difference Fouriers. The platinum positions were the largest peaks on the platinum difference Fourier using gold phases as were the gold sites on the gold difference Fourier with platinum phases. One of the two ways of combining isomorphous and anomalous differences gave a noticeably superior difference map for each derivative, thus enabling the 'hand' to be obtained readily. Figure I shows one section from each map which includes the major site of each derivative.

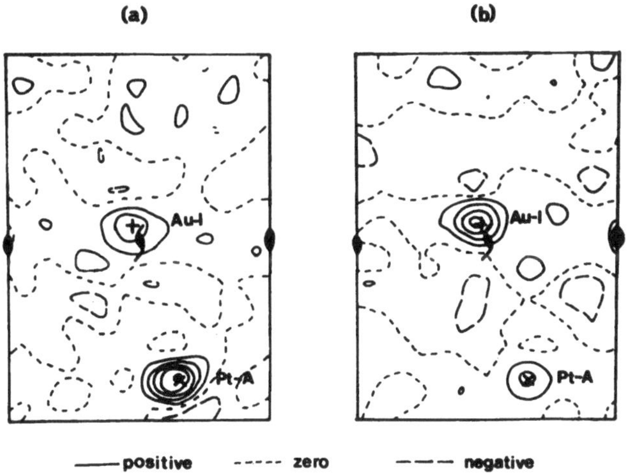

Fig. I

Comparison of section y = 6/72 b of difference Fouriers.
(a) Platinum difference Fourier with gold phases showing site Pt(A) and ghost of Au(1).
(b) Gold difference Fourier with platinum phases showing site Au(1) and ghost of Pt(A).

The heavy atom parameters were refined by the method of Least Squares using the centric data only, some 23% of the terms. The function $M = \sum\omega(|F_{HOBS}|-|F_{HCALC}|)^2$ was minimised where ω is an experimental weighting factor and $|F_{HOBS}|$ and $|F_{HCALC}|$ are the 'observed' and calculated structure amplitudes for the heavy atom components alone. The final heavy atom parameters are given in Table I and were used to calculate phases for the native structure factors. A 'best' electron density map (Blow and Crick, 1959) with terms $m|F_P|\exp(i\alpha_{best})$ was calculated.

The electron density map shows the dimer to be centered at approximately $0,\frac{3}{4},\frac{1}{4}$ with the dimer axis the crystallographic 2-fold axis parallel to b. The dimer was readily isolated and the individual subunits could be distinguished with ease. Two views of a balsa wood model of the subunit

TABLE I FINAL PARAMETERS USED FOR PHASING

				Positional Parameters				r.m.s.E*		r.m.s.
	B_{ov}**	Site	Z^+	x	y	z	B_i^x	centric	all	F_H
$K_2Pt(CN)_4$	-3.5	A	54	0.3330	0.0875	0.4453	15	277	200	314
		B	41	0.0301	0.1153	0.2368	15			
		C	48	0.276	0.2065	0.5036	15			
$KAu(CN)_2$	-3.5	1	78	0.2293	0.0792	0.2291	15	241	174	328
		2	39	0.4969	0.2228	0.1428	15			

$^+$Z is given in electrons, assuming full occupancy of site 1 of the gold derivative.

$^x B_i$, the individual heavy atom temperature factors, were not refined.

*r.m.s. E is the lack of closure error, $|F_{PH}| - |F_H+F_P|$. $|F_{PH}|$ is the heavy atom derivative structure amplitude, $|F_P|$ the protein structure amplitude, F_H the calculated heavy atom component.

**B_{ov} is the overall temperature factor to be applied to the derivative data for best fit with the native data.

are shown in Figure II. The protein subunit is ellipsoidal with a large cleft on one surface distant from the 2-fold axis. Although the subunit is not bilobal, the cleft divides it into two unequal parts. The major site of platinum substitution is on one edge of the cleft on the smaller part of the subunit, and the minor gold site is not far distant from the other edge on the larger part of the subunit. The other most interesting feature of the subunit is that a large part of it is made up

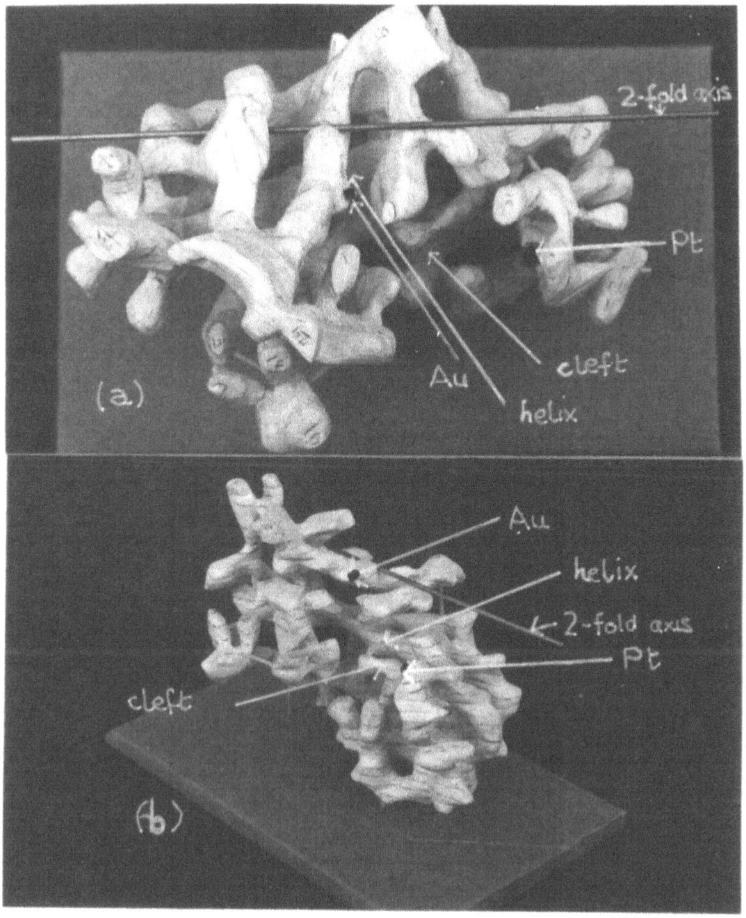

Fig. II

Views of the balsa wood model. The molecular 2-fold axis is indicated by the metal rod.

(a) Viewed from above, looking down (a).

(b) View showing outside of the dimer.

The minor gold site, the major platinum site, the cleft and a part of one long helix are arrowed.

of columns of density of the right dimensions for α-helix. Many of these columns run for up to 20Å but two are more than 60Å long, almost the length of the subunit, and form its backbone. They underlie the cleft region and one approaches the subunit interface.

BINDING OF THE CO-ENZYME $NADP^+$ AND THE ANALOGUE 3-IODOPYRIDINE ADENINE DINUCLEOTIDE PHOSPHATE TO CRYSTALS OF 6-PGDH

Attempts have been made to find suitable conditions for binding of $NADP^+$ to crystals of 6-PGDH. Experiments were carried out in the buffer used for growing crystals, 65% ammonium sulphate, 50mM potassium phosphate at pH 6.5. In these conditions, low concentrations of $NADP^+$ (5μM) had no effect on the intensity distribution. Use of 5mM $NADP^+$ gave small intensity changes which were maximal after 3 weeks soaking. These differences were reproducible and some of the intensity changes were similar to those for the platinum derivative.

A sample of 3-iodopyridine adenine dinucleotide phosphate ($IADP^+$) was kindly prepared for us by Dr. J-F Biellmann and his colleagues. Enzyme crystals, transferred to solutions containing 5mM $IADP^+$ showed reproducible intensity changes which were maximal after 24 hours. These changes were similar to those given by $NADP^+$ after 3 weeks. With $IADP^+$, longer soaking did not enhance the differences.

Diffractometer data sets to 6Å resolution were collected for both $NADP^+$ and $IADP^+$ soaked crystals. Data were processed by a technique which analysed the peak profile and is particularly suitable for small crystals with large numbers of weak reflections. 3000 measurements, 1500 independent reflections were collected from a single crystal for each derivative. The data were corrected for Lorentz-polarisation factors, absorption and radiation damage in the normal way.

Difference electron density maps were computed for the two series of differences using 'best' phases. Each of the two difference maps showed a large peak which was in a similar position in each. The peak was 12Å long and one end was adjacent to the major platinum site. This peak was twice as high as any other feature on each map. There was no very large negative density. Since the peak was close to the platinum site, it was considered that it might be sensitive to errors in the phases causing ripple around heavy atom sites. A second set of difference maps was thus calculated using phases derived only from the gold derivative making use of anomalous and isomorphous differences. Although these maps would be noisier since the phase sets were less reliable, there should be no bias at the platinum site. The major peak on each of these maps was in the same position and of the same size as before and again there was no other feature above background level. The two peaks must therefore correspond to $NADP^+$ and $IADP^+$ density. Composites of the difference maps viewed down both a and c are shown in Figure III(a) and (b). Figure IV shows a view of the balsa wood model of the protein subunit with the density corresponding to $NADP^+$ arrowed.

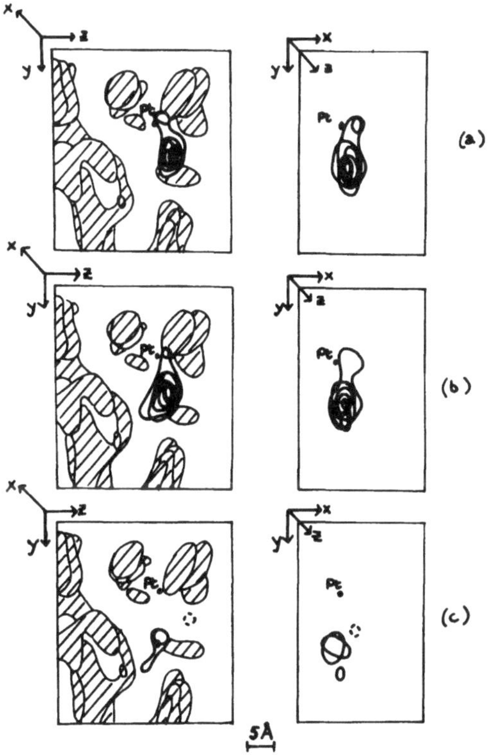

Fig. III

Composites of difference Fouriers for $NADP^+$ and $IADP^+$. Protein density (in views down a only) is shaded.

(a) $NADP^+$ - native protein.

(b) $IADP^+$ - native protein.

(c) $IADP^+$ - $NADP^+$

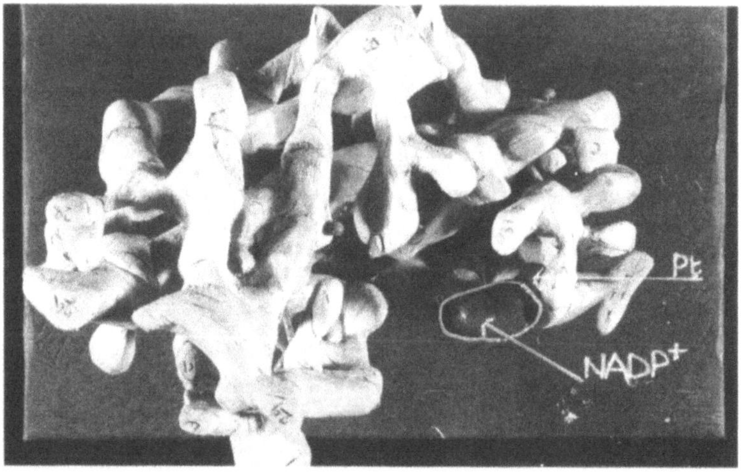

Fig. IV

View of the balsa wood model from above as in Fig. II(a) with $NADP^+$ density arrowed.

A further difference map with terms $m(F_{IADP^+} - F_{NADP^+})\exp(i\alpha_{best})$ showed a smaller positive region at one end of the co-enzyme density. This region is shown in Figure III(c). The peak, which should correspond to the difference between the iodine atom and the amide group at position 3 of the pyridine ring, is about 10Å from the site of platinum substitution and serves to orient the co-enzyme. The adenine part of the co-enzyme should thus be close to the platinum binding site. The binding of platinum compounds at the co-enzyme binding sites has been observed for lactate dehydrogenase (LDH), malate dehydrogenase (MDH) and liver alcohol dehydrogenase (LADH). In LADH the site is a general anion binding site and is at the position of the pyrophosphate. (Brändén et al. 1975).

The size and shape of the peak corresponding to $NADP^+$ indicates that while the co-enzyme certainly does not have the bases stacked, it is probably in a slightly less open conformation than that chosen by NAD^+ in LDH, MDH, LADH and glyceraldehyde phosphate dehydrogenase. The orientation of the molecule at one edge of and pointing into the cleft is interesting, it suggests an area of the subunit which might provide a substrate site, at the other side of the cleft on the larger part of the subunit. Any conclusion of this nature at this stage of the crystal structure analysis must, however, be extremely tentative.

It is even more difficult to compare the nature of the co-enzyme binding site with the 'nucleotide binding site' common to the four NAD^+ dependent dehydrogenases and maybe also to other enzymes (Rossmann et al., 1974) and the conclusions of this paragraph are entirely speculative. The smaller part of the enzyme contains four short columns of density which are within 60° of being co-axial. It is in principle difficult to see pleated sheet at 6Å resolution, and indeed, this map provides neither evidence for nor against its existence in this area. It may be argued that the elongation of the co-enzyme density should be in a rather different direction from that observed if the helices have been correctly identified as those of a 'nucleotide binding site'. The matter cannot be settled, however, without more data. A high resolution structure, in which the main chain can be traced is required before any informed decision can be made about the nature of the $NADP^+$ binding site.

EXPERIMENTS WITH DIFFERENT SUPPORTING MEDIA

The experiments described in this section arose from a chance observation that occasional crystals show a quite different intensity pattern but very similar cell dimensions and the same spacegroup. The occurrence of this form was unusual and unpredictable.

When the ammonium sulphate medium was changed to 3M potassium tartrate at pH 8 for 12 hours, crystals became very fragile, but those which could be handled showed the transformed pattern though they were less well ordered than the normal native enzyme crystals. The same transformed pattern could be produced reliably with 1.8M sodium citrate at a measured pH of 7.2, and small crystals give diffraction patterns which were well ordered. Figure V compares the intensity patterns of the normal 'sulphate' type (a) and the 'citrate' type (b) in the (h0ℓ) zone. The cell dimensions of the transformed crystals changed by a maximum of 3% and are $\underline{a}$ = 71.9Å, $\underline{b}$ = 146.8Å, $\underline{c}$ = 100.6Å. The nature of the transformation is of considerable interest. The intensity changes are much too large to be accounted for without some movement of the molecule but the molecular 2-fold axis must still be exact. Unfortunately there has been rotational disorder in every crystal investigated which is large enough for collection of diffractometer data and so the nature of the change has not been further elucidated yet.

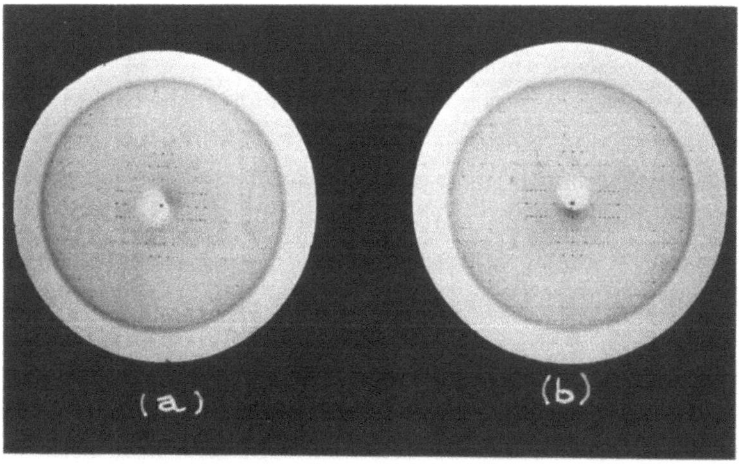

Fig. V

Comparison of h0ℓ photographs of (a) native 6-PGDH in ammonium sulphate, (b) crystals transferred to citrate.

When crystals were left in a supporting medium of 3M sodium/potassium phosphate buffer for seven days at pH 6.5, much smaller intenstiy changes were observed, not suggestive of any large-scale movement of the protein. The maximum change in cell dimensions was 0.5Å with a = 73.1Å, b = 147.7Å, c = 102.9Å. Diffractometer data have been collected to 6Å for a crystal in phosphate and a difference map between the phosphate and the sulphate forms has been calculated.

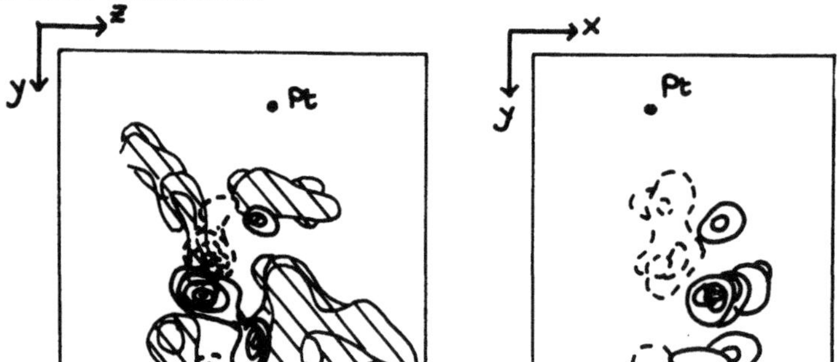

Fig. VI

Composite of difference Fouriers, phosphate crystal - sulphate crystal. Protein density (in view down a only), is shaded.

The major features of this map are all in one area 10 x 20 x 12$Å^3$ which is illustrated in Figure VI. There are two peaks and two holes which are up to 7Å long in their longest dimensions and a third small peak. The holes are on pieces of density at the edge of the cleft, the one closer to the major platinum site lies 15Å from it and is a little beyond the end of the co-enzyme site attributed to the nicotinamide. The second hole lies about 26Å from the platinum site. The peaks lie between the holes and may indicate movement of electron density. The further small peak is 10Å from the platinum site at a position where a part of the co-enzyme binds. The positions of the peaks and holes are indicated on a view of the balsa wood model on Figure VII.

It is extremely interesting to note that the changes take place in this part of the cleft. It would seem that either protein movement has occurred in this area or that bound sulphate ions have been removed and phosphate ions have bound in slightly different places. In view of the changed kinetics in the presence of phosphate and the fact that it is a competitive inhibitor (as in sulphate), it is tempting to suggest that some of the changes label the substrate site. The possibility seems strengthened since the peaks are not far from the nicotinamide portion of the co-enzyme. The substrate site would, of course, be best marked by

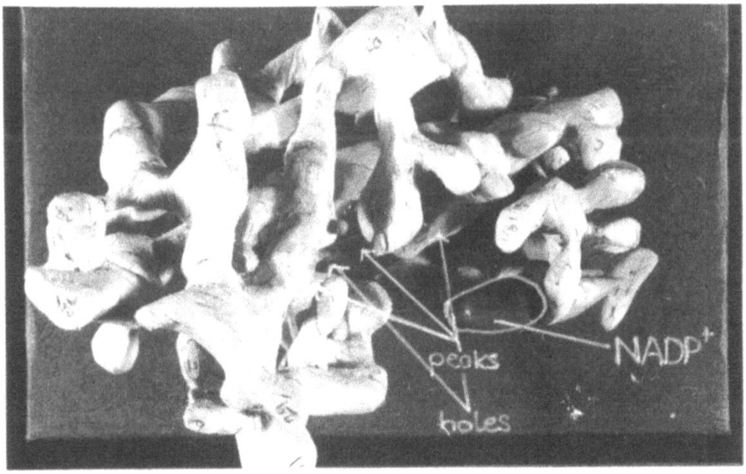

Fig. VII

View of the balsa wood model from above as in Fig. II(a) with position of phosphate-sulphate differences arrowed. The co-enzyme is in place and is indicated.

binding 6-phosphogluconate, but it has proved difficult to do this in the crystals in ammonium sulphate and in the absence of co-enzyme. Attempts are being made to bind 6-phosphogluconate in presence of NADPH and it is hoped that this will give a clearer indication of the substrate binding site. Extension of this work to higher resolution will give more definitive results.

ACKNOWLEDGEMENTS

We thank Professor D.C. Phillips and Professor Dorothy Hodgkin for their help and encouragement and Professor Don Voet for assistance and for helpful discussions. We are indebted to the Oxford Enzyme Group Preparation Laboratory for enzyme preparations. We acknowledge financial support from the Medical Research Council. One of the authors (IGA) holds an M.R.C. research scholarship, a second (JRH) a Science Research Council research scholarship. The third author (MJA) is a fellow of Somerville College, Oxford and a member of the Oxford Enzyme Group.

REFERENCES

Adams, M.J., Helliwell, J.R., Bugg, C.E.: The structure of 6-phosphogluconate dehydrogenase from sheep liver at 6Å resolution. Manuscript submitted to J. Mol. Biol. (1977).

Blow, D.M., Crick, F.H.: The treatment of errors in the isomorphous replacement method. Acta Cryst. 12, 794-801 (1959).

Brändén, C-I., Jörnvall, H., Eklund, M., Burugren, B.: The Enzymes 3rd edn. ed. P. Boyer (Academic Press) Vol. XI. Oxidation-Reduction Chap. 3 Alcohol Dehydrogenases p. 103-190 (1975).

Dyson, J.E.D., D'Orazio, R.E.: Sheep liver 6-phosphogluconate dehydrogenase - inhibition by nucleoside phosphates and by other metabolic inhibitors. J. Biol. Chem. 248, 5428-5435 (1973).

Dyson, J.E.D., D'Orazio, R.E., Hanson, W.H.: Sheep liver 6-phosphogluconate dehydrogenase, isolation procedure & effect of pH, ionic strength and metal ions on the kinetic parameters. Arch. Biochem. Biophys., 154, 623-635 (1973).

Rossmann, M.G., Moras, D., Olsen, K.W.: Chemical and biological evolution of a nucleotide binding protein. Nature 250, 194 (1974).

Silverberg, M.: Studies of 6-phosphogluconate dehydrogenase. D.Phil. Thesis, Oxford University.

Silverberg, M., Dalziel, K.: Crystalline 6-phosphogluconate dehydrogenase from sheep liver. Eur. J. Biochem., 38 229-238 (1973).

Silverberg, M., Dalziel, K., Adams, M.J.: Crystalline 6-phosphogluconate dehydrogenase from sheep liver. Biochem. Soc. Trans. 1, 1132-1134 (1973).

Villet, R.H., Dalziel, K.: Studies of 6-phosphogluconate dehydrogenase from sheep liver II Kinetics of the oxidative-decarboxylation reaction, co-enzyme binding and analyses for metals. Eur. J. Biochem. 27 251-258 (1972).

Received January 31, 1977

DISCUSSION

Sund: Is your enzyme inhibited by heavy metal ions and are the crystallized heavy metal derivatives still enzymatically active after dissolving?

Adams: The enzyme is inactivated on binding two moles of p-hydroxymercuribenzoate per mole subunit (Silverberg and Dalziel, Eur. J. Biochem. 38, (1973) 229-238), but our crystals crack before mercury is bound. The heavy atom derivatives were prepared by soaking crystals and we have not tested the activity of redissolved crystals of heavy atom derivatives.

Brändén: What happens when you try to cocrystallize complexes of enzyme and substrate from solution?

Adams: We have not been able to grow such crystals.

Jaenicke: What is the evidence that the dimer is the enzymatically active entity of 6-phosphogluconate dehydrogenase?

Adams: The dimer is the naturally occurring form. I cannot quote a paper which shows the subunit to be inactive.

Mannervik: I have been given the privilege to present data on the structure of human erythrocyte glutathione reductase obtained by X-ray diffraction at 6 Å resolution in the group of G.E. Schulz in Heidelberg (H.A. Zappe, G. Krohne-Ehrich and G.E. Schulz, J. Mol.Biol., in press). This enzyme is a dimer consisting of two apparently identical subunits of 50,000 molecular weight. The crystallographic asymmetric unit contains one subunit and the dimer interface is on a crystallographic twofold axis. The dimensions of a subunit are about 50 Å x 45 Å x 55 Å and each subunit can be divided into three domains, which are arranged in such a way that deep clefts are formed on opposite sides of the subunit. One of the clefts can accommodate the oxidized substrate glutathione disulfide (GSSG) and the other one can bind the oxidized product $NADP^+$. Thus, there is one site for glutathione and one site for pyridine nucleotide in each subunit. Both the two binary complexes and the ternary complexes involving GSSG and $NADP^+$ have been studied. The shortest distance between the disulfide bond of GSSG and the $NADP^+$ is 18 Å, and a connecting line between them passes through a region of high electron density, which may contain the prosthetic group FAD. A structure involving two topologically separated substrate-binding sites is clearly completely different from structures previously published on pyridine nucleotide-dependent dehydrogenases. This structure of glutathione reductase is also fully consistent with the proposed branching reaction scheme involving a combination of a ping-pong and a sequential mechanism (B. Mannervik (1976) in Flavins and Flavoproteins (Singer, T.P., ed.), pp. 485-491, Elsevier, Amsterdam).

Adams: This is a very interesting structure of a different type of $NADP^+$ dependent dehydrogenase. It is of interest to note that the $NADP^+$ conformation is compatible with the NAD^+ conformation in LDH, GPDH and liver ADH. At low resolution, it would be unjustifiable to assert that the $NADP^+$ conformation in this enzyme is different from that in 6-PGDH.

Veeger: Would you not expect that the $NADP^+$ is bound in a NAD^+-like conformation, since glutathione reductase is also rather active with NAD^+. Thus the lack of coenzyme specificity illustrates little influence of the 2'-phosphate, accepting the view that NAD^+ is bound in enzymes is rather similar conformations.

Mannervik: At 6 Å resolution the conformation of $NADP^+$ can evidently not be established in any detail, but Dr. Schulz claims that the conformation appears very similar to that found for NAD^+ bound to e.g. MDH, ADH, LDH, or malate dehydrogenase.

Sund: Do you have information about the distance between the flavin and the disulfide bridge and the nicotinamide moiety of NADP, respectively? Based on the model of the enzyme molecule, is the direct hydrogen transfer at all possible?

Mannervik: The only information available is the distance between the disulfide bond of glutathione disulfide and the nicotinamide ring of $NADP^+$ which is 18 Å. This distance was determined by the difference-fourier technique. The position of FAD on the enzyme could not be determined at 6 Å resolution. It is known that the enzyme in addition to FAD contains a disulfide, which participates in catalysis, and it is believed that these two redox-active groups are situated between the $NADP^+$- and the GSSG-binding clefts in a subunit. Direct transfer of reducing equivalents from NADPH to GSSG cannot take place when the substrates are bound to the enzyme.

Perham: I should just like to amplify a point of Dr. Mannervik'in reply to Dr. Sund. In glutathione reductase, a protein disulphide bridge participates in the mechanism, becoming alternately oxidized and reduced. This might help to span the distance of 18 Å between the nicotinamide ring and the disulphide of the substrate, in association of course with the cofactor FAD. Protein conformational changes could accompany intra-chain disulphide bridge breakage, which would help to reduce the distance to be spanned still further. The geometry of the binding of NADP and GSSG in active site as envisaged at present by Dr. Schirmer and Dr. Schulz is both novel and intriguing.

Section II. Symmetry and Coenzyme Binding

Chairmen: R.N. Perham and H. Gutfreund

CONFORMATIONAL CHANGES AND NON-EQUIVALENCE IN THE BINDING OF NAD^+ TO CYTOPLASMIC MALATE DEHYDROGENASE

M. Weininger, J.J. Birktoft and L.J. Banaszak
Department of Biological Chemistry, Division of Biology and Biomedical Sciences, Washington University School of Medicine, St. Louis, MO 63110, USA

Although pig heart cytoplasmic malate dehydrogenase (s-MDH) is a dimer of chemically identical subunits, the crystal structure as well as chemical analysis has suggested that the co-factor, NAD^+, binds to s-MDH crystals in both a non-stoichiometric as well as a non-equivalent manner (1,2). The so-called native crystals used in the x-ray analysis were grown in the presence of a 10-fold molar excess of NAD^+ corresponding to concentrations of the order of 0.7 mM (1). Chemical analysis of the crystals after extensive washing indicated the presence of 0.94 moles of NAD^+ per mole of s-MDH dimer (2). If the same "native" crystals were first soaked in 20 mM NAD^+ and then quickly washed, 1.2 moles of NAD^+ per mole of s-MDH dimer was found (2). Kinetic analysis of the enzyme in solution indicated that there was no apparent cooperativity in coenzyme binding in the absence of substrates (3). The non-equivalence in the binding of NAD^+ seemed, therefore, to be related to the crystal structure. In the case of the x-ray analysis of s-MDH crystals, the dimeric molecule is contained in the asymmetric unit and consequently such differential binding behavior can be observed.

The differences or non-equivalance of binding of NAD^+ to the two subunits was supported by two observations. First, when the "native" crystals were soaked in 20 mM NAD^+, additional binding of the coenzyme was observed in only one of the two subunits, the one which will be referred to as subunit 2 (SU2) (1). Secondly, analysis of the electron density map of the "native" enzyme at 2.5 Å resolution revealed a somewhat higher level of electron density in the coenzyme binding region of the subunit which cannot undergo additional substitution by NAD^+ (SU1) (4). While x-ray analysis of protein crystals is quite sensitive to changes in the contents of the unit cell, it is difficult to obtain accurate estimates of such problems as differential substitution when they occur in the native crystalline form. Frequently electron density is reduced in regions of a map because of temperature effects or varying kinds of disorder. Thus, the native electron density map of s-MDH even at high resolution could not be used directly to estimate the magnitude of the differential coenzyme binding.

In order to facilitate a discussion of the NAD^+-free form of s-MDH, a brief description of the current state of the interpretative model building from the 2.5 Å electron density map is necessary. At this time, a Kendrew-Watson model of the dimer has been constructed. Since the chemical sequence is not yet completed, the current model includes the polypeptide backbone as well as a speculative interpretation of the side chain electron density (the x-ray amino acid sequence). By comparing the electron density and model, residue by residue in each of the two subunits, we are hoping that an improved interpretation can be obtained. This approach should also enable us to define those parts of the s-MDH structure where conformational differences between the two subunits exists. Such differences are apparent in several regions of the s-MDH molecule, but they generally occur on the molecular surface where the electron density is frequently weaker. Without a chemical amino acid sequence, interpretations of these differences would be speculative.

This problem can be illustrated by a segment of polypeptide chain especially pertinent to the binding of NAD^+. This segment, called the loop region in s-MDH and other dehydrogenases, is a portion of the polypeptide chain connecting the secondary structural elements <u>βD</u> and <u>αD</u>. In lactate dehydrogenase (LDH), this turn has been shown to undergo a conformational change in going from the apoenzyme to the abortive ternary complex (6). A similar change could account for the differential binding of NAD^+ to s-MDH if, for example, crystal packing forces prevented identical movements of the loop region from occurring in both subunits. Because of the lack of chemical sequence data and weak density in the βDαD region of the high-resolution electron density map, minor or even no conformational differences may exist between the loop regions in the two subunits of s-MDH.

In order to further investigate the differential binding of NAD^+ to s-MDH, we have prepared crystals of the apo-form of s-MDH and measured the x-ray data from these crytals to a resolution corresponding to 5 Å. The unit cell dimension and a summary table of the x-ray measurements are given in Table I. The large changes observed in the x-ray intensities may be due, in part, to the changes in the unit cell dimensions and not solely to the removal of NAD^+.

Table I

Apo-Form of s-MDH

		Å	
A. Unit Cell Dimensions*	a	b	c
apo-	140.7	87.1	58.6
holo- (1.2 moles NAD^+/mole enzyme)	139.2	86.6	58.8

B. X-ray Data	
Mean $\Delta F/F_p$ (centric)	0.273
Mean $\Delta F/F_p$ (non-centric)	0.185
Total number of reflections 5.0 Å resolution**	3200

*Both crystal forms have the space group $P_{2_1 2_1 2}$.

**The data was measured by diffractometry in shells of about 900 reflections and scaled to holo-enzyme F's as previously described for the heavy atom derivatives (7).

Two sections of the resulting difference Fourier map are shown superimposed in Figure 1. The boundary of the two subunits comprising the dimeric s-MDH is also shown with SU1 to the left and SU2 to the right.

The average noise level on the map was estimated to be near the first (lowest) positive contour level which is drawn. In general terms,

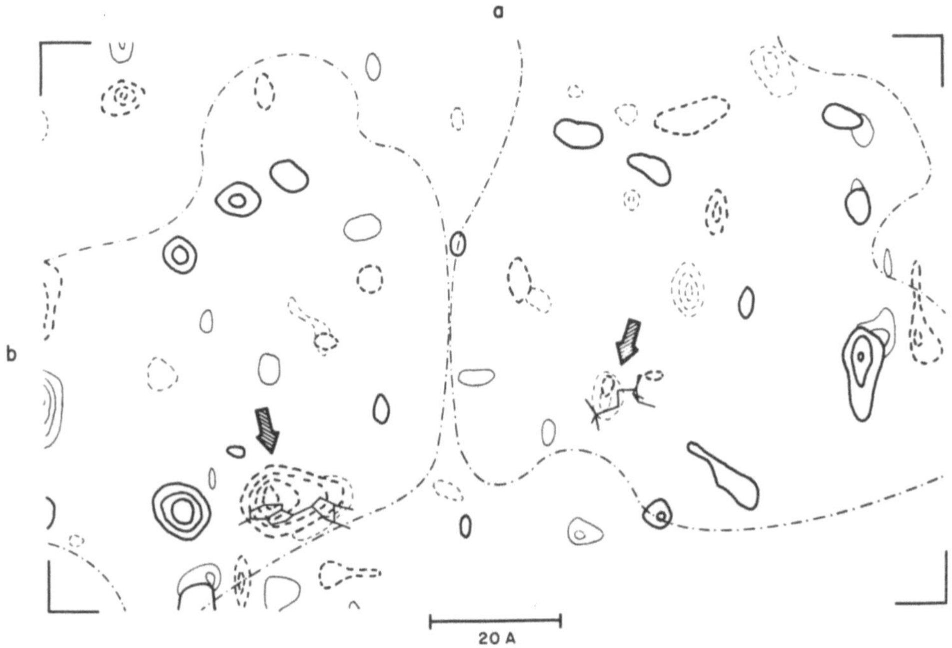

Figure 1 Coenzyme Difference Electron Density Map of s-MDH.
Negative contours are dashed; the line —·— represents the subunit boundaries with SU1 on the left and SU2 on the right. Two sections of the map are shown with the one closest to the reader drawn with a heavier line.

Figure 2 Several sections of the difference map in the region of the NAD^+ bound to SU1 as well as the atoms of the bound coenzyme are shown.

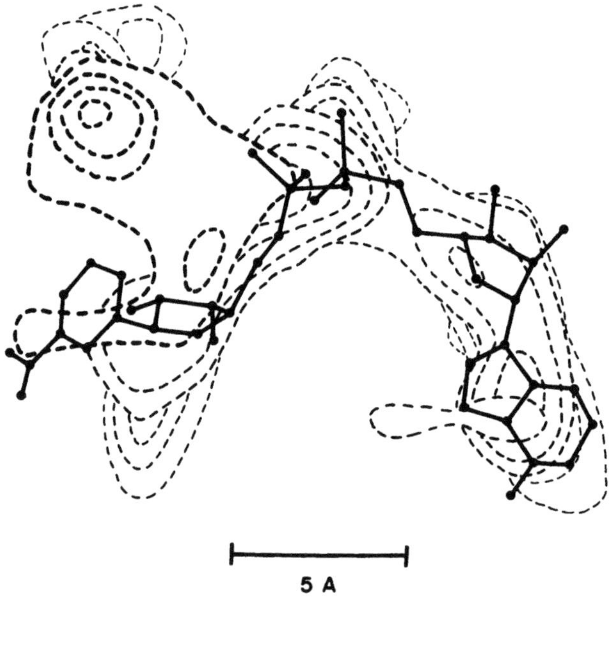

therefore, the map can be described as being relatively noisy. However, the major changes are quite striking and stand out well above the noise level.

Some of the more important features in the electron density difference map appear in part on the sections shown in Figure 1. Near the bottom of SU1, an arrow indicates a large negative peak which is continuous through several more sections. Superimposed on this negative electron density are the atoms of the adenine ring and the adenine ribose which all fall on the map sections which are shown. This negative region is illustrated in more detail in Figure 2 where the atoms of NAD^+ that are bound to SU1 in the native crystalline form (8) have been superimposed upon this negative electron density. As can be seen, the correspondence between the model and the electron density is quite good. Only the negative electron density associated with the nicotinamide ring is somewhat weak. The large negative peak in the upper left corner, which does not contain atoms belonging to the NAD^+, is associated with residues in the α2Gα3G corner. The changes in this segment of the polypeptide chain will be considered later together with other conformational changes in the polypeptide structure.

There is some weak negative electron density in the coenzyme-binding region in SU2. Part of these electron density changes can also be seen in Figure 1, where a weak negative region is observed at a position near where the pyrophosphate bridge of the bound NAD^+ is located as is indicated with an arrow. The negative peak at the coenzyme-binding site in SU2 is significantly weaker than that associated with the coenzyme-binding site in SU1 and is, in fact, only observed near the pyrophosphate region. Together, these two features support our earlier reports that SU1 contains slightly more bound NAD^+ than SU2 and that some additional coenzyme can be soaked into the active site of SU2.

Adjacent to the large negative peak in SU1, a large and extensive positive peak is observed, suggesting that some segments of the polypeptide have moved into this position. Close to this positive peak, high negative electron density is observed in a position correponding to several residues in the "loop" region. Thus, it appears that the βDαD segment, the "loop", moves towards the NAD^+ binding cavity upon removal of the coenzyme. We estimate that the movement of the polypeptide backbone could be as much as 7-9 Å and possibly more for side chain atoms. The weak negative density at the nicotinamide moiety in SU1 could be accounted for by a movement of a segment of the "loop", a side chain for example, into the space vacated by the nicotinamide ring. On the other hand, no movement is indicated for the corresponding "loop" region in SU2.

In both SU1 and SU2, no density is observed at the anion (dicarboxylic acid substrate) site, indicating that, just as in the apo-form of LDH (9), this site remains filled whether or not coenzyme is present. However, the above-mentioned movement of the "loop" in SU1 could well mask any ion displacement in this subunit.

A side chain near the end of the βF strand of s-MDH also appears to move during the transition to the apo-enzyme. In LDH, a serine (#163) is found in the homologous position and is thought to hydrogen bond to the carboxyamide side chain of the nicotinamide ring (10). Based on the structural homology between LDH and s-MDH, as well as the appearance of the side chain density, this residue is also thought to be a serine in s-MDH. The appearance of matching negative and positive peaks near this residue suggests a movement of the serine away from the NAD^+ and toward the surface of SU1. A relatively large positive peak and a somewhat weaker negative peak are located near what we believe is the active site histidine in SU1 in the βGβH corner. A movement similar to that described for the serine is believed to occur for this histidine. The weaker negative density at this residue could be accounted for by the movement of the "loop" which has been previously described. Perhaps some part of the "loop" is replacing the histidine.

In the active site of SU2, a relatively high negative electron density peak is found near the serine, and positive density is observed near the histidine, in the space between the serine and histidine. In this case, the difference electron density map is somewhat more difficult to interpret, but does suggest that, with respect to those two residues, similar changes occur in both subunits.

These conformational changes, as well as others to be described later, are illustrated in a somewhat schematic manner in Figures 3a and 3b. Figure 3a is a diagrammatic representation of the structure of one subunit of s-MDH (SU1). In Figure 3b, non-relevant segments of the structure have been stripped away, and only those parts of the s-MDH structure which are undergoing conformational changes are shown.

The conformational changes which take place upon going from the holo- to the apo-form of s-MDH are indicated with arrows. The size of the arrows are only intended to give a qualitative representation of the movements or conformational changes; no quantitative estimate is intended. However, we do believe that, with the exception of the movement of the "loop", the <u>βDαD</u> region, these movements are of the order of 2-4 Å for polypeptide backbone atoms, with larger changes possible for side chain atoms.

Despite the relatively noisy character of the difference map, a large number of other significant positive and negative peaks suggest that a number of other conformational differences may exist between the apo- and holo-crystals. Given the resolution of the difference map and the complexity of the changes, it is impossible in a limited space to describe them in detail. However, in a general, albeit speculative way, we feel that all can be linked to the removal of NAD^+ from the binding site in SU1.

Paired negative and positive peaks of electron density are observed around the structural segments of SU1 which are shown in Figure 3b. The movements which are indicated seem to occur in the same general direction -- namely, away from the coenzyme binding site and subunit interface, and towards the solvent surface of SU1. The most reasonable explanation for these conformational changes is that the movement of the "loop" into the NAD^+ cavity forces the s-MDH molecule to expand somewhat. The general direction of this expansion is along the direction of the crystallographic <u>a</u>-axis. In fact, this particular axis does increase 1.5 Å

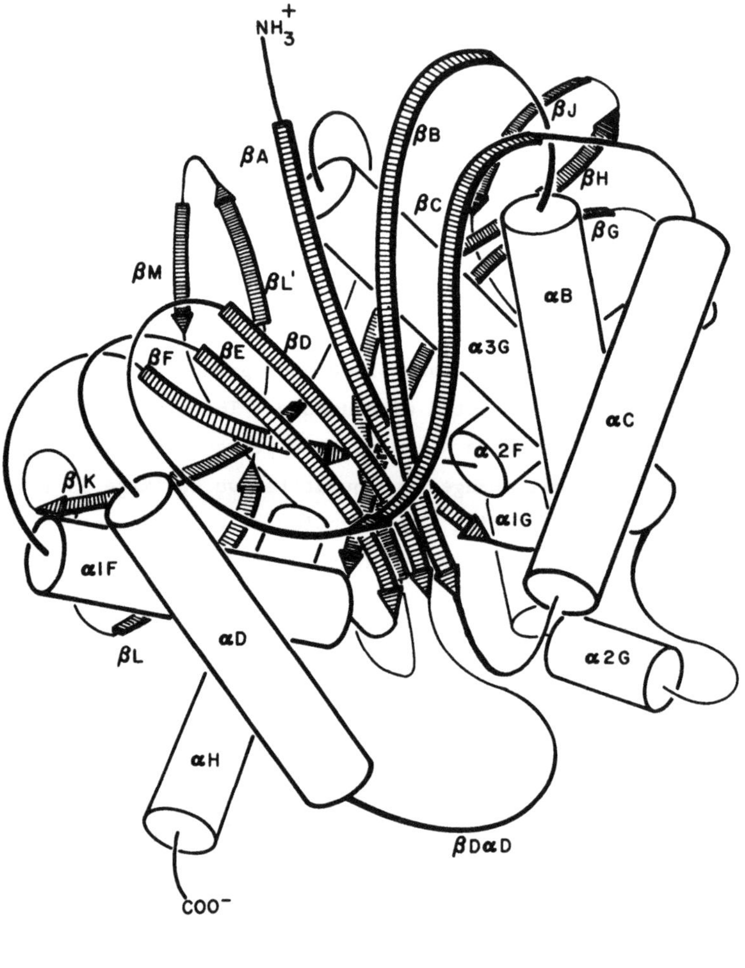

Figure 3A Schematic Representation of the Structure of s-MDH.

The segments of secondary structure are labelled with the standard dehydrogenase nomenclature (5). Although the coenzyme is not shown, this drawing is meant to represent the "holo-form" of the enzyme.

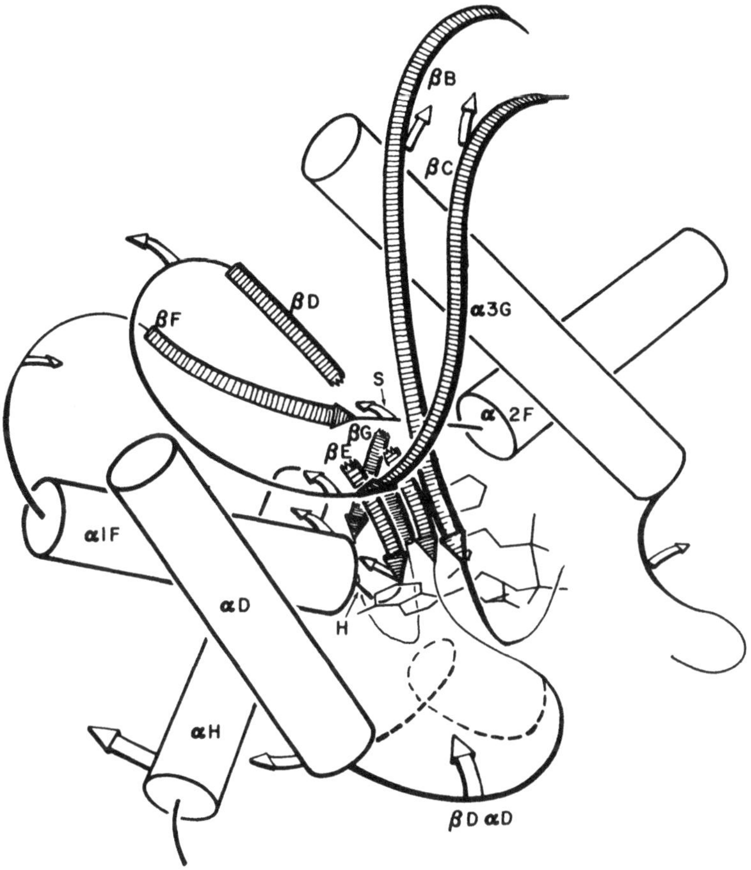

Figure 3B The NAD$^+$ Free Form of s-MDH.

The schematic drawing of s-MDH is the same as shown in Figure 3A except secondary structure which is uneffected by the transition to the apo-enzyme has been omitted. In addition, we have attempted to show the coenzyme in its perspective to the overall structure. Conformational differences between the apo- and holo-enzyme are shown schematically by the arrows except for the loop region changes for which the apo-form is drawn with a dotted line. The letters H and S indicate the approximate locations of the active site histidine and serine side chains.

on going from the holo- to the apo-enzyme.

Furthermore, the major conformational changes occurring in SU1 are transmitted to SU2, but not across the molecular two-fold axis. The subunit interface region is relatively unaffected, and only a movement of the <u>α2Gα3G</u> corner is suggested. Rather, intermolecular contacts resulting from crystal packing causes <u>αH</u> of SU1 to be in close contact with <u>αH</u> of SU2 of a symmetry related molecule. The movement of <u>αH</u> in SU1, as indicated in Figure 3b, is transmitted across to <u>αH</u> in SU2 in another molecule, and this movement is further propagated through part of SU2 with the following segments of the SU2 being affected: <u>αH</u>, <u>βMαH</u>, <u>βKβL</u>, <u>βH</u> and <u>α2F</u>. All of these chain segments are quite far removed from the NAD^+ binding site in SU2, and it is reasonable to suggest that the events observed in SU2 are changes caused by crystal packing.

In conclusion, the electron density difference map of the apo- versus the holo-form of s-MDH has shown that the largest negative peak appears at the location of the bound coenzyme in SU1, with a considerably weaker peak at the corresponding position in SU2. Two amino acids, believed to be the active site histidine and possibly a serine, both of which are interacting with the nicotinamide end of the bound coenzyme, appear to move upon removal of the coenzyme. Surprisingly, these residues appear to move in both subunits. The largest conformational change takes place in the "loop" region in SU1. During the transition from the holo- to the apo-enzyme, the "loop" moves approximately 7-9 Å towards or into the coenzyme binding cavity, with some part of this segment positioned at or near the location occupied by the nicotinamide ring in the holo-enzyme. On the other hand, no conformational changes in the "loop" region are observed in SU2.

The changes in the "loop" conformation, as clearly seen in SU1, are the opposite to that observed for LDH. In that enzyme, the "loop" moves away from the coenzyme binding region when NAD^+ is removed (6).

The "loop" movement in s-MDH is accompanied by a series of minor conformational changes in SU1. All of these movements occur in the same general direction, resulting in what can be described as a slight expansion of SU1. This complex set of changes could never be described in any more detail from the difference map. It would be necessary to independently determine the native structure of the apo-enzyme, an approach which we are now seriously considering. The conformational changes in SU1 are transmitted across an inter-molecular contact interface to SU2 in a symmetry related molecule and not through the subunit interface, resulting in several minor changes in the structure of SU2. The subunit interface region of the s-MDH dimer is essentially featureless in the electron density difference map. Therefore, there is no reason to believe that these crystallographic observations are related to coenzyme binding in solution.

The reason for the observed non-equivalence in NAD$^+$ binding in the crystalline state is, at this stage, obscure. The concentration of NAD$^+$ as employed in the preparation of the native crystals is 0.7 mM, which is approximately the same as the association constant for NAD$^+$ binding to s-MDH (3). Even if the high salt concentration employed during the crystallization, 63% saturated ammonium sulphate, is taken into consideration, it is not surprising that full saturation of s-MDH with NAD$^+$ is not observed in the native crystals. On the other hand, we do not have any satisfactory explanation for the non-equivalence in binding additional NAD$^+$. Since s-MDH crystallizes with a dimer in the asymmetric unit, intermolecular interactions and crystal packing forces generate different environments for SU1 and SU2. In SU2, the subunit which can bind additional NAD$^+$, the coenzyme binding region, including the "loop", is facing a large solvent cavity. The corresponding region in SU1 is, on the other hand, involved in intermolecular contacts with a symmetry related molecule. Specifically, there are several residues in the beginning of <u>αD</u> and in the <u>βBαC</u> region interacting with the <u>βMαH</u> region of SU2 of the adjacent molecule. So contrary to what might have been expected, NAD$^+$ is found with the highest occupancy in the subunit which is involved in the most extensive intermolecular interactions around the coenzyme binding site. This subunit would be the most likely to have the most serious lattice constraints on any protein conformational changes which might accompany coenzyme binding. Thus, the study of the apo-form of s-MDH has described an important conformational change accompanying the binding of NAD^+ but has also left unresolved the problem of non-equivalent binding in the native form of the crystals. We suggest that there is a difference in binding affinity in the active site of each of the subunits in these crystals, but this is due to intermolecular rather than intramolecular contacts. In a sense, it is cooperativity generated by a higher aggregation state, one which is only present in the crystal lattice.

ACKNOWLEDGEMENTS: We are grateful to Mr. Gary Barbarash for his help in the purification of the pig heart s-MDH and to the NIH and NSF (grant numbers GM-13925 and P-CM 76-81481) for support of these studies.

REFERENCES

1. Tsernoglou, D., Hill, E., Banaszak, L.J.; Cold Spring Harbor Symp. Quant. Biol. XXXVI, 171 (1971).

2. Glatthaar, B.E., Banaszak, L.J., Bradshaw, R.A.; Biochem. Biophys. Res. Comm. 46, 757 (1972).

3. Mueggler, P.A., Dahlquist, F.W., Wolfe, R.G.; Biochemistry 14, 3490 (1975).

4. Webb, L.E., Hill, E.J., Banaszak, L.J.; Biochemistry 12, 5101 (1973).

5. Rossmann, M.G., Liljas, A., Branden, C.-I., Banaszak, L.J.; "The Enzymes", 3rd Ed., 11A, 61 (1975).

6. White, J.L., Hackert, M.L., Buehner, M., Adams, M.J., Ford, G.C., Lentz, Jr., P.J., Smiley, I.E., Steindel, S.J., Rossmann, M.G.; J. Mol. Biol. 102, 759 (1976).

7. Tsernoglou, D., Hill, E., Banaszak, L.J.; J. Mol. Biol. 69, 75 (1972).

8. Banaszak, L.J., Webb, L.E.; "Structure and Conformation of Nucleic Acids and Protein-Nucleic Acid Interactions", ed. by M. Sundaralingan and S.T. Rao, University Park Press, Baltimore, p. 375 (1975).

9. Adams, M.J., Liljas, A., Rossmann, M.G.; J. Mol. Biol. 76, 519 (1973).

10. Eventoff, W., Rossmann, M.G., Taylor, S.S., Torff, H.-J., Meyer, H., Keil, W., Kiltz, H.-H.; Proc. Nat. Acad. Sci. USA, in press (1977).

Received March 15, 1977

DISCUSSION

Trommer: I should like to stress again an argument brought up by Dr. Hinz this morning. At 2 Å resolution the actual position of a given atom may be off by about 0.5 Å. Conformational changes in this order may not be observed in the X-ray analysis but may change binding constants of the ligands by orders of magnitudes. What I want to point out is that you cannot exclude conformational changes across the subunit interfaces leading to negative or half-of-the-sites reactivites. You may just not observe them.

Banaszak: Correct! On a difference electron density map at 5.5 Å resolution, changes of the order of 0.5 Å would not be observed.

Walk: Studies of the cytoplasmic and mitochondrial pig heart and watermelon MDH isozymes support the idea of a "closed" NAD-binding site in s-MDH, since the cytoplasmic isozymes pass through 5'AMP-Sepharose at low ionic strength whereas the mitochondrial isozymes bind to the matrix-bound ligand.

Banaszak: This does indeed correltate with our interpretation of the "loop" location in the apo-form of s-MDH.

Bernhard: Can you differentiate between crystal-lattice induced anticooperativity and inherent asymmetry in the dimer structure?

Banaszak: Not until the total structure is known. At this time we are still missing the amino acid sequence and therefore we are uncertain about parts of the molecule. At the stage when the complete molecular structure is known, if asymmetry is found one can attempt to trace it's origin. That is the conformational difference might be traced to the subunit-subunit contact or alternatively to an intermolecular contact. With very subtle differences in conformation, however, the resolution may never be good enough to determine their origin.

Brändén: If the loop in your binary complex represents the active conformation of the loop in the ternary complex your crystals should show activity without being shattered. Have you checked the activity of your crystals of the binary complex?

Banaszak: No, but that would certainly be worth doing. I am not sure, however, that the results would be unambiguous. In SU2, the loop density is weak indicating that it may be moving somewhat even in the native crystal.

Holbrook: Your interpretation of the moved loop position in s-MDH when compared to LDH is, as I understand it, that the enzyme binds an anion at what you presume to the active site in the absence of coenzyme. This would be consistent with the observation of Dr. Lodola in my laboratory, that added oxaloacetate alters the thermal stability of the enzyme in the absence of cofactor. There is of course no certainty that the changed thermal stability is due to binding at the active site.

Dunn: Is it possible that the crystallized enzyme is actually a dimer of dimers - that is a tetrameric enzyme with two classes of sites?

Banaszak: The intermolecular contacts in the crystal state lead to a continuous chains of dimers. As best I remember the contacts do not correspond to any inter-subunit interactions in the active dehydrogenases.

THE EFFECT OF NUCLEOTIDE BINDING ON SUBUNIT INTERACTIONS IN GLYCERALDEHYDE 3-PHOSPHATE DEHYDROGENASE AS DETERMINED BY THE KINETICS AND THERMODYNAMICS OF SUBUNIT EXCHANGE.

M.R. Hollaway, H.H. Osborne and G.M.L. Spotorno.
Department of Biochemistry, University College London, Gower Street, London WC1E 6BT.

Hybridization studies, involving measurement of the rate and extent of subunit interchange between identical or homologous oligomeric proteins, can provide information about the strength of subunit interactions and how these interactions change on binding specific ligands. This approach to the study of intersubunit binding has the advantage that it is not necessary to employ denaturing conditions to measure dissociation, although some care is necessary in interpreting results as the rate of subunit exchange gives information about the activation energy for the dissociation processes involved rather than the free energy of interaction between the subunits. We describe here the results of a study of the kinetics and thermodynamics of subunit interchange in the system comprising rabbit (R_4), yeast (Y_4) and hybrid (R_2Y_2) glyceraldehyde 3-phosphate dehydrogenases. The effect of the binding of specific ligands, such as NAD^+ and NADH, on the hybridization reaction will be discussed.

THE NATURE OF THE HYBRIDIZATION PROCESS

Mixing samples of the homotetrameric rabbit and yeast enzymes leads to the formation of the hybrid species R_2Y_2, comprising two subunits from each of the two types of molecule. (1,2). The rate and extent of formation of the R_2Y_2 species can be followed by densitometric analysis of starch gel electrophoretograms carried out at different times of incubation (3). Fig. 1 shows a typical analysis of a 'hybridization mixture' and illustrates the isolation of the hybrid species by ion exchange chromatography. The isolated hybrid allows 'hybrid reversion' experiments to be carried out: that is the measurement of the rate of reversion of hybrid to the equilibrium mixture, in the presence and absence of specific ligands.

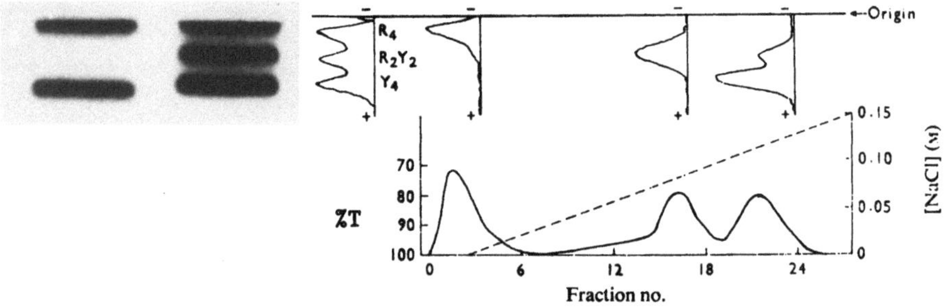

Fig. 1 Hybrid formation and isolation. A mixture of the yeast (Y_4) and rabbit enzymes (10 mg/ml each) was incubated for 14 h at 4°C in a 0.01 M phosphate buffer, pH 7.8. The photographs at the left are of starch gel electrophoretograms (carried out for 3 h at 25 V/cm in a Tris-borate-EDTA buffer, pH 8.7 - see ref. 3) of the mixture before and after incubation. 2 mls of the mixture was applied to a column (18x1.1 cm) of DEAE-Sephadex A-50 equilibrated with the phosphate buffer and the elution profile of %T at 254 nm recorded (solid line). The densitometric traces above each peak are of starch gel electrophoretograms of concentrated samples of the peak, and the one at the left is of the applied sample.

The results described here concern the process given in Scheme 1:

Scheme 1

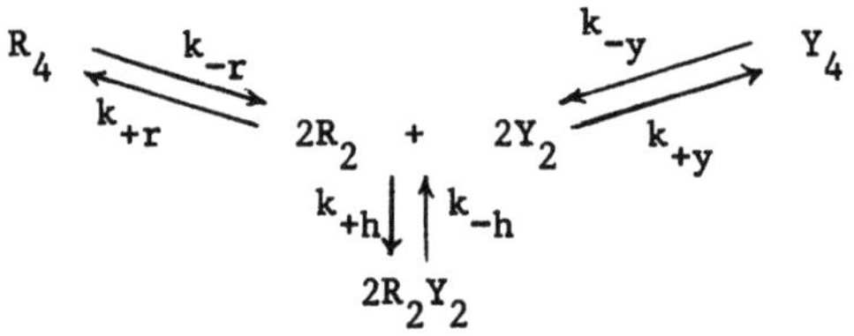

That is, on the time-scale of our experiments we are concerned only with the exchange of dimers between molecules (for a discussion see ref. 3). Given certain assumptions (see refs. 3 and 9), and under conditions where the enzymes are mostly in their associated forms (e.g. with at most a few percent of dimers present), the formation of hybrid from equal concentrations of species R_4 and Y_4 will be a pseudo first order process with an apparent first order rate constant, k_1^{obs}, given by eqn (1):

Eqn. 1 $$k_1^{obs} = (k_{-r} + k_{-h}) / 2$$

with the concentration of hybrid at equilibrium (A) given by eqn. (2):

Eqn. 2 $$A = 2k_{-r} \cdot [R]_o / (k_{-r} + k_{-h})$$

So values of k_{-r} and k_{-h} can be calculated from determinations of A and k_1^{obs}. (One of the assumptions made in deriving these equations is that $k_{-r} \approx k_{-y}$).

Only a single hybrid band, (R_2Y_2), is observed under the conditions of the experiments described here but it has been shown by Suzuki and Harris, (4) and subsequently confirmed in our laboratory (3), that under more stringent dissociating conditions the asymmetric" hybrids (RY_3 and R_3Y) can be detected, although only in much smaller amounts than the R_2Y_2 species. We have argued (3) from this greater thermodynamic stability of the R_2Y_2 species that the glyceraldehyde 3-phosphate dehydrogenase molecule can be represented as a "dimer of dimers", a conclusion confirmed by X-ray crystallographic studies (5-7).

The close homology between the amino acid sequences of the pig, rabbit, yeast and lobster GPD molecules (ref. 7 and J.I. Harris, personal communication), as well as the possibility of forming hybrids between the enzymes from different species (1,2,4,8,9) indicate that the three-dimensional structures of these enzymes are probably similar, and so can be discussed in terms of the detailed X-ray crystallographic structure for the lobster muscle enzyme presented by Buehner et al (5,6), and the human enzyme, given by Watson et al (1972) (7). The diagrammatic representation of the lobster enzyme molecule according to the colour coding of Buehner et al(5) is given in Fig. 2.

By consideration of the X-ray crystallographic structures (5,7) it was proposed that (3) the dissociation of tetramers (e.g. $R_4 \longrightarrow 2R_2$ in Scheme 1) involved cleavage about the plane generated by the Q and R axes, to give red-blue and green-yellow dimers (see Fig. 2). The refined X-ray structure of the lobster holoenzyme molecule (6) provides confirmation that the strongest subunit interactions are across the P axis, with 68 amino acid residues per subunit containing atoms less that 0.5 nm from an atom in the adjoining subunit, whereas there are only 32 such residues per subunit in the R axis related contacts, and 7 in the Q axis related contacts. The strong P-axis related interactions involve mainly the extensive anti-parallel pleated sheet arrangements of the catalytic domains. The greater number of dissimilarities between P-axis, subunit-contact amino acid residues in the rabbit (equals pig) and yeast enzymes (8) than between Q and R axis contact residues probably accounts for the instability of the R_3Y and RY_3 hybrids. The rabbit and lobster enzymes form asymmetric hybrids (RL_3 and R_3L) (10) and have fewer differences in P-axis contact amino acid (8).

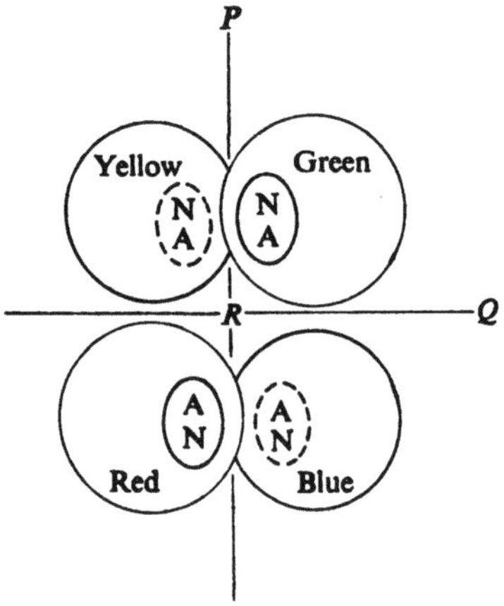

Fig. 2 Diagrammatic representation of the arrangement of subunits in the glyceraldehyde 3-phosphate molecule. The labelling of axes and colour coding of subunits is that given by Buehner et al (5). The P-axis dimers comprise the red-blue and green-yellow subunits respectively. The dissociation of the tetramer gives P-axis dimers by a process involving cleavage about the QR plane. Thus a R_2Y_2 hybrid comprises one P-axis dimer from the rabbit enzyme and one from the yeast. The nicotinamide and adenine moieties of NAD are designated as A and N respectively and the dashed lines indicate that two of the NAD-binding sites lie on the far side of the molecule as drawn. The X-ray crystallographic studies indicate that the red and yellow subunits have a different conformation from the blue and green subunits.

THE EFFECT OF NUCLEOTIDES ON SUBUNIT INTERACTIONS

The time-courses of hybrid formation, hybrid reversion (i.e. $2R_2Y_2$ $R_4 + Y_4$) and equilibrium perturbation were followed in the presence of different ligands and either relationships such as those given in eqns 1 and 2 or computer simulation studies (3) used to evaluate the dissociation rate constants k_{-r}, k_{-y} and k_{-h} for the liganded enzymes. The effect of ligand binding on these rate constants is taken to reflect changes induced in the interactions of the P-axis dimers across the QR dissociation plane of the molecules.

(a) The Effect of NADH on Subunit Interactions.

The results of the four different types of experiment shown in Fig. 3 demonstrate that the binding of NADH greatly increases the rate of subunit exchange between the species R_4, Y_4 and R_2Y_2. The ligand-induced rate of subunit exchange is so greatly enhanced that electrophoresis of a mixture of R_4, R_2Y_2 and Y_4 in a medium containing NADH gives only two bands, R_4 and Y_4 as the hybrid species 'melts out' as electrophoretic migration proceeds (Fig. 3(d)).

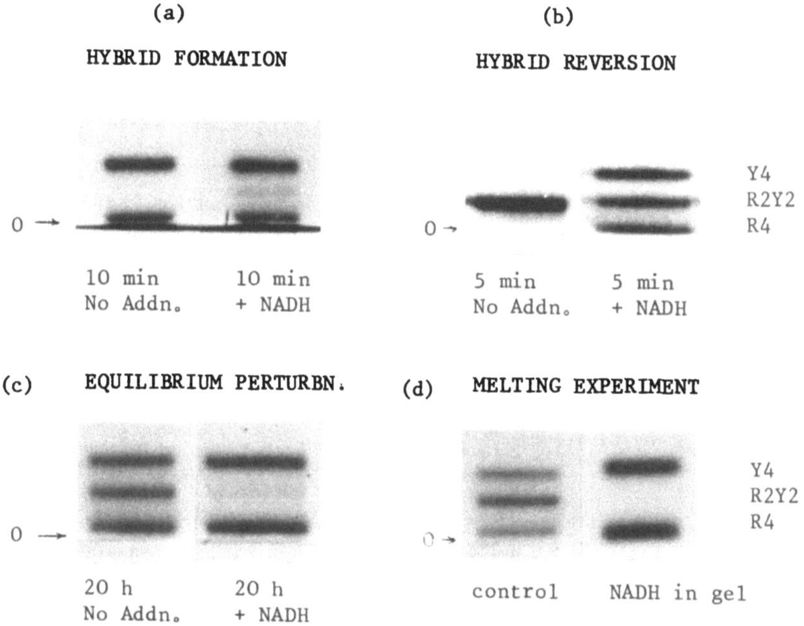

Fig. 3 The effect of NADH on subunit exchange. Experiments (a) - (c) were carried out using the 0.01M phosphate buffer pH 7.8 at 0°C. The concentration of NADH where indicated was 6 mM (the same results were obtained with 0.6 mM NADH) (a) incubation of R_4 and Y_4 for 10 mins (b) incubation of the isolated hybrid for 5 min (c) an equilibrium between R_4, R_2Y_2 and Y_4 was established at 37°, NADH then added and a further incubation carried out for 20 h at 0°C. (c) 37° equilibrium mixture was subjected to electrophoresis in the Tris-borate buffer with and without 1 mM NADH.

It is apparent also from Fig. 3 that there is much less of the R_2Y_2 species at equilibrium in the presence of NADH, suggesting that there are different NADH-induced changes in the structures of the rabbit and yeast enzymes. The results of these experiments illustrate the importance of carrying out hybrid reversion (decomposition of R_2Y_2) as well as hybrid formation experiments to avoid confusion of a relative thermodynamic instability of the hybrid with a slow rate of hybrid formation.

From a detailed study of the kinetics of the hybrid formation and reversion reactions we have shown that (9) the binding of 4 molecules NADH to the rabbit or yeast enzyme molecules greatly increases the rate constants for tetramer to dimer conversion (k_{-r} and k_{-y} - see Scheme 1). For example the value of k_{-r} increases by more that 80-fold in the presence of NADH corresponding to a decrease in Arrhenius activation energy for the dissociation process of at least 10 kJ/mol. We have assigned this destabilization of the tetramer to a weakening of the interaction between the two P-axis dimers across the QR plane of the molecule. This could possibly involve changes in the R-axis related contacts of 11 amino acids (between residues 9 and 49) in the nucleotide binding domain and /or of 16 residues (between residues 178 and 202), comprising mainly the residues of the S-shaped loop that is responsible mainly for red-green subunit interactions across the R-axis.

(b) The Effect of NAD on subunit interactions.

Hybrid formation, reversion, and equilibrium perturbation experiments were conducted with NAD at a concentration sufficient to saturate the four binding sites in each molecule. The results, shown in Fig. 4, demonstrate clearly that the binding of NAD 'freezes' the tetrameric structures of the R_4, Y_4 and R_2Y_2 species. The failure to form hybrid (Fig. 4(a)) in the presence of NAD is not due to an unfavourable equilibrium position arising from a destabilized, liganded hybrid, (i.e. $(RN)_2(YN)_2$), as the result of Fig. 4(b) shows that the structure of the hybrid is also 'frozen' in the presence of NAD.

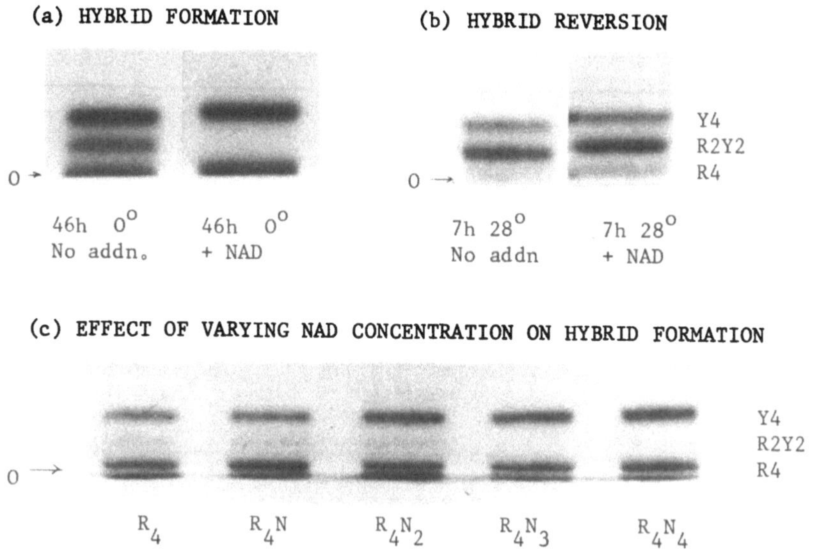

Fig. 4 The effect of NAD on subunit exchange. (a) Hybrid formation in the pH 7.8 phosphate buffer (both enzymes at 6 mg/ml). (b) Hybrid reversion: a 7 h incubation of the isolated hybrid (see Fig. 1) in the absence and presence of NAD. (c) Extent of hybrid formation in mixtures of yeast and rabbit apoenzymes containing different amounts of NAD. The designation R_4N refers to a mixture containing one mole of NAD per mole of rabbit enzyme, and so on. The higher affinity of the first 3 sites of the rabbit enzyme ensure that little Y_4N complex will form (see text).

From the kinetics of the hybridization reaction at 37° it was shown that NAD binding decreases the value of the rate constants for tetramer to dimer conversion (k_{-r} and k_{-y}, see scheme 1) by at least 100-fold (9), i.e. the $\Delta\Delta G^{\ddagger}$ for the dissociation increases by at least 8 kJ/mol. We ascribe this increase in activation energy to stronger interactions across the QR plane in the NAD holoenzyme compared with the apoenzyme.

In order to determine the number of bound NAD molecules per tetramer required to cause the tightening effect, hybrid formation experiments were conducted in which the yeast apoenzyme was incubated with samples of the rabbit enzyme containing 0, 1, 2, 3 and 4 molecules of NAD per enzyme molecule (designated R_4, R_4N and so on). The dissociation constants for the binding of successive NAD molecules to the rabbit enzyme are about 0.01, 0.09, 4.0 and 36μM respectively (11,12), whereas the corresponding values for the yeast enzyme are 220, 7, 13 and 280μM (13), so the added NAD in the hybridization experiments of Fig. 4(c)will bind almost exclusively to the sites on the rabbit enzyme.

It can be seen from the electrophoretograms of Fig. 4 that the rates of hybridization of R_4N and R_4N_2 are similar to that of R_4, but the R_4N_3, and R_4N_4 species do not form hybrids in the time of incubation. Therefore the tightening of the R_4 structure occurs on binding the third NAD molecule. The implications of this result for the interpretation of conformation changes giving rise to negative cooperativity of NAD binding will be discussed below.

(c) The Effect of Other Nucleotides on the Subunit Interactions

The effect on the hybridization process of a number of nucleotides that could be regarded as NAD or NADH analogues was investigated in an attempt to gain a further understanding of the basis of the NADH-induced weakening, and NAD strengthening of interactions between P-axis dimers across the QR plane of the glyceraldehyde 3-phosphate dehydrogenase molecule.

Adenosine alone (3 mM) or a mixture of adenosine and phosphate had no discernable effect on the rate of hybridization. However AMP, ADP, ATP, ADP-ribose, cyclic AMP and ITP each caused a very large increase in the rate of subunit exchange. The results obtained with each of these nucleotides were essentially the same (see Table 1): the hybrid formation and the hybrid reversion reactions reach equilibrium within 5 min in the presence of the ligand whereas in the absence of ligand the half-life for the reaction is about 28 h. It was also possible to 'melt out' the hybrid species from a mixture of R_4, R_2Y_2 and Y_4 by electrophoresis in a medium containing 1 mM ATP (cf the experiment with NADH shown in Fig. 3). It was possible to show that the binding of each of these ligands causes a decrease of at least 10 kJ/mol in the free energy of activation of the dissociation of the tetrameric enzyme. Therefore we conclude that the binding of AMP, ADP and ADPR, cause a weakening of the interaction between the P-axis related dimers. Thus the NADH-induced loosening of the interactions across the QR plane can be almost completely accounted for by the binding of the adenine-ribose-phosphate moieties in the nucleotide binding domains (5). Note that the phosphate group has to be joined covalently to the adenine-ribose moiety to give rise to the 'loosening' effect.

Table 1 The effect of specific ligands and combinations of ligands on the subunit excange parameters. The quotient $[R_2Y_2]_{eq}/[Y_4]_{eq}$ is the quotient of the areas under the hybrid and yeast peaks in the densitometric scans of electrophoretograms of equimolar mixtures of R_4 and Y_4 (6 mg/ml each), allowed to come to equilibrium in the presence of the indicated ligand. Essentially the same results were obtained for each ligand in 0.01 M phosphate buffer, pH 7.8 (shown in the table) and 0.1 M Tris, pH 7.8. All data is for 4°C. The rate constants are defined according to scheme 1 and were determined by use of eqn. 1; the corresponding eqn. for hybrid reversion, and computer simulation studies (see refs. 3 and 15)

LIGAND	$\frac{[R_2Y_2]_{eq}}{[Y_4]_{eq}}$	$10^5.k_1^{obs}(s^{-1})$	$10^5.k\ (s^{-1})$		
			k_{-r}	k_{-y}	k_{-h}
Control (P_i)	0.65	0.7	0.4	1.2	1.0
Control (Tris)	0.62	0.7	0.4	1.2	1.0
5mM ATP	1.0	200	133	133	267
5mM ADP	1.0	200	133	133	267
5mM AMP	1.0	200	133	133	267
3mM cyclic AMP	1.0	200	133	133	267
1mM ADP-ribose	1.2	200	150	150	250
3mM ITP	0.3	200	51	51	350
3mM NMN	0.6	0.7	0.4	1.2	1.0
3mM NAD	0.05	0.2	0.2	0.2	0.2
3mM NADH	0.22	200	32	32	360
1.5mM NAD + 5mM ATP	0.05	0.2	0.2	0.2	0.2
1.5mM NADH + 5mM ATP	0.2	200	32	32	360

If the 'loosening' effect of NADH is caused by the occupation of the nucleotide binding domain by the ADP-ribose moiety it seems axiomatic that the NAD-induced 'tightening' of the tetrameric structure must result from occupation of the substrate-binding domain (5) by the nicotinamide group. Thus it was somewhat surprising to find that NMN had no effect on the rate of hybridization (see table 1). However, it can be seen from the hybridization time-courses of Fig. 5 that a mixture of AMP and NMN greatly decreases the rate of subunit exchange compared with AMP alone so there is a residuum of the NAD tightening effect in this case. Therefore there must be a covalent bond between the NMN and AMP moieties before the full tightening effect is expressed. Therefore, the effect may depend on: (i)some sort of strain in the binding of NAD, whereby the conformation change required in order to bind the third NAD molecule overcomes the tendency for the occupancy of the nucleotide binding domain to weaken the interactions across the QR plane of the molecule; (ii) the correct charge on the diphosphate moiety; (iii) a rigorous steric requirement at the diphosphate moiety.

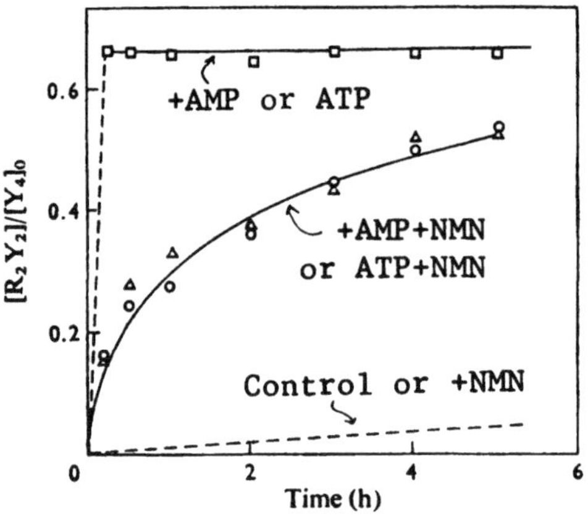

Fig. 5 The effect of specific ligands on the rate of hybrid formation. Mixtures of the Y_4 and R_4 species, each at 6 mg/ml in the pH 7.8 phosphate buffer at 4°C were incubated for different times in the presence of the indicated ligands. The relative amounts of hybrid formed at these times, (expressed as $[R_2Y_2] / [Y_4]_o$), were determined by the electrophoretic-densitometric analysis procedure.

DISCUSSION OF THE NEGATIVE COOPERATIVITY ON BINDING NAD AND NADH IN TERMS OF LIGAND-INDUCED CHANGES IN SUBUNIT INTERACTIONS.

There have been a large number of studies of the binding of NAD to rabbit muscle glyceraldehyde 3-phosphate dehydrogenase (e.g. 11,12,16,17) and there is general agreement that the enzyme contains two high affinity sites and two sites with considerable lower affinity. The most recent work by Bell and Dalziel (12) gives the dissociation constants at pH 7.6

and 25°C as 0.01, 0.09, 4 and μM. The large difference in affinity for the second and third sites was taken to suggest that(12) half-saturation of the tetramer produces a major conformational change in the molecule.

However, our results indicate that the major conformation change which leads to tightening of the structure occurs only on binding the 3rd NAD molecule and this may explain the low affinity of the third site. As the binding of this third molecule gives a stabilization of the interactions between the P-axis dimers and yet the binding is weaker than at sites one and two it is necessary to postulate that there is some compensating unfavourable change in the enzyme structure: e.g. a weakening of interactions between the P-axis related subunits, perhaps related to the movement of residues 149 to 166 away from the PR plane (see ref. 6). In any case it seems likely that the enzyme is forced to take up the asymmetric arrangement seen in the X-ray structure (6) only on occupation of the third site. Other studies have also indicated that there is a structural change on binding the third NAD molecule. Thus, the relative reactivity of the enzyme thiol groups towards the D and L-enantiomers of α-iodopropionate is inverted on binding the third NAD molecule (18), and the reactivity of the enzyme towards maléic anhydride is unaffected by the occupation of the 2 'tight' sites but greatly decreased by occupation of the third site (19). The foregoing considerations taken in conjunction with the structural studies (6) suggest that the order of binding of NAD to successive subunits is red, yellow, green (or blue) and blue (or green).

Rossmann and his coworkers (6) have suggested that the difference in conformation in the red and green subunits of the lobster NAD holoenzyme provides a basis for understanding the negative cooperativity of NAD binding. By inspection of the X-ray structure it can be seen that almost all of the interactions between the P-axis dimers across the QR plane (i.e. red-green and blue-yellow subunit interactions) are provided by Gly 9, Arg 10, Arg 13 and the amino acid residues in two loop structures. The first of these loops contains the 15 residues between 35 and 49, out of which 8 provide red to green subunit interactions across the R axis and are conserved in the known amino acid sequences (8). The second of the loops is the S-shaped configuration comprising residues 185 to 201 (6), and 11 of these

17 residues make interactions across the R-axis (red-green subunit interactions) and are also highly conserved in the known amino acid sequences (8). It is tempting to associate the weakening effect of ADP-ribose on red-green subunit interactions with changes involving the first loop and the residues around Arg 10, as the ADP-ribose moiety of bound NAD is closer to these residues (5,6). (This of course assumes that ADP-ribose occupies the same site as the ADP-ribose moiety in NAD). The NAD tightening effect could then involve mainly increased interactions of the S-shaped loops (see also ref. 10). Note that there is likely to be cooperativity between these two sets of interactions as indicated by the observation that NMN + AMP does not mimic NAD (see above).

Our conclusions about the tightening effect of binding NAD only apply to the muscle enzyme. However it has been shown that NAD also prevents hybridization of glyceraldehyde 3-phosphate dehydrogenases from different strains of yeast (19). It was not shown whether the coenzyme decreased the rate of dissociation to dimers or altered the equilibrium constant for hybridization, but the results presented here would favour the former interpretation.

In contrast to NAD-binding, our results show that the binding of NADH gives a weakening of the red-green (or blue-yellow) subunit interactions. However the rabbit muscle enzyme has similar affinities for NADH and NAD (12): the dissociation constants for NADH are 0.008 μM, 0.06 μM, 4 μM and 35 μM and are closely similar to the values for NAD binding (0.01, 0.09, 4 and 36 μM respectively.) In order to have such similar sets of dissociation constants and yet give different changes in subunit interactions, the respective conformation changes giving these altered subunit interactions must be energetically compensating.

CONCLUSIONS

It seems that the respective processes of binding NAD and NADH to glyceraldehyde 3-phosphate dehydrogenases are coupled to conformation changes that are constrained in different ways by interactions across the QR plane of the enzyme molecule. The binding of NADH weakens these interactions and this can explain readily the negative cooperativity of the binding of this ligand. However the binding of the third molecule of NAD (and in particular the positively charged nicotinamide moiety) to the enzyme molecule gives a strengthening of the interactions across the QR plane, yet NAD binding also shows negative cooperativity. In this case there must be a compensating, energetically unfavourable change elsewhere in the molecule, e.g. changes around the catalytic domain giving rise to changed P-axis contacts as well as the observed asymmetry of the holoenzyme structure (8). These interactions give the enzyme similar affinities for NAD and NADH, (an unusual feature in dehydrogenases - NADH usually binds much more tightly), and this seems likely to be associated with the role of the enzyme *in vivo*.

ACKNOWLEDGEMENT

H.H. Osborne thanks the Science Research Council for a Research Studentship.

REFERENCES

1. Spotorno, G.M.L. & Hollaway, M.R.: Hybrid molecules of yeast and rabbit GPD containing native and modified subunits. Nature (Lond), 226, 756-757 (1970).
2. Kirschner, K. & Schuster, I.: Recent studies on the allosteric glyceraldehyde 3-phosphate dehydrogenase from yeast, in 'Pyridine Nucleotide -Dependent Dehydrogenases' (Sund H., ed), Springer-Verlag, Berlin and New York, 217-228 (1970).
3. Osborne, H.H. & Hollaway, M.R.: The hybridization of Glyceraldehyde 3-phosphate dehydrogenases from rabbit muscle and yeast. Biochem. J., 143, 651-662 (1974).
4. Suzuki, K. & Harris J.I.: Hybridization of glyceraldehyde 3-phosphate dehydrogenase. J. Biochem. (Tokyo) 77, 587-593 (1975).
5. Buehner, M., Ford, G.C., Moras, D., Olsen, K.W. & Rossmann, M.G.: Three-dimensional structure of D-glyceraldehyde 3-phosphate dehydrogen-

ase. J. Mol. Biol., 90, 25-49 (1974).

6. Moras, D., Olsen, K.W., Sabesan, M.N., Buehner, M., Ford, G.C. & Rossmann, M.G.: Studies of asymmetry in the three-dimensional structure of lobster D-glyceraldehyde 3-phosphate dehydrogenase. J. Biol. Chem. 250, 9137-9162 (1975)

7. Watson, H.C., Duee, E. & Mercer, W.D.: Low resolution structure of glyceraldehyde 3-phosphate dehydrogenase. Nature (Lond), 240, 130-139 (1972).

8. Olsen, K.W., Moras, D., Rossmann, M.G. & Harris J.I.: Sequence variability and structure of D-glyceraldehyde 3-phosphate dehydrogenase. J. Biol. Chem., 250, 9313-9321 (1975).

9. Osborne, H.H. & Hollaway, M.R.: The investigation of substrate-induced changes in glyceraldehyde 3-phosphate dehydrogenases by measurement of the kinetics and thermodynamics of subunit exchange. Biochem. J. 151, 37-45 (1975).

10. Harris, J.I. & Water, M.: Glyceraldehyde 3-phosphate dehydrogenase in 'The Enzymes' (Boyer, P.D., ed), Academic Press, 13(C), 1-49, (1976).

11. Velick, S.F., Baggot, J.P. & Sturtevant, J.M.: Thermodynamics of NAD addition to the glyceraldehyde 3-phosphate dehydrogenases of yeast and rabbit muscle. An equilibrium and calorimetric analysis over a range of temperatures. Biochemistry, 10, 779-786 (1971).

12. Bell, J.E. & Dalziel, K.: Studies of coenzyme binding to rabbit muscle glyceraldehyde 3-phosphate dehydrogenase. Biochim. Biophys. Acta. 391, 249-258 (1975).

13. Cook, R.A. & Koshland, D.E.Jr.: Positive and negative cooperativity in yeast glyceraldehyde 3-phosphate dehydrogenase. Biochemistry 9, 3337-3342 (1970).

14. Bell, J.E. & Dalziel, K.: Conformational changes of glyceraldehyde 3-phosphate dehydrogenases induced by the binding of NAD. Biochim. Biophys. Acta., 410, 243-251 (1975).

15. Osborne, H.H. & Hollaway, M.R.: An investigation of the NAD-induced 'tightening' of the Structure of glyceraldehyde 3-phosphate dehydrogenase. Biochem. J. 157, 255-259 (1976).

16. Conway, A. & Koshland, D.E.Jr. : Negative cooperativity in enzyme action. The binding of diphosphopyridine nucleotide to glyceraldehyde 3-

phosphate dehydrogenase. Biochemistry, 7, 4011-4022 (1968).

17. De Vijlder, J.J.M. & Slater, E.C.: The reaction between NAD and rabbit -muscle glyceraldehyde 3-phosphate dehydrogenase. Biochim. Biophys. Acta. 167, 23-34 (1968).

18. Eisele, B. & Wallenfels, K.: The stereoselective inhibition of functional -SH groups of dehydrogenases. In 'Pyridine Nucleotide-Dependent Dehydrogenases' (Sund H., ed), Springer-Verlag, 91-101 (1970).

19. Ovadi, J., Telegdi, M., Batke, J. & Keleti, T.: Functional non-identity of subunits and isolation of active dimers of D-glyceraldehyde 3-phosphate dehydrogenase. Eur. J. Biochem. 22, 430-438 (1971).

20. Stallcup, W.B. & Koshland, D.E.Jr.: Half-of-the-sites reactivity and negative cooperativity: the case of yeast glyceraldehyde 3-phosphate dehydrogenase. J. Mol. Biol. 80, 41-62 (1973).

Received March 26, 1977

DISCUSSION

Woenckhaus: Complete inactivation of GAPDH from rabbit muscle is produced by covalent binding of [3-(3-bromoacetylpyridinio)-propyl]-adenosine pyrophosphate, a structural analog of NAD. Only two moles of the inactivator are bound to one mole of tetrameric enzyme. The completely inactivated enzyme (E_4-A_2) is able to bind two moles NAD (K_D = 12 μM). The inactivated enzyme forms hybrids with the native or inactivated enzyme from yeast. These hybrids show the same mobility in the electric field as the hybrids formed from the two native enzymes.

Hollaway: This is a very interesting observation and relates to our findings. We prepared a sample of rabbit GPDH modified at four sites per tetramer with 2-bromoacetamido -4-nitrophenol to give a catalytically inactive product: designated $(RK)_4$ (see Kirtley, M.E., & Koshland, D.E. Jr., Biochem.Biophys.Res. Comm. 23, 810 (1966)). This inactive species was then hybridized with native yeast enzyme: (Y_4). On electrophoresis a hybrid species: $(RK)_2Y_2$, was obtained which was catalytically active, with a specific activity of about 45 % of that of Y_4, i.e. the yeast chains retain their activity within a tetramer containing two inactive rabbit chains (Spotorni, G.M.L., & Hollaway, M.R., Nature 226, 756-757 (1970)). At first sight this result would seem to be in conflict with your finding. However I would like to propose a way in which the two results could be reconciled.

Your inhibitor, (let us call it W rather than A), resembles the coenzyme NAD and, like NAD, may react in a "negatively cooperative" way to give a species with one subunit in each P-axis dimer substituted at Cys-149 i.e.:

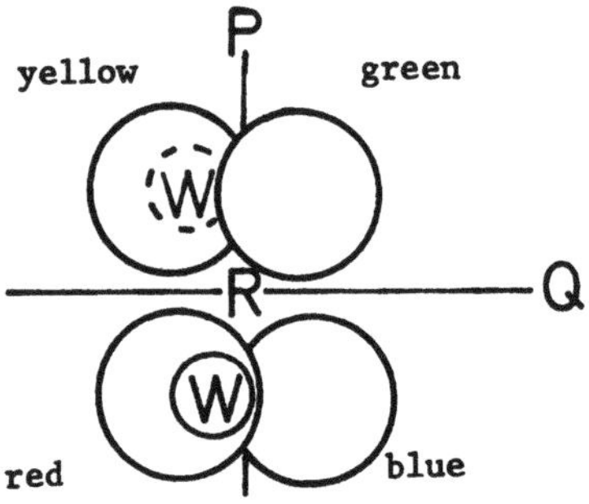

where the inhibitor is in the red and yellow subunits for the reasons given in the lecture (i.e. the inhibitor reacts in a "half-of-the-sites" way)

Our hybridization process can be represented as:

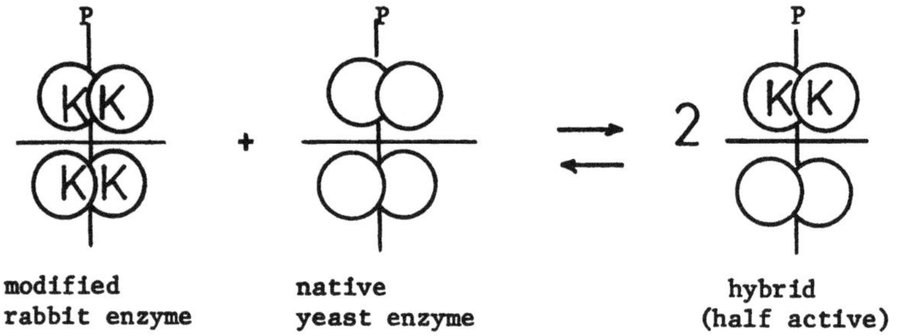

where the hybrid product retains something like a half of the activity of the yeast tetrameric species. In this case, the difference between the two disubstituted tetramers is that yours contains two modified P-axis dimers, whereas ours contains one modified P-axis dimer and one that is unmodified. All you have to do is to assume that modification of a P-axis dimer at one site with a reagent that resembles NAD gives an inactive species, whereas modification with a reagent that is not an NAD-analogue does not, and the apparent discrepancy is resolved.

This explanation hinges on the assumption that the reaction at one subunit by a modifier that resembles the coenzyme can prevent reaction at the remaining subunit within the P-axis dimer. The constraint could arise from an inhibition of reaction by virtue of intersubunit interactions (communication) across the R and P axes.

Perham: Dr. Hollaway has neatly resolved the apparent conflict between his results on the specific activity of the R_2Y_2 tetramer and Dr. Woenckhaus's report that his bis-alkylated tetramer is inactive. There remain the hybridization experiments carried out in Dr. Harris's laboratory, already referred to by Dr. Hollaway, in which alkylation (S-carboxymethylation of cysteine-149) of subunits successively reduces the catalytic activity of the tetramer. It must be that simple carboxymethylation of cysteine-149 in one subunit does not communicate inactivation to the neighbouring subunit whereas alkylation with Dr. Woenckhaus's analogue of NAD^+ does not. We should note that the alkylating agents are chemically very different and we should not be surprised therefore if different results ensue. Once again the conflict is more apparent than real. Having said that, let me ask Dr.

Bernhard to comment.

Jaenicke: The hybrid pattern suggests the dimer to be the only relevant dissociation product in the case of GAPDH. Is there independent evidence for the dissociation of GAPDH into (inactive) dimers which upon reassociation cause the formation of the active tetramer?

Hollaway: Yes, we have carried out the following experiment, part of which has been published (Hollaway, M.R., & White, H.A., Biochem. J. 149, 221-231 (1975)). Rabbit muscle GAPDH was reacted with an excess of DTNB (Ellman's reagent) to give a species that gave clear signs of dissociation to a dimer in the ultracentrifuge. This can be represented as: R_4 + nDTNB $\rightleftharpoons$ $2(RTNB_n)_2$. We then took the dimeric TNB derivative, which is catalytically inactive, and put it in one syringe of a stopped-flow apparatus. In the second syringe we put NAD, glyceraldehyde 3-phosphate and arsenate, together with β-mercaptoethanol. On mixing, the TNB groups were rapidly removed ($(RTNB_n)_2 \rightarrow R_2 + 2nTNB^-$) to give the R_2 dimer. At this time there was no enzymic activity, but activity then returned in a clearly second order reaction. The rate constant for the reappearance of enzymic activity was about 5×10^5 $M^{-1}s^{-1}$ which corresponds to the rate constant for combination of inactive dimers to give active tetramers (Hollaway, Osborne & Rosemeyer, unpublished work).

Hinz: I would like to indirectly support your conclusion, that you cannot mimic the effects resulting from binding of NAD by providing equimolar mixtures of NMN and AMP (at the same concentration) by similar evidence from investigations on pig heart muscle LDH. The heat of reaction on binding is considerably smaller when the equimolar mixture is employed and of the same magnitude as the ΔH determined when binding AMP alone.

Engel: On the question of whether NAD^+ can be replaced by its fragments, I should like to add a comment about bovine glutamate dehydrogenase. Fisher et al. showed that NMNH/NMN serves as a coenzyme for GDH at a very low rate. We tried to boost this activity by adding ADP, AMP or adenosine without any success, so it seems that for this enzyme too it is essential for the two moieties of the coenzyme to be covalently linked.

DINUCLEOTIDE DEPENDENT CONFORMATIONAL AND CHEMICAL BONDING CHANGES IN MUSCLE GLYCERALDEHYDE-3-PO_4 DEHYDROGENASE

Sidney Bernhard and Oswald Pfenninger
Institute of Molecular Biology and Department of Chemistry, University of Oregon, Eugene, Oregon 97403, USA

Om P. Malhotra
Department of Chemistry, Banaras Hindu University, Varanasi 5, India

Bernard Schwendimann
Department of Biochemistry, University of Geneva, 1211 Geneva, Switzerland

INTRODUCTION

Consequent upon the binding of nicotinamide-adenine dinucleotides to muscle and yeast glyceraldehyde-3-PO_4 dehydrogenase (GAPDH), kinetically first-order isomerizations have been observed (1,2). In this report we shall attempt to demonstrate that wherever it is possible to make the appropriate observations, these "conformational" changes are directly related to changes in the reactivity of the enzyme towards acylation and/or to changes in the reactivity of the metastable acyl enzyme intermediate according to all of the catalytic reaction pathways of Eq. 1.

$$(1)\quad \underset{\substack{\text{(glyceraldehyde)}\\ 3\text{-}PO_4}}{RCHO} + NAD^{\oplus} + ECH_2SH \underset{NAD^+}{\rightleftharpoons} ECH_2S\overset{\overset{O}{\parallel}}{C}R + NADH + H^{\oplus}$$

$$ECH_2S\overset{\overset{O}{\parallel}}{C}R + HPO_4^{2\ominus} \underset{NAD^+}{\overset{NAD^+}{\rightleftharpoons}} ECH_2SH + R\overset{\overset{O}{\parallel}}{C}OPO_3^{2\ominus}$$

It has already been shown by us that the chromophoric pseudo-substrate acyl enzyme, β-(2-furyl)acryloyl-GAPDH (FAE) undergoes a variety of spectral changes, characteristic of covalent perturbations in the acyl-enzyme (thiol ester) linkage (3-5). In Eq. 1, we have indicated those steps in the reaction pathway which are dependent on the presence of enzyme-bound NAD^+ for catalysis, but for which the chemical change being catalyzed does not involve the covalent participation of NAD^+. We and others have recently presented evidence that the equilibrium thermodynamic affinity of the enzyme for NAD^+ varies with both the extent of acylation and the extent of prior ligand occupancy by NAD^+ (6-9). Since these changes in affinity are substantial, and since bound ligand at particular types of enzyme sites is essential for enzyme catalysis, it is tempting to speculate that particular, ligand-effected, conformational states of the enzyme are associated with distinctive catalytic species along the reaction pathway: Each conformer is presumably selected for its efficiency in catalyzing one of the steps in the pathway of Eq. 1. In this regard, allosteric enzymes may turn out to be of great utility in reaction-mechanism studies since they may serve as trapping reagents for particular covalent enzyme-substrate intermediates by trapping (stabilizing) an appropriate conformational state.

In the reaction sequence (Eq. 1), the acylation of the enzyme by the pseudo-substrate analog, β-(2-furyl)acryloyl phosphate (FAP), occurs with a maximal stoichiometry of two acyl groups per enzyme tetramer (2,10). We now present further evidence that the enzyme functions as a dimer of dimers in which the two subunits per dimer are structurally non-equivalent, and especially, that the diacyl enzyme utilizes the non-equivalence of effector ligand sites in directing the catalytic pathway.

At 25°, in the presence of excess NAD^+, the two acyl groups per enzyme tetramer are transferred from the active site thiol to some particular amine residue [presumably lys 181 by analogy with other muscle GAPDH (11)]. The rate of transfer of the acyl group is the same as the rate of inactivation of the enzyme, as judged by the activity assay of Ferdinand (12) in the presence of NAD^+ and arsenate. Hence not only is there half-site reactivity, but there is specific half-site inactivation effected by the S to N transfer of two acyl groups per tetramer (13). In the studies which follow, we have investigated the kinetics and equilibria of the various acylation $\rightleftarrows$ deacylation reactions (Eq. 1). We shall attempt to interpret these results structurally in terms of the dimer-of-dimers model (14) and in terms of ligand-mediated conformational changes which occur therein.

MATERIALS AND METHODS

Nucleotide coenzymes were purchased from Sigma Chemicals and were of the highest purity available. DTND was recrystalized as described elsewhere (15). β-(2-furyl)acryloyl phosphate (FAP) was prepared according to the methods of Malhotra and Bernhard (2). $NADH \cdot H_2$ was prepared by the method of Bielmann (16); buffer solvents were all at pH 7.0 and 25°.

Sturgeon muscle GAPDH was prepared by the method of Seydoux et al. (6). Specific activities were in the range 280-320 units as assayed by the method of Ferdinand.

Acyl enzyme was prepared as previously described. In procedures involving time-consuming purification of the acyl enzyme, low temperatures (0-5°) were utilized for the purification of the acyl enzyme. These purification steps involve the removal of the excess of FAP, and the removal of bound nucleotides from the enzyme by adsorption to activated charcoal (4).

Spectrophotometric and slow kinetic studies were carried out utilizing a Cary Model 14 Spectrophotometer equipped, when necessary, with computerized signal averaging for the determination of spectra at low O.D. (the order of 0.01 maximal O.D.). Rapid kinetic measurements were all carried out utilizing the absorbance stopped-flow apparatus previously described (14). The phototube voltage, suitably amplified, was stored on analog tape and could subsequently be digitized for computer analysis.

RESULTS AND DISCUSSION

All experiments described herein involve the monitoring of changes in the electronic spectrum of the furylacryloyl group (18). These changes occur in the 320-390 nm wavelength region. Hence they do not overlap appreciably with the absorption spectra of the protein tyrosine and tryptophan constituents. These spectra do, however, overlap various dinucleotide spectra; most notably those of the reduced coenzyme, NADH, and of the perturbed forms of the oxidized coenzyme, for example the "Racker band" (λ_{max} = 360 nm). In overlapping regions, the spectral contributions from two chromophoric sources could often be resolved by additional known spectral characteristics of one of the contributing chromophores. For example, the Racker band, which occurs when NAD^+ is bound to an enzyme site which has a free SH group, is a far broader absorption band than is the furyl-acryloyl (π-π^*) transition; at longer wavelengths, its contribution to the total increases.

Signals of Chemical Reaction and Ligand Occupancy: Structural Interpretations

The furylacryloyl chromophore has a single electronic transition in the wavelength range 290-360 nm, dependent on both the nature of the carbonyl linkage and on the polarity of the solvent environment (19). In aqueous solution, the furylacryloyl thiol ester absorbs maximally in the wavelength range 337-340 nm, dependent on the chemical (electronic) nature of the covalent environment. In aqueous solvent, these chromophores are as "red-shifted" as they can be. For example, strong salt solutions or the strong hydrogen bonding solutes (guanidine hydrochloride) have virtually no effect on the aqueous absorption maximum (4). On the other hand, solvents less polar than water do shift the spectra of such chromophores appreciably to the blue.

The absorption maximum of furylacryloyl-GAPDH under varying conditions is summarized in Table 1. In the presence of excessive NAD^+ (∿ 1 mM), the absorption maximum is far red-shifted to 360 nm. We are confident that this spectral shift represents a change in the nature of the chemical bonding between the acyl group and the sulfur atom of cysteine 149 of the enzyme (4,20). At low NAD^+ concentrations (below ∿ 5 μM), native acyl enzyme is still significantly red-shifted relative to that anticipated for the "maximally red-shifted" chromophore, despite the fact that the apparent dissociation constant for the NAD^+-dependent complex is 60 μM (7). The difference is not so marked, but is, nevertheless, a real perturbation since the denatured acyl enzyme has a normal aqueous furylacryloyl thiol ester absorption maximum (339-341 nm).

TABLE 1

λ_{max} of Various FA Thiol Ester Derivatives

Thiol Group	Solvent and Perturbant	λ_m (in nm)
$-CH_2\ C_6C_5$	H_2O	337
	H_2O, 2% SDS	334
	H_2O, 5 M guanidine HCl	340
	acctonitrile	326
-GPDH	H_2O (0.1 MKCl)	346
	0.5 mM NAD^+	360
	0.05 mM NAD^+	347
	0.05 mM NADH	330
Papain	H_2O	360
	8 M urea	340

The spectrum of the FA-enzyme in the presence of bound NADH is severely "blue-shifted" (Table 1) (5). Since this "blue-shift" is no greater than that observed by transfer of the thiol ester chromophore to solvents of low polarity, it is conceivable that a change in the microenvironment surrounding the furylacryloyl group upon binding of NADH is responsible for the blue-shift. That the observed effect is more structural than enviornmental, however, is suggested by the fact (Table 1) that the NADH analogue, NADH•H_2, which binds to the acyl sites of the acyl enzyme does not induce any color change (neither a red nor a blue shift). The NADH-induced "blue-shifted" acyl enzyme is an intermediate in the reduction reaction (5). In light of this, it is suggestive that the spectral shift reflects a change in the electronic structure of the acyl group such that the transfer of hydrogen to the carbonyl carbon of the thiol ester is favored. If these chemical inferences are reasonable, it follows that nucleophilic displacement (by HPO_4^{2-} + AsO_4^{2-}), which requires the red-shifted acyl enzyme (4), and hydrogen transfer occur via different mechanisms involving different electronic states of the thiol ester.

We have already presented substantial evidence indicating that both the sturgeon muscle holo- and apo-enzyme are asymmetric, in that they contain two classes of subunits. Hence the enzymes behave as a "dimer of dimers". A prime argument in regard to the holo-enzyme is that the very rapid reaction of active site cysteine with Elman's reagent (DTNB), which depends on the absence of coenzymes (NAD^+), shows two classes of NAD^+-dissociable sites in the holo-enzyme. The sites differ by a factor of approximately 10 in their NAD^+-dissociation rates. A demonstration of this biphasicity is given in the stopped-flow experiment of Figure 1. More recently, we have demonstrated that when the apo-enzyme is partially titrated with iodoacetate, a reagent specific for the active site cysteine, the remaining non-alkylated sites can be distinguished both by their tighter binding and greater fluorescence quenching (of constituent tryptophan) upon titration with NAD^+ (Fig. 2). Apo-sturgeon muscle enzyme binds NAD^+ in a manner consistent with the pre-existence of two classes of sites. Similarly (Fig. 2), partially alkylated (carboxymethylated) enzyme still shows two equal classes of NAD^+-binding sites among the non-alkylated sites. We have argued that the common finding in GAPDH of "negative cooperativity", as in the two classes of binding sites, and half-site reactivity, as in

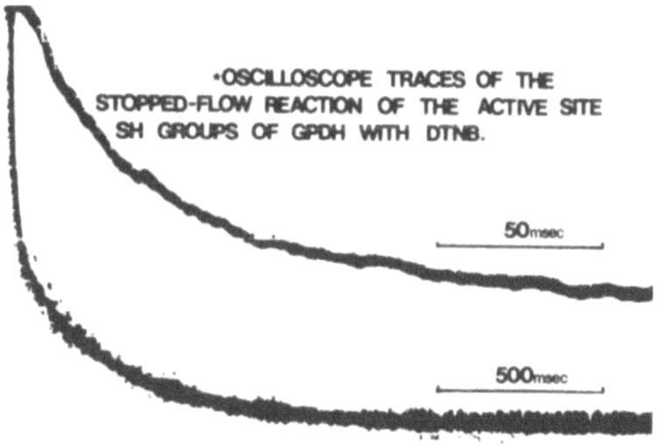

Fig. 1. The formation of the yellow thionitrobenzoate^{2-} (TNB) occurs very rapidly at NAD^+-free sites ($k \approx 10^6$ s^{-1} M^{-1}) but virtually not all at NAD^+-bound sites. The reaction rate is hence directly dependent on the rate of NAD^+ desorption; any heterogeneity in NAD^+ affinity will be reflected in heterogeneity of exponential decay rates under first order conditions. [DTNB] = 5 mM, [NAD^+] = 0.5 mM, [GAPDH] = .001 mM in 10 mM ethylene diamine buffer, 100 mM KCl, pH 7.0, 25°. The transmission changes are linear with concentration of TNB.

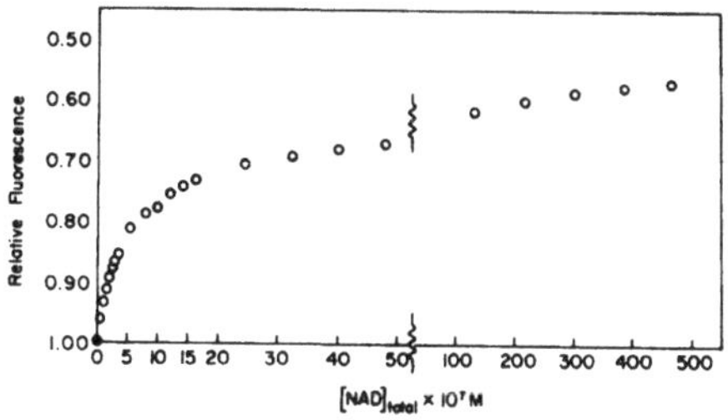

Fig. 2. NAD^+ titration of sturgeon enzyme randomly carboxymethylated among 1.68 sites. Conditions: 0.076 μM sturgeon enzyme. EDA buffer, pH 7.0, 25° C.

acylation by FAP, are explainable on the basis of a common structural mechanism (2). In this paper we shall consider some functional consequences of this structural mechanism. In this regard it is advantageous to have the capability of identification of as many different types of bound coenzyme species as are generated by the asymmetry of the enzyme. As illustrated in Table 1, we can identify NAD^+ bound to acylated sites, and NADH bound to acylated sites. We can distinguish NAD^+ bound to non-acylated sites from the NAD^+ bound to acylated sites by the occurrence at non-acylated sites of the "Racker band", a broad band with λ_{max} at 360 nm but extending into the visible absorption range. There is no Racker band when NAD^+ binds to the acylated sites as is evident by NAD^+ titration of the diacylated, dicarboxymethylated enzyme tetramer (4). We can identify NADH binding to non-acylated sites by the quenching of the fluorescence of the constituent tryptophanes of the protein and/or by the effect of binding on the intrinsic NADH fluorescence. Due to the unusually wide variety of signals available to us, it is possible in most cases described below to assess the extent, and the asymmetric location, of coenzyme and substrate ligand occupancy.

Kinetic Studies of the Dependence of GAPDH Reaction on NAD^+ Concentration

The time dependence of a variety of different GAPDH reactions are illustrated in Figure 3. Note that each reaction illustrated shows a biphasic

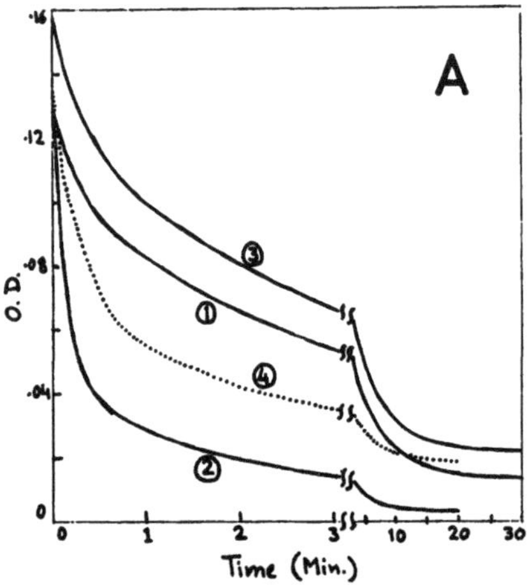

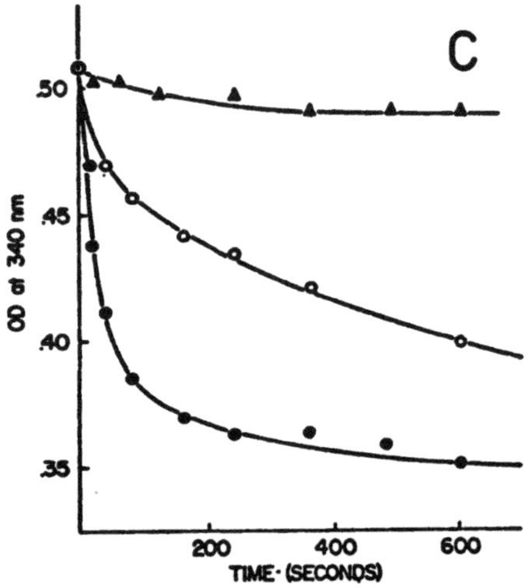

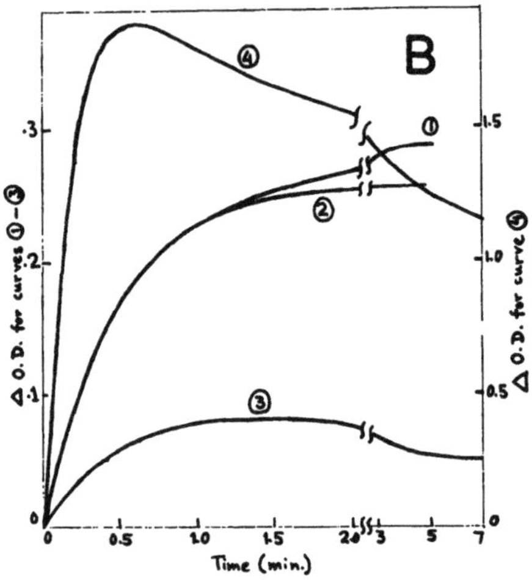

Fig. 3a. Kinetics of deacylation of FA-GAPDH. In every case the reaction was started with a small volume of acceptor and monitored at 360 nm. Protein and NAD^+ were pre-incubated. Concentration conditions are as follows: (1) 2.4 µM protein with 2.08 FA groups per mole enzyme, 25 µM NAD^+, and 0.5 mM arsenate; (2) 2.4 µM protein with 2.08 FA groups per mole enzyme , 250 µM NAD^+ and 0.5 mM arsenate; (3) 4.1 µM protein with 2.08 FA and 2 carboxymethyl groups per mole enzyme, 65 µM NAD^+, and 0.5 mM arsenate; (4) 3.1 µM protein with 1.83 FA groups per mole enzyme, 131 µM NAD^+, and 0.5 mM phosphate (some of the curves have been shifted along the vertical scale for clarity).

Fig. 3b. Time-dependent O.D. changes in the reaction of FAP and GAPDH monitored at 350 (1), 360 (2), and 390 nm [(3) and (4)]. For curves (1), (2), and (3), the concentrations are 3.8 µM protein, 65 µM NAD^+, and 1.2 mM FAP. For curve (4), they are 82 µM holo-enzyme (with 4 moles NAD^+ bound per mole enzyme), and 4.4 mM FAP. In every case the reaction was started by adding enzyme to a solution of FAP.

Fig. 3c. Reduction of alkylated furylacryloyl enzyme by NADH.
-▲-▲- Di-furylacryloyl enzyme with 1.5 NAD^+ molecules per tetramer; 9.7 NADH molecules per furylacryloyl site.
-●-●- Dicarboxymethyl-di-furylacryloyl enzyme with 1.5 NAD^+ molecules per tetramer; 9.7 NADH molecules per furylacryloyl site.
-o-o- Di[-S-(ethylsuccinimido)]-di-furylacryloyl enzyme; 202 µM NAD^+; 2.72 NADH molecules per tetramer.

time dependence.

The rates and amplitudes of fast and slow processes in the biphasic reactions (Fig. 3) are affected by the concentration of NAD^+. The slow and the fast rates and amplitudes of acylation, nucleophilic displacement and spectral perturbation processes reach saturating values at high NAD^+ concentrations. Moreover, at high NAD^+ concentrations (∿ 0.2 mM), all of the reactions we have studied (acylation, phosphorolysis, arsenolysis, reduction by NADH, and NAD^+-induced perturbations of the furyacryloyl thiol

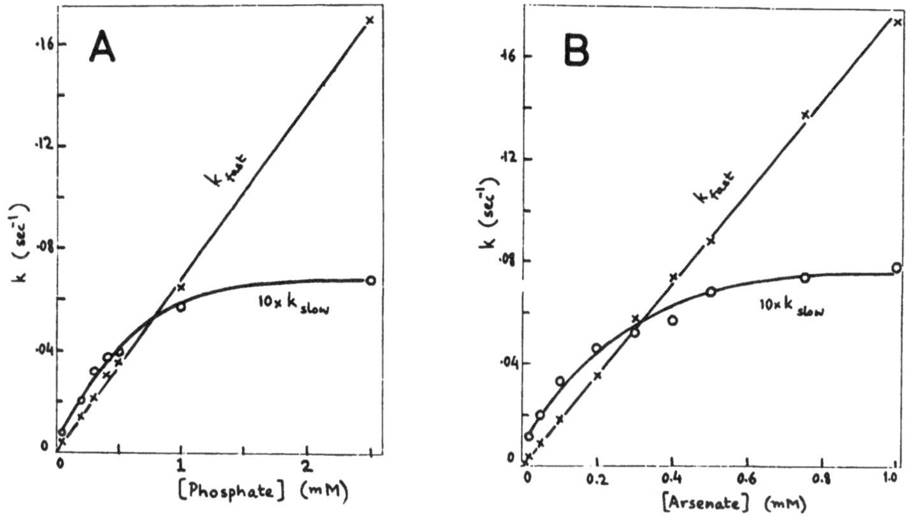

Fig. 4a. Phosphate concentration dependence of the kinetics of deacylation of FA-GPDH. Protein concentration is 4.0 μM has 1.75 FA-groups per mole enzyme. NAD⁺ concentration is 68,5 μM, λ = 360 nm.

Fig. 4b. Arsenate concentration dependence of the kinetics of deacylation of FA-GPDH. Protein concentration is 4.2 μM and has 1.63 FA-groups per mole enzyme. NAD⁺ concentration is 72 μM.

ester spectrum) saturate with a limiting common slow specific rate of approximately 0.01 sec^{-1} under otherwise near-optimal conditions of acceptor or NADH concentrations. The fact that the perturbation in the furylacryloyl spectrum (the red-shift) shows approximately the same slow kinetic parameter as the phosphorolysis and arsenolysis reactions, suggests that it is the rate of formation of this spectrally-perturbed species which, at sufficiently high acceptor concentrations, governs the rate of arsenolysis and phosphorolysis. At lower concentrations of acceptor (phosphate or arsenate) (Fig. 4), both the slow and the fast rates of deacylation are dependent on the acceptor concentration. At high concentrations, however, the slow kinetic parameter, but not the fast one, becomes independent of the acceptor concentration. In correspondence with our suggestion that rate-control is at the step of activation of the thiol ester linkage, acylation of the enzyme by FAP in the presence of NAD^+ also exhibits kinetic biphasicity. Previously (3), we had reported that in the presence of NAD^+ the acylation reaction exhibits precise first-order kinetics. These two reports are not in conflict: Previously we had measured the rate of the acylation reaction by the change in absorbance at

360 nm, the isosbestic point of the unperturbed vs. NAD^+-perturbed acyl enzyme spectrum. At other wavelengths, the biphasicity in the kinetics of acylation, which is apparent in the data of Figure 3 reflects at least two processes; the acylation of the enzyme site and its subsequent NAD^+-induced spectral perturbation. Clearly, this spectral perturbation must occur at a rate which is substantially slower than the acylation reaction itself. The absolute magnitude of this slow isomerization of acyl enzyme ($\sim$ 0.01 sec^{-1}) is noteworthy; it does not involve the formation of any new chemical bond, and hence it must involve a protein conformational change concomitant with thiol ester bond perturbation. No such slow isomerization is detectable in any reaction of the true substrate, 3-phosphoglyceroyl-GAPDH, although much faster rate-limiting isomerizations have been reported with this acyl enzyme as well (2). The catalytically linked isomerization of 3-phosphoglyceroyl-GAPDH has a half-life of about 50-100 msec.

An interesting common characteristic of the phosphorolysis, arsenolysis, and spectral perturbation reactions is that the kinetic biphasicity is not completely abolished at either very high or at minute NAD^+ concentrations. Hence whatever the detailed mechanism of protein isomerization, it seems to be more in accord with the mechanism proposed by Monod et al. (20): A reactive and an unreactive conformation are both a priori states of the acyl enzyme. The equilibrium distribution between these conformations is, however, affected by the presence of bound NAD^+. If indeed this interpretation is correct, the spectrum of the "apo"-acyl enzyme should correspondingly reflect the presence of some spectrally-perturbed acyl sites. The perturbation of the acyl enzyme spectrum in the absence of NAD^+, vis-à-vis the spectrum of the denatured acyl enzyme, is in fairly good quantitative agreement with the kinetic data on reactions of the apo-acyl enzyme. Approximately 20-25% of active (and hence red-shifted) acyl sites is indicated by the amplitude of the fast phase in reactions of the apo-acyl enzyme. The precise percentage of spectral shift anticipated in the apo-acyl enzyme depends, of course, on our estimate of the unperturbed acyl enzyme spectrum. We estimate that of the λ_{max} for the furylacryloyl thiol ester linked to the active cysteine is 340-342 nm.

In the reduction-reaction with NADH, the kinetic situation is more complex due to the significant competition at equivalent concentrations between NAD^+ and NADH at non-acylated sites, and due to competitions between the two nucleotides at the acyl sites when NAD^+ concentration is high relative to NADH. Consequently, it is difficult to obtain good extrapolations to saturating conditions for the two essential dinucleotides in this reaction (5). The data presented in Figure 3 are under conditions of minimal competition between nucleotides, yet still nearly saturating concentrations. Nevertheless, we estimate a 50% decrease in the rate of the slow process due to the competition between NAD^+ and NADH at the acylated sites.

All of the above NAD^+-dependent biphasic kinetic results can be explained on the basis of the minimal model given in Eq. 2.

(2)

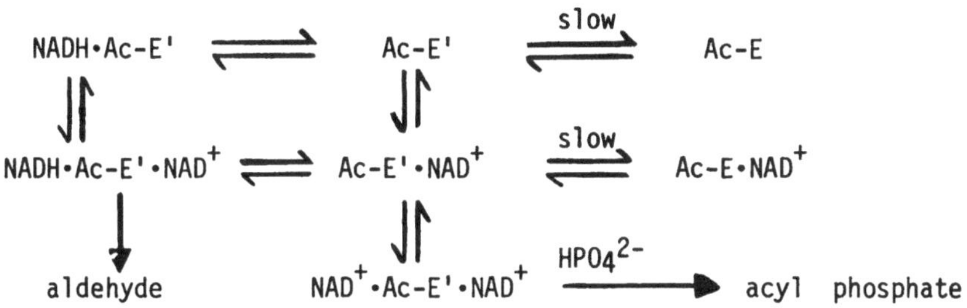

According to this scheme, the only slow step is the conformational change indicated in the horizontal. The capacity for the protein to undergo this conformational change is dependent on the presence of bound NAD^+ at the non-acylated site. The equilibrium distribution between the two states, however, is dependent on bound coenzyme ligand (NAD^+ or NADH) at the acylated site. Hence, the dependence of reaction velocity on NAD^+ will vary according to which of the various chemical reactions are being monitored: the reduction reaction will depend on bound NAD^+ at the non-acylated site exclusively, whereas the nucleophilic displacement reaction will depend on the low concentrations of NAD^+ required for driving the equilibrium toward the reactive conformation by binding at the acylated site. Previously we had suggested that the red-shifted acyl enzyme spectrum was dependent solely on the binding of NAD^+ to the acylated sites. Our arguments were based on the fact that carboxymethylation of the non-acylated sites did not change the NAD^+ concentration dependence of the red-shift, or of the arsenolysis reaction. However, more recent information (Fig. 2) indicates that carboxymethylated sites still bind NAD^+ more tightly than do acylated sites. Therefore under the conditions of our former experiments, the saturation of reaction rates and amplitudes would depend on the NAD^+ concentration required for saturation of the acylated sites, whether or not binding at the non-acylated sites was a prerequisite for catalysis.

The results on the reduction of acyl enzyme by NADH clearly indicate that the binding of NAD^+ at the two non-acylated sites is required for catalytic reduction. The correspondence in magnitude of the slow isomerization rate for this reaction and for the phosphorolysis and arsenolysis reactions suggests that the same conformational isomerization is rate limiting in all cases. An indication that NAD^+ binding to non-acylated sites is a ubiquitous requirement for catalysis is the fact that alkylation of the non-acylated sites with an alkylating reagent which leads to a less affine site for NAD^+ binding (viz., N-ethyl maleimide), reduces the amplitude of the fast phosphorolysis and arsenolysis step at concentrations of NAD^+ which are optimal for the unmodified enzyme but below saturation for the alkylated sites. Similarly, alkylation with N-ethyl maleimide inhibits the rate of reduction of acyl enzyme by NADH (5).

The function of this critical enzyme in the glycolytic pathway can be modulated by a variety of different effectors, notably, those affecting the binding of coenzyme dinucleotide to each type of site in the functioning enzyme molecule. Among the potential effector molecules for this system, we have investigated the adenine nucleotides, ATP and ADP. Both are competitive ligands vs. NAD^+ and NADH. In further communications we shall report the relative effectiveness of these nucleotides in displacing NAD^+ and NADH from the four classes of sites (acylated, non-acylated, tight and loose sites). The effect of these nucleotides on the reaction is most pronounced in the displacement of NAD^+ from acylated sites. In this regard it is noteworthy that magnesium-nucleotide complexes are far weaker competitive inhibitors than are the free nucleotides.

CONCLUSIONS

It is noteworthy that as in the reactions of furylacryloyl acyl enzyme, the phosphorolytic, hydrolytic, and reductive reactions of true substrate all proceed at nearly identical rates, under optimal conditions for all substrates. In some of these processes, a rate-limiting protein conformational isomerization has been detected by rapid-flow techniques. The isomerization rate constant is very similar to the specific rate of turnover, as calculated from steady state data. This rate is approximately 50-100 sec^{-1}, pH 7, 25°. Hence, it differs from the furylacryloyl isomerization rate by a factor of approximately 10^4. Since this isomerization does not involve a making and/or breaking of new covalent bonds, it is noteworthy that the conformational isomerization is so tightly linked to a change in the nature of the thiol ester bond. The enzyme's catalytic specificity resides not only in the increased affinity of enzyme for the specific substrate over the pseudo-substrate (a factor of about 10^3), but still more in the coupling of isomerization to bond perturbation. We believe it is reasonable to assume that the conformational change is a driving force in the catalytic process. Accepting this view, it would seem likely that inhibitors can act to stabilize a conformation, thus preventing isomerization to successive catalytically active states.

It is noteworthy that all of the results which we have found with the model acyl enzyme system demonstrate the structural asymmetry in pairs, of the enzyme tetramer. The mechanisms of regulation of catalysis all appear to require this type of asymmetry. In an accompanying paper, Dr. Seydoux will present arguments that similar asymmetry obtains in the case of the true substrate acyl enzyme and its reactions, modified however by the very much faster rates of conformational isomerization with this covalent enzyme-substrate intermediate.

ACKNOWLEDGMENTS

This work was supported by NSF Research Grant BMS-75-23297, PHS Research Grant GM-10451, and PHS Training Grant GM-00715.

REFERENCES

1. Kirschner, K., Eigen, M., Bittman, R., Voight, B.: The binding of nicotinamide-adenine dinucleotide to yeast GAPDH: temperature-jump relaxation studies on the mechanism of an allosteric enzyme. Proc. Nat. Acad. Sci. U.S.A. 56, 1661-1667 (1966).

2. Seydoux, F., Malhotra, O.P., Bernhard, S.A.: Half-site reactivity. CRC Critical Reviews of Biochemistry, March, pp. 227-257 (1974).
3. Malhotra, O.P., Bernhard, S.A.: Spectrophotometric identification of an active site-specific acyl GAPDH. J. Biol. Chem. 243, 1243-1252 (1968).
4. Malhotra, O.P., Bernhard, S.A.: Activation of a covalent enzyme-substrate bond by noncovalent interaction with an effector. Proc. Nat. Acad. Sci. U.S.A. 70, 2077-2081 (1973).
5. Bernhard, S.A, Malhotra, O.P., Ingbar, D., Schwendimann, B.: On the function of half-site reactivity: intersubunit NAD^+-dependent activation of acyl-GAPDH reduction by NADH. J. Mol. Biol. 108, 123-138 (1976).
6. Seydoux, F., Bernhard, S., Pfenninger, O., Payne, M., Malhotra, O.P.: Preparation and active-site specific properties of sturgeon muscle GAPDH. Biochemistry 12, 4290-4300 (1973).
7. Kelemen, N., Kellershohn, N., Seydoux, F.: Sturgeon GAPDH. Formation of binary complexes with coenzymes and substrates. Eur. J. Biochem. 57, 69-78 (1975).
8. Seydoux, F.J., Kelemen, N., Kellershohn, N., Roucous, C.: Specific interactions of 3-phosphoglyceroyl-GAPDH with coenzymes. Eur. J. Biochem. 64, 481-489 (1976).
9. Conway, H., Koshland, D.E.: Negative cooperativity in enzyme reaction: the binding of phosphopyridine nucleotide to GAPDH. Biochemistry 7, 4011-4023 (1968).
10. MacQuarrie, R.A., Bernhard, S.A.: Subunit conformation and catalytic function in rabbit-muscle GAPDH. J. Mol. Biol. 55, 181-192 (1971).
11. Olsen, K.W., Moras, D., Rossmann, M.G., Harris, J.I.: Sequence variability and structure of D-GAPDH. J. Biol. Chem. 250, 9313-9321 (1975).
12. Ferdinand, M.: The isolation and specific activity of rabbit-muscle glyceraldehyde phosphate dehydrogenase. Biochem. J. 92, 578-585 (1964).
13. Malhotra, O.P., Bernhard, S.A.: Submitted for publication.
14. Bernhard, S.A., MacQuarrie, R.A.: Half-site reactivity and the "induced-fit" hypothesis. J. Mol. Biol. 74, 73-78 (1973).
15. Seydoux, F., Bernhard, S.: On site heterogeneity in sturgeon muscle GAPDH: a kinetic approach. Biophys. Chem. 1, 161-174 (1974).

16. Biellmann, J.-F., Jung, M.J.: Mechanism of the alcohol dehydrogenases from yeast and horse liver. Eur. J. Biochem. 19, 130-134 (1971).
17. Bernhard, S.A., Lau, S.-J.: Spectrophotometric and structural evidence as to the mechanism of protease catalysis at chemical bonding resolution. Cold Spring Harbor Symp. Quant. Biol. 36, 75-83 (1971)
18. Bernhard, S.A., Lau, S.J., Noller, H.: Spectrophotometric identification of acyl enzyme intermediates. Biochemistry 4, 1108-1118 (1965).
19. Charney, E., Bernhard, S.A.: Optical properties and the chemical nature of acyl-chymotrypsin linkages. J. Am. Chem. Soc. 89, 2726-2733 (1967).
20. Bernhard, S.A., Malhotra, O.P.: Color, conformation and catalysis. Israel J. of Chem. 12, 471-481 (1974).
21. Monod, J., Wyman, J., Changeux, J.-P.: On the nature of allosteric transitions: a plausible model. J. Mol. Biol. 12, 88-118 (1965).

Received February 11, 1977

DISCUSSION

Werber: Are the different processes elicited by alkylation and acylation uniquely dependent on the chemical nature of the modifier, or can one obtain under some conditions "acylated" behavior upon alkylation of "alkylated" behavior upon acylation?

Bernhard: I am not certain. There are reported half-site NAD-dependent alkylations which I consider a characteristic of acylation. Unlike acylation which always has these characteristics, alkylation properties always seem to be solvent-, species-, and pH-dependent.

Mannervik: It was not clear to me how you could assign dissociation constants (K_D) for the binding of dinucleotides to specific subunits.

Bernhard: The different "colors" of the chromophoric acyl group indicate whether NAD^+ ("red"), NADH ("blue") or nothing ("neutral") is bound to the site. Since the extent of color change is linear with dinucleotide occupation, these K_D's can be determined. NAD^+ binding to non-acylated sites is indicated by the characteristic broad Racker band and NADH occupancy of non-acylated sites by the changes in spectrum in the 270 nm range and by the fluorescence quenching of NADH emission.

Heilmann: Do you regard the two dimers within the tetramer as binding the ligands completely independently from each other? In the yeast enzyme one obtains from titration studies four different NAD^+ binding constants; the first three ligand molecules apparently being bound cooperatively, whereas the forth one binds in an anticooperative manner.

Bernhard: In the fish muscle enzymes and in B. stearothermophilus enzymes, the subunits behave as independent heterologous dimers. Presumably the dimer-dimer constraints do not function allosterically. In yeast, at pH 7 and 25^{0} C, the situation appears to be quite similar (half-site, NAD^{+}-dependent acylation) but the dimer-dimer (p-axis) constraints do appear to function allosterically. The simplest model is that the yeast tetramer can exist in two differenct symmetry states ◆,●:

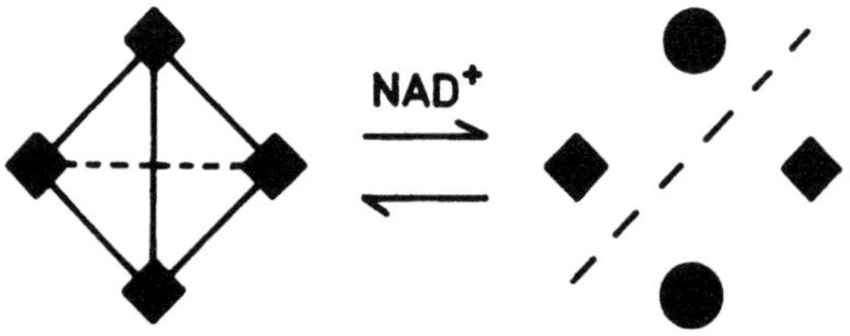

(2,2,2) (Dimer of Dimers)

This model predicts positive and negative cooperativity in ligand binding and is quantitatively consistent with our own data and that of Cook & Koshland, on the binding of NAD to yeast GPDH. It is also consistent with the enhanced cooperativity of NAD-binding in the presence of Michaelian ligands such as ADPR and ATP.

Plaut: Is the change of spectrum of the probe due to an acyl transfer from a thiol ester to, e.g. histidine group on the enzyme or due to a change in environment surrounding the acyl probe?

Bernhard: I don't think so, although it cannot be ruled out by spectroscopy. Arguments against the acyl transfer are as follows: (1) The acyl shift in papain is exactly the same. The evidence in papain (from the X-ray crystallography) argues against such a transfer. (2) Denaturation of the holo-acyl enzyme always yields the thiol ester (normal spectrum). (3) There are complementary "blue-shifts" upon the binding of NADH. This effect is highly specific and is obligatory for reduction. There is no apparent alternate acyl acceptor corresponding to this spectrum and reactivity.

NON EQUIVALENT ACTIVE SITES IN TRANSIENT KINETICS OF STURGEON GLYCERALDEHYDE-3-PHOSPHATE DEHYDROGENASE.

N. Kellershohn and F.J. Seydoux.
E.P.C.M., CNRS, Bat. 430, Université Paris-Sud - ORSAY 91405 - France.

Tetrameric muscle GPDH*'s from various species exhibit well documented anticooperative properties which are revealed in two distinct phenomena :

i) The binding of the oxidized (or reduced) coenzyme NAD^+ to apoGPDH which can be described quantitatively on the basis of two classes of binding sites in the case of sturgeon GPDH (1).

ii) The half site reactivity in the acylation of an essential cysteine residue of Holo(NAD^+) GPDH by the substrate analog β-(2-Furyl)acryloyl phosphate (2).

Although these phenomena are indicative of peculiar structural features of the functional oligomer (3), it is not clear at the present time if anticooperativity is involved in the physiological reaction catalyzed by GPDH. It has been demonstrated that the physiological substrate DPG acylates (1) all the four active sites of the enzyme.On the other hand, previous kinetic analysis of various reactions of the fully acylated enzyme indicates that the tetramer behaves kinetically as a dimer of dimer, each functional dimer containing two non equivalent active sites with distinct reactivity (3). Further insight of the role of anticooperativity in the specific catalysis by GPDH can be obtained from the analysis of the presteady state kinetics of the reductive dephosphorylation of the normal substrate DPG which is strongly dependent on the condition of preincubation of the active enzyme with DPG. Such an analysis allows the determination of the number of active sites which are acylated under steady state conditions.

---

*Abbreviations : GPDH : Glyceraldehyde-3-Phosphate dehydrogenase.
DPG : 1,3 bisphosphoglycerate.

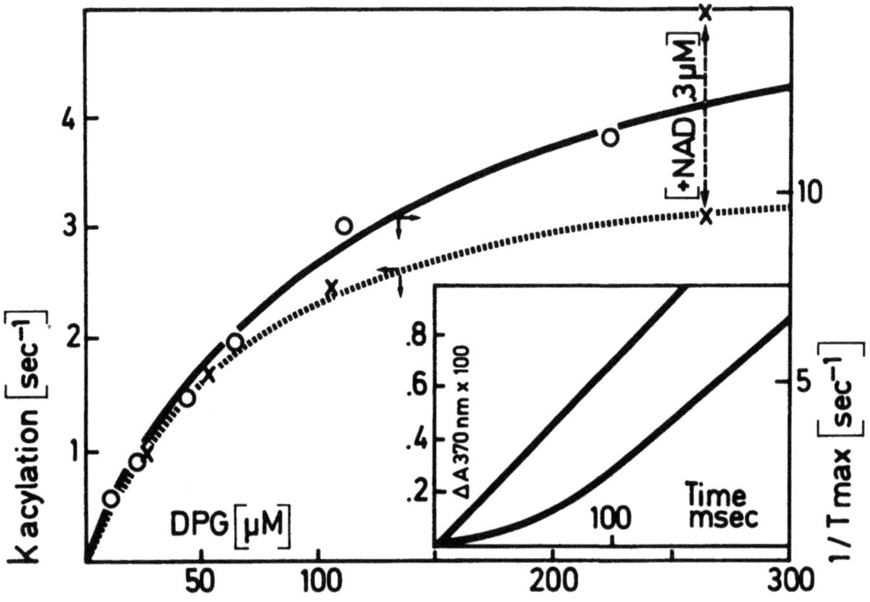

Fig. 1 Lag phase in the reductive dephosphorylation of DPG.
□ : lag period as function of the final DPG concentration, 0.06 µM enzyme, 48 µM NADH , + : pseudo first order rate constant of the acylation of apo-GPDH as function of the DPG concentration, 4 µM enzyme. The insert shows the progress curve of the reaction (0.035 µM apoGPDH, 53 µM NADH, 97 µM DPG) initiated with apoenzyme (lower curve) or with apoenzyme preincubated with DPG (upper curve).

EXPERIMENTAL

Sturgeon apoGPDH and DPG were prepared as previously described (1). All experiments were carried out at 10°, pH 7.0 in standard Tris sulfonate buffer. Kinetic measurements were performed with a Durrum stopped-flow apparatus equipped with a 2 cm optical path observation cell. After analogic to digital conversion in a Biomation 802 transient recorder, the kinetic data (1024 points per kinetic run) were stored on a magnetic Tape and analyzed with a Wang 2200 B calculator.

RESULTS AND DISCUSSION.

Lag phase in the kinetics of reduction of DPG.

When apoGPDH is mixed in the stopped-flow with an excess of NADH and DPG, the progress curve of the reaction monitored at 370 nm exhibits a well defined lag phase. This lag phase is invariant with respect to the final concentrations of NADH and enzyme in the reaction mixture but decreases with increasing concentrations of DPG and shows a michaelian saturation behavior as depicted in Fig. 1.

The lag phase results from the slow acylation of apoGPDH by DPG.

The kinetics of the acylation of apoGPDH by DPG, as monitored spectrophotometrically at 287 nm (4) shows michaelian saturation kinetics similar to that observed for the dependence of the lag phase with respect to DPG (Fig.1). Addition of very small amounts of NAD^+ in the reaction mixture results in a significant increase of the acylation rate. This confirms the specific "catalytic" function of NAD^+ which promotes greatly the acylation reaction (5) (activation of about two order of magnitude when the enzyme is saturated by NAD^+). This observation provides a simple explanation of the lag phase phenomenon. It is well accepted that the reductive dephosphorylation of DPG proceeds via at least two distinct, consecutive steps :

i) Acylation of the enzyme, this step being strongly and exclusively promoted by NAD^+ (not by NADH).

ii) Reductive deacylation of the thioacyl-enzyme by NADH with formation of NAD^+.

Thus, NAD^+ produced in step (ii) activates the first acylation step which is rate limiting in the apoenzyme reaction. Hence a lag phase appears in the evolution of the system toward its steady state. In agreement with this interpretation, the lag phase disappears completely upon incubation of the enzyme with NAD^+ or DPG which acylates the enzyme stoechiometrically (1,4) (see Fig. 1).

Extent of acylation required for a complete elimination of the lag phase.

The evolution of the reaction toward its steady state is strongly dependent on the conditions of the preincubation of the apoenzyme with DPG(or NAD^+). If the extent of acylation of the enzyme under given conditions of

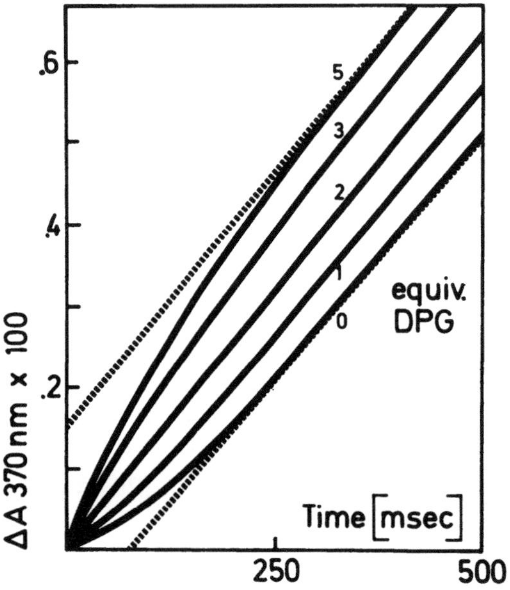

Fig 2 Presteady state kinetics of the reductive dephosphorylation of DPG as function of the preincubation of apoGPDH with increasing stoechiometric amounts of DPG as indicated : 0.035 μM enzyme, 53 μM NADH and 97 μM DPG. Dashed curves correspond to the limiting steady state.

preincubation is similar to that of the system in the steady state, no lag phase should be observed. This has been investigated by studying the presteady state kinetics obtained upon preincubation of the apoenzyme with various amounts of DPG. The final concentration of DPG in the reaction mixture (about 5 times the apparent Km) was chosen so that the progress curve remains linear after the lag phase over a sufficient period of time. In these conditions, as shown in Fig. 2 the lag phase decreases when the enzyme is incubated with increasing amount of DPG. With 2 equivalents of DPG, the lag phase is completely eliminated. With more than 2 equivalents of DPG in the incubation mixture (up to 4-5 eq), the evolution of the reaction to the steady state proceeds via a "burst" like kinetics. The amplitude of this pseudo-burst is about 2.5 times the active site concentration. This observation rules out any "burst" mechanism in which the excess of DPG in the incubation mixture is consumed before reaching the steady state.
Thus, the only class of mechanism which can account for this pseudo burst effect should involve a slow transition between two steady states : one in which the enzyme is fully acylated by DPG and the other in which only <u>two</u> of the <u>four</u> active sites are acylated. All the present data can be explained on the basis of the following scheme :

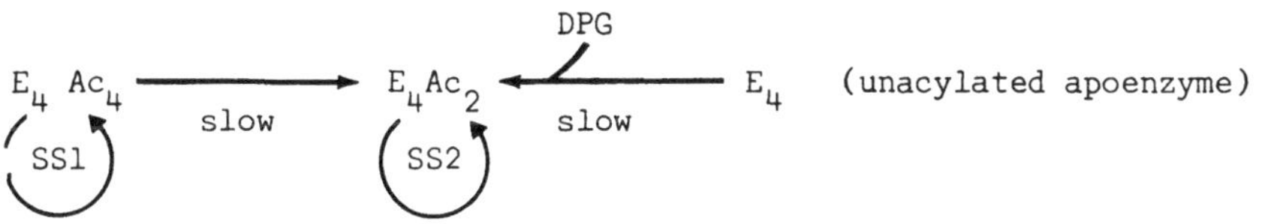

in which the turnover of steady state 1 (SS1) is about twice as fast as that of steady state 2 which corresponds to a "stable steady state" finally reached by the system.

Although we cannot provide yet a detailed molecular picture of the proposed mechanism, the present results are clearly consistent with the accumulation at the steady state of an enzymic species in which only half of the four active sites are acylated by the physiological substrate DPG. Thus, the extensively reported anticooperative properties of muscle GPDH are also apparent in the specific reaction of the enzyme.

ACKNOWLEDGMENTS

We are indebted to Professors S.A. Bernhard and J. Yon for stimulating discussions. This work was supported by grants from the C.N.R.S. (G.R. 13 and A.T.P. n° 1952).

REFERENCES

1- Seydoux, F.J., Bernhard, S.A., Pfenninger, O., Payne, M. & Malhotra, O.P. : Preparation and active site properties of Sturgeon Muscle Glyceraldehyde-3-phosphate dehydrogenase. Biochemistry 12, 4290-4300 (1973).

2- Malhotra, O.P. & Bernhard, S.A. : Spectrophotometric identification of an active site specific Acyl Glyceraldehyde-3-phosphate dehydrogenase. J. Biol. Chem., 243, 1243-1252 (1968).

3- Seydoux, F.J., Malhotra, O.P. & Bernhard, S.A. : Half site reactivity. C.R.C. Critic Rev. Biochem. 2, 227-257 (1974).

4- Kelemen, N., Kellershohn, N. & Seydoux, F.J. : Sturgeon Glyceraldehyde-3-phosphate dehydrogenase. Formation of binary complexes with coenzymes and substrates. Eur. J. Biochem. 57, 69-78 (1975).

5- Trentham, D.R. : Reactions of D-glyceraldehyde-3-phosphate dehydrogenase facilitated by NAD^+; Biochem. J. 122, 59-69 (1971).

Received February 19, 1977

DISCUSSION

Veeger: Is it absolutely sure that the lag phase observed in the reaction of 1,3 diphosphoglycerate + NADH is due to the autocatalytic formation of NAD^+? Since NADH always contains traces of NAD^+, one should carry out this reaction by adding NADH pre-treated with NADase (to destroy all NAD^+) in order to observe no disappearence of NADH, in case there is a requirement for NAD^+ to acylate the enzyme. Similarly NADase must be present in the reaction medium.

Seydoux: It is absolutely sure that NAD^+ activates the reductive dephosphorylation of 1,3 diphosphoglycerate. We don't know however, if the slow initial rate of reaction is due to trace amounts of NAD^+ in the reaction mixture, although addition of very small amounts of NAD^+ does increase this rate significantly.

Scheek: A possible explanation of the "pseudo-burst", seen if reductive dephosphorylation is started with fully acylated enzyme, might be that the four acyl groups don't react at the same rate, so that they are running slowly out uf phase during the first turnovers, until the real steady state is reached, in which on the average only two sites are acylated in the tetramer.

Seydoux: This is a likely explanation provided that the complete "phasing out" of two active sites takes place only after several turnovers of the fully reactive tetra-acylated enzyme.

Werber: How do you know that in the reaction of GPDH with excess of 1,3 diphosphoglycerate - which can give a tetraacylated enzyme - the species turning over at the steady state is a diacetylated enzyme? Were you able to separate by gel-filtration this species from the reaction mixture and prove chemically that it is diacetylated? Or what other evidence do you have in favor of this suggestion?

Seydoux: It is impossible to isolate diacylated enzyme from the reaction mixture because of the reversibility of the overall reaction. The only evidence we have for the accumulation of diacylated enzyme comes from our kinetic results. It should be, however, possible to characterize the diacylated species in a rapid quenching apparatus.

Dubied: Your steady-state cycles I and II are they eventually reversibly connected, or is the cycle II the only system working in the real final steady-state of reaction?

Seydoux: The fact, that starting with E_4Ac_2 we found the steady-state right form the beginning, seems to provide evidence that cycle II is the final and stable cycle. However, a certain equilibrium may be real.

Hollaway: Is it possible that the time-course in the inset of Fig. 1 showing the acceleration phase, could arise from a combination of inactive dimers to give active tetramers on mixing with substrates? The curve seems to have the shape expected for such a second order combination and the final rate does not equal that of the reaction where the enzyme was preincubated with 1,3 diphosphoglycerate. Are you absolutely sure that the shape of the acceleration phase is independent of the enzyme concentration?

Seydoux: The enzyme concentration was varied by a factor of 2.5 and the lag period did not change significantly. Also the time-course can be fitted by a first order gain of enzyme activity, and a zero-order catalytic process. The sturgeon muscle apoenzyme is more stable than the rabbit enzyme and would be less likely to dissociate.

Mosbach: I would suggest the following approach in case you wish to differentiate between the E and E-Ac form in the sequence:

$$E + NAD^+ + \text{glyceraldehyde 3-phosphate} \longrightarrow E\text{-}Ac + NADH$$

Run the reaction on solid phase with NAD^+ covalently bound to e.g. Sepharose. Subsequently you apply your mixture, which is free of soluble NAD(H), on another NAD^+-column where because of the pronounced difference in affinity to NAD^+ of E and E-Ac, the two forms could be easily separated.

Seydoux: This is a very interesting suggestion and I hope we can discuss this point later on in detail.

Südi: In your "simplest possible scheme" in which you interpret the arsenolysis reaction, you represent arsenolysis itself with a single irreversible reaction step. Do you derive this idea from the irreversibility of the overall reaction?

Seydoux: From 1 to 10 mM arsenate, the overall rate of the reaction is linearly increasing with the arsenate concentration. Thus, the overall rate of reaction is not limited under these conditions by the decomposition of an eventual arsenate-acyl group. Adduct and the arsenolysis can be treated kinetically as a single irreversible step.

SYMMETRY AND NAD$^+$-DEPENDENT STRUCTURAL CHANGES IN D-GLYCERALDEHYDE-3-PHOSPHATE DEHYDROGENASE

A.J. Wonacott & G. Biesecker*
MRC Laboratory of Molecular Biology, Hills Road, Cambridge, U.K.
*Scripps Clinic and Research Foundation, 476 Prospect Street, La Jolla, CA 92037, U.S.A.

INTRODUCTION

D-glyceraldehyde-3-phosphate dehydrogenase (GPDH) is a tetramer of molecular weight 140,000, which catalyses oxidative phosphorylation of its substrate to D-1,3 diphosphoglyceric acid in the presence of the coenzyme NAD$^+$. The catalytic mechanism and cooperative properties have been studied extensively (for review see ref. 1). It is evident that bound ligands act as allosteric effectors of adjacent subunits in the tetramer and that these substrate-induced conformational changes may play an important role in the regulation and specificity of the enzyme in *vivo* (2, 3,4).

Although the tetramer consists of chemically identical subunits and binds four molecules of NAD$^+$, their affinity for NAD$^+$ is not the same. Thus the enzyme from muscle sources displays negative cooperativity (2,5,6) while that from yeast is essentially positively cooperative (7,8,9). Several models have been proposed to account for the behaviour of GPDH. Some of the data are best explained by a symmetrical conformation change of the tetramer (8,10,11). On the other hand sequential ligand-induced conformational changes (5,7,12) or pre-existing asymmetry among the subunits (13,14,15) gives a simpler explanation of other data.

The crystallographic studies of GPDH from the thermophile *Bacillus stearothermophilus* were undertaken in an attempt to explain these properties on a molecular basis. The bacterial enzyme has properties very similar to those of the muscle enzymes (16,17,18). However the bacterial enzyme remains quite stable when NAD^+ is removed. Different forms of the enzyme with NAD^+ stoichiometries from zero to four molecules per tetramer have been crystallized (19). The structure of the fully saturated holo-enzyme has been determined at high resolution and compared with that of the lobster holoenzyme (20). A low-resolution electron density map of the apoenzyme has been calculated. In addition, the effect of removing NAD^+ from holoenzyme crystals has been investigated.

These studies indicate that the thermophile holoenzyme possesses accurate 222 symmetry; significant conformational changes occur on binding NAD^+ to the tetramer which are not confined to the binding of the first molecule of NAD^+ (cf. ref 21). Taken together with the data on the crystals forms these results point to a mechanism of sequential ligand-induced conformational changes.

MOLECULAR STRUCTURE OF HOLO-GPDH

The structure of lobster holo-GPDH has been described in detail (22,23) and compared with that of the thermophile holoenzyme (20). The overall structures are closely similar, both in sequence (51% of the residues being identical) and in the folding of the polypeptide chain. Three orthogonal two fold axes P, Q and R relate the subunits to one another (Fig. 1). NAD^+ binding sites are located near the centre of the tetramer, close to the interface between subunits related by the R-axis dyad.

A single subunit, shown in Fig. 2, is divided into two folding domains. The N-terminal or 'coenzyme-binding' domain, residues 0-148, contains most of the residues involved in NAD^+ binding, while the C-terminal or 'catalytic domain', comprising residues 149-333, contributes most of the residues implicated in catalysis. The same secondary and tertiary structural features are present in both structures; differences are confined for the most part to external loop regions between segments of secondary structure.

The C-terminal helix, α_3, is associated with the coenzyme binding domain, where it fits into a groove. Its function appears to be that of a hinge allowing the relative rotation of the two domains (see below).

Each subunit makes contact with every other subunit by virtue of the three twofold axes in the tetramer. The three kinds of subunit interface present are very different in nature. The P-axis interface involves extensive contact between the anti-parallel β-sheet region of the catalytic domain with an equivalent region of the related subunit (Fig. 3). The residues involved in this contact are highly conserved in all known sequences (24).

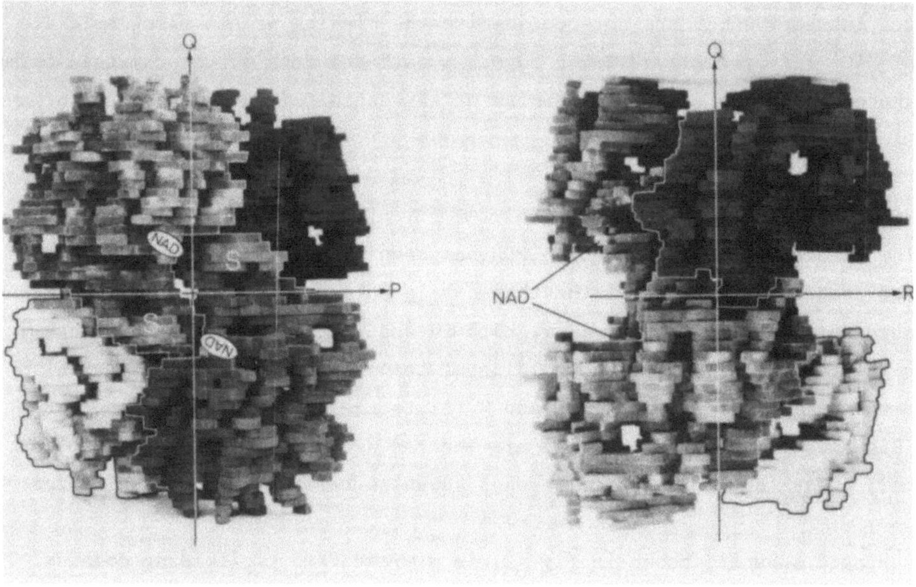

Figure 1

Schematic diagram of the GPDH tetramer showing four subunits related by three orthogonal 2-fold axes, labelled P,Q and R according to the nomenclature adopted for lactate dehydrogenase and lobster GPDH. The four molecules of NAD^+ bind at a molecular 'waist'.

(a) The tetramer viewed down the R-axis with the S-loop regions labelled 'S'.

(b) View down P axis.

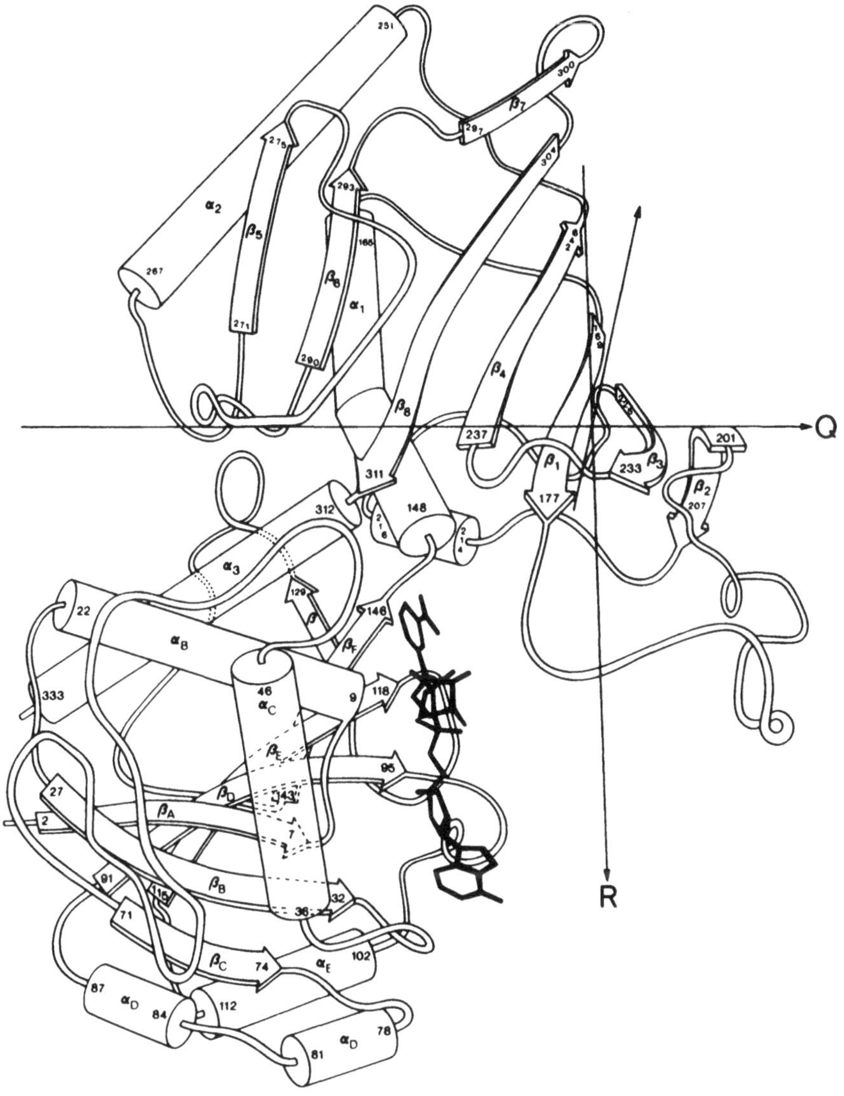

Figure 2

Folding diagram of one subunit of B. stearothermophilus GPDH. Nomenclature for sheets and helices of first domain is that used for lactate dehydrogenase and lobster GPDH. The binding site for NAD^+ (shown in bold) is at the end of the β-sheet region of the N-terminal domain.

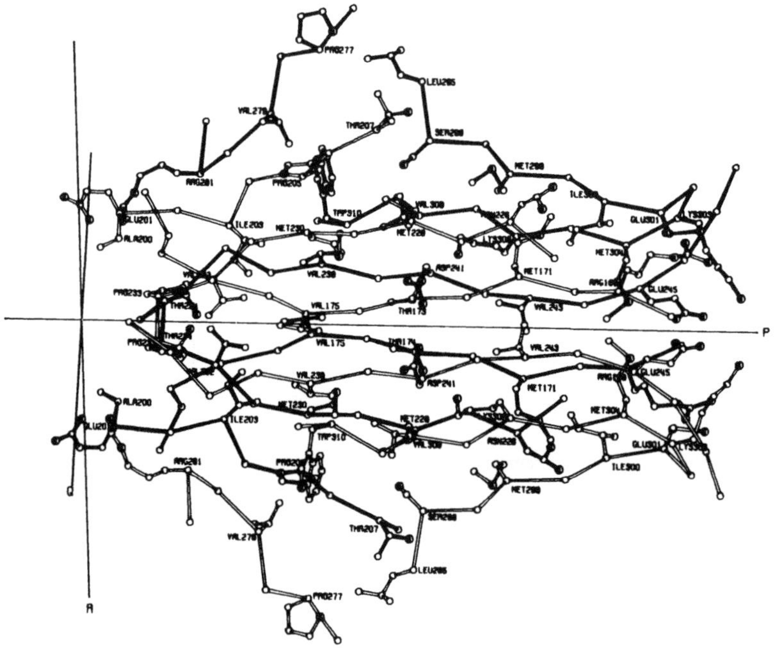

Figure 3

The extensive anti-parallel β-sheet interface between subunits which are related by P-axis dyad. Residues sandwiched between sheets are very highly conserved in all GPDHs.

A significant feature of the C-terminal domain is an irregular loop of polypeptide chain (residues 178-201) termed the S-loop which is in Van der Waals contact with the molecule of NAD^+ bound to the R-axis related subunit (see Fig. 4). An interface is formed by the contact of one S-loop with the S-loop and α_c helix of the subunit related by the dyad axis R. This contact contains a number of interacting charged residues which is an unusual feature of subunit interfaces. The S-loops of the tetramer form the core of the molecule; most of the residues are internal and make important interactions with other subunits. It is noticeable that this region of the molecule is not highly conserved between the mesophile and thermophile dehydrogenases (see Harris & Walker, this volume), in addition to which, many main-chain hydrogen bonding groups appear to be unsatisfied in both the lobster and thermophile enzymes. These observations have led to the suggestion (20) that the S-loop is a flexible region, whose structure is determined by the packing of side chains and by the interactions that it makes with adjacent subunits through contacts across the P and R axes.

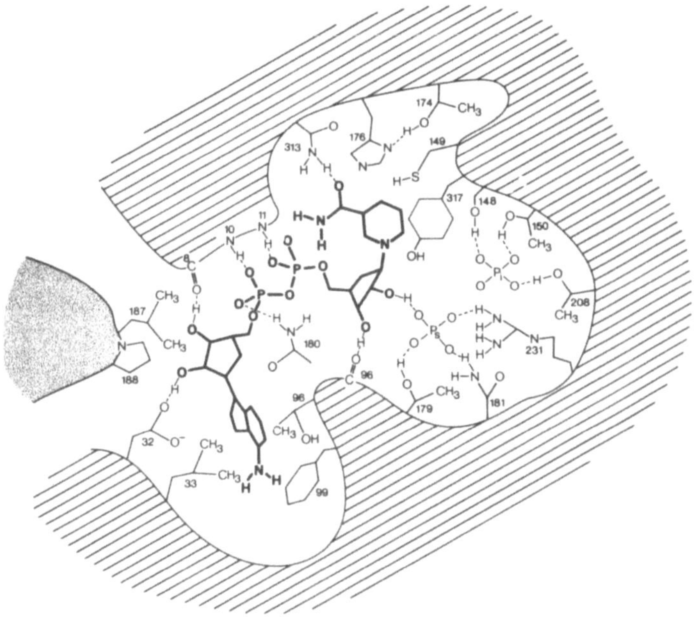

Figure 4

Schematic diagram of the NAD$^+$ binding site in B. stearothermophilus GPDH. Two anion-binding sites, labelled Ps and Pi, are shown; residues leucine-187 and proline-188 are from the R-axis related subunit (shown stippled).

The protein interactions involved in NAD$^+$ binding are shown in Fig. 4. These are almost identical to those found for lobster GPDH apart from the residues on one side of the adenine binding pocket and the additional interaction of asparagine 180 with the phosphate group. However an additional residue, leucine-187 from the R-axis related subunit, is in contact with the adenosine ribose.

Two non-protein electron dense features, interpreted as bound anions, have been observed in the active site pocket of both lobster and B. stearothermophilus GPDH. These are considered to be the probable sites of the phosphate group of the substrate and the inorganic phosphate in the catalytic reaction (21,23). One of these sites bridges between the nicotinamide ribose and side chains in the S-loop region, thus linking the catalytic site of one subunit to the adenine site of the R-axis related subunit.

MOLECULAR SYMMETRY

Both lobster and B. stearothermophilus holo-GPDH crystallise with the whole molecule in the asymmetric unit of the crystal so that the molecular symmetry is not expressed directly. The molecule of human holoenzyme (25) is known to have one exact twofold axis since the Q-axis dyad is incorporated in the crystal symmetry.

Moras et al (23) have averaged their map of lobster GPDH about the Q dyad. In interpreting this map they have suggested that the molecule is nonsymmetrical, the coenzyme being bound differently in subunits related by the R-axis dyad. This asymmetry has been advanced as a partial explanation of the negative cooperativity of NAD^+ binding. Other differences particularly in the active site region and the S loops of R-axis related subunits are described.

The procedure used to obtain the electron density map of B stearothermophilus GPDH has been described (20). An averaged electron density map was not used for interpretation. Obviously there are differences in the electron density of different subunits due to noise, but it was found unnecessary to invoke structural differences between the four subunits in order to fit the molecular model.

We have examined carefully the possibility that the coenzyme takes up different conformations in each subunit. We conclude that the electron density can only be fitted satisfactorily by having an NAD^+ conformation in which both ribose rings have a 2'-endo condormation with the adenine ring in an anti-conformation. To make the comparison between NAD^+ conformations in different subunits more objective, the molecule was refined into the electron density using Diamond's realspace refinement procedure (26). The fit of the molecule to the electron density for two subunits related by R-axis symmetry can be seen in Fig. 5. There is a slight difference in the orientation of the adenine rings of the NAD^+ bound to the R-axis related subunits corresponding to a rotation of $\sim 20^o$ in the dihedral angle about the glycosyl bond.

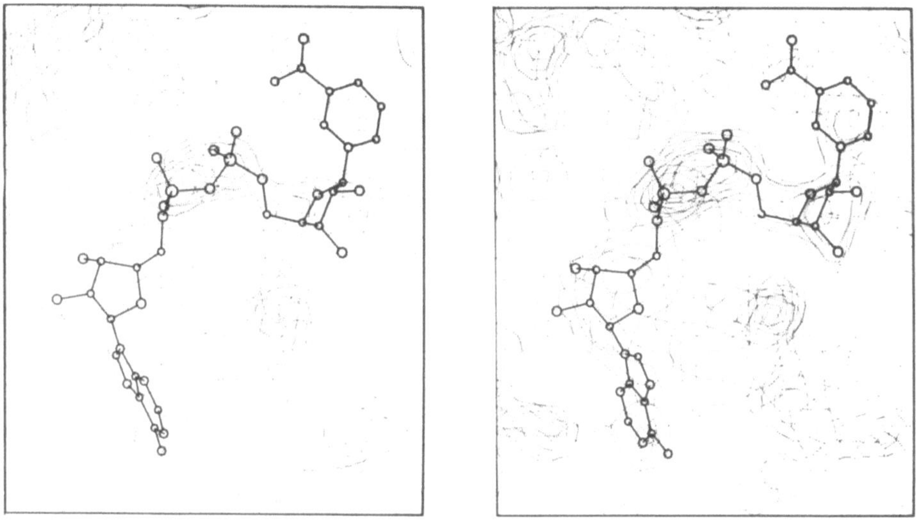

Figure 5

Molecules of NAD^+ from R-axis related subunits superimposed on their respective electron densities.

The r.m.s. difference in the refined coordinates of NAD^+ in the four subunits is < 0.25 Å.

The conformation of the NAD^+ molecule in the thermophile enzyme is closely similar to that for the NAD^+ bound to the 'red' subunit in lobster GPDH, although there are differences of at least 1 Å in the individual atomic coordinates.

The orientation of the adenine ring would be very sensitive to small movements of leucine 33 with which it is in contact. No significant difference in the position of this residue can be detected at the level of accuracy attainable with a 2.7 Å electron density map.

We are satisfied that no significant differences between subunits exist within the limits of the presently available electron density map. A more definitive answer can only be provided by extending the structural analysis to higher resolution.

NAD^+-DEPENDENT STRUCTURAL CHANGES

Crystal Forms of GPDH

Three distinct but closely related crystal forms of GPDH from B. stearothermophilus have been crystallized. The unit-cell dimensions, space-group and NAD^+ stoichiometries are given in Table 1. Further details of crystallization conditions and crystal properties will be given elsewhere (19).

Table 1 Crystalline Forms of GPD from B. stearothermophilus

Form	Space group	Unit Cell A				NAD^+ content
		a	b	c	β	moles/tetramer
Apo	$P2_1$	83.6	130.2	82.9	107.25°	<0.2
Holo I (half-holo)	$P2_12_12$	133.2	126.8	99.3	90°	1 - 2.5
		≡(83.1	126.8	83.1	106.6°)	
Holo II (holo)	$P2_1$	82.6	124.4	82.6	109.1	2.7 - 4.0

For the holo (Holo II) and half-holo (Holo I) crystal forms the unit cell dimensions vary depending on the NAD^+ content or NAD^+ concentration in the growth medium of the crystals (the values given in Table 1 are for Holo I, ∿ 2 moles/tetramer and Holo II, 4 moles/tetramer).

When apo-enzyme crystals are soaked in solutions containing NAD^+ at concentration of 10^{-6} - 10^{-5}M, the binding of the coenzyme causes cracking and eventual disintegration of the crystals (27). With higher concentrations of NAD^+, 10^{-4} - 10^{-3}M, Holo I crystals bind NAD^+ causing fine cracks to appear; the extent of cracking depends on the original NAD^+ saturation of the crystal. The resulting crystal gives a diffraction pattern corresponding to that of Holo II crystals.

Conformational Changes

An electron density map of apo-GPDH has been obtained at 6 Å resolution by isomorphous replacement and compared with the map of the holo enzyme at a similar resolution. Fig. 6 shows superimposed sections of the two electron density maps for a region of the coenzyme-binding domain. The main structural features are clearly delineated. Comparison shows that the coenzyme-binding domain in apo-GPDH is rotated outwards, away from the NAD^+ binding site in the holo enzyme, about an axis which corresponds approximately to that of the C-terminal helix, α_3. The positions of helices α_D and α_E differ by about 5 Å. As expected there is no electron density at the position of the NAD^+ binding site. In addition the conformation of

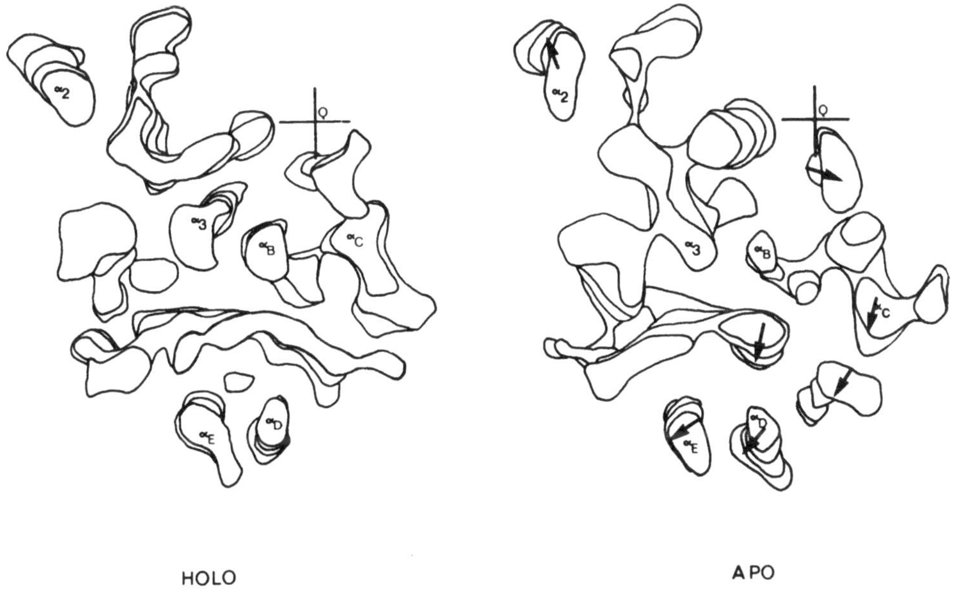

Figure 6

Superimposed sections of equivalent regions of the low resolution electron density maps of holo- and apo-GPDH in the region of the coenzyme-binding domain. Main structural features are labelled as in Fig. 2. Sections are perpendicular to the Q-axis of the tetramer. Arrows indicate approximate direction and magnitude of movements of helices and sheet region in transformation from the holo to apo conformation.

the S-loop appears to be modified, which together with the rotation of the coenzyme binding, results in the S-loop being more accessible to solvent.

Symmetry of the holoenzyme is only maintained in the presence of saturating concentrations of NAD^+. In crystals grown in lower concentrations of NAD^+, or soaked in NAD^+-free mother liquor, 0.5 moles of NAD^+ are lost. Data have been collected for crystals with approximately 3.5 moles/tetramer. A difference map against holo-GPDH shows clearly that NAD^+ is lost from only one of the four subunits and that structural changes in the protein occur. Although there is a slight change in the packing of the tetramer in the unit cell which complicates the analysis, the major conformational changes are associated with the coenzyme-binding domain of the subunit which loses NAD^+. These changes are towards the conformation observed in apo-GPDH. The fact that NAD^+ is lost from one subunit only might be taken to imply that NAD^+ is bound less tightly to this subunit. A more likely explanation is that molecule packing in the crystal is such that changes accompanying loss of NAD^+ are more easily accommodated by removal of NAD^+ from this particular subunit. We are fortunate in this particular case to be able to observe uniquely the species with only 3 molecules/tetramer.

DISCUSSION

The movement of the coenzyme-binding domain on binding NAD^+ to apo-GPDH produces a contraction of the tetramer. This is consistent with the decrease in molecular volume detected by sedimentation studies (28,29) and by small angle scattering (30,31) as well as the observed changes in protein hydration (32) in the formation of the holoenzyme.

The S-loop region of the subunit appears to be flexible. Binding of NAD^+ could result in changes in the packing of parts of the protein chain, in particular the residues of the S-loop, giving rise to negative cooperativity. In the transition to apo-enzyme the molecule expands and opens up this region, possibly allowing greater flexibility of the S-loop. The instability of apo-enzyme from mesophile sources could be a direct consequence of this. The hypothesis of a flexible S-loop is given further support by the observed acyl-transfer in muscle apo-GPDH (33,34) in which an acyl group is transferred specifically from the essential cysteine-149 to N^{ε} of lysine-183. The groups involved are too far apart in the holoenzyme for this reaction to occur.

Studies of the effect of binding NAD^+ analogues to crystals of apo-GPDH indicate that the ribose of the nicotinamide plays an essential role in effecting the structural changes associated with NAD^+ binding (27). It has been suggested that the anion binding site which bridges between the nicotinamide ribose and residues of the S-loop is the crucial interaction which triggers the conformational changes.

The relationship between crystal forms of the B. stearothermophilus enzyme and their behaviour in the presence of added NAD^+ implies that:

(a) The conformation of the GPDH tetramer depends on the degree of saturation with NAD^+, structural changes occurring for each molecule of NAD^+ bound.

(b) There is a virtual continuum of structural states for the molecule of GPDH with at least one molecule of NAD^+ bound since Holo I crystals can be converted to the Holo II form without loss of the diffraction pattern.

(c) The largest conformational change takes place when the first molecule of NAD^+ is bound to apo-enzyme crystals since the crystals are unable to accommodate the structural change.

These data fit the prediction made by Herzfield and Schlesinger (35) from their analysis of previously reported data on GPDH, 'that the binding of each NAD^+ molecule will result in a conformational change, but that the binding of the first, and possibly the second, NAD^+ molecule will result in much more extensive changes than will subsequent NAD^+ binding'.

CONCLUSIONS

An allosteric mechanism is proposed in which the primary effect of NAD^+ binding is to cause a rotation of its coenzyme binding domain. This in turn induces structural changes in the flexible core of the protein which is capable of transmitting structural information from one subunit to another.

ACKNOWLEDGEMENTS

We thank Dr D.M. Blow for his encouragement in the course of this work and Mrs J.M. Smith for her excellent technical assistance.

REFERENCES

1. Harris, J.I. & Waters, M.: The Enzymes, Vol XIII, 3rd Edition, Academic Press, New York, 1-50, (1976).
2. Seydoux, F., Bernhard, S.A., Pfenninger, O., Payne, M. & Malhotra, O.P.: Biochemistry, 12, 4290-4300 (1973).
3. Byers, L.D. & Koshland, D.E.: Biochemistry, 14, 3661-3669 (1975).
4. Schwendimann, B., Ingbar, D. & Bernhard, S.A.: J. Mol. Biol. 108, 123-138 (1976).
5. Conway, A. & Koshland, D.E., Jr.: Biochemistry, 7, 4011-4023 (1968).
6. de Vijlder, J.J.M., Hilvers, A.G., van Lis, J.M.J. & Slater, E.C.: Biochim. Biophys. Acta. 191, 221-228 (1969).
7. Cook, R.A. & Koshland, D.E., Jr.: Biochemistry, 9, 3337-3342, (1970).
8. Kirschner, K.: J. Mol. Biol. 58, 51-68 (1971).
9. Gennis, L.S.: Proc. Nat. Acad. Sci. U.S.A. 73, 3928-3932, (1976).
10. Kirschner, K., Gallego, E., Schuster, I & Goodall, D.: J. Mol. Biol. 58, 29-50 (1971).
11. von Ellenrieder, G., Kirschner, K. & Schuster, I.: Eur. J. Biochem. 26, 220-236 (1972).
12. Koshland, D.E., Jr., Nemethy, G. & Filmer, D.: Biochemistry, 5, 365-385 (1966).
13. Malhotra, O.P. & Bernhard, S.A.: J. Biol. Chem. 243, 1243-1252 (1968).
14. MacQuarrie, R. & Bernhard, S.A.: J. Mol. Biol. 55, 181-192 (1971).
15. Malhotra, O.P. & Bernhard, S.A.: Proc. Nat. Acad. Sci. U.S.A. 70, 2077-2081 (1973).
16. Singleton, R., Kimmel, J.R. & Amelunxen, R.E.: J. Biol. Chem. 244, 1623-1630 (1969).
17. Suzuki, K. & Harris, J.I.: FEBS Letters, 13, 217-220 (1971).
18. Allen, G. & Harris, J.I.: Biochem. J. 151, 747-749 (1975).
19. Thierry, J.C., Biesecker, G., Wonacott, A.J., Butler, R. & Sweet, R. M.: in preparation.
20. Biesecker, G., Harris, J.I., Thierry, J.C. & Walker, J.E.: Nature, 266, 328-333 (1977).

21. Olsen, K.W., Garavito, R.M., Sabesan, M.N. & Rossmann, M.G.: J. Mol. Biol. 107, 577-584 (1976).

22. Buehner, M., Ford, G.C., Moras, D., Olsen, K.W. & Rossmann, M.G.: J. Mol. Biol. 90, 25-40 (1974).

23. Moras, D., Olsen, K.W., Sabesan, M.N., Buehner, M., Ford, G.C. & Rossmann, M.G.: J. Biol. Chem. 250, 9137-9162 (1975).

24. Olsen, K.W., Moras, D., Rossmann, M.G. & Harris, J.I.: J. Biol. Chem. 250, 9313-9321 (1975).

25. Watson, H.C., Duée, E. & Mercer, W.D.: Nature New Biol. 240, 130-133 (1972).

26. Diamond, R.: Acta Cryst. 21, 253-266 (1966).

27. Biesecker, G. & Wonacott, A.J.: Biochem. Soc. Trans. 5, 649-654 (1977).

28. Fuller Noel, J.K. & Schumaker, V.W.: J. Mol. Biol. 68, 523-532 (1972).

29. Smith, G.D. & Schachman, H.K.: Biochemistry, 12, 3789-3801 (1973).

30. Simon, I.: Eur. J. Biochem., 30, 184-189 (1972).

31. Durchschlag, H., Puchwein, G., Kratky, O., Schuster, I. & Kirschner, K.: Eur. J. Biochem. 19, 9-22 (1971).

32. Sloan, D.L. & Velick, S.F.: J. Biol. Chem. 248, 5419-5423 (1973).

33. Harris, J.I. & Polgar, L.: J. Mol. Biol. 14, 630-633 (1965).

34. Park, J.H., Agnello, C.F., Matthew, E.: J. Biol. Chem. 241, 769-771 (1966).

35. Herzfield, J. & Schlesinger, P.A.: J. Mol. Biol. 97, 483-517 (1975).

Received March 25, 1977

DISCUSSION

Dunn: It seems to me that the observation of negative cooperative phenomena, for example NAD^+ binding isotherms that are multiphasic, is rather temperature dependent. Dr. Seydoux has shown us that the sturgeon GPDH exhibits hyperbolic NAD^+ binding isotherms at temperatures well above that of the environment of the fish. This also is known for the yeast enzyme. Perhaps the lack of structural asymmetry in the $E(NAD^+)_4$ structure is a consequence of the temperature conditions used in the structure determination?

Wonacott: Certainly NAD^+ binding appears to be temperature dependent, the onset of hyperbolic binding being associated with increasing temperature. Negative cooperativity of binding is observed at room temperature, in this case well below that at which the organism normally grows. Crystals have been examined over a wide range of temperatures from -10° up to $+50^o$ C and no changes in their diffraction pattern were apparent.

Hinz: Is it possible that the reason for not seeing any conformational changes in X-ray investigations at different temperatures resides in the crystal forces and not in an intrinsic invariance of the protein conformation with regard to temperature?

Wonacott: Although crystal forces should not be overlooked, we do not believe that they are a very strong constraining influence on the protein in this case. NAD^+ can be dialysed into and out of the crystals with associated changes in both the unit cell dimensions and the diffraction pattern, so that it appears that the molecule is capable of undergoing conformational changes in the crystal.

Seydoux: I think that it is important in general to distinguish between two types of conformational changes: a very large one which arises from the filling of the site by NAD^+ and a more subtle one which can be transmitted from one subunit to an other. Anticooperativity can result only from second type of conformational change which is more difficult to detect in crystallographic studies.

Wonacott: I think that the two kinds of changes which you refer to are closely related. Negative cooperativity of NAD^+ binding could result from the large conformational changes. Binding of substrates at one active site which results in half-site reactivity is likely to produce smaller conformational changes but these would be transmitted to other subunits by much the same mechanism, that is changes in the packing of the S-loop core of the tetramer.

Bernhard: The ligand-binding properties of *B. stearothermophilus* GPDH is like that for sturgeon muscle as contrasted with rabbit muscle. Thus far we have found only Michaelian binding for NAD^+ analogs to sturgeon muscle GPDH with ε-NAD^+, 8-BrNAD^+ and ADPR. Extrapolating from sturgeon to *B. stearothermophilus*, the anticooperative properties of NAD^+-binding do not reside in ligand-induced structural changes which are equally well-induced by both NAD^+ and the above mentioned NAD^+-analogs.

Wonacott: From measurements of NAD^+ binding to *B. stearothermophilus* GPDH it appears to be similar to that from rabbit muscle. So that I would not agree with the extrapolation that you are suggesting.

Saenger: We should note that the conformation of NAD^+ in this holoenzyme displays significant defferences with respect to NAD^+ in other holoenzymes. While the adenosine moiety is similar as in the other holoenzymes, the nicotinamide-riboside conformation is *gauche, gauche* about the C(4')-C(5') bond instead of *trans, gauche*. Further the amide group is rotated by 180° with the C-NH_2 bond being *cis*-planar to the C(2)-C(3) bond. This could imply that a hydrogen bonding interaction between the amide amino group and the phosphate group attached to nicotinamide-riboside becomes possible.

Harris: Quite apart from the crystallographic evidence there are interesting chemical differences between apo and holo GPDH that are indicative of significant conformational changes in three-dimensional structure. Thus for example with muscle and yeast enzymes there is the S → N migration of acyl groups, from Cys-149 to Lys-183 in the S-loop, that occurs in apo- but not in holoenzyme. Then again the *B. stearothermophilus* holoenzyme can be extensively succinylated without loss of activity, whereas the apoenzyme is rapidly inhibited by small amounts of succinic anhydride due to specific reaction with Thr-150 which is adjacent to the active site cysteine.

Rossmann: 1.) I would like to acknowledge my co-workers (referred to frequently by Dr. Wonacott). They are Manfred Bühner, Ken Olsen, Dino Moras and Mike Garavito.

2.) I would also like to comment on the asymmetry of lobster GPD found by alkylation with trifluoro acetonyl (TFA):

$$-S-CH_2-\overset{\overset{\displaystyle O}{\|}}{C}-CF_3$$

In difference electron density maps between the alkylated crystals and the native holo crystals a different conformation of the TFA adduct is clearly seen between subunits related by the R-axis. This is consistent with Bode et al. (J. Bode, M. Blumenstein & M.A. Raftery, Biochemistry (1975) 14, 1153-1110) observation of two environments for the CF_3 groups and with the observation that co-operativity is across the R-axis.

3.) The finding by Dr. Wonacott that there is some asymmetry of the adenine group in the holoenzyme across the R-axis is partially consistent with our tentative results. It also agrees with the results of Alex Levitzky that the co-operativity resides in the adenine end of the coenzyme. We have suggested that the mechanism of information transfer between subunits is by way of Pro-188 in the S-shaped loop. This residue in the green subunit closes the hydrophobic pocket of adenosine in the red subunit.

Perucho: I would like to mention some results that we have on glyceraldehyde 3-phosphate dehydrogenase. This work was done at the Center of Molecular Biology, C.S.I.C., U.A.M., Madrid, Spain, by Manuel Perucho, José Salas and Maria L. Salas. We were studying the synthesis of DNA-binding proteins in cultured mammalian cells. One of these proteins (P8) had several interesting properties. Thus, it was synthesized in large amounts in growing fibroblasts of line nil but its synthesis decreased considerably in resting cultures. In contrast to the behavior of P8 in normal cells, the synthesis of this protein was kept at high levels in dense cultures of virus transformed cells. It was also found that P8 was retained on columns of single-stranded DNA-cellulose, but not of double-stranded DNA-cellulose.

In order to investigate the nature of P8 we purified to homogeneity this protein from a virus-transformed cell line, using chromatography on DNA-cellulose and phosphocellulose. Much to our surprise, we identified P8 as glyceraldehyde 3-phosphate dehydrogenase by a number of criteria. Since the purification method of P8 is based on the finding that this protein is retained on columns of denatured but not native DNA-cellulose, we tested the affinity of P8 and hamster muscle GPDH, purified by conventional methods, for DNA using a millipore filtration technique. Both P8 and hamster muscle GPDH bind with the same efficiency to denatured DNA. In contrast, very little binding of these proteins to native DNA is found. We also tested a commercial preparation (from Sigma) or rabbit muscle GPDH and found that it also binds to single-stranded DNA.

We then tested whether NAD, NADH, NADP, ATP, AMP and glyceraldehyde 3-phosphate could inhibit the binding of DNA to the enzyme. When these compounds are tested at a concentration of 0.01 mM, NAD and NADH, but not the other compounds, strongly inhibit the binding of denatured DNA to GPDH. ATP and AMP at higher concentrations (1 mM) also inhibit the binding of DNA to the enzyme. In contrast, with 1 mM glyceraldehyde 3-phosphate or 5 mM inorganic phosphate no inhibition was observed. These results are compatible with the possibility that the interaction of GPDH with DNA takes place through the NAD binding sites of the enzyme.

The Unimer Model of Glutamate Dehydrogenase: A Verification Using Chemical Modification

I. Rasched, A. Bohn, D. Peetz, and H. Sund
Fachbereich Biologie, Universität Konstanz
D - 7750 Konstanz, West Germany

INTRODUCTION

One of the relevant problems of enzymology is the relation between the catalytic properties of an enzyme and its chemical structure, and, consequently, the folding of its polypeptide chain. An additional problem of enzymes having a quaternary structure is the geometrical arrangement of the polypeptide chains, the nature of the noncovalent bonds responsible for the stabilization of the quaternary structure and the change of the conformation of the polypeptide chains upon association. In the last 15 years the X-ray analysis of many enzymes and other proteins has provided us with detailed insight into the nature of these problems. Unfortunately, this analysis is restricted to those proteins forming suitable crystals.

In the case of glutamate dehydrogenase from beef liver it was not possible to obtain crystals suitable for X-ray analysis. This enzyme exhibits in solution an association-dissociation equilibrium with the association in the direction of the major axis of the elongated unimer (for a description of the unimer, see below) forming rodlike particles [1]. The particles form helical tubes before they are incorporated into the crystals, i.e. crystal growth is achieved mainly by association of the helical tubes and not by single unimers. As the tubes are morphologically heterogeneous it is assumed [2] that the crystals are inhomogeneous which may limit the growth of crystals large enough to be used in X-ray analysis [2].

An alternative (but much less unambiguous) approach to solve the problem mentioned above is the selective chemical modification of functional groups. Using the appropriate chemical reagent one may characterize particular functional groups of the enzyme with respect to its kinetic properties, its topography [3] or even more to elucidate the arrangement and symmetry of the polypeptide chains in the quaternary structure [4, 5]. In this context the amidination reaction has proved to be the method of choice for the following reasons:

1.) the reaction is specific for amino groups and can be performed under mild conditions [6]

2.) the amidinated protein obtained does not change its overall charge, and proves to maintain its native conformation and

3.) the reagents are easy to prepare, the monofunctional as well as the bifunctional ones with different carbon chain lengths.

The unimer of glutamate dehydrogenase (molecular weight 336 000) contains six identical polypeptide chains (molecular weight 56 000). It is the smallest enzymatically active form of the enzyme [7]. On the basis of small-angle X-ray experiments [8], light-scattering measurements [9] and electron microscope studies [1] a physical model for the unimer has been proposed (Fig. 1) consisting of two layers, each composed of three polypeptide chains arranged in a triangular fashion. The two layers form an elongated particle (unimer), which associates in the direction of the major axes of the prolate ellipsoids.

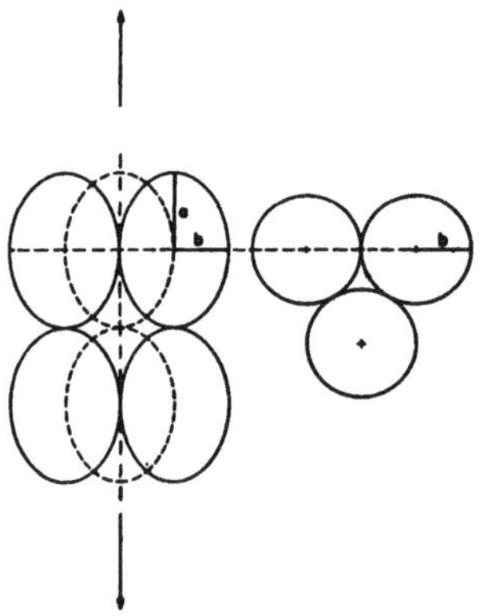

Fig. 1 Macroscopic schematic model of the unimer of glutamate dehydrogenase as proposed by Eisenberg and Reisler [9]. a = 33.25 Å, b = 21.50 Å. The cross-section shows a three-fold symmetry arrangement of the half-unimer.

This model implies three different interface domains:

1.) interfaces between the three polypeptide chains within a layer (half of the unimer),

2.) interfaces between the two layers within the unimer and

3.) interfaces between the associated unimers, which we may call the "super quaternary structure" of this enzyme.

In fact, as will be shown in the following, the modification with bis-(imido esters) discriminates between two different interface domains within the unimer. The data from the quantitative and sequence analyses of the cross-linked peptides within one layer reveal a striking feature of the geometrical arrangement of three polypeptide chains; these can hardly be arranged around a three-fold axis of rotational symmetry.

We report finally some preliminary results concerning the modification of a side chain group using a "hydrophobic sites" directed label. This modification does not alter the enzymic activity but is concomitant with the dissociation of the "super quaternary structure" suggesting a critical role of the modified residue for the association properties of the unimer.

RESULTS AND DISCUSSION

Cross-Linking with [^{14}C]Dimethyladipimidate

Fig. 1 shows the schematic model proposed by Eisenberg and Reisler [9] for the unimer of glutamate dehydrogenase. The assembly of the polypeptide chains in two layers of three (Fig. 1) is supported by cross-linking experiments reported by Hucho and Janda [10], which show that the trimeric species is preferencially cross-linked. Our work has been directed toward the elucidation of the arrangement of the three polypeptide chains within the half-unimer. The scheme in Fig. 1 proposes a symmetric arrangement of the three chains, meaning identical and heterologous (for definition see [11]) interface domains between them.

The time course of the cross-linking reaction with dimethyladipimidate

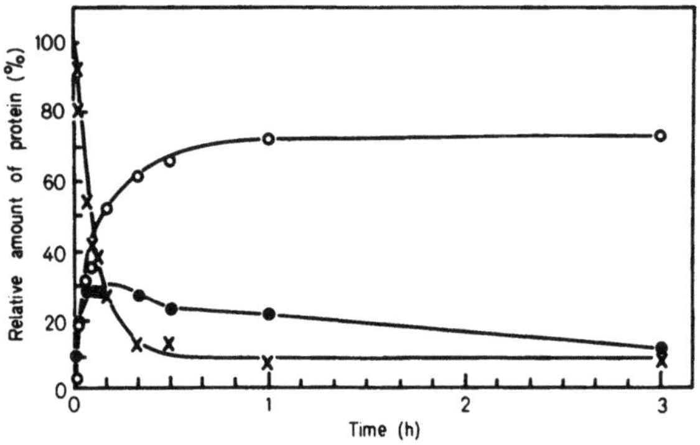

Fig. 2 Time course of the cross-linking reaction of glutamate dehydrogenase with dimethyladipimidate. 0.5 mg/ml protein was incubated with 3 mg/ml (12.2 mM) dimethyladipimidate in 0.1 M sodium pyrophosphate/HCl buffer, pH 9.4. The reaction was stopped at the indicated times as described in [5]. 5 % sodium dodecylsulfate/polyacrylamide gels were run for every sample, stained and scanned at 550 nm. Percentages of the relative amount of total protein are calculated by integration of the peak areas. Relative amounts of trimer (o), dimer (•) and monomer (x).

(Fig. 2) shows that the trimeric species is formed immediately at the beginning of the reaction. This means that at least two interface domains are simultaneously cross-linked, suggesting an identical structure of at least two interface domains or a very similar reactivity of the involved lysine residues in structurally different interface domains.

The sequence analysis of the peptides involved in the cross-links may help to solve the problem. Fig. 3 shows the quantitative determination of the number of modified amino groups and incorporated [^{14}C]dimethyladipimidate. The number of cross-links/polypeptide chain can be calculated from the difference, which is about 0.9 mol/polypeptide chain after one hour of reaction. This means that about 0.9 mol dimethyladipimidate has reacted bifunctionally. For the formation of a trimer two interchain cross-links are needed, i.e. 0.67 mol reagent/polypeptide chain.

Fig. 4 shows the tryptic and chymotryptic fingerprints of a sample prepared as mentioned in the legend. Noteworthy is the specificity of the

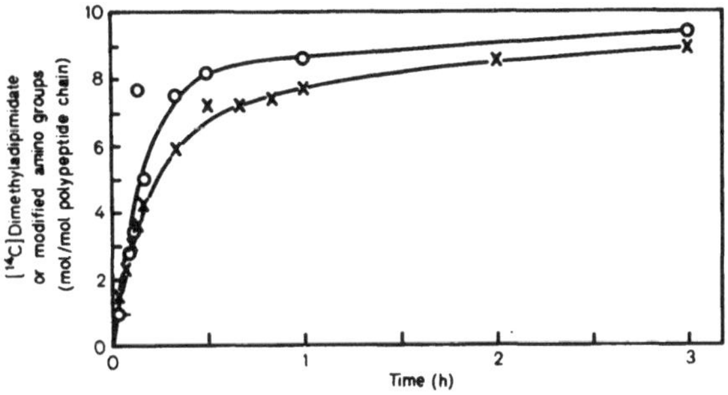

Fig. 3 Modification of amino groups and incorporation of [^{14}C]dimethyladipimidate during the cross-linking reaction. Reaction conditions: 0.5 mg/ml protein in 0.1 M sodium pyrophosphate/HCl buffer, pH 9.4. The [^{14}C]dimethyladipimidate concentration was 3 mg/ml, further details see [5]. (x) The number of mol of [^{14}C]dimethyladipimidate/mol polypeptide chain incorporated during the cross-linking reaction: (o) the number of mol modified amino groups/mol polypeptide chains.

amidination reaction of glutamate dehydrogenase as displayed by the relatively few labeled peptides. Ammonolysis removes almost all of the label from the peptides. Peptides 3 and 6 subsequently split each in two peptides

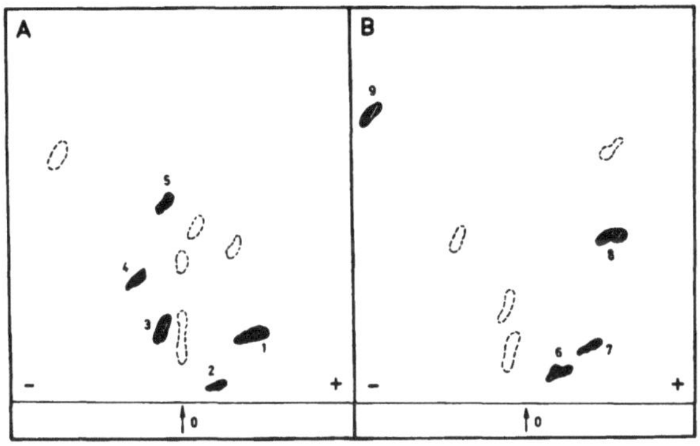

Fig. 4 Peptide maps of glutamate dehydrogenase reacted with [^{14}C]dimethyladipimidate. (A) Map from tryptic digest of carboxymethylated protein, (B) map from chymotryptic digest of carboxymethylated protein. Only radioactive peptides (detected by autoradiography) are shown. Strong spots are cross-hatched, faint spots are enclosed with broken lines. The sample was applied to the origin (0), subjected to electrophoresis at pH 6.5, and then chromatographed in pyridine/butanol/acetic acid/water.

with opposite charge, as can be detected by high voltage electrophoresis at pH 6.5. The other peptides were not altered in their electrophoretic mobility, see Fig. 9 in [5].

A discrimination between isologous (cross-linking of two identical peptides) and monosubstituted peptides is possible either by quantitative determination of the incorporated label/mol peptide (or lysine) or by identification of N^{ε},N^{ε}-adipamidino-bis-lysine in the peptide. The determination of the bis-lysine derivative turned out to be ambiguous due to its destruction during acid hydrolysis and its poor color yield with ninhydrin. Much more reliable was the quantitisation of the label/mol peptide by scintillation counting. The results of which are shown in Table 1. Peptide 1 carries 0.5 mol label/mol, which means an isologous cross-link. The same conclusion can be drawn for peptides 7 and 2. The peptides 3 and 6 carry also half an

Table 1 Distribution of the radioactivity among the tryptic and chymotryptic peptides of glutamate dehydrogenase reacted with [^{14}C]dimethyladipimidate.

Peptide	Lysine residue/ peptide*	Radioactivity/ lysine**	[^{14}C]Dimethyl-adipimidate***
		cpm/nmol	mol/mol peptide
1	2	140	0.48
2	1	270	0.46
3	4	65	0.45
4	1	595	1.02
5	0	550	0.95
6	3	110	0.57
7	1	295	0.51
8	0	610	1.05
9	1	560	0.96

* Number of lysine residues as expected from primary structure [12]

** Scintillation counting was performed with aliquots prior to ammonolysis

*** 1 nmol [^{14}C]dimethyladipimidate contains 580 counts/min.

Table 2 Sequence analysis of the carboxymethylated tryptic and chymotryptic peptides from [^{14}C]dimethyladipimidate-reacted glutamate dehydrogenase after ammonolysis. ⟶ indicates sequence analysed by the dansyl-Edman method.
This information together with the amino acid composition of the peptides [5] is sufficient for their identification.

Peptides	
1	Phe-Gly-Ala-Lys(269)-Cys(Cm)-Val ....Lys
2	Ile-Ile-Ala-Glx-Gly-Ala ....Lys(358) ....Lys
3	Tyr-Ser-Thr-Asx-Val ....Lys(105) ....Lys
3'	Phe-Thr-Met-Glx ....Lys(154)-Lys(155)
4	Leu-Thr-Phe-Lys(399) ....Arg
5	Ala(1)-Asp-Arg
6	Ser(96)-Thr-Asx-Val-Ser-Val ....Lys(105) .....Tyr
6'	Thr(149)-Met-Glx-Leu-Ala ....Lys(154)-Lys(155) ....Phe
7	Gly(267)-Ala-Lys(269)-Cys(Cm)-Val-Ala-Val ....Trp
8	Ala(1)-Asx-Arg-Glx-Asx ....Phe
9	Lys(399)-Tyr

equivalent cross-linker/mol, but they split after ammonolysis into two different peptides, suggesting a heterologous cross-link.

The cross-linker occurs in all other peptides in a 1:1 ratio, which means that these peptides are not really cross-linked but substituted monofunctionally. Table 2 shows the identification of the lysine residues involved in the reaction with [^{14}C]dimethyladipimidate.

The results presented can be summarized as follows:

1. Three peptides are involved in the cross-linking reaction, two of them are obviously cross-links between identical (isologous) lysine residues in different polypeptide chains, Lys 269 and Lys 358 whereas Lys 114 is cross-linked to Lys 154 or 155 (heterologous).

2. In case of the heterologous cross-link no differentiation can be made between interchain and intrachain cross-links, since two linkages are sufficient to build the trimeric species.

Conclusions: The fact that two isologous cross-linked peptides have been isolated from the trimeric species of glutamate dehydrogenase reacted with [^{14}C]dimethyladipimidate may be due to different reasons:

1. The interface domains within the half-unimer are really not identical, which implies that it does not possess a three-fold axis of rotational symmetry. In other words the three polypeptide chains in the half-unimer are not symmetrically arranged, and therefore isologous cross-links may occur [5].

2. The reactivity as well as the distance between two Lysines 269 and two Lysines 358 in different polypeptide chains of the half-unimer are particularly fitted for the reaction with dimethyladipimidate, regardless of the specific geometrical arrangement of the polypeptide chains. Thus, any interpretation of the results with respect to the symmetry in the half-unimer would be fortuitous.

3. The isologous cross-linked peptides are a by-product of the reaction, and the heterologous cross-linked peptide is the expected main-product (according to the symmetrical arrangement of the polypeptide chains shown in Fig. 1). This is not probable because the isologous peptides as well as the heterologous peptide each accounts for approximately 15 % of the radioactivity incorporated, see Table 4 in [5].

Investigation of the "Super Quaternary" Structure of Glutamate Dehydrogenase

The ability of the enzyme to associate along the three-fold symmetry axis of the unimer has been mentioned above. Eisenberg et al. [1] assumed that the rate-determining step in self-assembly involves breakdown of ordered water-structure associated with a hydrophobic region of the protein surface. This presumption is supported by the temperature dependence of the association and the role of aromatic hydrocarbons, which promot the association. Furthermore the property of association has been related to some specific conformation of the associating end surfaces. The association behavior is highly sensitive to certain biologically active compounds as nucleotides, amino acids and hormones. Those which enhance association also enhance the activity and vice versa. Therefore one may argue that the unimer domains involved in association have a specific structure, very probably hydrophobic as proposed in [1], stabilized by a specific conformation different from that of not associating particles. A shift in the accessibility of functional groups of the unimer to specific chemical reagents concomitant with a shift of the association behavior would monitor the existence of different conformations of the unimer. The modification reaction proposed in the following section is an approach to check the validity of this presumption.

A modification of -SH groups seemed to be the most straight-forward approach because these groups have been shown to be inaccessible to all common alkylating reagents [13]. This indicates that they are probably buried or located in a hydrophobic environment of the polypeptide chains. Using an alkylating reagent adapted to hydrophobic sites should circumvent this barrier. We propose N-(1-pyrene)maleimide (Fig. 5A) as the reagent of choice for this purpose. It was first synthesized by Haugland [14]. It does not fluoresce in aqueous solution but forms fluorescent covalent adducts with proteins which contain -SH groups (Fig. 5B) [15]. The fluorescence emission spectrum is shown in Fig. 6, the derivative N-(1-pyrene)succinimide (PS) (Fig. 5C) shows the same spectrum.

In order to eliminate the loss of quantum yield by absorbance effects of the reactants at the excitation wave lengths (342 nm), the following equation has been developed (Fig. 7)

$$\ln \frac{F}{c} = \ln \frac{F_o}{c} - B \cdot E_{exc.} \quad \text{when } E_{emission} \approx 0$$

where F is the measured fluorescence at 379 nm, F_0 the absolute fluorescence

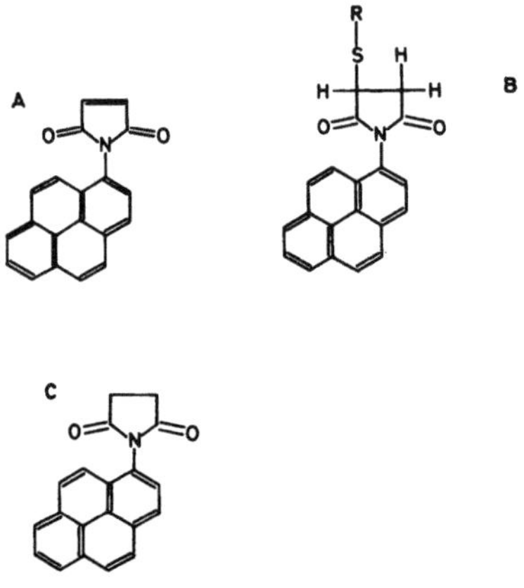

Fig. 5 Structures of N-(1-pyrene)maleimide (PM, A), adduct of PM with an -SH group in proteins or with cysteine, (B), and N-1(1-pyrene)succinimide (PS, C).

(extrapolated to $E_{exc.} = 0$), c the concentration of the fluorescence material and $E_{exc.}$ the absorbance of the solution at 342 n. B is a constant dependent on the measuring instrument.

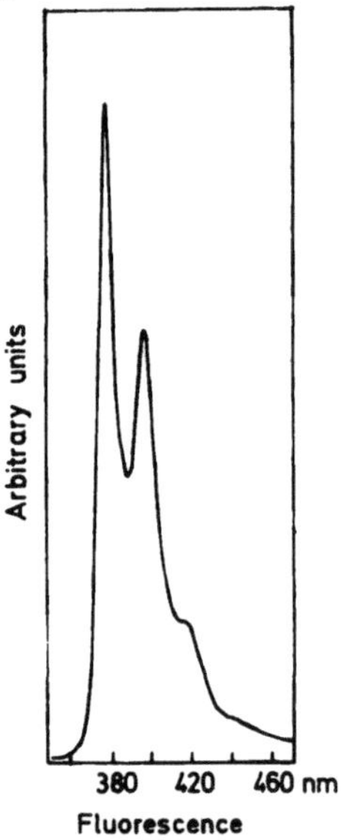

Fig. 6 Fluorescence emission spectrum of the PM-glutamate dehydrogenase adduct. The same spectrum is displayed by PS. The excitation wavelength was 342 nm in 0.067 M phosphate buffer, pH 7.6 (1% aceton) at 20° C.

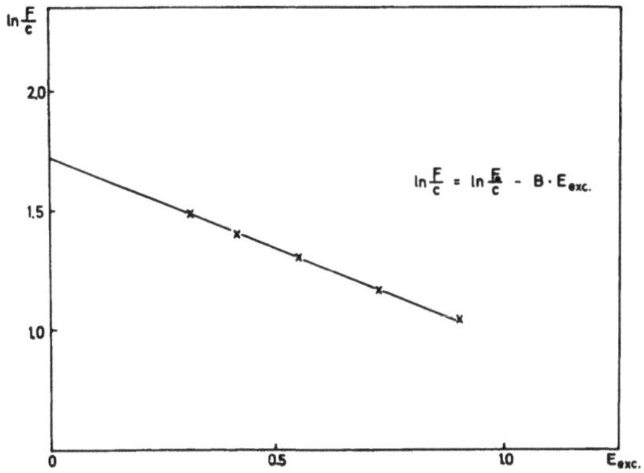

Fig. 7 Fluorescence F/c (in arbitrary units) at different solute concentrations plotted against the corresponding absorbance ($E_{exc.}$) of the solution at 342 nm, in order to determine the absolute fluorescence ($E_{exc.} = 0$).

Fig. 8 shows that the rate constant of glutamate dehydrogenase with PM is less than first order, which means that the rate of the noncovalent addition of PM to the binding site is higher than the rate of covalent binding.

The results in Fig. 9 support this finding as PS, which cannot react covalently, completely inhibits the covalent incorporation of PM with glutamate dehydrogenase.

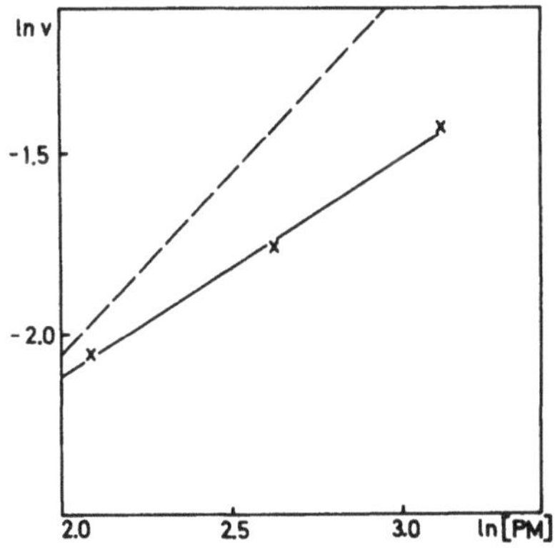

Fig. 8 Determination of the rate constant of the PM-glutamate dehydrogenase reaction with respect to PM. The logarithm of the initial rate v is plotted against the PM concentration. Calculated slope for first-order reaction rate (---), reaction of glutamate dehydrogenase with PM (x).

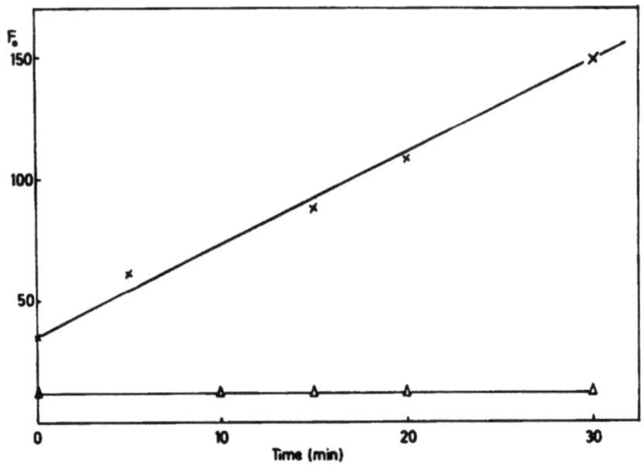

Fig. 9 Time course of the reaction of glutamate dehydrogenase with PM in the presence and absence of PS. Reaction conditions: 4 μM glutamate dehydrogenase, 0.067 M phosphate buffer, pH 7.6, 1 % aceton at 20° C.

x-x-x = reaction with 10 μM PM

Δ - Δ = the solution was made 10 μM in PS and than 10 μM in PM

F_o = fluorescence in arbitrary units in presence of PS, corrected for its background fluorescence.

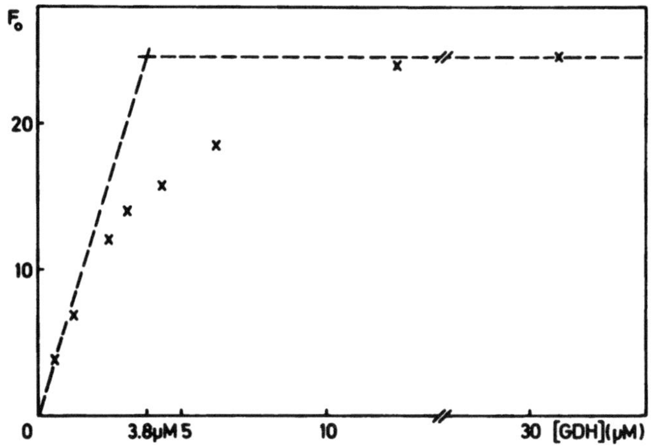

Fig. 10 Stoichiometry of the reaction of glutamate dehydrogenase with PM in 0.067 M phosphate buffer, pH 7.6 (1% aceton) at 20° C. PM 2.5 μM, measured fluorescence at different enzyme concentrations (x), extrapolation to the point of equivalence corresponding to 3.8 μM polypeptide chain (---).

Fig. 10 shows that with 2.5 μM PM about 3.8 μM polypeptide chain of glutamate dehydrogenase is covalently labeled. This excludes that more than one PM molecule per polypeptide chain has reacted. Neither the enzymic activity nor the allosteric properties (ADP activation, GTP inhibition) of the enzyme are altered by the modification.

The sedimentation pattern of PM and PS treated glutamate dehydrogenase show that dissociation of the associated particles into the 12.9 S component (unimer) occurs only with PM indicating that dissociation is coupled with the covalent modification of one amino acid residue (Fig. 11).

The isolation and identification of the PM labeled residue in the amino acid sequence of glutamate dehydrogenase is the object of our current research.

Conclusions: Using the alkylating reagent N-(1-pyrene)maleimide we were able to show that glutamate dehydrogenase can be dissociated into its unimer without alteration of either its catalytic or its regulatory proper-

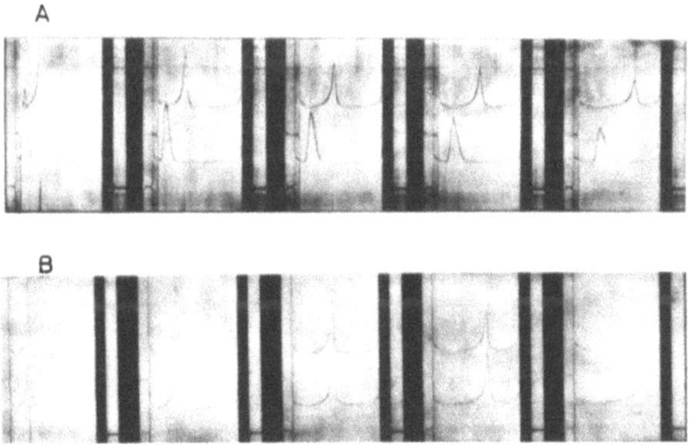

Fig. 11 Sedimentation pattern of PM reacted (A) and PS treated (B) glutamate dehydrogenase. Measurements in 0.067 M phosphate buffer, pH 7.6 and 1 % aceton at 68 000 rev/min and 20° C in 12 mm normal and wedge-window cells, angle of the schlieren diagram 60°. Pictures were taken 16 min after the rotor had reached full speed. Upper curve in A and B (wedge cell) native enzyme (6 mg/ml), lower curve in A glutamate dehydrogenase (6 mg/ml) reacted with 2-fold molar excess PM, in B glutamate dehydrogenase (6 mg/ml) in the presence of 2-fold molar excess PS.

ties; thus proving that this association phenomenon is independent of its other enzymic properties.

The suppression of the reaction of glutamate dehydrogenase with PM in presence of PS proves that PM specifically reacts with the residue located in the hydrophobic environment to which both reagents very probably are directed by their specific molecular structure, i.e. their hydrophobic moiety, the pyrene.

Acknowledgment: The authors thank Mrs. U. Markau for her excellent technical assistance. This work was supported by grants from the Deutsche Forschungsgemeinschaft (Sonderforschungsbereich 138: Biologische Grenzflächen und Spezifität) and the Fonds der Chemischen Industrie.

REFERENCES

1. H. Eisenberg, R. Josephs, and E. Reisler: Bovine Liver Glutamate Dehydrogenase. Adv. Protein Chemistry 30, 101-181 (1976).

2. R. Josephs: An Analysis of the Mechanism of Crystallization of Glutamic Dehydrogenase. J. Mol. Biol. 97, 127-130 (1975).

3. J.M. Lambert and R.N. Perham: Folding Domains and Intramolecular Ionic Interactions of Lysine Residues in Glycerinaldehyde-3-Phosphate Dehydrogenase. Biochem. J. 161, 49-62 (1977).

4. F. Hucho, H. Müllner and H. Sund: Investigation of the Symmetry of Oligomeric Enzymes with Bifunctional Reagents. Eur. J. Biochem. 59, 79-87 (1975).

5. I.R. Rasched, A. Bohn and H. Sund: Studies of Glutamate Dehydrogenase. Analysis of Quaternary Structure and Contact Areas between the Polypeptide Chains. Eur. J. Biochem. 74, 365-377 (1977).

6. M.J. Hunter and M.L. Ludwig: Amidination. Methods in Enzymology 25, 585-596 (1972).

7. H. Eisenberg and G.M. Tomkins: Molecular Weight of the Subunits, Oligomeric and Associated Forms of Bovine Liver Glutamate Dehydrogenase. J. Mol. Biol. 31, 37-49 (1968).

8. H. Sund, I. Pilz and M. Herbst: The X-Ray Small-Angle Investigation of Beef Liver Glutamate Dehydrogenase. Eur. J. Biochem. 7, 517-525 (1969).

9. H. Eisenberg and E. Reisler: A Physical Model for the Structure of Glutamate Dehydrogenase. Biopolymers 9, 113-115 (1970).

10. F. Hucho and M. Janda: Investigation of the Quaternary Structure of Beef Liver Glutamate Dehydrogenase with Bifunctional Reagents. Biochem. Biophys. Res. Commun. 57, 1080-1088 (1974).

11. A.J. Cornish-Bowden and D.E. Koshland: The Quaternary Structure of Proteins Composed of Identical Subunits. J. Biol. Chem. 246, 3092-3102 (1971).

12. K. Moon and E.L. Smith: Sequence of Bovine Liver Glutamate Dehydrogenase. VIII. Peptides Produced by Specific Chemical Cleavages; The Complete Sequence of the Protein. J. Biol. Chem. 248, 3082-3088 (1973).

13. F. Hucho, I. Rasched and H. Sund: Studies of Glutamate Dehydrogenase. Analysis of Functional Areas and Functional Groups. Eur. J. Biochem. 52, 221-230 (1975).

14. R.P. Haugland, Ph.D. Thesis, Stanford University, Palo Alto (1970).

15. F.Y. Wu, L.R. Yarbrough and C.-W. Wu: Conformational Transition of Escherichia coli RNA Polymerase Induced by the Interaction of σ Subunit with Core Enzyme. Biochemistry 15, 3254-3258 (1976).

Received March 11, 1977

DISCUSSION

Werber: What is the evidence that the cysteine residue modified with the maleimide derivative is in a hydrophobic environment?

Rasched: It is an indirect evidence. The fact is that water soluble ethylmaleimide does not react with any of the -SH groups of the enzyme neither does it compete with the hydrophobic pyrene maleimide which reacts with one -SH group per polypeptide chain.

I think that the pyrene moiety of the molecule plays a crucial role in piloting the maleimide part of the molecule toward the -SH groups. As pyrene is extremely hydrophobic it is easy to argue that the site to which it directs the maleimide moiety is lipophilic.

Veeger: Can you completely exclude the possibility of intra-peptide crosslinking? It could be the source of induced asymmetry of the trimer.

Rasched: I cannot exclude intra-chain cross-links; but homologous crosslinks can only be due to inter-chain cross-linking.

Boos: Is it possible that heterologous peptides arise from cross-links from within the trimer and the homologous peptides from between the trimer layer?

Rasched: This cannot be completely excluded. Although it is not probably because cross-links between the trimer layers only occur when suberimidate is used, i.e. two carbon atoms longer than the dimethyladipimidate used in our experiments.

Kiltz: As you mentioned the non identified peptide, how sure are you, that you are using the same enzyme in respect to animal race, organ etc., as it has been sequenced?

Rasched: All I can say is that we are using the bovine liver enzyme purchased from Boehringer, Mannheim. E.L. Smith used for the sequence the enzyme from the same source and the same producer.

Walker: Is the maleimide reagent soluble in water? What is the molar excess of reagent over protein?

Rasched: The pyrene maleimide is not soluble in water, but partly soluble in buffer containing 1 % or more aceton. The molar excess of reagent over -SH groups per polypeptide chain is generally 1.5.

Harris: Could you elaborate on the difference between the peptide that you have isolated and the published sequence of glutamic dehydrogenase? I presume that you are working with the enzyme from the same species and tissue? What's the size of the peptide and was it isolated in high yield? Is the yield of the peptide compatible with the level of radioactivity incorporated into the native enzyme?

Rasched: As I just answered Dr. Kiltz, we are working with the same enzyme as E.L. Smith and collaborators. The size of the peptide is between 12 and 14 amino acids, with a high content of hydrophobic amino acids. Almost all the incorporated radioactivity is recovered with this peptide, when normalized to the overall yield which is relatively low (about 8 %).

Brändén: You are making the assumption when you interpret your data that the subunits within the trimer cannot be arranged in such a way that you can get homologous cross-linking. Have you checked if this is the case in the recently published X-ray structure of a trimeric aldolase by Tulinsky?

Rasched:You mean the folding of the trimeric 2-keto-3-deoxy-6-phosphogluconic aldolase. This enzyme has the first well-documented quaternary structure containing three polypeptide chains. But as far as I understood the published structure of this enzyme, only heterologous, cross-links between the polypeptide chains could be expected as it has a three-fold rotation axis and the intersubunits contacts are between heterologous regions of the polypeptide chains.

THE IMMOBILIZATION TECHNIQUE AS AN AID IN THE STUDY OF THE QUATERNARY STRUCTURE OF DEHYDROGENASES WITH SPECIAL REFERENCE TO SUBUNIT ASSOCIATION AND ALLOSTERIC REGULATION

K. Mosbach and L. Andersson
Biochemistry 2, Chemical Center, University of Lund, S-220 07 Lund 7, SWEDEN

The immobilization of enzymes, as primarily utilized in enzyme technology (1), and that of small ligands, as used in affinity chromatography (2), has received great attention during the last few years. In the following we wish to focus the reader's attention to an additional aspect of immobilization, namely, the use of this technique per se in helping to answer questions in fundamental biochemistry. The examples that will be discussed here concern pyridine nucleotide-dependent dehydrogenases and in particular the question as to what effects the quaternary structures of these enzymes have on activity.

A. Rabbit Muscle Lactate Dehydrogenase

The two first examples to be discussed concern lactate and alcohol dehydrogenase and the question whether isolated subunits are enzymically active or whether they can regenerate activity by interacting with subunits in solution. The immobilization technique has proven of value in such studies since for many enzymes the native oligomeric structures are so stable that severe conditions are necessary to cause dissociation. Subsequent return to non-dissociating conditions then leads to spontaneous reassociation into the native structure. Therefore the oligomeric and monomeric forms cannot be compared under normal assay conditions.

The first example on the use of this technique as a tool to answer such questions (for a general discussion of the technique see reference 3) involved studies on rabbit muscle aldolase in which context it could be demonstrated that the immobilized subunit in fact is active (4).

Also pyridine nucleotide-dependent dehydrogenases such as lactate dehydrogenase (LDH) have been studied using this approach in part because of apparently conflicting results found in the literature concerning LDH monomer activity (5-7).

The following approaches, as depicted in Fig. 1 a and b, were chosen (8). In one case rabbit muscle lactate dehydrogenase was coupled to weakly CNBr-activated Sepharose in such a way that each molecule is expected to be attached via only one subunit. Subsequent dissociation of the bound active enzyme by several methods including dissociation at pH 3.0 and use of 6 M guanidine hydrochloride all yielded immobilized subunit derivatives that were inactive. However, these derivatives were capable of regenerating activity by interacting specifically with "refolded" subunits in solution formed transiently during renaturation.

In the other alternative approach LDH was attached to Sepharose via disulfide bonds and again, following dissociation, the immobilized subunit was found to be inactive. After reduction with mercaptoethanol, leading to detachment of the immobilized subunit, activity was regenerated and as judged from the kinetics of this reactivation process, reassociation is required for appearance of activity. All these results strongly indicate that subunit interactions are essential for LDH activity and as a by-product of the studies above it was found that the immobilized subunits had retained their capacity to reassociate with soluble subunits.

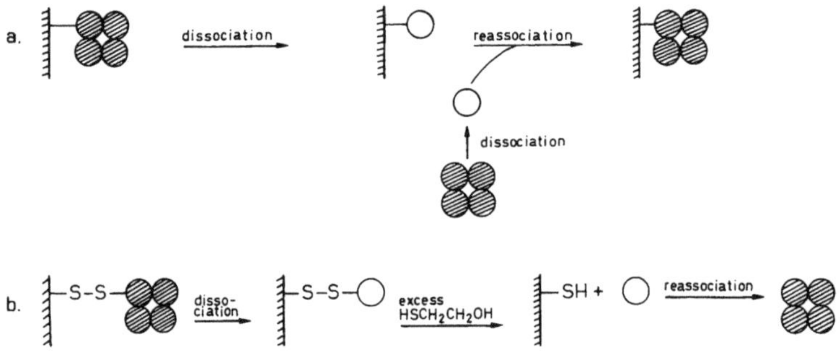

Fig. 1 (a) Reaction scheme for the coupling of LDH to weakly CNBr-activated Sepharose followed by dissociation and reassociation with soluble monomers. (b) Reaction scheme for the coupling of LDH to Sepharose via disulfide bonds and the subsequent dissociation of the bound enzyme and detachment of the subunits. Shaded circles denote activity.

B. Liver Alcohol Dehydrogenase

The possible requirement of subunit interaction for enzyme activity has also recently been investigated for alcohol dehydrogenase from horse liver (HLADH) (9). As known, this dimeric enzyme occurs in the EE, ES or SS forms (10). All three are active toward ethanol as substrate, but only the ES and SS forms are active toward 3β-hydroxysteroids (11). The same principle procedure as employed for LDH was used as outlined in Fig. 2.

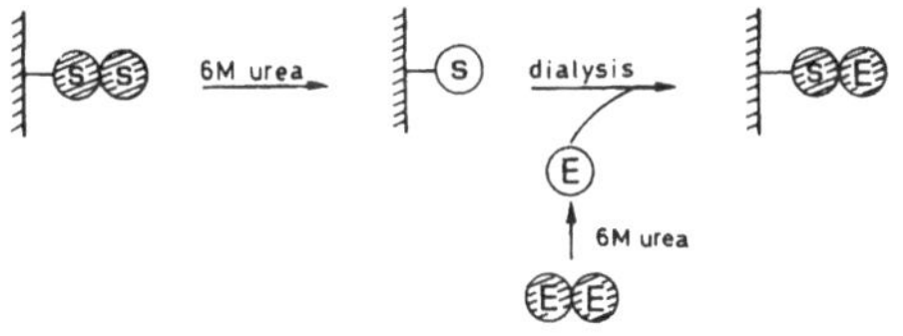

Fig. 2 Reaction scheme for the coupling of the steroid active SS isozyme of HLADH to weakly CNBr-activated Sepharose followed by dissociation with 6 M urea at 4°C and regeneration of steroid activity at room temperature with soluble subunits of the EE form of HLADH. Shaded circles denote activity.

Pure SS alcohol dehydrogenase, obtained by affinity chromatography on immobilized AMP (12), was bound to weakly CNBr-activated Sepharose. After dissociation with 6 M urea, the preparation had lost practically all its activity. On subsequent addition of soluble subunits of the E-form, the immobilized S-subunit regained its 3β-hydroxysteroid activity. The ratio between ethanol and 3β-hydroxysteroid activity for the bound "redimerized" enzyme was 5.9, a value close to that found for the corresponding native, soluble ES isozyme (about 4 (10)). The successful hybridization on a solid-phase shows that not only is the immobilized S-subunit, although inactive in itself, capable of specifically picking up soluble subunits, but in addition, in doing so, the immobilized subunit regains its original steroid activity.

C. Beef Liver Glutamate Dehydrogenase

In the following an additional example on the usefulness of the immobilization technique per se for studies relating to the quaternary structure of dehydrogenases is given. Glutamate dehydrogenase is known as an enzyme in which ligand binding is accompanied by association or dissociation of the oligomeric enzyme. Here immobilization has been shown to offer a means for "uncoupling" these interactions so that those changes associated with ligand binding can be assessed independently from those resulting from protein subunit association-dissociation (13).

Beef liver glutamate dehydrogenase is composed of six identical polypeptide chains corresponding each to a molecular weight of 56.000 and associated to the smallest enzymically active subunit, the unimer (monomer), which therefore has a molecular weight of 336.000 (14). The unimer associates reversibly in the direction of the major axis forming rodlike particles of molecular weights exceeding 2.000.000. It is known that the reversible aggregation is dependent on a variety of factors. Thus, ADP, which allosterically increases the glutamate dehydrogenase activity, similarly increases its degree of polymerization and GTP, known to allosterically inhibit the enzyme, also effects depolymerization of the enzyme to its unimer.

In order to obtain direct experimental evidence concerning the effectiveness of ADP and GTP in allosterically activating and inhibiting glutamate dehydrogenase, respectively, in the absence of perturbation through polymerization and depolymerization interactions, the enzyme was covalently bound to a rigid matrix, porous succinamidopropyl-glass beads using water-soluble carbodi-imide (13). The immobilized enzyme showed an identical pH-optimum as the native enzyme and the K_m value for NADPH after correction for diffusional effects corresponded to that for native enzyme. The enzyme preparation was then examined for its sensitivity to activation by ADP and inhibition by GTP. As seen in Fig. 3, allosteric activation and inhibition are reversible and apparently do not depend on linear association of unimers (since there was no free enzyme in solution) or dissociation of aggregated forms.

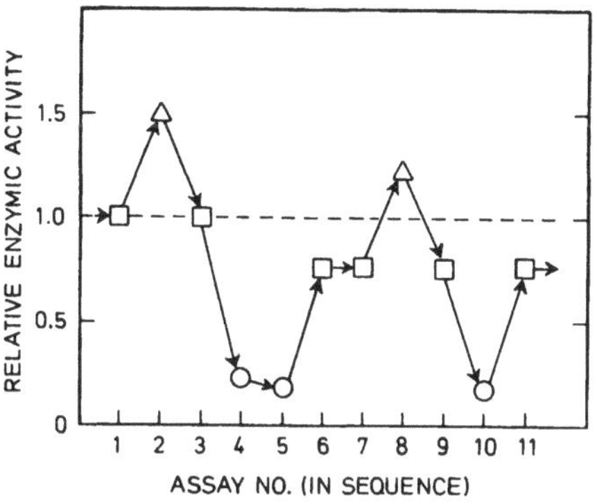

Fig. 3 Effects of ADP and GTP on activity of immobilized glutamate dehydrogenase at pH 7.75. Relative enzymic activity of 1.0 represents 1.93 µmoles NADH oxidized/min · g of beads. After each consecutive assay, the beads were washed and immediately resuspended in the appropriate medium for the next assay. Assays 1-3, beads which had never been exposed to GTP were assayed in the absence and presence of ADP as shown; they were subsequently assayed in the presence of GTP (4 and 5), and then in the absence and presence of ADP and GTP, respectively. □ = standard assays. Δ = 75 µM ADP. ○ = 20 µM GTP.

Closer examination revealed that following initial exposure to GTP, the immobilized enzyme lost about one-fifth of its total activity. Repeated exposure to GTP did not lead to further losses in activity. It appears that this was the result of depolymerization of partially polymerized enzyme by the action of GTP to result in immobilized unimer enzyme.

As demonstrated here, the technique of immobilization provides a convenient way of assessing allosteric effects when these are accompanied by association-dissociation. Although other techniques have also been applied to this particular enzyme establishing a lack of correlation between association and enzymic activity (14, 15), the immobilization technique seems to be highly suitable for such investigations (In this context it should be mentioned that the enzyme had previously been immobilized to formaldehyde-"tanned" collagen (16) or covalently connected by reaction with glutaraldehyde to clusters (17). The objectives with these studies referred to, however, did not focus on the theme discussed here with the exception of the last study indicating that the enzymic properties are independent of molecular weight.). Due to the relative simplicity of the technique and conclusiveness of the results obtained, provided the enzyme's properties are not changed appreciably through coupling, it will probably be applied in the future to the elucidation of other similar systems. In addition, the technique also provides a means for the direct determination of the reversibility of allosteric ligand induced changes, since the same enzyme molecules can conveniently be alternately exposed to various solutions.

REFERENCES

1. Mosbach, K. (ed.): Methods in Enzymology, vol. 44 (1977), Academic Press, New York.

2. Jakoby, W.B., Wilchek, M. (eds.): Methods in Enzymology, vol. 34 (1974), Academic Press, New York.

3. Chan, W.W.-C.: in Methods in Enzymology (Mosbach, K., ed.), vol. 44, pp. 491-503 (1977), Academic Press, New York.

4. Chan, W.W.-C.: Matrix-bound protein subunits. Biochem. Biophys. Res. Commun. 41, 1198-1204 (1970).

5. Markert, C.L.: Lactate dehydrogenase isozymes: Dissociation and recombination of subunits. Science 140, 1329-1330 (1963).

6. Cho, I.C., Swaisgood, H.E.: The reactivation of an unfolded subunit enzyme covalently linked to a solid surface. Biochim. Biophys. Acta 258, 675-679 (1972).

7. Jaenicke, R.: Reassociation and reactivation of lactate dehydrogenase from the unfolded subunits. Eur. J. Biochem. 46, 149-155 (1974).

8. Chan, W.W.-C., Mosbach, K.: Effects of subunit interactions on the activity of lactate dehydrogenase studied in immobilized enzyme systems. Biochemistry 15, 4215-4222 (1976).

9. Andersson, L., Mosbach, K.: To be published.

10. Pietruszko, K., Theorell, H.: Subunit composition of horse liver alcohol dehydrogenase. Arch. Biochem. Biophys. 131, 288-298 (1969).

11. Cronholm, T., Larsén, C., Sjövall, J., Theorell, H., Åkeson, Å.: Steroid oxidoreductase activity of alcohol dehydrogenase from horse,

rat and human liver. Acta Chem. Scand. B29, 571-576 (1975).

12. Andersson, L., Jörnvall, H., Mosbach, K.: Preparative purification of homogeneous steroid-active isozyme of horse liver alcohol dehydrogenase by affinity chromatography on an immobilized AMP-analog. Anal. Biochem. 69, 401-409 (1975).

13. Horton, H.R., Swaisgood, H.E., Mosbach, K.: Reversible allosteric modulation of activity of immobilized hepatic glutamate dehydrogenase. Biochem. Biophys. Res. Commun. 61, 1119-1124 (1974).

14. Sund, H., Dieter, H., Koberstein, R., Rasched, I.: Glutamate dehydrogenase: Chemical modification and ligand binding. J. Mol. Catalysis 2, 1-23 (1977).

15. Frieden, C., Colman, R.F.: Glutamate dehydrogenase concentration as a determinant in the effect of purine nucleotides on enzymatic activity. J. Biol. Chem. 242, 1705-1715 (1967).

16. Julliard, J.H., Godinot, C., Gautheron, D.C.: Some modifications of the kinetic properties of bovine liver glutamate dehydrogenase (NAD(P)) covalently bound to a solid matrix of collagen. FEBS Lett. 14, 185-188 (1971).

17. Josephs, R., Eisenberg, H., Reisler, E.: in Protein-Protein Interactions (Jaenicke, R., Helmreich, E., eds.), pp. 57-90 (1972), Springer-Verlag, Berlin, Heidelberg, New York.

Received February 7, 1977

DISCUSSION

Sund: What is the experimental evidence that the covalently bound enzymic inactive polypeptide chain of LDH is active upon reassociation (Fig. 1)? How did you determine the protein concentrations and the specific enzymatic activity?

Mosbach: As to your first question I agree that in the case of LDH there exists no direct quantitative or qualitative (as in the case of horse liver ADH) evidence to what extent the bound, inactive polypeptide chain of LDH becomes active on reassociation with soluble subunits.

The protein content of matrix-bound derivatives was determined after acid hydrolysis and subsequent amino analysis of lyophilized Sepharose-bound enzyme.
In order to assay the activity of the immobilized enzyme, a spectrophotometer (Zeiss PMQ 3) with a stirring device mounted in the cuvette house was used. The matrix-bound activity was measured by following the decrease in 340 nm absorbance after the addition of an aliquote of a suitably diluted gel suspension and substrates to the buffer solution. On assaying, the incubation mixture was stirred with a small magnetic ball. For a general discussion of the method see e.g. Chan in "Methods in Enzymology", Vol. 44, (ed. Mosbach), 1977, page 491; or Chan and Mosbach, Biochemistry (1976), 15, 4215.

Trommer: In the case of LDH you may encounter an additional problem. You dissociate your matrix-bound enzyme under denaturing conditions i.e. pH 3. If I remember correctly Dr. Jaenicke has shown that monomers of LDH are indeed partially denatured. Refolding then seems to occur upon reassociation to oligomers. In other words, even when you can raise the pH of the solution of your immobilized monomer it may still be partially unfolded and therefore inactive, this could still leave open the question of a catalytically active "native" monomer.

Mosbach: Your question is related to that of Dr. Sund. However, on dissociation from the matrix made up of Sepharose-S-S-LDH-subunit using excess 2-mercaptoethanol, activity graudally appeared in free solution. This would be consistent with the postulated existence of correctly folded but inactive subunits on the matrix which are then detached by thiol treatment.

Engel: Have you tried the same sorts of experiment that you describe for ADH and LDH with glutamate dehydrogenase? In other words have you had any success in refolding immobilised GDH monomers with or without free subunits present?

Mosbach: I am afraid we did not try such experiments with glutamate dehydrogenase.

Gafni: Do the immobilized Sepharose bound subunits of horse liver ADH or LDH bind coenzyme (NAD or NADH) molecules?

Mosbach: We did never investigate whether the immobilized subunits of either LDH or horse liver ADH were able to bind coenzyme.

Shore: You mentioned that addition of native E isomer dimers to monomeric immobilized S isomer resulted in the formation of immobilized hybrid dimers with a regeneration of steroid activity. How long did the activity regeneration take and what was the extent? We have estimated that the subunit dissociation constant is at least 10^{-12} M. With a diffusion limited subunit association constant of 10^7 $M^{-1}sec^{-1}$, the rate constant for dissociation of the dimer would be 10^{-5} sec^{-1}, with a half life of 1 000 minutes. This is the fastest possible rate for the hybridization reaction.

Mosbach: We mentioned this experiment since the fact that native EE isozyme did regenerate steroid activity in the immobilized S-subunit did puzzle us indeed and we welcome any suggestion. As to your specific question: About 500 µg of immobilized S-subunit were mixed with 1.5 mg of dimeric, native EE isozyme in 2 ml 0.1 M sodium phosphate, pH 7.5, and 14 mM β-mercaptoethanol. The incubation mixture was left, under slow rocking, for about 4 hours at room temperature. This treatment yielded bound steroid activity that was about 15 % of the steroid activity found for the immobilized SS isozyme.

Jaenicke: How does fixation of enzymes to the matrix affect the catalytic properties?

Mosbach: This is still a question of controversy and I must admit that no definite answer can be given. The sometimes low specific activity of immobilized enzymes may be due to diffusion effects exerted by the matrix but also the derivatization per se taking place in covalent binding may affect the system.

Jaenicke: Considering the contradictions between your LDH data and the data presented by Cho and Swaisgood (BBA 258 (1972) 675-679) I would like to ask you, what are the pitfalls of your approach regarding subunit activity?

Mosbach: I refer to a recent paper in which this question is discussed in greater details (Chan and Mosbach, Biochemistry 15 (1976) 4215).

Jaenicke: Has zinc been removed in your urea treatment of liver ADH?

Mosbach: We never did investigate the zinc contents of our immobilized liver ADH-subunit preparations. Although we have every reason to believe that zinc was still available in sufficient amounts, we should do this experiment.

Perham: One of the problems associated with experiments of the sort you describe is the difficulty in ensuring that immobilized subunits are not bound sufficiently close together on the matrix to interact when refolding is allowed to take place. If you recover, say, 2 % activity, that may represent the true activity of the monomer or it may be due to reformation of small amounts of oligomers that also might be expected to show activity. I am not saying it can't be done, but that interpretation of the experiments should be made with great care. I make this comment only as a general point of caution.

Mosbach: I fully agree on this point. The possible risk of interaction in the refolding step between bound subunits close together on the matrix should be considered as has been done in another context in a thorough study by Gree (Biochem. J. 133 (1973) 687-700). In our experiments, using a "diluted" matrix, this risk is very small. In addition, in the experiments referred to, the specific activity was far higher than 2 %.

Section III. Chemical Mechanism and Coenzyme Binding

Chairmen: J.I. Harris and M. Rossmann

On the mode of hydrogen transfer and catalysis in nicotinamide-dependent oxidoreduction

Gunter Blankenhorn
Fachbereich Biologie der Universität Konstanz, Postfach 7733
D-7750 Konstanz, Federal Republic of Germany

In order to classify reactions and possible mechanisms involved in biological nicotinamide-dependent oxidoreduction, let us consider electron flow in the respiratory chain: two electrons plus one proton are transferred from alcoholic substrates to nicotinamide defining the input reaction (scheme 1). In the output reaction $2e^-$-equivalents are transferred to flavin, which acts as a center where the incoming $2e^-$-equivalents are transformed into radical electrons.
Currently, the possible mechanisms by which $2e^-$-equivalents are transferred to and from (dihydro)nicotinamides are heavily disputed. For the input reaction, hydride (H^++$2e^-$) transfer as the unique mechanism has been challenged and the al-

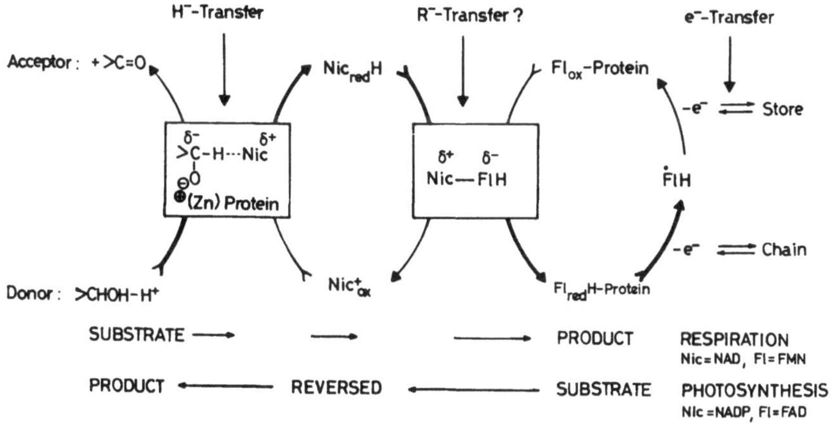

Scheme 1 Transfer and transformation of redox equivalents in the respiratory chain (from left to right). Reversal of these processes describes photosynthetic electron transport.

ternative, hydrogen atom (H^++e^-) transfer,has been suggested instead (1-3). It will be one aim of this article to demonstrate that the energy required for this latter process is biologically not attainable. For flavin-nicotinamide oxidoreduction, the output reaction in scheme 1, a group transfer mechanism has been postulated, involving σ-bonded intermediates. Both the validity of this latter postulate, along with the closely related problem of structure and function of possible catalytic intermediates, will be discussed in the second part of this article (4).

$1e^-$ AND $2e^-$ OXIDATION REDUCTION POTENTIALS AND RADICAL STABILITY OF NICOTINAMIDES

Dihydronicotinamides contain the kinetically stable sp^3-C (4)H bond. Hence, only the upper nicotinamide $1e^-$-shuttle (E_1) $NAD^+ \rightleftharpoons NAD^\cdot$ represents a reversible, pH-independent redox system. Both the lower $1e^-$-shuttle (E_2) $NAD^\cdot \rightleftharpoons NADH$ and the $2e^-$-shuttle $NAD^+ \rightleftharpoons NADH$ (E_m) represent pH-dependent oxidation reduction couples which are not rapidly equilibrated. Use of mediators is therefore necessary to catalyze equilibration between oxidized and reduced forms. At half reduction, thermodynamic stability of the radical state is defined by the equilibrium position of radical dismutation:

$$2\ NAD^\cdot + H^+ \rightleftharpoons NAD^+ + NADH$$

At a given pH, the semiquinone stability constant is given by

$$E_1 - E_2 = 0.059 \log K$$

Hence, at half reduction the radical state is only thermodynamically stable if $E_1 > E_2$.

Apart from thermodynamic radical stability, kinetic radical stability is important. In the case of nicotinamides the latter primarily concerns the rate of radical dimerization

$$2\ NAD^\cdot \rightleftharpoons NAD\text{-}NAD$$

When nicotinamide radicals are generated by either pulse radiolysis (5,6) or electrochemical methods (7,8) rapid dimerization is obersved ($k=10^7\ M^{-1}s^{-1}$). Only very recently has it been possible therefore, to generate NADH from NAD^+ electrochemically. Using liquid crystal membrane electrodes radical dimerization can be prevented (9). Because of radical dimerization, experimentally determined E_1 potentials of nicotinamides represent kinetic potentials. The corresponding thermodynamic potentials (oxidation reduction potentials), however, can only be estimated. Kinetic and thermodynamic potentials diverge increasingly with increasing rates of dimer formation.

In table 1 estimated thermodynamic potentials of some nicotinamide analogs (Elving, P.J., personal communication) derived from polarographically determined E_1-potentials (10) have been summarized. The corresponding E_m-potentials have been

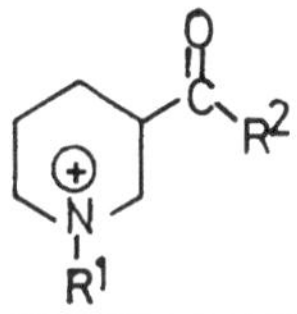	E_1 (mV)	E_2 (mV)	E_m (mV)	E_1-E_2(mV)
$R^1 = CH_3$, $R^2 = NH_2$	- 965	+ 153	- 406	- 1118
$R^1 = CH_3$, $R^2 = OC_2H_5$	- 980	+ 158	- 411	- 1138
$R^1 = CH_3$, $R^2 = CH_3$	- 760	+ 86	- 337	- 846
R^1 = R-P-P-A $R^2 = NH_2$ (NAD^+)	- 850	+ 210	- 320	- 1060
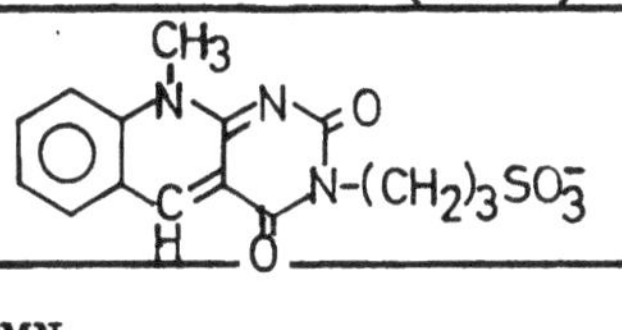	- 770	+ 130	- 320	- 900
FMN	- 238	- 172	- 205	- 66

Table 1 Oxidation reduction potentials of nicotinamide analogs and FMN. E_1 and E_2 represent estimates, derived from polarographically determined kinetic potentials (E_1) and potentiometrically determined thermodynamic potentials (E_m) (10).

determined independently (10), making quite reliable estimates of nicotinamide E_2-oxidation reduction potentials possible. E_1-oxidation reduction potentials of nicotinamides are lower than the corresponding E_2-potentials by more than 1000 mv which expresses an extreme thermodynamic radical instability. Relative to nicotinamides, flavin radicals are thermodynamically stable (table 1). 5-deazaflavins, however, which have been widely used as flavin analogs in enzymatic reactions show the high thermodynamical instability characteristic for nicotinamides.

The data shown in table 1 demonstrate that $2e^-$-oxidation of dihydronicotinamides by consecutive $1e^-$-steps with biological $1e^-$-acceptors is energetically feasible; however, <u>$1e^-$-reduction of nicotinamides is energetically blocked.</u> If one considers NAD radicals as possible catalytic intermediates in enzyme reactions it is of primary importance to describe means of overcoming this barrier.

Model studies have shown that pyridinyl radicals are thermodynamically stabilized by electron withdrawing substituents, placed at the conjugated positions 2,4, and 6 (3). Stabilization is attained only if the substituent allows further delocalization of spin density. None of these structural factors is present in nicotinamides; it appears, therefore, that nicotinamides represent a pyridinyl redox system characterized by maximal thermodynamic radical instability. Even in the enzyme-bound state no significant stabilization is achieved. In view of this fact, reversible $1e^-$-transfer in biological nicotinamide-dependent oxidoreduction can be eliminated as a feasible mechanism. This is in contrast to flavins, which are characterized by a thermodynamically stable radical state and have been demonstrated to be involvolved in biological $1e^-$-transfer (11).

FREE ENERGY RELATIONSHIPS

Linear free energy relationships have frequently been used

to correlate structure-reactivity patterns. The slopes obtained in such diagrams have been interpreted in terms of transition state geometry, for example as to whether transition states are more substrate or more product like (12). We have tried to use this approach as a tool allowing one to differentiate between rate limiting hydrogen atom transfer and rate limiting hydride transfer. From the basic free energy equations

$$G = -RT \ln K \quad \text{and} \quad G = -nEF$$

one obtains $$E = \frac{RT}{nF} \ln K$$

and since $K = \frac{k_1}{k_{-1}}$ and $\frac{RT}{F} \ln 10 = 0.059$ V at 20° C

then $$E = \frac{0.059}{n} (\log k_1 - \log k_{-1})$$

where k_1 and k_{-1} represent the individual rate constants for rate limiting hydrogen transfer. From this one should expect two sets of conditions in which the slope of a linear free energy relationship E versus log k should only be dependent on n, the number of electrons transferred in the rate limiting hydrogen transfer step:

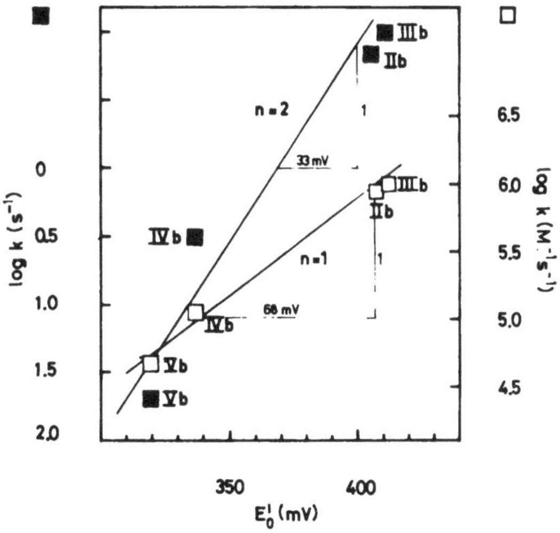

Figure 1 Linear free energy relationships for the oxidation of NADH analogs to NAD^+. □ spirocyclohexylporphyrexide (1e⁻-oxidant), lumiflavin (2e⁻-oxidant). ■ Data taken from (10).

a. The energy of transition states depends on substituent effects to the same extent as that of the corresponding ground states.
b. The energy of transition states is independent of substituent effects.

We have tested this hypothesis in the following model system (10): Dihydronicotinamide analogs were oxidized to the corresponding nicotinamidium salts by both a mandatory $1e^-$-oxidant, spirocyclohexyl porphyrexide, and by lumiflavin, which can react as either a $1e^-$ on a $2e^-$-oxidant. The resulting linear free energy relationships (fig. 1) showed slopes of 68 mv and 33 mv per unit of log k respectively, indicating that the first reaction proceeded via hydrogen atom transfer but the second via hydride transfer. If both reactions had proceeded via hydrogen atom transfer very similar if not identical slopes would have been expected.

For several nicotinamide-dependent dehydrogenases free energy relationships have been reported (13-16); however, in most examples insufficient information has been given to apply the concept discussed in this report. The one notable

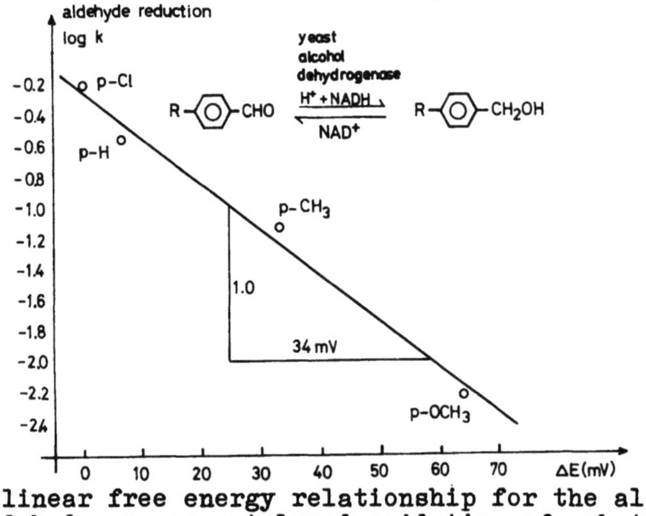

Figure 2 linear free energy relationship for the alcohol dehydrogenase-catalyzed oxidation of substituted benzaldehydes. Data taken from (16).

exception so far, is Klinman's elegant work on yeast alcohol dehydrogenase (16), which demonstrated that for a given set of substrates C-H cleavage was at least partially rate limiting. Binding constants for aldehyde substrates and alcohol products were also reported, thus allowing application of the free energy concept (fig. 2). A linear free energy relationship is obtained for the conversion of enzyme-bound aldehyde to enzyme-bound alcohol. The slope of 34 mv per unit of log k indicates that this enzyme-catalyzed reaction proceeds via a hydride transfer mechanism. The reverse reaction, enzyme-catalyzed alcohol dehydrogenation was found to be independent of substituent effects.

We believe to have demonstrated that linear free engery relationships can be a useful tool for distinguishing rate limiting hydrogen atom and hydride transfer. It will be very interesting to investigate further enzymes and model systems to get a fuller understanding of the scope, and limitations, of the concept.

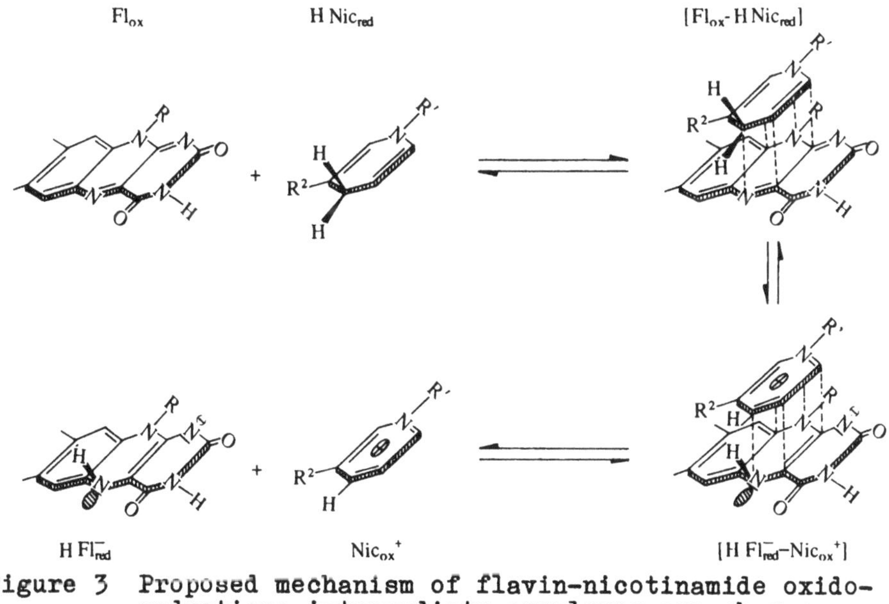

Figure 3 Proposed mechanism of flavin-nicotinamide oxidoreduction: intermediate complexes are charge transfer complexes. (10).

SUBSTRATE ACTIVATION AND POSSIBLE INTERMEDIATES IN NICOTINAMIDE-DEPENDENT OXIDOREDUCTION

There are two types of biological substrates of nicotinamide-dependent (de)hydrogenases (see also scheme 1):

a. with (dihydro)flavins kinetically labile N-H bonds are formed or cleaved,

b. with alcohols and most other substrates kinetically stable C-H bonds at sp^3 carbon have to be cleaved or formed.

These substrates differ in their properties to a degree that their interactions with nicotinamides have to be discussed separately. Flavin-nicotinamide oxidoreduction: Flavins and nicotinamides in the oxidized state represent electron deficient π-acceptors whereas they are electron excess π-donors in the reduced state. Thus, it is not surprising that the respective complexes Fl_{ox}-$Nic_{red}H$ and $Fl_{red}H^-$-Nic^+_{ox} possess charge transfer properties. Due to their long wavelength absorption they have been detected early in enzymic systems and implied as catalytic intermediates (17).
A better understanding as to the catalytic function and possible geometry of flavin-nicotinamide charge transfer complexes was derived from model studies (10,18). Flavin-nicotinamide biscoenzymes, linked by three methylene groups through flavin position N(10) and nicotinamide N(1), have been shown to mimic the properties of enzyme-bound flavin-nicotinamide couples very well. Both types of charge transfer complexes are formed and oxidation-reduction rates (345 s^{-1}) are higher than in most of the enzymic counterparts. Results from the study on flavin-nicotinamide biscoenzymes suggest that in the active complex C(4) of nicotinamide and C(4a)-N(5) of flavin are juxtaposed in a similar way as shown in figure 3.
The mechanism outlined here accomodates the minimal requirements known from model and enzyme studies. Obviously it is an open question whether yet another intermediate, having escaped detection so far, may be involved in catalysis.

Such an intermediate has in fact been postulated because product isotope effects observed in the reaction of dihydronicotinamides with trifluoroacetophenone or N-methyl-acridinium ion differ from those predicted by the corresponding kinetic isotope effects (19,20).

$$\text{substrate} + \text{Nic}_{red}\text{H(D)} \underset{k_{-1}}{\overset{k_1}{\rightleftharpoons}} [\text{intermediate}] \begin{cases} \xrightarrow{k_2} \text{product(H)} \\ \xrightarrow{k_2} \text{product(D)} \end{cases}$$

Product isotope effect: $\dfrac{k_2^H}{k_2^D}$

Kinetic isotope effect: $\dfrac{k^H_{rate}}{k^D_{rate}}$, $k_{rate} = \dfrac{k_1 k_2}{k_{-1} + k_2}$

The observed difference has lead to the postulate of an intermediate which could partition between substrate and product, i.e. that k_{-1} had to be in the order of k_2.
Although a yet unknown intermediate might well be formed in these reactions there is one trivial possibility that could also account for the observed deuterium isotope effects.
If substrate and product reacted with each other at rates compatible with the overall reaction scrambling of the deuterium label could be a result of a substrate-product equilibrium isotope effect rather than partitioning of an intermediate.

$$\text{substrate} \overset{K^H/K^D}{\rightleftharpoons} \text{product}$$

For nicotinamide-5-deazaflavin interaction validity of this latter assumption can actually be demonstrated: the rate of hydrogen exchange between oxidized and reduced deazaflavin has been determined as $k=22M^{-1}s^{-1}$ (21), whereas reduction of deazaflavin by N-methyl-dihydronicotinamide is characterized by a $k=12M^{-1}s^{-1}$ (22).

If in flavin-nicotinamide oxidoreduction another intermediate than the respective charge transfer complexes was involved, its most probable structure would be that of a covalent adduct.

Evidence suggesting a possible catalytic role for covalent flavin-substrate adducts stems from experiments with the flavoenzymes D-amino acid oxidase and lactate oxidase as well as model studies (23,24). Very recently, spectral evidence compatible with covalent adduct formation in a flavin-nicotinamide model system at -40°C has been obtained Knappe (W.-R. Knappe, personal communication) in our laboratory.

Our present knowledge of flavin-nicotinamide oxidoreduction can be summarized as follows: preorientation of substrate and product and prepolarization of the dihydronicotinamide C(4)H bond is achieved in a charge transfer complex oriented in a way that C(4) of nicotinamide and C(4a)-N(5) of flavin are juxtaposed. Direct hydride transfer appears as the most likely mechanism of oxidoreduction although participation of a covalent intermediate cannot be excluded. Hydrogen atom transfer does not appear as a feasible mechanism.

In alcohol-nicotinamide oxidoreduction activation of substrate cannot be achieved via charge transfer complexation, for the simple reason that alcohols have no tendency to form such complexes. Binding and proper orientation of coenzyme and substrate to alcohol dehydrogenases appear to be regulated by ion pair interactions and hydrogen bonding. These processes are accompanied by protein conformational changes in the binding loop region (25). Activation (prepolarization) of substrate probably involves acidbase catalysis and/or metal ion complexation (Lewis acid catalysis by Zn^{2+}). In reactions catalyzed by yeast alcohol dehydrogenase participation of a base with a pK of 8.25 and of Zn^{2+} has actually been demonstrated (26).

Different catalytic pathways in flavin nicotinamide oxidoreduction as compared to the yeast alcohol dehydrogenase reaction also become apparent from free energy versus reaction pathway diagrams. In yeast alcohol dehydrogenase substituents affect transition state energies to a very

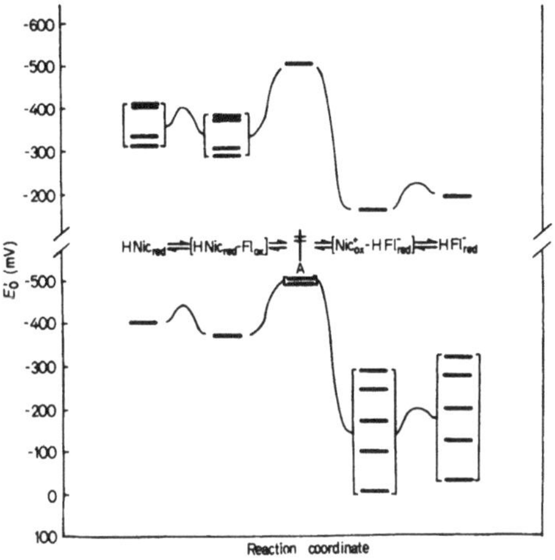

Figure 4 Substituent effects in flavin-nicotinamide oxidoreduction. Upper part: reaction of dihydronicotinamide analogs with lumiflavin. Lower part: reaction of N-Methyldihydronicotinamide with flavin analogs. Data taken from (10).

similar if not the same extent as the corresponding ground state energies of the alcohol NAD^{+}-enzyme complex. Hence, it may be assumed that transition state geometry resembles that of the alcohol-NAD^{+}-enzyme complex rather than that of the aldehyde-NADH-enzyme complex (16). This result can be readily rationalized as reflecting the very high kinetic stability of the alcohol C-H-bond. For flavin nicotinamide oxidoreduction a completely different pattern evolves: transition state energies appear to be independent of nicotinamide substituent effects and only marginally dependent on flavin substituent effects (fig. 4). Thus, the transition state complex apparently cannot be described as either substrate or product like.

As recently as the last symposium on "Pyridine Nucleotide Dehydrogenases" Schellenberg, on account of the lack of experimental tools for differentiation of possible modes of hydrogen transfer, made the suggestion(p.16)that "distinction between hydride transfer and hydrogen atom transfer might ultimately prove to be only semantic". It is hoped that the recent developments in nicotinamide-dependent oxidation reduction studies presented above allow a more optimistic outlook.

References

1. Williams, R.F., Shinkai, S., and Bruice, T.C.: Radical Mechanisms for 1,5-Dihydroflavin Reduction of Carbonyl Compounds. Proc.Natl. Acad. Sci. U.S.A. 72, 1763-1767 (1975)
2. Fisher, J., Spencer, R., and Walsh, C.: Enzyme-Catalysed Redox-Reactions with the Flavin Analogs 5-Deazariboflavin, 5-Deazariboflavin 5'-Phosphate, and 5-Deazariboflavin 5'-Diphosphate 5' → 5'-Adenosine Ester. Biochemistry 15, 1054-1064 (1976)
3. Kosower, E.M., in "Free Radicals in Biology" Vol.2 (Pryor, W.A. ed.): Pyridinyl Radicals in Biology, p. 1-54, Academic Press, New York, 1976
4. Hemmerich, P. in "Progress in the Chemistry of Organic Natural Products" (Herz, W., Grisebach, H., and Kirby, G.W. eds.): The Present Status of Flavocoenzyme Chemistry, p. 451-527, Springer Verlag Wien, New York, 1976
5. Land, E.J., and Swallow, A.J.: One Electron Reactions in Biochemical Systems as Studied by Pulse Radiolysis. I. Nicotinamide Adenine Dinucleotide and Related Compounds. Biochim. Biophys. Acta 162, 327-337 (1968)
6. Brühlmann, U. and Hayon, E.: One-Electron Redox Reactions of Water-Soluble Vitamins.I. Nicotinamide (Vitamin B_5) and Related Compounds. J.Am.Chem.Soc. 96, 6169-6175 (1974)
7. Burnett, J.M. and Underwood, A.L.: Electrochemical Reduction of Diphosphopyridine Nucleotide. Biochemistry 4, 2060-2064 (1968)
8. Schmakel, C.D., Santhanam, K.S.V., and Elving, P.J.: Nicotinamide Adenine Dinucleotide (NAD^+) and Related Compounds. Electrochemical Redoxpattern and Allied Chemical Behaviour. J.Am.Chem.Soc. 97, 5083-5092 (1975)
9. Alzawa, M., Suzuki, S., and Bubo, M.: Electrolytic Regeneration of NADH from NAD^+ with a Liquid Crystal Membrane

Electrode. Biochim. Biophys. Acta 444, 886-892 (1976)
10. Blankenhorn, G.: Nicotinamide-Dependent One-Electron and Two-Electron (Flavin) Oxidoreduction. Thermodynamics, Kinetics, and Mechanism. Eur. J. Biochem. 67, 67-80 (1976).
11. Mayhew, S.G. and Ludwig, M.L. in "The Enzymes", (Boyer, P.D. ed.) Vol.XII, 3rd ed. Flavodoxins and Electron--Transferring Flavoproteins. p. 57-118, Academic Press, New York 1975
12. Jencks, W.P.: Structure Reactivity Correlation and General Acid Base Catalysis in Enzymic Transacylation Reactions. Cold Spring Harbor Symp. Quant. Biol. 36, 1-13 (1971)
13. Hardman, M.J., Blackwell, L.F., Boswell, C.R., and Buckley, P.D.: Substituent Effects on the Pre-Steady-State Kinetics of Oxidation of Benzylalcohols by Liver Alcohol Dehydrogenase. Eur. J. Biochem. 50, 113-118 (1974)
14. Dworschack, R.T. and Plapp, B.V.: Isotope and Substituent Effects on the Interconversion of Aromatic Substrates Catalyzed by Hydroxybutyrimidilated Liver Alcohol Dehydrogenase. Fed. Proc. 35, 1748 (1976)
15. Brown, A. and Fisher, H.F.: A Comparison of Glutamate Dehydrogenase Catalysed Oxidation of NADPH by Trinitrobenzenesulfonate with the Uncatalyzed Reaction. J.Am.Chem. Soc. 98, 5682-5688 (1976)
16. Klinman, J.P.: Isotope Effects and Structure-Reactivity Correlations in the Yeast Alcohol Dehydrogenase Reaction. A study of the Enzyme-Catalyzed Oxidation of Aromatic Alcohols. Biochemistry 15, 2018-2026 (1976)
17. Massey, V., Matthews, R.G., Foust, G.P., Howell, L.G., Williams, C.H., Zanetti, G., and Ronchi, S. in "Pyridine Nucleotide-Dependent Dehydrogenases" (Sund, H.ed.): A new Intermediate in TPNH-Linked Flavoproteins. p. 393-409 Springer Verlag, Berlin-Heidelberg-New York (1970)
18. Blankenhorn, G.: Flavin-Nicotinamide Biscoenzymes: Models for the Interaction between NADH (NADPH) and Flavin in Flavoenzymes. Eur. J. Biochem. 50, 351-356 (1975)
19. Steffens, J.J. and Chipman, D.M.: Evidence for an Inter-

mediate in the Reduction of Trifluoroacetophenone by 1-Substituted Dihydronicotinamides. J.Am.Chem.Soc. 93, 6694-6696 (1971)

20. Creighton, D.J., Hajdu, J., Mooser, G., and Sigman, D.S.: Model Dehydrogenase Reactions. Reduction of N-Methylacridinium Ion by Reduced Nicotinamide Adenine Dinucleotide and ist Derivatives. J.Am.Chem.Soc. 95, 6855-6857 (1973)

21. Spencer, R., Fisher, J., and Walsh, C.: Preparation, Characterization, and Chemical Properties of the Flavin Coenzyme Analogues 5-Deazariboflavin, 5-Deazariboflavin 5'-Phosphate, and 5-Deazariboflavin 5'-Diphosphate, 5' → 5'-Adenosine Ester. Biochemistry 15, 1043-1053 (1976)

22. Blankenhorn, G.: Intermolecular Complexes between N-Methyl-1,4-dihydronicotinamide and Flavines. The Influence of Steric and Electronic Factors on Complex Formation and the Rate of Flavin-Dependent Dihydronicotinamide Dehydrogenation. Biochemistry 14, 3172-3176 (1975)

23. Porter, D.J.T., Voet, J.G., and Bright, H.J.: Direct Evidence for Carbanions and Covalent N(5)-Flavin-Carbanion Adduct as Catalytic Intermediates in the Oxidation of Nitroehtane by D-Amino-Acid Oxidase. J.Biol.Chem. 248, 4400-4416 (1973)

24. Massey, V., and Ghisla, S. in "Proceedings of the Tenth FEBS-Meeting": The Mechanism of Flavoprotein Catalyzed Dehydrogenation Reactions. p. 145-158 (1975)

25. Brändén, C.-J., Jörnvall, H., Eklund, H., and Furguren, B. in "The Enzymes" (Boyer, P.D. ed.) Vol.XI, 3rd ed.: Alcohol Dehydrogenases. p.104-190, Academic Press, New York 1975

26. Klinman, J.P.: Acid-Base Catalysis in the Yeast Alcohol Dehydrogenase Reaction. J.Biol.Chem. 250, 2569-2573 (1975)

Received February 22, 1977

DISCUSSION

Knappe: UV spectroscopic evidence for π-complexes in the redox reaction between NADH analogs and flavin has been reported by Blankenhorn (1). In this report evidence for σ-complexes in this redox reaction will be discussed.

When a solution of 3-benzyllumiflavin in acetonitrile or 2-methyltetrahydrofuran containing an equimolar amount of N-propyldihydronicotinamide was illuminated with blue light (λ = 420-480 nm) under anaerobic conditions, a smooth photoreduction took place, during which no CT absorption could be seen. When this photoreduction was performed at -40° C, the spectrum shown in curve 2 of Fig. 1 was observed.

It shows a maximum at 354 nm (ε ≈ 10,000). Upon warming the solution to room temperature, the spectrum shown in curve 3 appeared. This spectrum is very similar to the absorption spectrum of the anion of 1,5-dihydroflavin

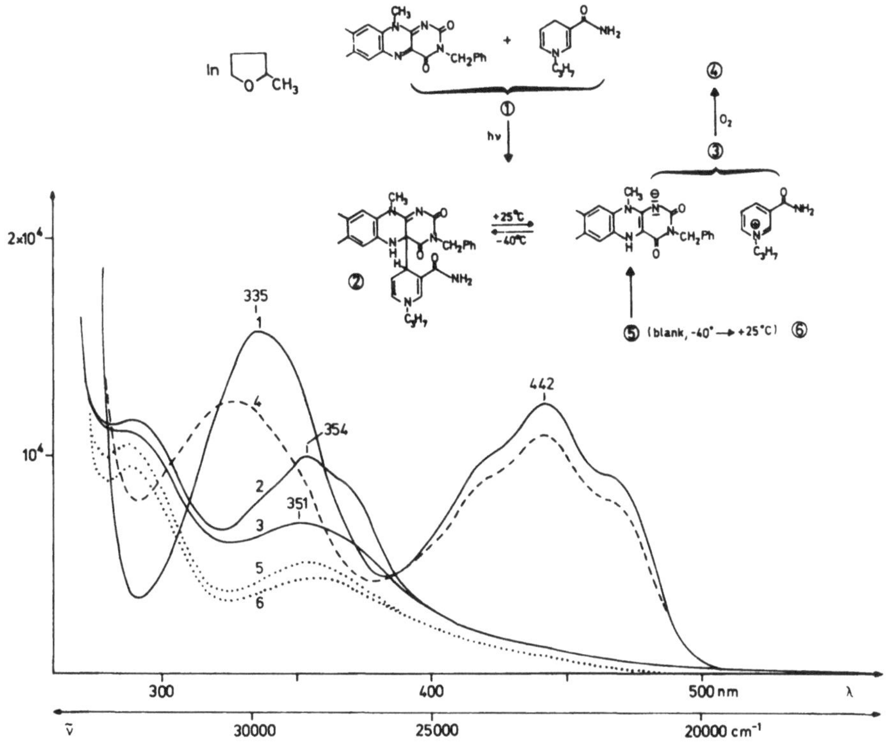

Fig. 1 UV spectra of the species involved in the redox reaction between 3-benzyllumiflavin and N-propyldihydronicotinamide in 2-methyltetrahydrofuran. 1: 1:1-mixture of the starting materials; 2: σ-complex after illumination with blue light at -40° C; 3: dihydroflavin anion - nicotinamide cation - ion pair at +25° C; 4: spectrum after admission of oxygen; 5: dihydroflavin anion at -40° C; 6. dihydroflavin anion at +25° C.

shown in curve 6. Upon cooling to -40° C, the spectrum shown in curve 2 reappeared. However, when the solution of the dihydroflavin anion was cooled, only a small increase in absorbance (< 10 %) due to volume contraction was observed. Upon admission of air, more than 90 % of the starting oxidized flavin was formed. These UV spectral results are interpreted in the following way:

In the low temperature reaction a σ-complex (a 4a-substituted 4a,5-dihydroflavin derivative) formed, in which the 4-position of the N-propyldihydronicotinamide is linked covalently to the 4a-position of the flavin nucleus. Many such derivatives with alkyl substituents in the 4a-position are known, and their structures have been confirmed (2, 3). The UV spectrum shown in curve 2 is a superposition of the 4a,5-dihydroflavin chromophore (maximum 340-360 nm, $\varepsilon \approx 6000$ (2, 3)) and the dihydronicotinamide chromophore (340 nm, $\varepsilon = 6200$ (4)), and it obeys the Beer-Lambert law. Upon warming, this σ-complex splits heterolytically, yielding an ion pair consisting of the anion of 1,5-dihydroflavin and the nicotinamide cation. Upon cooling, this ion pair recombines to the σ-complex.

Although this redox reaction must be induced photochemically in these nonprotic solvents, the same reaction sequence could occur in water under nonphotochemical conditions. Some similarities between flavin photochemistry and flavin biocatalysis are well established (5).

To support this UV spectroscopic evidence for a solvent intermediate, further models of the NADH system were investigated. In Fig. 2 the results of the photoreaction between 3-benzyllumiflavin and 1,3,10-trimethyl-1,5-dihydro-5-deazaisoalloxazine (6) are shown. According to Hemmerich (7, 8), this ring system should be regarded as a model for reduced nicotinamide despite its flavin shape.

At -40° C the UV spectrum 2 was obtained, showing an absorbance maximum at 380 nm ($\varepsilon = 5500$). However, when this reaction mixture was warmed, it behaved very differently from the N-propyldihydronicotinamide-adduct: 50 % of the oxidized flavin reappeared (curve 3), and there was no change on cooling to -40° C. On admission of oxygen, 100 % of the oxidized flavin was formed (curve 4).

The 4a-adduct splits homolytically, yielding flavin radical. The flavin radical then disproportionates to 50 % oxidized + 50 % reduced flavin and the radical of the deazaisoalloxazine derivative dimerizes. Indeed, the dimer of the deaza derivative was the only oxidation product of this ring system. It was identified by tlc comparison with an authentic sample which was a generous gift of Dr. Bauch in Berlin (6).

The results are summarized in scheme 1.

It should be emphasized that the results are very different depending on the NADH model compound used. Results obtained from model reactions, perhaps even those presented in this lecture, should be interpreted with great care.

The 1,5-dihydro-5-deazaisoalloxazine model compound, which is not blocked in the N(1)-position, yielded only 1,5-dihydroflavin and oxidized 5-de-

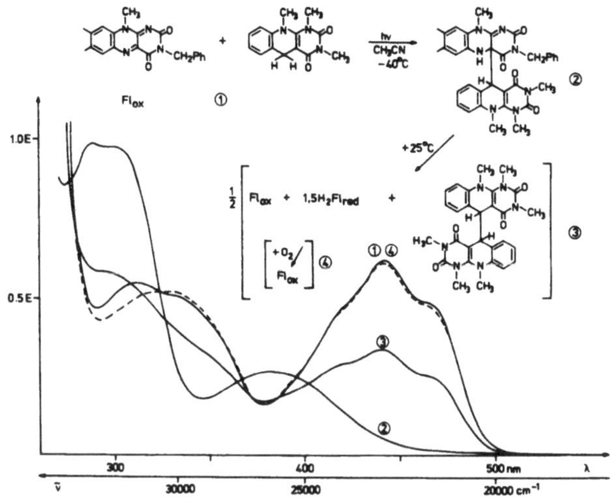

Fig. 2 UV spectra observed in the photoreaction between 3-benzyllumiflavin and 1,3,10-trimethyl-1,5-dihydro-5-deazaisoalloxazine in acetonitrile under anaerobic conditions. 1: 1:1-mixture of the starting materials (5.0×10^{-5}M); 2: 4a-substituted 4a,5-dihydroflavin derivative; 3: solution 2 after warming from -40° C to +25° C; 4: solution 3 after admission of oxygen.

azaisoalloxazine. The thermal decomposition of the postulated 4a-adduct occurred heterolytically even at -80° C. No dimer of the 1,5-dihydro-5-deazaisoalloxazine could be detected by TLC.

The N(1) blocked 5-deazaisoalloxazine derivative decomposed homolytically on warming. Thus, the deaza radical dimerized, whereas the flavin radical disproportionated.

Of course, both of these decomposition reactions with deazaisoalloxazine derivatives are strictly irreversible.

N-propyldihydronicotinamide, the best model for NADH, showed the most interesting behavior, since its dissociation reaction was reversible. The 4a-adduct formed at -40° C dissociated upon warming, yielding an ion pair which recombined again on cooling.

In summary, the redox reaction between oxidized flavin and reduced nicotinamide can occur via a covalent σ-complex in aprotic solvents. In protic solvents the CT- or π-complexes have been observed and their kinetic competence has been established by Massey and coworkers (9) and Blankenhorn (10). Nevertheless, covalent adduct formation might occur even in enzymes, being hardly detectable because of the short lifetime of the σ-complex under these conditions. The redox process consists of the following two steps: First, a σ-bond between the 4-position of the NADH and the C(4a) of the flavin nucleus is formed. This covalent adduct then cleaves heterolytically on warming in a reversible reaction, transferring the redox equivalents to the flavin. Of course, these reaction steps proceed in biochemical

Thermal decomposition of covalent adducts between lumiflavin and NADH models

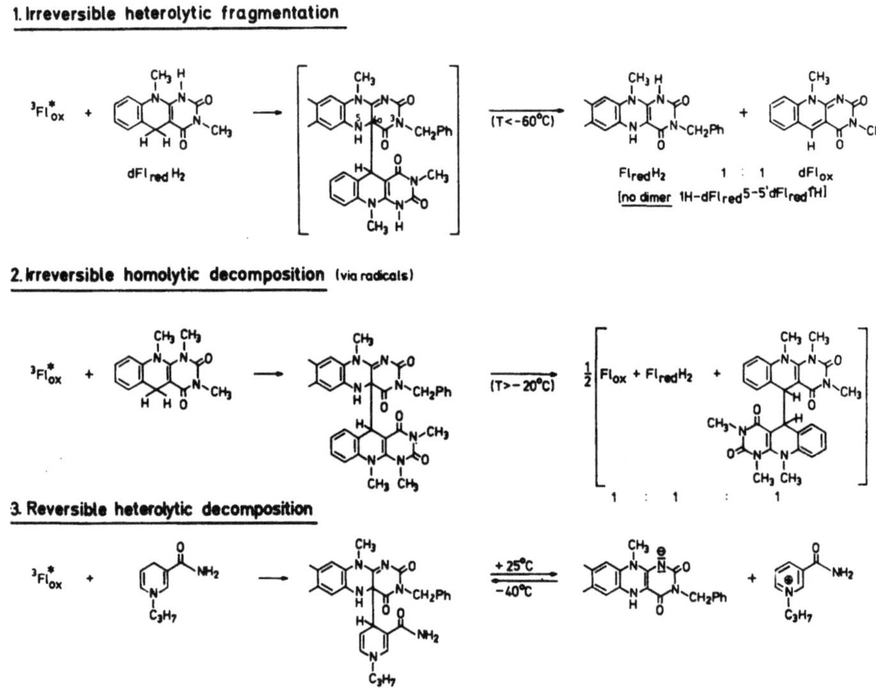

reactions in the dark via enzyme catalysis. However, in this model study the first step had to be performed photochemically. The subsequent reaction step in both cases should be identical. Further experiments, mainly NMR studies to characterize the σ-complex, are planned.

References

1. Blankenhorn, G., this volume, p. 185.
2. Walker, W.H., Hemmerich, P., and Massey, V., Reduktive Photoalkylierung des Flavinkerns und flavinkatalysierte Photodecarboxylierung von Phenylacetat. Helv. Chim. Acta 50, 2269-2279 (1967); Light-induced Alkylation and Dealkylation of the Flavin Nucleus. Eur. J. Biochem. 13, 258-266 (1970).
3. Knappe, W.-R. and Hemmerich, P., Reductive Photoalkylation of the Flavin-Nucleus: Structure and Reactivity of the Photoproducts. Liebigs Ann. Chem. 1976, 2037-2057.
4. Kornberg, A. and Horecker, B.L., in Snell, E.E., Biochemical Preparations, Wiley, New York 1953, Vol. 3, p. 24.
5. Hemmerich, P., The Present Status of Flavin and Flavocoenzyme Chemistry. Progress in the Chemistry of Organic Natural Products 33, 451-527 (1976); and literature quoted therein.

6. Bauch, W., Thesis, Berlin 1977

7. Hemmerich, P., and Schuman Jorns, M., Mechanisms of Hydrogen Transfer in Redox Enzymes, in Veeger, C., Drenth, J., and Oesterbaan, R.A., Enzymes: Structure and Function, North Holland Publ., Amsterdam 1972, p. 95-118

8. Hemmerich, P., this book

9. Massey, V., Matthews, R.G., Foust, G.P., Howell, L.G., Williams, Jr., C.H.,, Zanetti, G., and Ronchi, S., A New Intermediate in TPNH-Linked Flavoproteins, in Sund, H., Pyridine Nucleotide-Dependent Dehydrogenases, Springer Verlag, Berlin 1970, p. 393-409.

10. Blankenhorn, G., Nicotinamide Dependent One-Electron and Two-Electron (Flavin) Oxidoreduction: Thermodynamics, Kinetics, and Mechanism. Eur. J. Biochem. 67, 67-80 (1976), and literature quoted therein.

Hemmerich: I am challenged to judge between the two antagonists of my lab, and in order to make two long stories short, I want to draw the problem in this way: Four questions must be raised when the mechanism of redox equivalent "transfer" is at stake:

1) The $1e^-/2e^-$-question, which at my advice has been solved nonequivocally for nicotinamide by Dr. Blankenhorn in favor of $2e^-$-transfer.

2) If $2e^-$-transfer, we have to ask for the $2e^-$-"carrier", which yields the H^-/R^- - or hydride versus carbanion question. Here we have to distinguish the nicotinamide "input" (from alcohol), which is obviously hydride, from the "output" towards flavin, which at Dr. Knappe's and my advice is "carbanion", while Dr. Blankenhorn thinks, it is still hydride. (Neither H^- nor R^- will, of course, ever appear as "free" anions during catalysis.)
While this question cannot yet be decided with safety, it seems to me that nature is not as dull as to make use of two cofactors, nicotinamide and flavin, in sequence, if flavin was able to digest hydride equivalents. The very slow chemical reaction of flavin with BH_4^- is in agreement with this way of thinking.

What Dr. Blankenhorn has seen experimentally, concerns the 3. question: π- or σ-linkage between the interacting cofactors? Dr. Blankenhorn has characterized model π-complexes and shown that their spectra correspond to those of essential enzyme-substrate intermediates - for some flavoprotein dehydrogenases. Other ones do not show intermediate π-complexes and do still catalyze nicotinamide oxidation. Hence, Dr. Knappe's model seems to apply. In other words: Evidence for π-complexation is no evidence against "σ-transfer". On the contrary, as the Figure shows, σ-transfer can be shown to occur easily between position 4a (flavin) and 2 or 6 (nicotinamide), which is the answer to the 4. question: If σ-transfer, which positions of the heteroaromatic nuclei are concerned?

The picture of the Figure is at my opinion clearly preferable over H^--transfer towards nitrogen (flavin position 5), as involved in Dr. Blankenhorn's Fig. 3. Nucleophilic attack at an even very weakly basic pyridine type nitrogen seems to me chemically unsound if compared with the alternative of H^+-transfer to N(5) plus 2e-σ-transfer into C(4a).

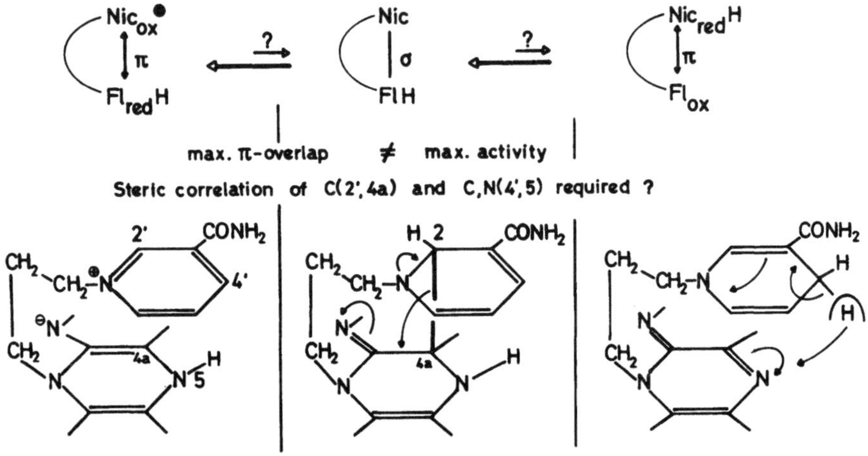

Nicotinamide-Flavin σ-Transfer proposed to occur between Positon 2 (Nicotinamide) and Position 4a (Flavin) with concomitant Transfer from Position 4 (Nicotinamide) to Position 5 (Flavin).

Veeger: Although I agree fully with a mechanism of H^- transfer in dehydrogenases which transfer reducing equivalents to a substrate like an aldehyde even if mediated via a flavoprotein. Nevertheless I think it might be possible that the combination of a low potential electron + proton gradient within a membrane could be the mechanism by which ATP is synthesized within a membrane.

In the NADH dehydrogenase of the respiratory chain the combination flavin and non-heme iron upon oxidizing the NADH could - by whatever mechanism - generate a single electron of about -700 mV. Furthermore in the respiratory chain the C_4-H^- NADH is not transferred but expelled.

Blankenhorn: I don't see why ATP synthesis in NADH dehydrogenase of the respiratory chain should be driven by both a low potential electron and a pH-gradient. To generate a potential of -700 mV for an iron-sulfur center you would need a pH-gradient of more than three pH-units which would be enough to drive ATP synthesis anyway.

Hemmerich: Dr. Veeger is concerned about the fact, that e.g. in mitochondrial respiration as well as in nitrogen fixation the "input" substrate nicotinamide does not mark the lowest potential reached, but there are apparently proteins (FeS-proteins or flavodoxins) working at lower level.

This has nothing to do with potential NAD-radicals. Instead, (Figure) we have a $2e^-$-input ($\approx$ -330 mV) from NADH into flavin, and the flavin "splits" the electron pair into one electron of low potential, which is transferred into the chain, and the second electron of higher potential, which is delivered to a store (another FeS-protein).

This is what we call a "transformase" unit, made up from one "dehydrogenase

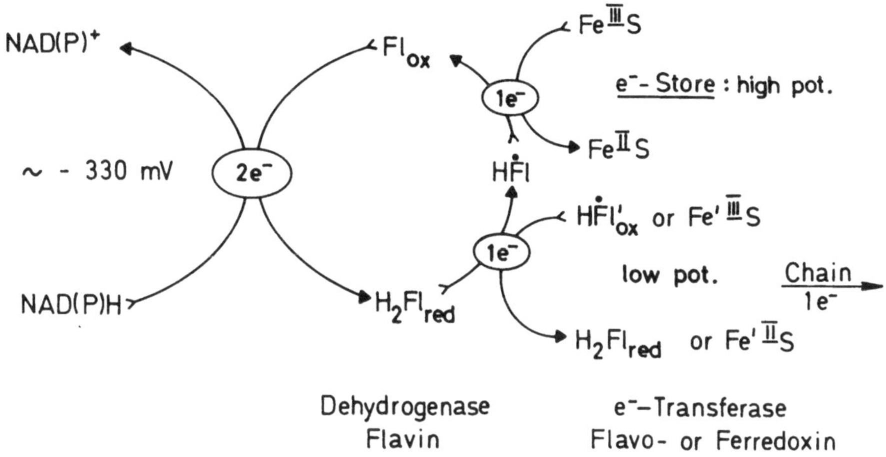

$$Fl + \begin{cases} Fl' \\ \text{or} \\ Fe'S \end{cases} + FeS = \text{TRANSFORMASE UNIT}$$

flavoprotein" running with 2e$^-$-input and 2 x 1e$^-$-output, and a "e$^-$-transferase", i.e. either ferredoxin or flavodoxin, and a third "storage" protein, i.e. high potential FeS-protein.

Südi: Dr. Blankenhorn, you seem to expect that, or would at least prefer, if NAD-dependent dehydrogenases showed neat linear free energy relations. I am afraid I can't follow you. Won't you agree that substrate specificity, as it is usually found in enzyme reactions, cannot obtain at the same time as a linear free energy relation?

Blankenhorn: If an enzyme is specific for a single substrate you will not be able to get a linear free energy relationship, obviously. If several substrates are turned over by the enzyme, a linear free energy relationship may be obtained if several conditions are met: the basic assumption is that ground states are in equilibrium with the transition state, then you have to be able to pick up the respective rates of the hydrogen transfer step, and finally you need to know the respective binding constants for substrates and products. In correcting for differential interactions, you actually account for substrate specificity. If no additional effects are important, you should get a linear free energy relationship telling you whether hydrogen was transferred as a hydride or a hydrogen atom as demonstrated in the case of yeast alcohol dehydrogenase.

SPECTROPHOTOMETRIC AND KINETIC IDENTIFICATION OF TRANSIENT INTERMEDIATES IN THE HORSE LIVER ALCOHOL DEHYDROGENASE CATALYZED REDUCTION OF SOME AROMATIC SUBSTRATES

M.F. Dunn, P. Schack, S.C. Koerber, A.M.-J. Au, G. Saliman and R.G. Morris
Department of Biochemistry, University of California, Riverside, California 92507, USA

INTRODUCTION

In 1973 Dunn and Hutchison (1) reported kinetic and spectrophotometric evidence documenting the formation of a transient chemical intermediate (I) during the equine liver alcohol dehydrogenase (LADH) catalyzed reduction of trans-4-N,N-dimethylaminocinnamaldehyde (DACA). In the original report and in subsequent studies (2,3), it was established that the intermediate (λ_{max} 464 nm, ε_{max} 6.2 x 10^4 M^{-1} cm^{-1}) is a relatively stable species above pH 9 in the presence of high concentrations of the enzyme-NADH complex and stoichiometric amounts of DACA (Fig. 1). At lower pH values, the intermediate decays to trans-4-N,N-dimethylaminocinnamyl alcohol and NAD^+. Detailed kinetic studies using the combined stopped-flow temperature-jump method indicated that intermediate formation is a simple, reversible pH-independent process (Eq. 1),

$$(1) \quad E(NADH) + DACA \underset{k_{-1}}{\overset{k_1}{\rightleftharpoons}} E(I) \rightleftharpoons \cdots$$

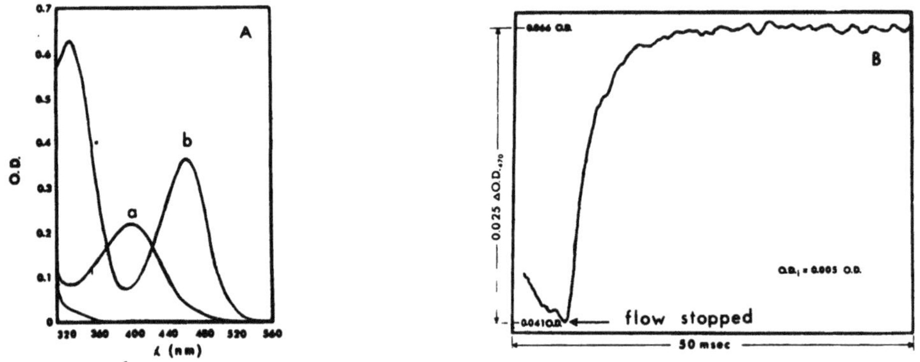

FIG. 1 (A) Comparison of the spectrum of 7.10 μM DACA (a) in a solution of 98.0 μN LADH with the spectrum of the intermediate (b) generated by the addition of 131 μM NADH to the cuvette at pH 9.36. (B) Stopped-flow time-course for the formation of the intermediate. Conditions: $[E]_0$ 7.3 μN; $[NADH]_0$ 78.0 μM; $[DACA]_0$ 2.50 μM; pH 8.75 sodium pyrophosphate buffer, 0.1 M; 25.0 ± 0.2°. Data taken from reference (1) with permission.

with $k_1 \simeq 4 \times 10^7\ M^{-1}\ sec^{-1}$ and $k_{-1} \simeq 280\ sec^{-1}$. It was also found that neither NAD^+, ADPR, ADP, AMP nor NMN will function in place of NADH in the formation of the intermediate, even though NADH is not chemically altered in this step (1,2).

The kinetics of intermediate formation, the spectral properties of the intermediate, and spectral and structural studies of model zinc-DACA complexes (3) are fully consistent with a structure for the intermediate involving a coordination bond between the zinc ion at the enzyme active site and the carbonyl oxygen of DACA (Scheme I).

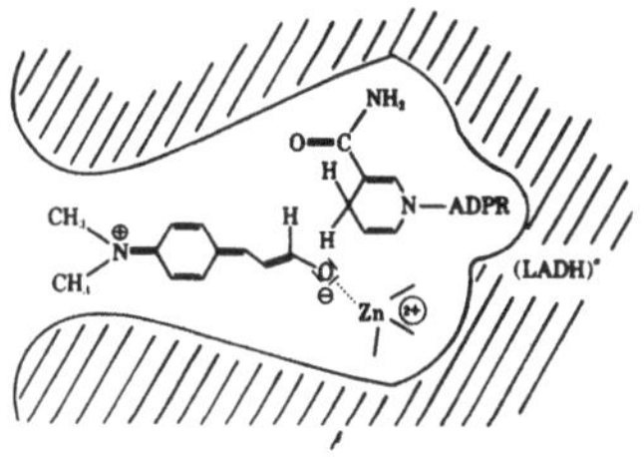

Scheme I

These findings led us to propose (a) that the LADH active site zinc ion functions as a Lewis acid in activating the aldehyde carbonyl for reduction (viz. Scheme I), and (b) that NADH functions as a noncovalent effector in this activation process.

The work described above provides a description of the catalytic mechanism prior to the redox process. In the studies described in this paper, we have used the following quasi-substrates and substrate analogues in combination with rapid kinetic techniques

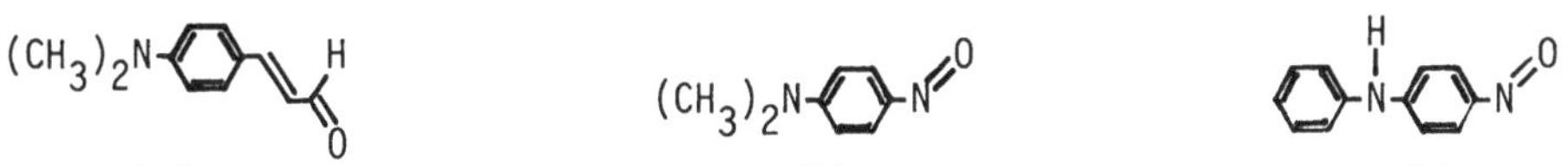

DACA NDMA NPA

to examine both the events which occur during the redox step(s) and the subsequent events which culminate in the desorption of products from the enzyme site. In addition, we report a novel LADH-catalyzed redox-elimination reaction involving arylnitroso compounds as substrate analogues.

RESULTS AND DISCUSSION

McFarland and Bernhard (4) have shown that under the conditions, [S], [NADH] >> [E], reaction can be limited to a single turnover of enzyme sites if a high concentration of pyrazole (i.e., 20 mM) is present. Termination of reaction after a single turnover is due to the quasi-irreversible formation of the E(NAD-pyrazole) adduct when alcohol dissociates from the E(NAD^+, Alc) product complex, as shown below. (Note that reaction at a single site is shown throughout for the sake of simplicity.)

$$(2)\quad \text{E(NADH)} + \text{Ald} \rightleftharpoons \text{E(NADH, Ald)} \rightleftharpoons \text{E(NAD}^+\text{, Alc)} \rightarrow \text{E(NAD}^+\text{)} + \text{Alc} \xrightarrow{\text{pyrazole}} \text{E(NAD-pyrazole)}$$

Under these single turnover conditions, the decay to products of the intermediate formed in the reaction of DACA with the E(NADH) complex obeys an apparent, first-order rate law when [DACA] >> [E(NADH)] (Fig. 2). Note also in Fig. 2 that when (4R)-4-deuterio-NADH (NADD) is used in place of NADH, the rate of the decay process is slowed. The magnitude of the observed rate constant (k_{app}) varies with substrate concentration, as shown in the inset to Fig. 2, and the best fit rectangular hyperbola to these data (solid line) is characterized by a hyperbolic constant, K_{app}, of 4×10^{-6} M. This value is identical (within experimental error) to the directly measured equilibrium constant for the process of intermediate formation (i.e., $k_{-1}/k_1 = K_D$) depicted in Eq. 1 (1). Values of K_{app}, determined as in Fig. 2, were found to exhibit little or no pH dependence over the pH range 4.5 to 9.

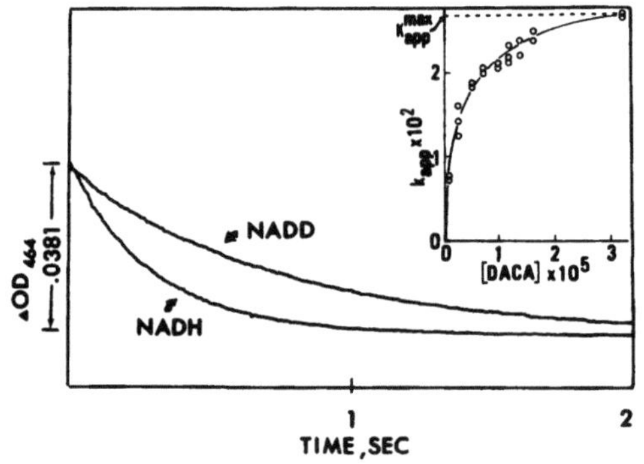

FIG. 2 Effects of deuterium substitution for the (4R)-4 hydrogen of NADH on the time-course for intermediate decay. Conditions: $[E]_0$ 9.5 µN; $[NADH]_0$ or $[NADD]_0$ 89 µM; $[DACA]_0$ 37.1 µM; $[pyrazole]_0$ 20 mM; pH 4.73 sodium acetate buffer, 0.1 M; 25.0 ± 0.2°. The inset shows the dependence of k_{app} on the concentration of DACA. Conditions: $[E]_0$ 2.39 µN; $[NADH]_0$ 98.0 µM; $[pyrazole]_0$ 8.2 mM; $[DACA]_0$ variable.

The preliminary rate studies reported by Dunn and Hutchison (1) indicated that the rate of the decay process is remarkably pH dependent. We have carried out a detailed study to establish the pH-rate profile for this process over the pH range 4.5 to 9. The data presented in Fig. 3 compare the pH-rate profiles for NADH and for NADD. Since LADH undergoes denaturation at a significant rate below pH 6, the k_{app}^{max} values at low pH were obtained by employing a pH-jump technique. The k_{app}^{max} values refer to the apparent first-order rate of intermediate decay measured in the presence of 20 mM pyrazole under the conditions, [E(NADH)] << [DACA] > K_{app}. Note in Fig. 3 that the pH dependencies for NADH and NADD indicate that the rate of intermediate decay is subject to a large, primary kinetic isotope effect at low pH ($k_H/k_D \simeq 2.5$), while at high pH, the isotope effect decreases and approaches a value of $k_H/k_D \simeq 1.0$. Furthermore, the rate of intermediate decay exhibits an apparent first-order dependency on the concentration of the protonated form of the enzyme. The data in Fig. 3 yield apparent pKa values of 6.3 and 6.7 respectively for NADH and NADD. Finally, note that the high pH form of the enzyme retains a residual k_{app}^{max} value which is 700-fold lower than the k_{app}^{max} value of the low pH form of the enzyme.

The pH dependencies of k_H/k_D and k_{app}^{max} together constitute evidence for a change in the nature of the rate limiting step as a function of hydrogen

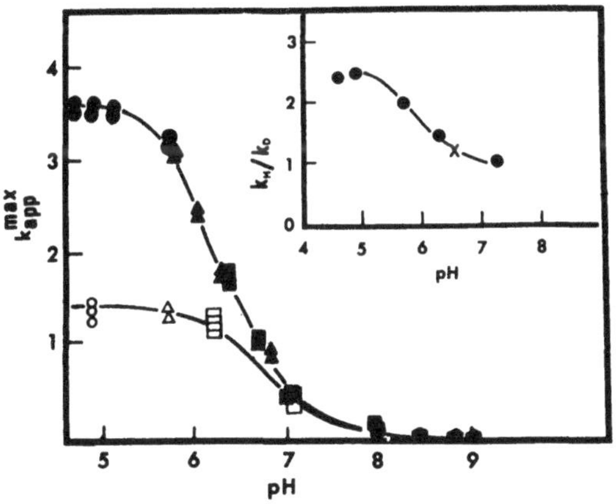

FIG. 3 Effects of deuterium substitution for the (4R)-hydrogen of NADH on k_{app}^{max} as a function of pH at 25.0 ± 0.2°. Buffers: (●,o) sodium acetate; (▲,△) sodium cacodylate; (■,□) sodium phosphate; (⬢,⬡) sodium pyrophosphate. NADH rates, solid symbols; NADD rates, open symbols. The inset shows the pH dependence of the kinetic isotope effect (k_H/k_D).

ion concentration. At low pH, the magnitude of the primary kinetic isotope effect establishes that hydride transfer is a component of the rate limiting step. At high pH, the absence of a kinetic isotope effect establishes that some subsequent step must be rate limiting.

These findings greatly restrict possible mechanisms for the process of intermediate decay. The spectrum of the intermediate and the kinetics of intermediate formation both are insensitive to pH changes over the pH range 4.5 to 10.5 as shown in previous work (1,2) and in the present studies. Therefore, the protonation event which influences the rate of intermediate decay (Fig. 3) does not involve protonation of the intermediate (Eq. 3); hence, all such schemes are ruled out.

$$(3)\quad E(I) + H^+ \rightleftharpoons E(IH^+) \longrightarrow \cdots$$

Furthermore, note that schemes which involve protonation of the enzyme in a pre-equilibrium step, i.e., Eq. 4,

$$(4)\quad E(I) + H^+ \rightleftharpoons H^+E(I) \longrightarrow \cdots$$

must also be discarded since all such mechanisms predict pH dependencies for k_H/k_D and k_{app}^{max} that do not correspond to the observed dependencies (Fig. 3).

The simplest reasonable family of mechanisms consistent with the observed pH dependencies of both k_{app}^{max} and k_H/k_D involves protonation in a step subsequent to the transfer of hydride (Schemes II and III below):

Scheme II

$$(5)\quad E(I) \underset{k_{-1}}{\overset{k_1}{\rightleftharpoons}} E(NAD^+, RCH_2O^-)$$

where R = 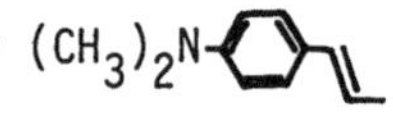

$$(6)\quad E(NAD^+, RCH_2O^-) + H^+ \underset{k_{-2}}{\overset{k_2}{\rightleftharpoons}} E(NAD^+, RCH_2OH) \qquad k_{-2}/k_2 = K_a$$

$$(7)\quad E(NAD^+, RCH_2OH) \xrightarrow{k_3} E(NAD^+) + RCH_2OH \xrightarrow[k_4]{\text{pyrazole}} E(NAD\text{-}pyr)$$

Scheme III

$$(8)\quad E(I) \underset{k_{-1}}{\overset{k_1}{\rightleftharpoons}} E(NAD^+, Alc)$$

$$(9)\quad E(NAD^+, Alc) + H^+ \underset{k_{-2}}{\overset{k_2}{\rightleftharpoons}} H^+E(NAD^+, Alc) \qquad k_{-2}/k_2 = K_a$$

$$(10)\quad H^+E(NAD^+, Alc) \xrightarrow{k_3} H^+E(NAD^+) + Alc \xrightarrow[k_4]{\text{pyrazole}} H^+E(NAD\text{-}pyr)$$

Assuming steady states for the products in Eqs. 5 and 8 and assuming mobile equilibria for Eqs. 6 and 9 and assuming the step, k_3, is quasi-irreversible (i.e., $k_4 >> k_3$), then the rate is given by the expression:

$$(11)\quad \text{rate} = -\frac{d[E(I)]}{dt} \simeq \frac{k_1[E(I)][H^+]}{K_a(k_{-1}/k_3) + [H^+]} = \frac{k_1[E(I)][H^+]}{K_{app} + [H^+]}$$

Both schemes assume that equilibrium in the first step favors E(I). Scheme II postulates that the protonation of a zinc-alcoholate product complex (Eq. 6) determines the rate of decay at high pH, while hydride transfer (Eq. 5) becomes rate limiting at low pH.

The mechanisms described by Scheme III differ from Scheme II only in that the group protonated is an enzyme residue. In one case the group is in close proximity to the active site and influences the affinity of the site for products via direct bonding interactions. Likely candidates include the zinc bound water molecule (assuming a pentacoordinate zinc ion) and the His(51) - Ser(48) proton relay system (5,6). The other case involves the protonation of a group that influences the affinity of the site for alcohol (and NAD^+) by modulating the conformation of the enzyme via an allosteric interaction.

As is obvious from Eqs. 5-7 and Eqs. 8-10, these mechanisms are kinetically indistinguishable. Note also that $K_{app} = K_a(k_{-1}/k_3)$, Eq. 11, and hence cannot be equated to the microscopic ionization constant of any group. Nevertheless, it is apparent from other studies (7-11) that the affinity of the enzyme for NAD^+ and for other LADH substrates is regulated by protonation equilibria. These studies demonstrate that the behavior of the DACA system is completely analogous to other substrates in this regard. Although the ionizable group involved has not been identified, this work provides strong evidence indicating that the role played by hydrogen ion in the catalytic mechanism is restricted to the modulation of the site affinity for NAD^+ and alcohol.

In 1971 Dunn and Bernhard (12) published a preliminary account of the LADH-catalyzed reaction of p-nitroso-N,N-dimethylaniline (NDMA) with NADH. We found that under single turnover conditions (i.e., limiting [NDMA]), the NDMA chromophore (λ_{max} 440 nm, ε_{max} 3.54 x 10^4 M^{-1} cm^{-1}) disappeared (Fig. 4B,D) in a rapid, apparent second-order reaction with the E(NADH) complex ($k_2 \simeq 4 \times 10^7$ M^{-1} sec^{-1}). In contrast, the optical density changes at 330 nm were found to lag considerably behind the 440 nm changes, and hence suggested the build up of stoichiometrically significant amounts of intermediate (Fig. 4C).

Later, in a serendipitous set of experiments in which Schack and Dunn (13) attempted to use phenolphthalein (λ_{max} 550 nm) as a probe of pH changes, it was discovered that the disappearance of NDMA is followed by the formation of a transient intermediate absorbing at 555 nm (Fig. 4A) having

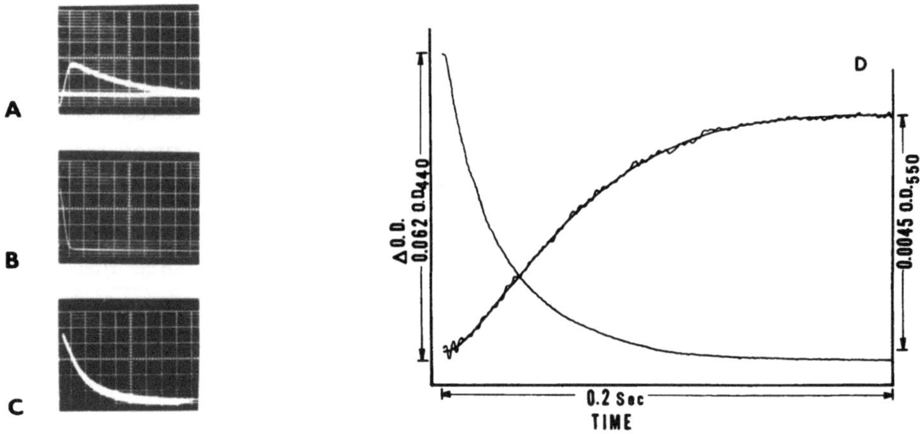

FIG. 4 (A,B,C) Stopped-flow kinetic traces for the reaction of NDMA with the E(NADH) complex at: (A) 550 nm; (B) 440 nm; (C) 330 nm. Ordinate, ∝ O./D.; abscissa, time (50 msec/div). The traces in (D) compare the 440 nm time-course with the 555 nm time-course under single-turnover conditions. $[E]_0$ 2.0 μN; $[NADH]_0$ 200 μM; [NDMA] 3.5 μM; $[pyrazole]_0$ 20 mM; pH 8.71 sodium phosphate buffer, 0.1 M; 25.3 ± 0.1°.

nothing to do with phenolphthalein. Further studies of the NDMA reaction and the reaction of a homolog, 4-nitroso-N-phenylaniline (NPA), provide a fairly detailed description of the LADH-catalyzed reactions of these arylnitroso compounds with NADH.

The NDMA reaction proceeds with the overall stoichiometry:

(12) 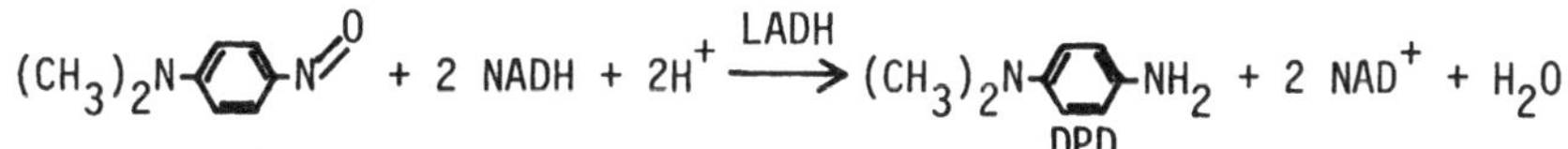

The products, NAD^+ and N,N-dimethyl-p-phenylenediamine (DPD) have been identified by isolation, derivitization, and comparison with known derivatives. The time course for the formation of the 555 nm intermediate is characterized by an initial lag phase (Fig. 4D), suggesting the intervention of at least one additional intermediate subsequent to the disappearance of NDMA but prior to the formation of the 555 nm species. The decay of the 555 nm species to DPD is first order with respect to the concentration of free NADH, and independent of the concentration of enzyme-bound NADH. Thus, the decay process is nonenzymatic and therefore one of the two moles of NADH consumed (viz. Eq. 12) involves a nonenzymatic process. Finally, the spectrum of the intermediate exhibits two long wavelength absorption bands, $\lambda_{max} \sim 515$ nm and $\lambda_{max} \sim 555$ nm (Fig. 5A).

We have tentatively identified the intermediate as the radical cation of DPD $(CH_3)_2N$--$\dot{N}^+H_2$, via spectral comparisons with an authentic sample

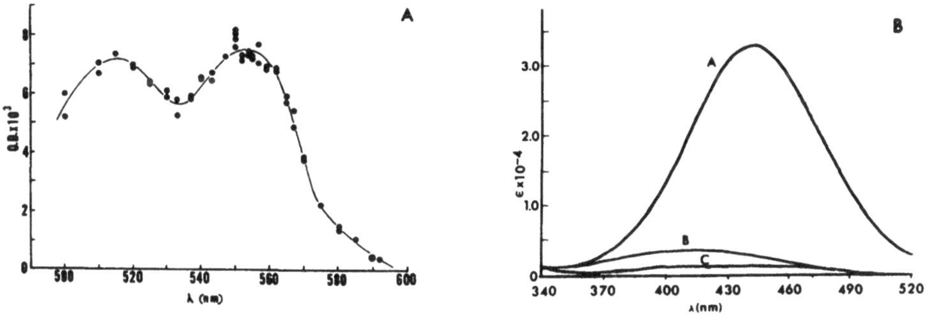

FIG. 5 (A) Spectrum of the 555 nm intermediate derived from the reaction of NDMA with the E(NADH) complex in 0.1 M pH 8.71 sodium pyrosphosphate buffer at 25.0 ± 0.2°. The spectrum is reconstructed from stopped-flow kinetic traces collected as a function of wavelength. (B) Comparison of the spectrum of NPA (trace A) with products derived from the reaction of NPA with the E(NADH) complex. Trace B, the initial product, 4-(phenylimino)-2,5-cyclohexadien-1-imine; trace C, the final hydrolysis product, 4-(phenylimino)-2,5-cyclohexadien-1-one.

generated by the $Fe(CN)_6^{-2}$ oxidation of DPD (14).

The behavior of the LADH-catalyzed reaction of NPA with NADH is in several respects analogous to the behavior of the NDMA system. Under single turnover conditions, NPA (λ_{max} 440 nm, ε_{max} 3.3×10^4 M^{-1} cm^{-1}) also disappears in a rapid, apparent second-order process ($k_2 = 4 \times 10^7$ M^{-1} sec^{-1}), yielding NAD^+ and a product absorbing at 430 nm, apparent ε_{max} 4.0×10^3 M^{-1} cm^{-1}, (Fig. 5B). This initially formed product undergoes a slow ($t_{1/2} \simeq$ 20 min.) nonenzymatic conversion to the quinonemonoimine, 4-(phenylimino)-2,5-cyclohexadien-1-one (15). Treatment of the initial product with $NaBH_4$ and then acetic anhydride yields N-acetyl-N'-phenyl-p-phenylenediamine. Treatment of the initial product with hydroxylamine regenerates NPA. These observations allow a tentative identification of the initial product as the quinonediimine, 4-(phenylimino)-2,5-cyclohexadien-1-imine (viz. Eq. 13):

(13)
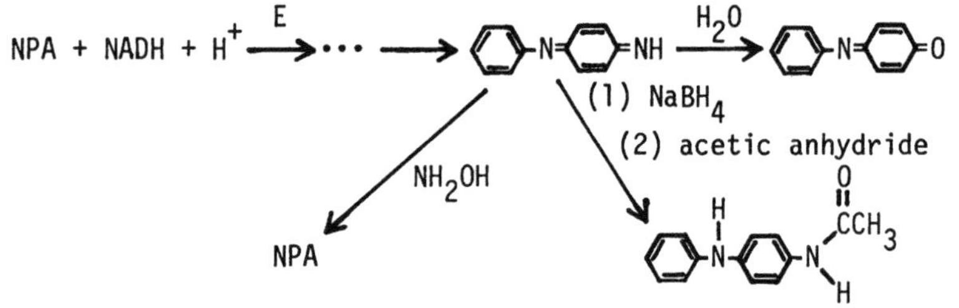

From these studies it is clear that both NDMA and NPA undergo a facile LADH-catalyzed redox-elimination reaction with NADH, yielding quinonediimine products which then decompose via nonenzymatic pathways. The following mechanism (Scheme IV) is proposed to explain these findings:

Scheme IV

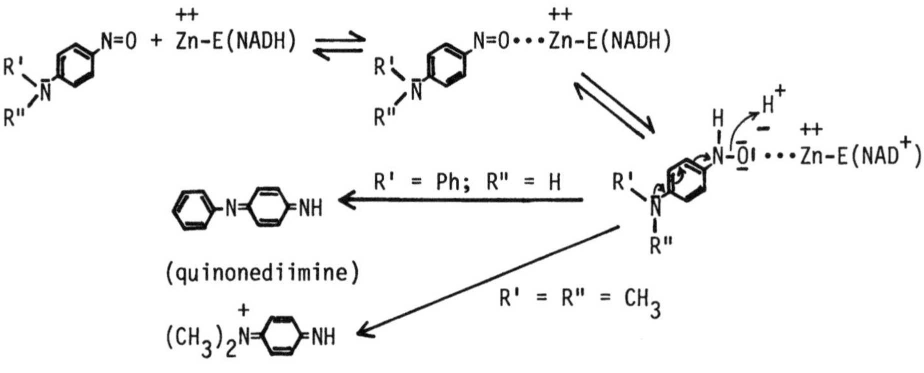

Scheme IV (continued)

When R' = R" = CH_3,

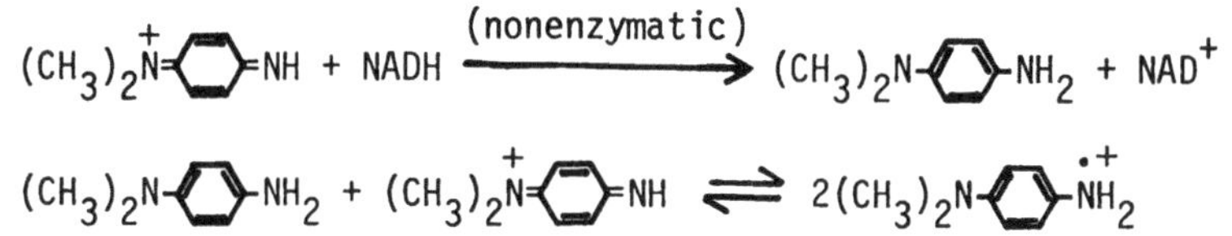

Here, the active site zinc ion is postulated to function as a Lewis acid both in the reduction of the nitroso group to the hydroxylamine oxidation state and in the elimination of the hydroxyl group to generate the quinonediimine product. The quinonediimine derived from NPA undergoes slow hydrolysis as indicated in Eq. 13. The decay of the quinonediimine derived from NDMA is more complicated. The radical cation intermediate (Fig. 5A) probably arises via the reduction of the quinonediimine to DPD by reaction with free NADH; then DPD undergoes a one electron transfer reaction to the quinonediimine to yield the radical cation. The work of Baetzold and Tong (14) has established that the (reversible) formation of this radical cation via the proposed one electron transfer is a facile process. Our preliminary studies show that NADH rapidly converts the radical cation-quinonediimine equilibrium mixture to DPD under conditions similar to those employed in the enzyme initiated reaction.

CONCLUSIONS

The only reasonable interpretation of the kinetic and spectral properties of the DACA-LADH system is that the intermediate formed is poised for hydride transfer via inner sphere coordination of the carbonyl oxygen to the active site zinc ion (5). NADH plays a critical, noncovalent role in facilitating the formation of this activated intermediate, presumably by binding to the catalytically active conformation of the enzyme. Similar noncovalent effector roles have been documented for NAD^+ and NADH in the D-glyceraldehyde-3-phosphate dehydrogenase system (see Bernhard et al., this volume). Following hydride transfer, protonation of the E(NAD^+, Alc) ternary complex triggers the release of products from the site. The novel redox-elimination reaction observed in the LADH catalyzed reaction of the p-nitrosoaniline substrate analogues (NDMA and NPA) serves to further emphasize the Lewis acid character of the LADH active site zinc ions.

ACKNOWLEDGMENTS

This work was supported by National Science Foundation Grants GB-31151 and BMS 75-20339. A.M.-J. Au and G. Saliman received support from the National Science Foundation Undergraduate Summer Research Program (GY 9922) during a portion of these studies. We thank Professor Sidney A. Bernhard for contributions in the form of theoretical discussions.

REFERENCES

1. Dunn, M.F., Hutchison, J.S.: Roles of zinc ion and reduced coenzyme in the formation of a transient chemical intermediate during the equine liver alcohol dehydrogenase catalyzed reaction of an aromatic aldehyde. Biochemistry 12, 4882-4892 (1973).
2. Dunn, M.F., Biellmann, J.-F., Branlant, G.: Roles of zinc ion and reduced coenzyme in horse liver alcohol dehydrogenase catalysis. The mechanism of aldehyde activation. Biochemistry 14, 3176-3182 (1975).
3. Angelis, C.T., Dunn, M.F., Muchmore, D.C., Wing, R.M.: Studies on selected Lewis acid complexes which exhibit striking spectroscopic similarities to a productive horse liver alcohol dehydrogenase ternary complex. Biochemistry, in press (1977).
4. McFarland, J.T., Bernhard, S.A.: Catalytic steps during the single-turnover reduction of aldehydes by alcohol dehydrogenase. Biochemistry 11, 1486-1493 (1972).
5. Dunn, M.F.: Mechanisms of zinc ion catalysis in small molecules and enzymes. Structure and Bonding 23, 61-122 (1975).
6. Brändén, C.-I., Jörnvall, H., Eklund, H., Furugren, B.: Alcohol dehydrogenases. Enzymes, 3rd. ed., Vol. XI, Part A, 103-190 (1976).
7. Taniguchi, S., Theorell, H., Åkeson, Å.: Dissociation constants of the binary complex of homogeneous horse liver alcohol dehydrogenase and nicotinamide adenine dinucleotide. Acta Chem. Scand. 21, 1903-1920 (1967).
8. Dunn, M.F.: A comparison of the kinetics and stoichiometry of proton uptake with aldehyde reduction for liver alcohol dehydrogenase under single turnover conditions. Biochemistry 13, 1146-1151 (1974).
9. Shore, J.D., Gutfreund, H., Brooks, R.L., Santiago, D., Santiago, P.: Proton equilibria and kinetics in the liver alcohol dehydrogenase reaction mechanism. Biochemistry 13, 4185-4191 (1974).

10. McFarland, J.T., Chu, Y.-H.: Effect of pH on the liver alcohol dehydrogenase reaction. Biochemistry 14, 1140-1146 (1975).
11. DeTraglia, M.C., Schmidt, J., Dunn, M.F., McFarland, J.T.: Liver alcohol dehydrogenase-coenzyme reaction rates. Biochemistry, in press (1977).
12. Dunn, M.F., Bernhard, S.A.: Rapid kinetic evidence for adduct formation between the substrate analog p-nitroso-N,N-dimethylaniline and reduced nicotinamide-adenine dinucleotide during enzymic reduction. Biochemistry 10, 4569-4575 (1971).
13. Schack, P., Dunn, M.F.: Direct spectrophotometric observation of a transient intermediate in the LADH-catalyzed reduction of 4-nitroso-N,N-dimethylaniline (NDMA) by NADH. 8th Meeting of the Federation of European Biochemical Societies, Amsterdam, Holland, August 20-25 (1972).
14. Baetzold, R.C., Tong, L.K.J.: Kinetics of redox reactions of oxidized p-phenylenediamine derivatives. I. J. Amer. Chem. Soc. 93, 1347-1353 (1971).
15. Bhatnagar, I., George, M.V.: Oxidation with metal oxides. III. Oxidation of diamines and hydrazines with manganese dioxide. J. Org. Chem. 33, 2407-2411 (1968).

Received February 2, 1977

DISCUSSION

Südi: In your studies of the reduction of dimethylamino cinnamaldehyde you have observed a very interesting pH dependence of the kinetic isotope effect at substrate saturation. In our interpretation of this pH dependence you have used a scheme to which you referred as describing "the decay of a (single) intermediate to products". Now, is this an assumption, or do you have any evidence on the point that maximal velocity is to be interpreted in this case in terms of a so called "one-central complex mechanism"?

Dunn: First of all, we have not been able to detect any additional stoichiometrically significant species involving E, coenzyme and substrate along the reaction path. Certainly the decay of the intermediate is a strict first-order process under single turnover conditions. Therefore, the spectral changes at this wavelength appear to reflect the decay of a single intermediate. Within the limitations of the temperature jump instrument used, the formation of the intermediate is adequately described as a bimolecular association-unimolecular dissociation process, E(NADH) + DACA $\rightleftharpoons$ E(I). The equilibrium constant for this process is numerically equal to the steady-state K_m value for DACA. The saturated apparent first-order decay rate is numerically equal to the steady-state k_{cat} value. Therefore, the kinetics do not justify the existence of additional, stoichiometrically significant intermediates. However, the pH dependence of the decay and the pH dependence of the kinetic isotope effect demonstrate the existence of one or more kinetically significant intermediates, as postulated in schemes II and III.

Hemmerich: I must congratulate Dr. Dunn, since he is one of the few biochemists of my knowledge, who were not fooled by finding a radical in his system, but recognized immediately that the radical arose from a secondary chemical, fast comproportionation reaction, and not from direct $1e^-$-transfer between biological redox-carriers.

Concerning the mechanism of substrate activation in nicotinamide dependent dehydrogenation, one must, however, ask, whether Dr. Dunn's mechanism is unique. I agree fully with his nice findings and interpretations, but I want to propose, that at least four alternative modes of alcohol activation are possible, namely

1) acidic, i.e. by Zn^{2+}, such as in alcohol dehydrogenase (Dunn)
2) basic, i.e. by histidine, such as in LDH (Rossmann)
3) π-complex, as postulated for chemical models by Chipman (1), and
4) "σ-complex", for which I want to give a chemical example, as presently studied by Dr. Duchstein and myself: Up to date, no chemical reaction has been described which might fulfil the pretention of being a model reaction for nicotinamide dependent biological dehydrogenation of "simple" alcohols (1, 2, 3). Such a reaction must cope at least with the following criteria:
   1) It should proceed at ambient temperature with at least a comfortably measurable, if not rapid rate.
   2) It should work with "simple", i.e. small, aliphatic, saturated alcohols.
   3) It should work with nicotinamide derivatives devoid of the nucleotide side chain, but otherwise as unmodified as possible.

If we consider the role of the apoprotein in NAD(P) dependent oxidoreduction - and not magics - as being responsible for the apparent distinction between chemical and enzymic reactions, it is easily seen that three prerequisites must be dealt with apart from the acid-base catalysis required to pre-activate the alcohol which the chemical and the biological reaction have in commen (4) (zero prerequisite):

o) B + >CHOH $\rightleftharpoons$ BH^+ + $>CHO^-$
1) The apoprotein must rule out attack of $-O^-$ at the NAD^+ nucleus, unless the alcoholate adduct is an intermediate.
2) The apoprotein must protect nicotinamide positions 2 and 6 against nucleophilic attack (in the oxidized state) (5) and position 5 against electrophilic attack (6) (in the reduced state).
3) The apoprotein must keep any α-CH groups of the carbonyl product apart from the reoxidized nicotinamide, in order to exclude carbanion additions (7) (Fig. 1).

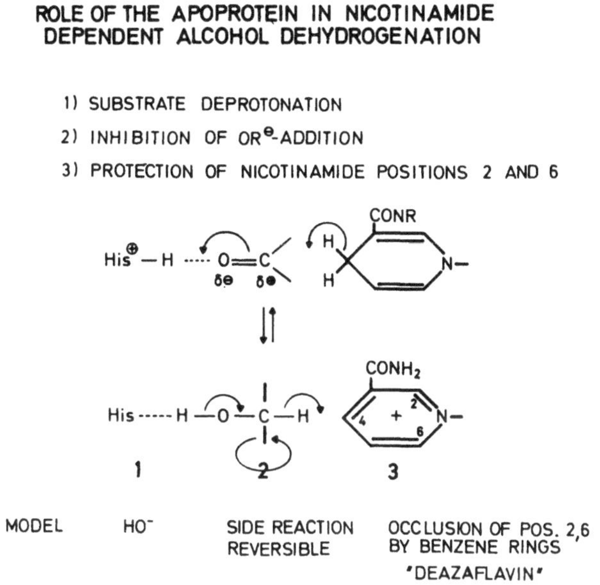

In the design of a model system, prerequisite 3 can be taken care of by keeping the substrate in high excess over the product, while prerequisite 1 may be fulfilled by the reversibility of alcoholate addition. This might decide whether the alcoholate adduct can be an intermediate in the reductive reaction, such as in the reduction of NAD^+ by dithionite (8).

The most serious difficulties arise from prerequisite 2 which demands the occlusion of the pyridine nucleus by at least two benzene nuclei. Preferably, however, the design should be such as to incorporate the $CONH_2$-function into one of the blocking groups, and this leads to a model once thought to be a "flavin analog":

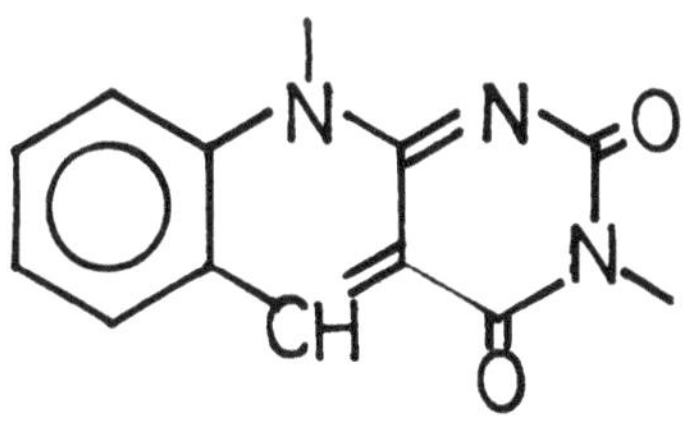

It turns out, however, that this "5-deazaflavin" unit behaves as a "flavin-shaped nicotinamide analog".

We have found, and we have joined a Japanese research group headed by Dr. Yoneda arriving at similar results independently (9) that 5-deazaflavin oxidizes "simple" alcohols (with the exception of methanol) at pH $\approx$ 13 at ambient temperature, if only slowly, but quantitatively. Fig. 2 shows the spectral course of this reaction, which involves the alcoholate adduct as

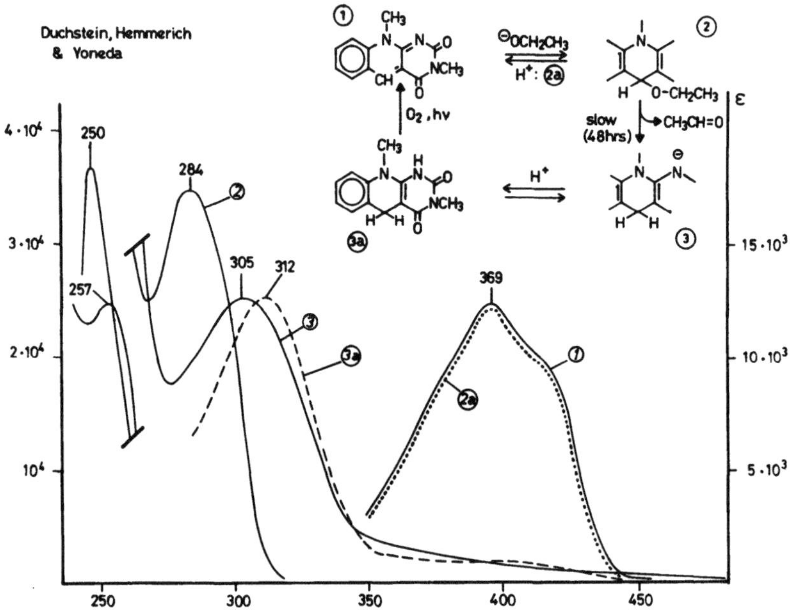

Figure 2

Tentative Interpretation of the Spectral Behaviour of Nicotinamide towards Dry Alcoholate

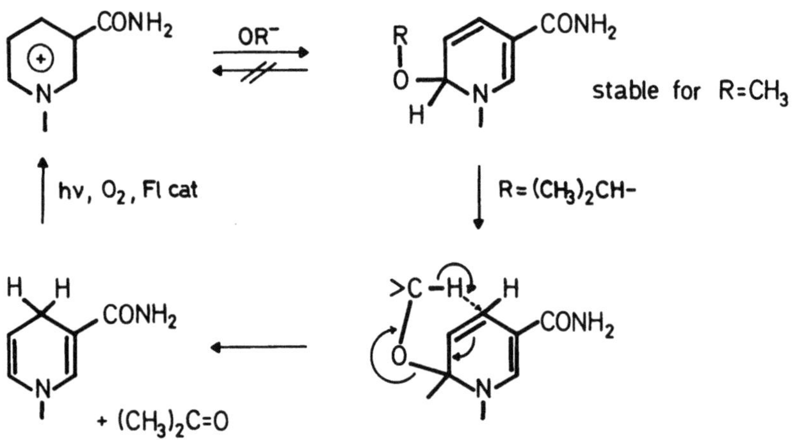

Figure 3

a rapidly and reversibly formed intermediate rather than a side product. The intramolecular hydride transfer is rate determining, as shown by the spectral shift from 284 to 305 nm. After this rearrangement, the formed product can quantitatively be reconverted to the starting deazaflavin by oxidation.

Since the properties and the reactivities of deazaflavin radicals - leading to stable covalent dimers - have been worked out by us, (10, 11) we can conclude that the present reaction does not involve $1e^-$-transfer steps in confirmation of Dr. Blankenhorn's results presented at this symposium.

Presently we are concerned with the application of these data to 2,5,6-"unblocked" N-alkyl nicotinamides. Preliminary results lead us to propose the reaction scheme of Fig. 3 based on an original proposal of Hamilton (12). There is no doubt that this model reaction is unefficient as compared with a hypothetic enzymic arrangement of e.g. histidine base, substrate and coenzyme being arranged "on line", but it shows that some kind of intermediate fixation of the alcoholate oxygen is chemically required in favor of an efficient hydride activation.

References:

1) J.J. Steffens, D.M. Chipman, J.Am.Chem.Soc. 93, 6694 (1971)
2) D.J. Creighton, J. Hajdu, G. Mooser, D.S. Sigman, J.Am.Chem.Soc. 95, 6855 (1973)
3) D.J. Creighton, J. Hajdu, D.S. Sigman, J.Am.Chem.Soc. 98, 4619 (1975)
4) M. Rossmann, this Volume
5) K. Wallenfels, H. Schüly and D. Hofmann, Ann.Chem. 621, 106 (1959)
6) A.G. Anderson and G. Berkelhammer, J.Am.Chem.Soc. 80, 992 (1958)
7) J. Ludowieg, N. Bhacca and A. Levy, Biochem.Biophys.Res.Comm. 14, 431 (1964)
8) J.F. Biellmann and H.J. Callot, Bull.Soc.Chim.Fr. 1154 (1969)
9) F. Yoneda, personal communication
10) H. Fenner, H.H. Rößler, H.-J. Duchstein, P. Hemmerich (1976), Fifth International Symposium on Flavins and Flavoproteins, Amsterdam, Elsevier Publ., p. 343
11) H.-J. Duchstein, H. Fenner, R. Grauert, G. Blankenhorn, P. Hemmerich, W.-R. Knappe, V. Massey, M. Goldberg, I. Pecht, submitted to FEBS Letters
12) G.A.Hamilton, Prog.Bioorg.Chem. 1, 83 (1971)

CONFORMATION OF NAD^+ IN SOLUTION, IN HOLOENZYMES AND IN THE CRYSTALLINE Li^+-COMPLEX

W. Saenger, B.S. Reddy, K. Mühlegger* and G. Weimann*
Abteilung Chemie, Max-Planck-Institut für experimentelle Medizin,
Hermann-Rein-Str. 3, D-3400 Göttingen, Germany
*Boehringer-Mannheim, Biochemica-Werk Tutzing, D-8132 Tutzing, Germany

INTRODUCTION

The coenzyme NAD^+ is composed of two nucleotides, adenylic acid and nicotinamide-ribosyl-5'-phosphate linked together in a head to head fashion by a pyrophosphate linkage. This chemical structure (fig. 1) was elucidated in 1936 by Warburg and Christian and since then the three dimensional geometry of the molecule was studied by spectroscopic methods in aqueous medium and by X-ray analysis in several crystalline holoenzyme complexes (1). Although a wealth of data was obtained, high resolution details of NAD^+ were not available until a single crystal analysis of the Li^+-complex of

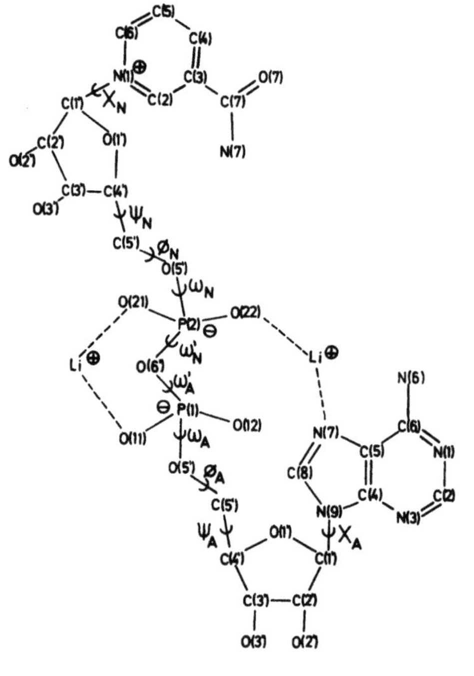

Fig. 1 The chemical structure of NAD^+ and numbering convention. The torsion angles are as defined in ref. 3. The Li^+-coordination is indicated by broken lines.

NAD^+ became available. In this report the conformation of NAD^+ in solution, in holoenzyme and in the Li^+-complex will be described, compared and discussed.

GENERAL PRINCIPLES OF NUCLEOTIDE STRUCTURE

Conformational analyses of nucleotides have been carried out by spectroscopic methods (NMR, ORD, CD), quantum chemical calculations and by X-ray crystallographic techniques. The main results obtained from these investigations can be summarized with a few statements (for review, see ref. 2-5).

The sugar (ribose) five membered ring is never planar but puckered in either an envelope form with atoms C(2') or C(3') out of the plane through the other four atoms or in a half chair form with both atoms C(2')- and C(3') on opposite sides of the plane formed by atoms C(1')-O(1')-C(4'). The conformations and the nomenclature (<u>endo</u> = on the same side as C(5'), <u>exo</u> = on the other side) displayed in fig. 2 represent the preferred puckering modes; other puckerings are possible but less frequently observed.

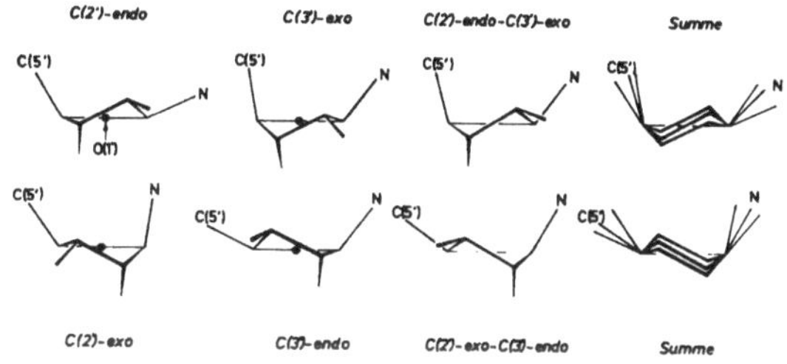

Fig. 2 The puckering modes of riboses. Note that the orientations of glycosyl linkage and C(4')-C(5') bond depend on the sugar conformation.

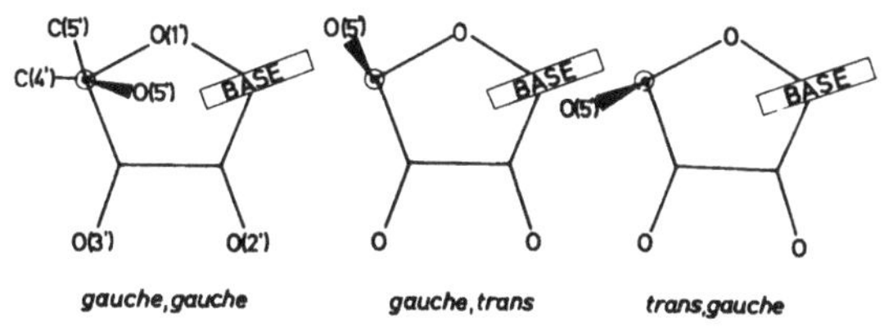

Fig. 3 Conformation about the C(4')-C(5') bond.

The heterocycles can occur in a syn or anti orientation relative to the sugar moiety. In the anti form which corresponds to the preferred one, the bulkier group of the heterocycles (pyrimidine ring in adenosine and amide group in nicotinamide) point away from the sugar ring (χ about 30°) while in the syn conformation they are "above" the sugar (χ about -150°).

A third structural parameter concerns the conformation about the C(4')-C(5') bond (fig. 3). The orientation of the C(5')-O(5') bond with respect to the ribose is generally gauche, gauche, i.e. the torsion angles C(3')-C(4')-C(5')-O(5') and O(1')-C(4')-C(5')-O(5') are at $+60^o$ and at -60° (ψ is at (+)gauche). This is true for all the double helical polynucleotides studied thus far, for most of the deoxyribonucleotides with a few exceptions but for all the natural ribonucleotides with only one exception, namely the modified 6-azauridine-5'-phosphate (6).

For the torsion angles involving the phosphate group, similar standard values can be given. Thus, the torsion angle ϕ (C(4')-C(5')-O(5')-P) is generally trans, and the angles about the O(5')-P and P-O(6') bonds, ω and ω' are either both (+)gauche or both (-)gauche or ω' can be trans with ω ($\pm$)gauche (3). However, the pyrophosphate linkage P-O-P possesses considerable greater torsional flexibility than P-O-C linkages because the interatomic angle at pyrophosphate oxygen, about 130°, is increased relative to the angle at the ester oxygen atom, 120° (7). Therefore, the torsion angles O-P-O-P (ω') can cover all ranges except cis-planarity (0°).

In conclusion, the preferred geometry of a nucleotide can be described as: sugar puckering mode either envelope (C(3')-endo and C(2')-endo) or twist (C(2')-endo-C(3')-exo and C(2')-exo-C(3')-endo); orientation about the C(1')-N linkage anti (χ around 30°) and about the C(4')-C(5') bond gauche, gauche (ψ at (+)gauche). The phosphate group is trans to C(4')-C(5') with ϕ near 180°, the orientation about the O(5')-P bond (ω) is ($\pm$)gauche and the pyrophosphate linkage P-O-P is quite flexible with a forbidden range at 0° (cis-planar) for the torsional angle ω'.

STRUCTURE OF $Li^+ \cdot NAD^+$ IN THE CRYSTALLINE STATE

The Li^+-complex of NAD^+ crystallizes from aqueous methanol as dihydrate. The structure was solved by conventional single crystal X-ray diffraction methods but could only be refined to a discrepancy index R = 10 % owing to experimental difficulties (minute crystals). However, the main features of the molecular geometry are clear (fig. 4) and the Li^+ cation with only two electrons could be located with certainty.

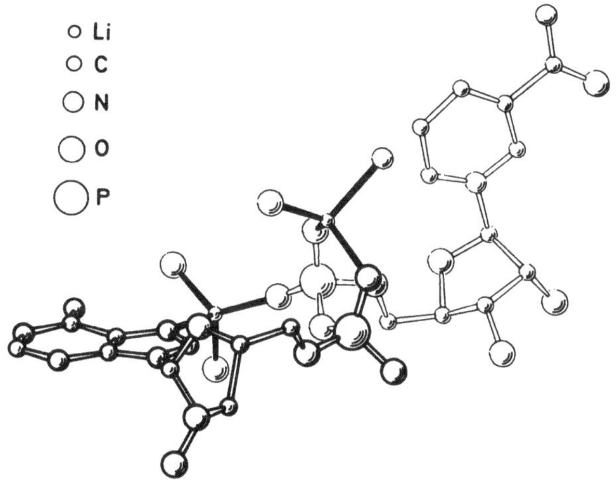

Fig. 4 Structure of the crystalline $Li^+ \cdot NAD^+$ complex. One Li^+ cation is chelated with two symmetry related NAD^+ molecules or (vice versa) one NAD^+ is chelated with two Li^+ cations.

In the Li^+-complex, the NAD^+ molecule exists in an "extended" form with adenine and nicotinamide heterocycles separated by about 11.5 Å and nearly perpendicular to each other (fig. 4,5,6,7). Both nucleotides are in the anti conformation, the torsion angles involving the C(4')-C(5') bonds, ψ, are (+)gauche and the sugar puckering modes are C(3')-endo for nicotinamide riboside but C(2')-endo-C(3')-exo for adenosine (table 1). This difference is probably brought about by the coordination with the Li^+-cation which involves adenine N(7) but not the nicotinamide heterocycle.

TABLE 1

Torsion angles (°) of NAD^+ in binary and ternary complexes with lactate dehydrogenae (LDH; 10) and in $Li^+ \cdot NAD^+$. Angles defined as in (3); α and ß are torsion angles N(9)···P(1)···P(2)···N(1) and $O(5')_A$-P(1)···P(2)-$O(5')_N$. A and N denote adenosine and nicotinamide-riboside.

Torsion angle	$LDH \cdot NAD^+$ A	$LDH \cdot NAD^+$ N	$LDH \cdot NAD^+ \cdot$pyruvate A	$LDH \cdot NAD^+ \cdot$pyruvate N	$Li^+ \cdot NAD^+$ A	$Li^+ \cdot NAD^+$ N
χ	30	74	37	40	52	15
ψ	-75	-88	-75	-47	48	47
φ	-173	180	-162	-172	163	179
ω	104	109	107	85	-125	79
ω'	-156	37	-156	49	133	72
α	-45		-41		172	
β	-115		-130		-164	

Sugar puckerings are all C(3')-endo except adenosine in $Li^+ \cdot NAD^+$, C(2')-endo-C(3')-exo.

Although standard deviations of bond angles and bond distances are rather large (∼0.01 Å to ∼0.02 Å and 1°) for the structure determination of $Li^+ \cdot NAD^+$, particuliarities involving the pyrophosphate bridge are worth mentioning. Similar as in other pyrophosphate structures (8) the P(1)-O(6')-P(2) angle, 133°, is widened as compared to "normal" diester bond angles, C-O-P, of 120° and the P(1)-O(6') and O(6')-P(2) distances, 1.56 Å and 1.66 Å, are significantly different from each other. Because the covalently bonded substituents to P(1) and P(2) are essentially the same (riboses), the difference in P-O(6') bond distances has to be attributed to the Li^+-coordination.

The Li^+ is coordinated with four ligands belonging to two different NAD^+ molecules (fig. 4,5); or, in other terms, one NAD^+ molecule is coordinated to two Li^+ cations. One of these bridges adenine N(7) and an unesterified oxygen atom of the phosphate group P(2) while the other bridges two unesterified oxygen atoms belonging to phosphates P(1) and P(2). The former coordination forces the ADP residue of NAD^+ to adopt a conformation with

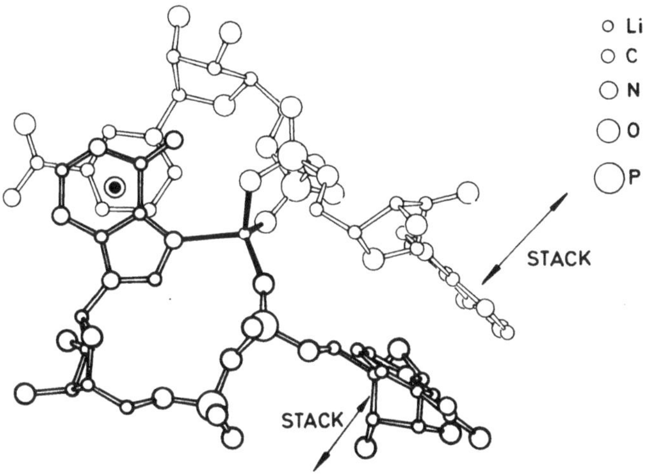

Fig. 5 Dimer formation in $Li^+ \cdot$ NAD crystals. Intermolecular heterocycle stacks are indicated by ⊙ and by arrows. Note the tetrahedral coordination of Li^+ cation.

the sugar puckered C(2')-endo-C(3')-exo and the torsion angles ω_A and ω'_A both in eclipsed orientation while these angles are usually gauche or trans in 5'-nucleotides and in holoenzyme $\cdot$ NAD^+ (table 1). The coordination of the second Li^+-ion forces the pyrophosphate group into a conformation with both P(1)-O(11) and P(2)-O(21) bonds nearly cis-planar because only then both phosphate groups can ligand the same Li^+-ion; in a nonplanar conformation the O(11)···Li···O(21) distances would be too wide to allow such complex formation. The overall shape of the $Li^+ \cdot NAD^+$ molecule is such that the O(5') atoms of the nucleoside residues are arranged trans with respect to the pyrophosphate linkage and contrasts the eclipsed orientation of the nucleosides in NAD^+ bound to enzyme (fig. 6 and angles ß in table 1).

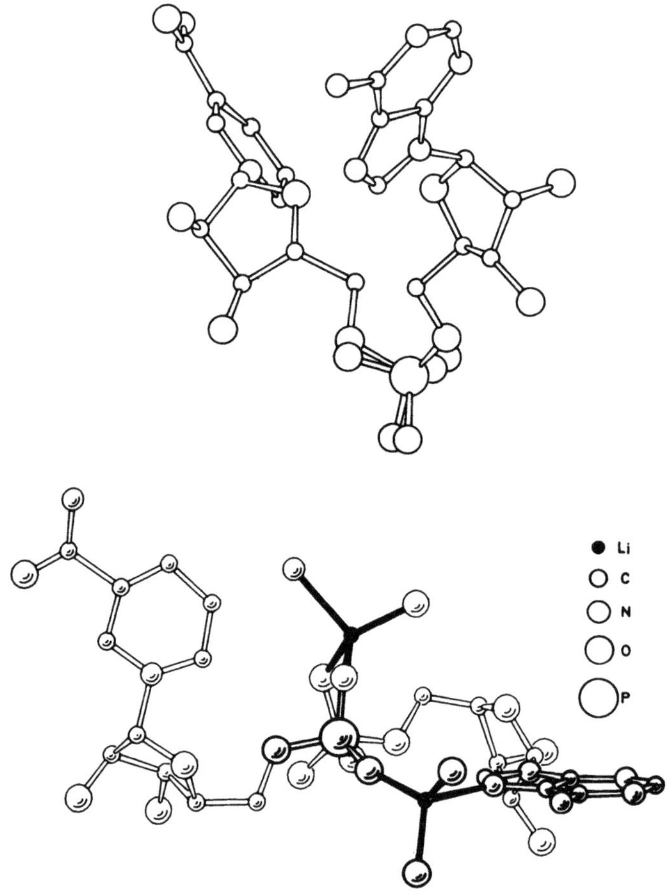

Fig. 6 Comparison of $Li^+ \cdot NAD^+$ (bottom) and the ternary LDH · NAD^+ · pyruvate complex (top; 10) in the projection along the P-P vector. $O(5')_A$-P···P-$O(5')_N$ torsion angles ß are in the eclipsed (top) and trans (bottom) ranges; see also angles α in table 1.

In the crystal lattice, the two Li^+-complexed NAD^+ molecules are in head-to-tail orientation with adenine of molecule I stacked at 3.4 Å distance on nicotinamide of molecule II but nicotinamide of molecule I and adenine of molecule II are stacked with symmetry related, adjacent NAD^+ molecules. The stack geometry is displayed in fig. 5. It is of interest to note that the bases overlap with the heterocycles themselves and not merely with the amino or amide substituents (9) and that the two glycosyl bonds are opposite to each other.

STRUCTURE OF NAD^+ IN THE HOLOENZYME COMPLEXES

In the holoenzyme binary complexes with LDH, LADH, GAPDH, s-MDH*, and in the ternary LDH · NAD^+ · pyruvate complex (1,10,11,12) the structure of the coenzyme NAD^+ could only be derived on the basis of difference electron density maps at rather low resolution (2.5 Å as compared to 1.09 Å in case of the Li^+ · NAD^+ complex). Although structural details such as bond angles and distances cannot be deduced from such experiments the main

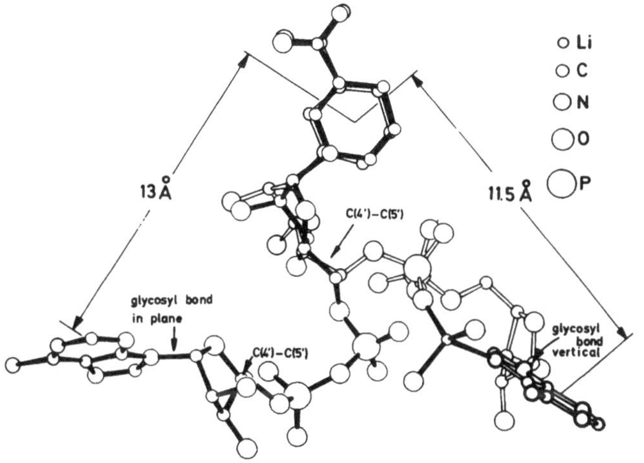

Fig. 7 Structures of the NAD^+ molecules in the Li^+ complex (open bonds) and in the ternary complex (10; solid bonds). For easy comparison, both molecules are projected on the nicotinamide plane. Note differences in conformation about the C(4')-C(5') bonds and nearly perpendicular orientations of the adenine and nicotinamide heterocycles. Further, the adenines in the two molecules are related by a 90° rotation (glycosyl linkages in-plane and vertical-to-plane).

* Footnote: LDH (lactate dehydrogenase), LADH (liver alcohol dehydrogenase), GAPDH (glyceraldehyde-3-phosphate dehydrogenase), s-MDH (soluble malate dehydrogenase).

conformational features can be obtained with some confidence. In all the holoenzyme binary and ternary complexes the NAD^+ molecule adopts a similar conformation owing to the comparable structures of the nucleotide binding sites (13). Therefore, it is sufficient here to describe NAD^+ in one binary and in the ternary complex, see table 1 and figures 6,7.

When bound to dehydrogenases, NAD^+ adopts an extended structure similar as in the Li^+-NAD^+ crystal form. The adenine and nicotinamide heterocycles are again nearly perpendicular to each other and about 13 Å apart and they are in an anti orientation*. However, the structural features forced by the Li^+-coordination are not present in the holoenzyme complex: Thus, the adenosine ribose displays C(3')-endo puckering, the torsion angles ω_A and ω'_A are in the (+)gauche and trans ranges (instead of being eclipsed) and the torsion angle $O(5')_A$-P(1)···P(2)-$O(5')_N$ (ß) is (-) eclipsed rather than trans as in Li^+-NAD^+ (fig. 6).

From a structural point of view, the most salient difference is found in the conformations about the C(4')-C(5') bonds - trans, gauche in holoenzyme - NAD^+ (ψ at (-)gauche, table 1) as compared to gauche, gauche in all the other natural ribonucleotides (ψ at (+)gauche). This unusual conformation corresponds to a "high energy" state of NAD^+ and can be explained by several contributing factors. First, NAD^+ becomes dehydrated when diffusing from the aqueous solution to the enzyme surface. Then, the liganded cation is expelled and replaced by (charged) groups of amino acid side chains which also form hydrogen bonds to the heteroatoms such as hydroxyl- and amino groups of NAD^+, and, further, adenine and nicotinamide heterocycles fit into hydrophobic pockets of the enzyme (1,10,11,12).

In total, NAD^+ molecules in the Li^+ and in the holoenzyme complexes differ in detail. If the overall features are compared such as relative separation and orientation of heterocycles (fig. 7), they are similar and one could conclude that the NAD^+ conformation as found in the Li^+ complex is related to the conformation of NAD^+ in aqueous solution and that it represents the form recognized by and bound to the enzyme active site.

*Footnote: The nicotinamide heterocycle is anti in A-type and syn in B-type dehydrogenase complexes (1).

STRUCTURE OF NAD^+ IN SOLUTION

A similar "extended" structure was also proposed for NAD^+ in aqueous solution at pH < 4, after addition of 30 % methanol or ethanol or 4 M urea and at elevated temperatures (14-16). Under "normal" conditions, i.e. neutral pH and room temperature, the "extended" form of NAD^+ exists to about 60 % and is in equilibrium with two "folded" forms with adenine and nicotinamide heterocycles stacked intramolecularly at about 3.4 Å distance. Because adenine can stack "on top" or "below" nicotinamide, left- and right handed helices result, the "extended" form representing a transition intermediate (fig. 8).

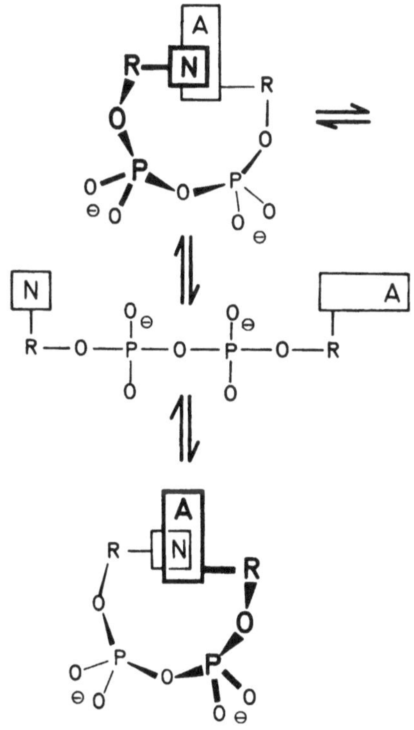

Fig. 8 The "folded" and "extended" forms of NAD^+ in aqueous solution. Note that the two "folded" forms describe right- and left handed helices.

In more detail, in aqueous solution the adenine and nicotinamide heterocycles of NAD^+ are in *anti* and *syn* orientations respectively, the conformations about the C(4')-C(5') bonds are *gauche*, *gauche* and the sugar puckering corresponds to the standard envelope forms outlined previously (14). The conformational parameters are all "normal", the *syn* and *anti* orientation of the nicotinamide heterocycle being nearly equally probable because the substituents in *ortho* position to the glycosyl linkage, C(2)-H and C(6)-H, are the same. Owing to the folding of NAD^+, the glycosyl linkages are arranged on opposite sides of the stack, similar as observed in the stack formed in the $Li^+ \cdot NAD$ crystals (fig. 5).

It should be noted here that in most of the publications on spectroscopic studies of NAD^+ the counterions were not defined. However, it is clear from model building that the folded form of NAD^+ is only possible if adenine N(7) and P(2) phosphate are not connected via a metal ion. Therefore, we conclude that small cations like Li^+ or Mg^{2+} would cause NAD^+ to adopt an "extended" structure similar as observed in the Li^+ NAD^+ complex and that the "folded" form is possible only in the presence of cations unable to link N(7) and P(2) simultaneously.

SUMMARY AND CONCLUSIONS

The results obtained from the different studies suggest that NAD^+ is chameleonic in character, its structural flexibility being brought about by the torsional freedom about the P-O-P linkage. In aqueous solution the molecule exists in a "folded", helical structure in equilibrium with an "extended" one which dominates at pH <4, at elevated temperatures or if alcohols are added to the solution. When crystallized as Li^+-salt from aqueous methanol, NAD^+ adopts an "extended" form in which both nucleotides are in standard (preferred) conformations. The adenine plane is nearly perpendicular to the nicotinamide plane and the distance between these planes is about 12 Å. In the holoenzyme complexes, NAD^+ displays a similar extended conformation. However, the orientations about the C(4')-C(5') bonds correspond to a high energy form not observed previously for naturally occurring nucleotides and nucleic acids.

This latter form of NAD^+ will not be present to a significant amount in aqueous solution but a population of "folded" and "extended" forms with standard conformations will exist. As the overall geometrical features of the enzyme-bound and "extended" NAD^+ conformations are comparable - the heterocycles have similar distance and orientation with respect to each other - we assume that the enzyme-coenzyme complex formation is initiated by the recognition of the adenine and nicotinamide heterocycles in the "extended" NAD^+ structure by the enzyme. After recognition, the enzyme binds the NAD^+ molecule and during the "tightening process", NAD^+ becomes distorted because energetically favourable interactions with amino acids and side chains at the enzyme surface compensate for the increase of conformational energy of the coenzyme.

REFERENCES

1) "The Enzymes Vol. XI, Oxidation-Reduction", Academic Press, New York, edit. Boyer, P.D. (1975).
2) Sundaralingem, M.: Structure and Conformation of Nucleosides and Nucleotides and Their Analogs as Determined by X-Ray Diffraction, in "Chemistry, Biology and Clinical Uses of Nucleoside Analogs", Annals of the New York Academy of Sciences, edit. Block, A., 255, 1-42 (1975).
3) Sundaralingam, M.: Stereochemistry of Nucleic Acids and Their Constituents. IV. Allowed and Preferred Conformations, Biopolymers 7, 821-860 (1969).
4) Saenger, W.: Structure and Function of Nucleosides and Nucleotides. Angew. Chem. internat. edit. 12, 591-601 (1973).
5) Pullman, B. and Saran, A.: Quantum-Mechanical Studies on the Conformation of Nucleic Acids and Their Constituents. Progr.Nucl.Acid Res. and Mol.Biol. 18, 215-322 (1976).
6) Saenger, W., Suck, D.: 6-Azauridine-5'-phosphoric Acid: Unusual Molecular Structure and Functional Mechanism. Nature 242, 610-612 (1973).
7) Saenger, W., Reddy, B.S., Mühlegger, K., Weimann, G.: X-Ray Study of the Lithium Complex of Nicotinamide-Adenine-Dinucleotide. To be published (1977).
8) Viswamitra, M.A., Hosur, M.V., Shakked, Z., Kennard, O.: X-Ray Study of the Rubidium Salt of ADP. Nature 262, 234-236 (1976).
9) Bugg, C.E., Thomas, J.M., Sundaralingam, M., Rao, S.T.: Stereochemistry of Nucleic Acids and Their Constituents X. Solid-State Base-Stacking Patterns in Nucleic Acid Constituents and Polynucleotides. Biopolymers 10, 175-219 (1971).
10) Chandrasekhar, K., McPherson, Jr.,A., Adams, M.J., Rossmann, M.G.: Conformation of Coenzyme Fragments when Bound to Lactate Dehydrogenase. J.Mol.Biol. 76, 503-518 (1973).
11) Banaszak, L.J., Webb, L.E.: Nicotinamide Adenine Dinucleotide and the Active Site of Cytoplasmic Malate Dehydrogenase, in "Structure and Conformation of Nucleic Acids and Protein-Nucleic Acid Interactions", ed. Sundaralingam, M., and Rao, S.T., University Park Press, London

pp. 375-386 (1974).

12) Webb, L.E., Hill, E.J., Banaszak, L.J.: Conformation of Nicotinamide Adenine Dinucleotide Bound to Cytoplasmic Malate Dehydrogenase. Biochemistry 12, 5101-5109 (1973).

13) Rossmann, M.G., Moras, D., Olsen, K.W.: Chemical and Biological Evolution of a Nucleotide-Binding Protein. Nature 250, 194-199 (1974).

14) Sarma, R.H., Mynott, R.J.: Manifestation of the Conformation of Pyridine Coenzymes in ^{31}P and ^{1}H Fourier Transform Nuclear Magnetic Resonance Spectroscopy, in "Proceedings of the Jerusalem Symposia on Quantum Chemistry and Biochemistry", ed. Bergmann, E.D. and Pullmann, B, Academic Press, New York, V, 591-626 (1973).

15) Catterall, W.A., Hollis, D.P., Walter, C.F.: Nuclear Magnetic Resonance Study of the Conformation of NAD^+ and NADH in Solution. Biochemistry 8, 4032-4036 (1969).

16) Blumenstein, M. and Raftery, M.A.: Natural Abundance ^{13}C Nuclear Magnetic Resonance Spectra of Nicotinamide Adenine Dinucleotide and Related Nucleotides. Biochemistry 12, 3585-3590 (1973).

Received January 31, 1977

DISCUSSION

Blankenhorn: The conformations derived by X-ray crystallography for molecules capable of stacking interactions have to be taken with great care. In aqueous solution intramolecular stacking is important both in NAD^+ and in the flavin-nicotinamide biscoenzymes discussed earlier. In the crystal, however, intermolecular stacking becomes more important, giving rise to conformations different from those present in solution.

Saenger: Your comment includes two statements, a general and a special one. Generally, it is true that molecules can crystallize in conformations with potential energy well above the energy minimum. One of the best examples in nucleoside structure analysis is 4-thio-uridine, a substance crystallizing in the unusual syn-conformation but adopting the anti form when in aqueous solution (Saenger & Scheit, J.Mol.Biol. 50, 153 (1970)). Therefore, it is not allowed to draw conclusions on conformations of molecules from one crystal structure alone. If, however, a great number of X-ray analyses of one class of substances, like the nucleosides and nucleotides, are known, the minimum energy conformation can be assessed reliably and the spectrum of conformations with higher potential energy also becomes evident. Further, the X-ray data are generally corroborated by potential energy calculations. Let me now discuss your second statement. In aqueous solution, NAD^+ occurs in two conformations, one "folded", with intramolecular stack and the other "extended" with no stack. These two forms are in equilibrium, with about 60 % for the "extended" one. If alcohol is added to the NAD^+ solution as we have done to crystallize NAD^+, the "folded" form opens up and NAD^+ exists almost exclusively in the "extended" form which we observe in the crystal structure. Thus solution-spectroscopic and solid-state X-ray data do not contradict each other.

Luisi: I suspect that intermolecular stacking interactions in the lattice can be a major contribution to the conformation in the crystal. Coordination of Li^+ can be another important factor for determining the crystal conformation. It follows that the conformation you have established - although very interesting and useful per se - may have little bearing to the conformation of NAD^+ free in solution or bound to enzymes.

Saenger: The coordination of Li^+ certainly is an important factor stabilizing this particular conformation of NAD^+. I have stressed this point in greater detail in the article than it was possible in this oral presentation. In the light of this study it would be interesting to repeat the spectroscopic investigations with different cations in order to establish the influence of the counterions on the NAD^+ conformation.

Wonacott: On checking the conformation about the C(4')-C(5') bond for the nicotinamide-ribose of NAD^+ bound to lobster glyceraldehyde 3-phosphate dehydrogenase I note that it too has the gauche, gauche conformation that we have interpreted for NAD^+ bound to the enzyme form B. stearothermophilus. Thus results for glyceraldehyde 3-phosphate dehydrogenase are consistent but differ from the conformation reported for LDH and MDH.

Saenger: Obviously, I misinterpreted the glyceraldehyde 3-phosphate dehydrogenase results because I thought NAD^+ to be in a similar trans, gauche conformation as in the other dehydrogenase holoenzyme complexes. It is very interesting to see that this is not the case and that we have now three different conformations for the "extended" form of NAD^+: the one with both C(4')-C(5') conformations gauche, gauche, the one with both trans, gauche and the one with nicotinamide-ribose gauche, gauche but adenine-ribose trans, gauche. The one that is missing in this series is nicotinamide-ribose trans, gauche, but adenine-ribose gauche, gauche.

Rossmann: How does the orientation of the carboxamide group in the Li^+ complex agree with that on nicotinamide as seen in other compounds such as those studied by Jan Paul?

Saenger: In the crystal structure of the unsubstituted nicotinamide, C-NH_2 is oriented trans to the C(3)-C(4) bond while to my knowledge it is cis in most or all other, N(1) substituted nicotinamide derivatives.

Wonacott: I want to give a statement regarding the orientation of carboxamide in the nicotinamide residue as interpreted for glyceraldehyde 3-phosphate dehydrogenase.

There is no way of distinguishing between the two possibilities from the electron density map. We have chosen the orientation in which the C(2) and -NH_2 of the nicotinamide residue are cis. This allows us to form a hydrogen bond from phosphate to $-NH_2$ of nicotinamide.

Banaszak: Is the carboxamide group involved in hydrogen bonding?

Saenger: Both oxygen and the $-NH_2$ group are involved in hydrogen bonding in the crystal structure.

Brändén: The conformation you observe in the Li^+ salt of NAD^+ seems to me to be rather similar to that we observe for the coenzyme analogue pyridine-adenine dinucleotide when it binds to liver ADH. This analogue has a very different conformation than bound NAD^+. In particular the C(4')-C(5') bond of the adenosine ribose could be gauche, gauche, as you observe.

Bernhard: Can you make an estimate of the energy differences among conformations, and if so, are these differences of a magnitude useful in regulating protein conformations?

Saenger: The conformational change from gauche, gauche to trans,gauche involves an increase in energy in the order of about 5 kcal/Mole. This energy is compensated by several factors, including entrance of NAD^+ into the hydrophobic pocket of the enzyme. We assume that NAD^+ is recognized by the enzyme in the "extended" form as observed in the crystal structure and that the conformational change of NAD^+ occurs upon tighter binding to the enzyme. The initial recognition of NAD^+ could already trigger conformational changes of the enzyme and additional regulation could be involved when the NAD^+ molecule is finally bound in the trans, gauche conformation.

Hinz: The discussion gives the impression to me that it is considered to be highly improbable that dogfish muscle LDH could accommodate an NADH molecule which adopts a conformation being by approximately 5 kcal/Mole higher in energy than the conformation occurring in solution. From a purely energetic point of view it is easily possible that on binary complex formation sufficient energy is provided to convert the gauche, gauche conformation of the NADH riboses into the trans, gauche apparently found in the LDH-NADH-oxamate complex. Although the ΔH-values for the dogfish enzyme are not known, the enthalpies for analogous binary complex formation between LDH H_4 or LDH M_4 (pig) and NRDH are certainly large enough, not to speak of the binding enthalpies associated with ternary complex formation.

CONFORMATION OF εNAD^+ IN SOLUTION AND BOUND TO DEHYDROGENASES REVEALED BY FLUORESCENCE DECAY KINETICS

A. Gafni
Department of Chemical Physics, Weizmann Institute of Science,
Rehovot, Israel

INTRODUCTION

Nicotinamide $1,N^6$-ethenoadenine dinucleotide (εNAD^+) is a fluorescent analog of NAD^+ in which much of the biochemical activity of the parent compound is retained (1,2). This analog and similar derivatives have been used as fluorescent probes in studies of enzymes which require adenine derivatives as coenzymes, effectors or substrates (1-5). In εNAD it is the modified adenine ring which fluoresces in contrast to NADH where the fluorophore is localized on the reduced nicotinamide ring. These two dinucleotides thus probe different areas of the coenzyme binding sites in dehydrogenases. The interaction of εNAD with a few dehydrogenases has been investigated in several laboratories using various fluorescence techniques (2,4,6) and the large changes in the fluorescence intensity which accompany the binding were used to monitor the binding process and to gain information on the conformation of the bound coenzyme analog (4,6).

In the present work fluorescence decay measurements were used to study the conformation of εNAD^+ in aqueous solution and when bound to several dehydrogenases.

MATERIALS AND METHODS

Horse liver alcohol dehydrogenase (LADH), beef heart lactate dehydrogenase (LDH) and phosphodiesterase (from Crotalus terr terr) were all purchased from Boehringer-Mannheim. Rabbit muscle glyceraldehyde 3-phosphate dehydrogenase (GPDH) and εNAD^+ were kindly donated by Professor A. Levitzki and Y. Henis (The Hebrew University, Jerusalem). εADP was purchased from Sigma. Apo-GPDH was prepared by treating the holoenzyme with HCl treated Norit A as described by Schlessinger and Levitzki (4). The apoenzyme had an A280/A260 ratio of 1.85 indicating that it was substantially free of NAD^+.

εNAD^+ was cleaved to εAMP and NMN by adding 10μg of phosphodiesterase to 2.5ml of 40μM εNAD^+ solution in 0.1M phosphate buffer pH=7.4 . The reaction mixture was then incubated for four hours at 23°C.

Fluorescence decay measurements were done with an instrument of the type described by Hundley et al (7) and modified in our laboratory to overcome problems due to drift which are associated with the repeated averaging procedures used in fluorescence decay measurements (8). The excitation wavelength was selected using a Jarrell-Ash double monochromator (Model 82-410), the wavelength bandwidth being 8-10nm. The fluorescence decay was observed at a right angle through a Schott KV394 cut-off filter transmitting light of wavelengths above ∿400nm. The time profile of the excitation light pulse was collected using a scattering suspension. The decay curve and the excitation light profile were stored in the two halves of a multichannel analyzer.

In the experiments done using enzyme solutions small amounts of stray light were present in the decay curves due mainly to scattered excitation light. This background was eliminated by subtracting from the observed decay curve the "decay" curve obtained for the enzyme solution (in the absence of εNAD^+) under exactly the same experimental conditions.

The fluorescence decay curves were analyzed using the method of nonlinear least squares (9). The fluorescence decay mechanism was assumed to be monoexponential or multiexponential function of the type: $I(t)=\sum_i \alpha_i \exp(-t/\tau_i)$ where $I(t)$ is the decay function and α_i and τ_i are the amplitude and lifetime of the i-th component. $I(t)$ was convoluted with the time profile of the excitation pulse, $G(t)$, to give the calculated decay curve $F_c(t)$ (9). The latter was compared with the experimental decay curve, $F(t)$. The decay parameters (amplitudes and lifetimes) are used by the method of nonlinear least squares as free parameters to be changed so as to minimize the value of the root mean square of the deviations between the calculated and experimental curves (RMS) defined as:

$$RMS = \left\{\frac{1}{n} \sum_{j=1}^{n} [F_c(t_j) - F(t_j)]^2\right\}^{1/2}$$

The summation is done over all n data points (for a typical experiment n=500).

RESULTS AND DISCUSSION

The fluorescence decay of εNAD^+ in neutral aqueous solution is presented in Fig. 1 and in Table I which summarizes the experimental results obtained in the present study. The decay kinetics markedly deviate from a single exponential decay law. The fit for two exponential components is excellent as evidenced by the low RMS value and by the shapes of the deviations and of the autocorrelation function of the deviations. εNAD^+ is known to exist, in aqueous solution, in a state of equilibrium between folded and extended conformations. One may therefore try to relate these two species with the two components observed in the fluorescence decay. Thus the extended conformation may be assigned the longer lifetime of 24.6nsec while the folded conformation in which stacking of the nicotinamide and ε-adenine rings results in fluorescence quenching assigned the shorter lifetime of 2.2nsec. This simple scheme requires that the rate of interconversion between the folded and extended conformations be slow compared with the fluorescence decay rates so that the two emitting species decay independently. The fluorescence quenching, by folding, is thus assumed to be purely static.

An alternative explanation for the two components observed in the fluorescence decay might be presented based on an excited state conversion of one conformation into the other. Such excited state reactions might be either reversible or irreversible and in both cases the lifetime of the species originally excited becomes shorter than in the absence of the reaction (in the case of a reversible excited state reaction the decay mechanism of each emitting species is no longer monoexponential and the decay components observed do not represent directly the emitting species) (10-12). The mechanism for εNAD^+ quenching may thus be clarified by studying the fluorescence decay of the fully folded and fully unstacked conformations. The decay of the latter was studied using εADP, and shown in Fig. 2, and also by using phosphodiesterase treated εNAD^+ (see Table I). In both cases single lifetimes were observed whose values were identical to the longer component obtained for εNAD^+. A very similar value for the fluorescence lifetime of εAMP has been reported by Spencer and coworkers (13). Both the reversible excited state folding-unfolding reaction, and the excited state irreversible folding reaction (originating in the unstacked conformation)

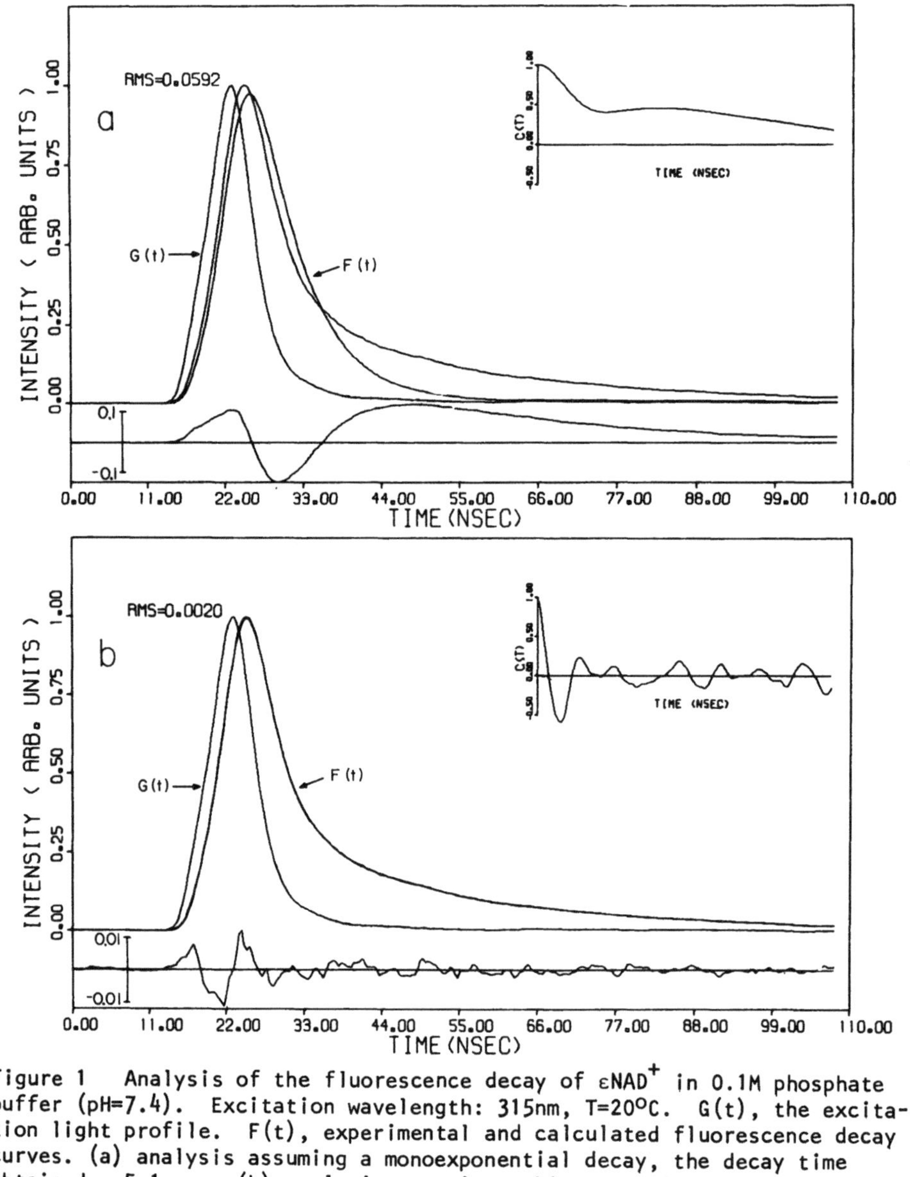

Figure 1 Analysis of the fluorescence decay of εNAD^+ in 0.1M phosphate buffer (pH=7.4). Excitation wavelength: 315nm, T=20°C. G(t), the excitation light profile. F(t), experimental and calculated fluorescence decay curves. (a) analysis assuming a monoexponential decay, the decay time obtained τ=5.1nsec. (b) analysis assuming a biexponential decay. The parameters obtained: τ_1=2,2nsec, τ_2=24.6nsec, α_1=0.90, α_2=0.10. The noisy curves at the bottom of the two plots are the traces of the deviations between the experimental and the calculated curves. The inserts describe the autocorrelation function of the deviations. The RMS values obtained for each analysis are also included. The fit to a single exponent is very poor while that for two components is excellent.

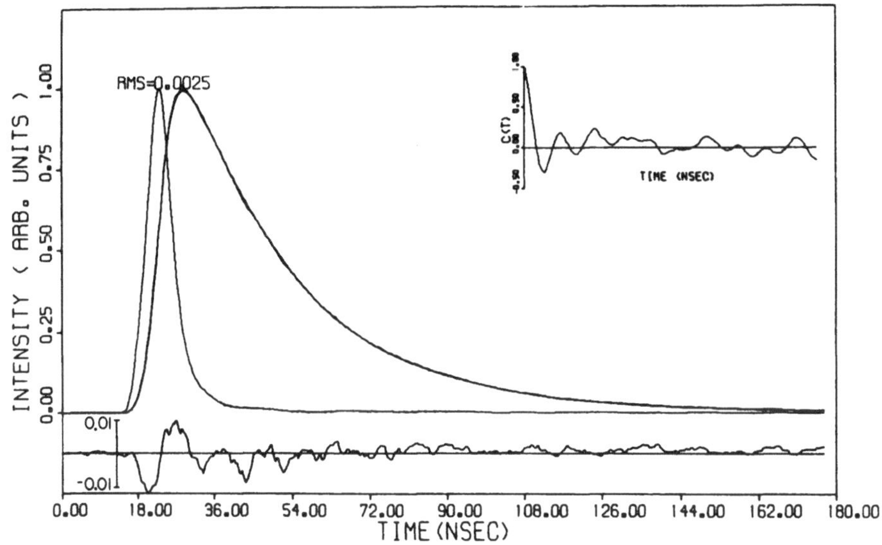

Figure 2 Analysis of the fluorescence decay of εADP in 0.1M phosphate buffer (pH=7.4). Excitation wavelength: 315nm, T=20°C. A single exponential decay was assumed and the decay time obtained τ=24.3nsec. The fit is very good as seen from the low RMS value, the random distribution of the deviations and the shape of the autocorrelation function.

---

thus have negligible rates compared to that of light emission and do not affect the decay mechanism. Since the fluorescence lifetime of εNAD$^+$ in the fully folded conformation could not be studied,an irreversible opening of the folded conformation in the excited state cannot be ruled out without additional information. This was obtained from steady state absorption and fluorescence spectra which reveal that the unfolding reaction is even slower than the folding reaction (14). In view of the discussion presented above we conclude that εNAD$^+$ in neutral aqueous solution exists in equilibrium between open and folded conformations, the latter being quenched by a factor of eleven relative to the first. The rate of interconversion of these two forms is slow compared with the fluorescence decay rate so that the quenching due to stacking is solely static.

The decay amplitudes associated with the two conformations of εNAD$^+$ reflect the relative concentrations of these species in the electronically excited state. The corresponding concentrations in the ground state may be deduced if their relative absorbances are known. No change in the absorption at

315nm (the excitation wavelength) was detected upon cleavage of εNAD$^+$ with phosphodiesterase and therefore the relative concentrations of the folded and extended conformations in the ground and electronically excited states are the same. Thus it may be concluded that in neutral aqueous solution 90% of the εNAD$^+$ molecules are in the stacked conformation. This value is higher than the one reported by Lee and Everse, namely about 40% stacked conformation (2). The latter value was derived from NMR studies of the cyanide adduct of εNAD$^+$, a procedure which involves several assumptions regarding the kinetics of cyanide addition to the extended and folded conformations of the coenzyme analog (15).

Table I Summary of Fluorescence Decay Data Obtained for εNAD$^+$ Under Various Conditions

Sample composition and buffer used	One component analysis τnsec	RMS·10^4	Two components analysis $\tau_1$nsec	$\tau_2$nsec	α_1	α_2	RMS·10^4
εNAD$^+$(40μM), 0.1M phosphate (pH=7.4)	5.1	590	2.2	24.6	0.90	0.10	20
εNAD$^+$(40μM) cleaved by phosphodiesterase to 5'-εAMP+NMN,0.1M phosphate (pH=7.4)	24.5	17					
εADP(30μM)0.1M phosphate (pH=7.4)	24.3	25					
εNAD$^+$(30μM);LDH(14μM) oxalate (0.1M)*,0.1M phosphate (pH=7.4)	25.6	59					
εNAD$^+$(10μM);LADH (10μM) pyrazole (1.3mM)*,0.1M phosphate (pH=7.4)	28.6	41					
εNAD$^+$(80μM);GPDH 80μM)*, 0.05M HEPES, 0.01M EDTA(pH=7.5)	8.6	91	7.3	22.5	0.91	0.09	18
εNAD$^+$(240μM);GPDH (80μM)*,0.05M HEPES, 0.01M EDTA(pH=7.5)	8.6	119	7.0	22.5	0.89	0.11	16

* The coenzyme analog is fully bound to the enzyme under these conditions.

Since the fluorescence decay parameters of εNAD^+ have proven to be a sensitive indicator of its conformation, this technique may be used to gain detailed information on the coenzyme conformation when bound to dehydrogenases. Figure 3 presents the fluorescence decay of εNAD^+ bound to LDH

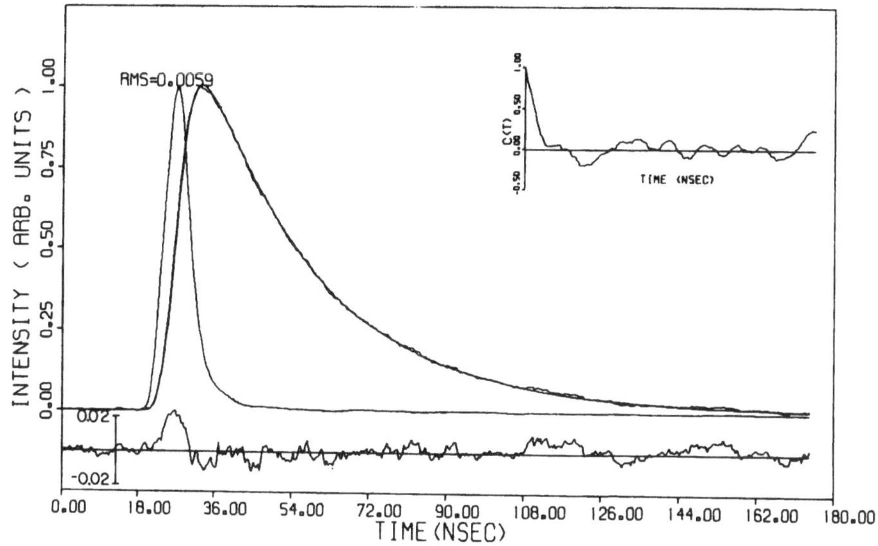

Figure 3 Analysis of the fluorescence decay of εNAD^+ in the ternary complex with LDH and oxalate. Excitation wavelength: 320nm, T=20°C. A single exponential decay was assumed and the decay time obtained τ=25.6nsec.

in the presence of oxalate. A single exponential decay is observed whose value is very close to that assigned before to the unfolded conformation in aqueous solution. Similar results were obtained for εNAD^+ in the ternary complex with LADH and pyrazole (see Table I). Thus the εNAD^+ molecule, when bound to these two dehydrogenases in the presence of substrate analogs, is fully in the extended conformation. This conclusion is in agreement with results obtained by other techniques (16,17).

The decay kinetics of εNAD^+ bound to GPDH were found to be more complex. The results obtained for the $E(\varepsilon NAD)_1$ complex are shown in Fig. 4. Analysis assuming a single component yields a poor fit while the fit obtained for two components is very good. The decay amplitudes in the latter case are similar to those obtained for the dinucleotide in aqueous solution. The lifetimes however are quite different; τ_1 is considerably longer than the one observed in aqueous solution while τ_2 is somewhat shorter in the GPDH

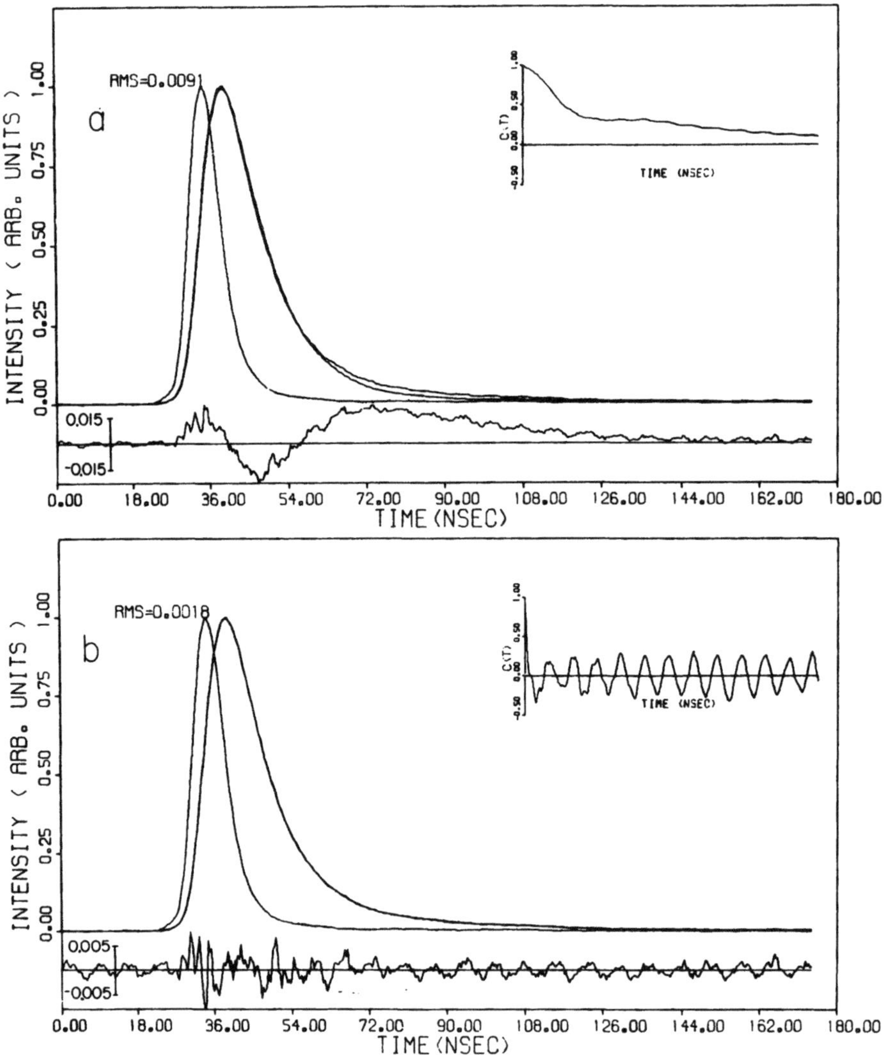

Figure 4 Analysis of the fluorescence decay of εNAD^+ in the 1:1 complex with GPDH. Excitation wavelength: 315nm, T=20°C. (a) analysis assuming a single exponential decay, the decay time obtained τ=8.6nsec. (b) Analysis assuming a biexponential decay. Parameters obtained: τ_1=7.3nsec, τ_2=22.5 nsec, α_1=0.91, α_2=0.09 .

bound εNAD^+. It is tempting to correlate the two decay amplitudes of the GPDH bound εNAD^+ with those observed in the absence of enzyme and conclude that the coenzyme analog is bound predominantly in the folded conformation. Folded conformations for NADH and NAD^+ when bound to GPDH have been proposed by several investigators based on fluorescence excitation spectra and NMR studies (18-20). However, X-ray crystallographic studies clearly show that the coenzyme is bound to GPDH in the unfolded form (21). A different origin for the two components observed in the decay must therefore be sought. The partial quenching of fluorescence of the ethenoadenine ring may arise from interactions with aromatic amino acid residues in its binding domain. As discussed above such interactions may lead to static or dynamic quenching depending on the rate constant for interaction. While no conclusive mechanism for the proposed interactions can be deduced from the data obtained in the present study it seems likely that excited state interactions play a role in fluorescence quenching. This is based on the observation that τ_2 is smaller than the corresponding lifetime found for the unstacked species in aqueous solution and when bound to other dehydrogenases (see Table I). One should thus be cautious in interpreting the two decay components as representing two species of bound εNAD^+ since this might not be valid in the case of excited state interactions (10-12).

εNAD^+ binding to GPDH shows negative cooperativity similar to that observed for the native coenzyme (4), and it has been shown by several investigators that binding of NAD^+ is associated with changes in the conformation of the protein (22,23). It was therefore of interest to study the decay mechanisms of εNAD^+ bound to GPDH at various degrees of binding sites saturation. Results obtained for two complexes, $E(\varepsilon NAD^+)_1$ and $E(\varepsilon NAD^+)_3$ are summarized in Table I. No significant differences among the decay parameters of these complexes were detected. This observation is in full agreement with the results reported by Schlessinger and Levitzki who found no change in the quantum yield of εNAD^+ fluorescence as the number of occupied binding sites of GPDH was varied between one and four (4). It has been previously shown by a variety of experimental techniques that the major structural changes in the tetrameric enzyme occur in response to the binding of the first NAD^+ molecule (see p. 34 in Ref. 21 for a list of references). Conformational changes induced by subsequent coenzyme binding appear to have no effect on the fluorescence lifetime or quantum yield of the bound εNAD^+.

REFERENCES

1. Barrio, J.R., Secrist, J.A., Leonard, N.J.: A Fluorescent Analog of Nicotinamide Adenine Dinucleotide. Proc.Nat.Acad.Sci.USA 69, 2039-2042 (1972).
2. Lee, C.Y., Everse, J.: Studies on the Properties of 1,N^6-Ethenoadenine Derivatives of Various Coenzymes. Arch.Biochem.Biophys. 157, 83-90 (1973).
3. Secrist, J.A., Barrio, J.R., Leonard, N.J.: A Fluorescent Modification of Adenosine Triphosphate With Activity in Enzyme Systems: 1,N^6-Ethenoadenosine Triphosphate. Science 175, 646-647 (1972).
4. Schlessinger, J., Levitzki, A.: Molecular Basis of Negative Cooperativity in Rabbit Muscle Glyceraldehyde-3-Phosphate Dehydrogenase. J.Mol. Biol. 82, 547-561 (1974).
5. Secrist, J.A., Barrio, J.R., Leonard, N.J., Weber, G.: Fluorescent Modification of Adenosine Containing Coenzymes. Biological Activities and Spectroscopic Properties. Biochemistry 11, 3499-3506 (1972).
6. Luisi, P.L., Baici, A., Bonner, F.J., Aboderin, A.A.: Relationship Between Fluorescence and Conformation of εNAD^+ Bound to Dehydrogenases. Biochemistry 14, 362-368 (1975).
7. Hundley, L., Coburn, T., Garwin, E., Stryer, L.: Nanosecond Fluorimeter. Rev.Sci.Inst. 38, 488-492 (1967).
8. Hazan, G., Grinvald, A., Maytal, M., Steinberg, I.Z.: An Improvement of Nanosecond Fluorimeters to Overcome Drift Problems. Rev.Sci.Inst. 45, 1602-1604 (1974).
9. Grinvald, A., Steinberg, I.Z.: On the Analysis of Fluorescence Decay Kinetics by the Method of Least Squares. Anal.Biochem. 59, 583-598 (1974).
10. Steinberg, I.Z., Katchalski, E.: Theoretical Analysis of the Role of Diffusion in Chemical Reactions, Fluorescence Quenching, and Nonradiative Energy Transfer. J.Chem.Phys. 48, 2404-2410 (1968).
11. Loken, M.R.: Determination of Rates of Excited State Reactions Using Nanosecond Fluorometric Techniques. Ph.D. Thesis, Johns Hopkins University (1973).
12. Gafni, A., Brand, L.: Fluorescence Decay Studies of Reduced Nicotinamide Adenine Dinucleotide in Solution and Bound to Liver Alcohol Dehydrogenase. Biochemistry 15, 3165-3171 (1976).

13. Spencer, R.D., Weber, G., Tolman, G.L., Barrio, J.R., Leonard, N.J.: Species Responsible for the Fluorescence of 1:N^6-Ethenoadenosine. Eur. J.Biochem. 45, 425-429 (1974).

14. Gafni, A.: To be published.

15. Oppenheimer, N.J., Arnold, L.J., Kaplan, N.O.: A Structure of Pyridine Nucleotides in Solution. Proc.Nat.Acad.Sci.USA 68, 3200-3205 (1971).

15. Holbrook, J.J., Liljas, A., Steindel, S.J., Rossmann, M.G.: Lactate Dehydrogenase, in "The Enzymes" Vol. II, P.D. Boyer, Ed., pp. 191-292, Academic Press (1975).

17. Brånden, C.I., Jörnvall, H., Eklund, H., Furugnen, B.: Alcohol Dehydrogenase, in "The Enzymes", Vol. II, P.D. Boyer, Ed., pp. 103-190, Academic Press (1975).

18. Velick, S.F.: Spectra and Structure in Enzyme Complexes of Pyridine and Flavin Nucleotides, in "Light and Life", W.D. McElroy and B. Glass, Eds., Johns Hopkins Press, Baltimore, pp. 108-143 (1961).

19. Eby, D., Kirtley, M.E.: Interaction of Nicotinamide-Adenine Dinucleotide and Its Analogs with Glyceraldehyde 3-Phosphate Dehydrogenase. Biochemistry 10, 2677-2682 (1971).

20. Lee, C.Y., Eichner, R.D., Kaplan, N.O.: Conformations of Diphosphopyridine Coenzymes Upon Binding to Dehydrogenases. Proc.Nat.Acad. Sci.USA 70, 1593-1597 (1973).

21. Harris, J.I., Waters, M.: Glyceraldehyde 3-Phosphate Dehydrogenase, in "The Enzymes", Vol, 13, P.D. Boyer, Ed., pp. 1-50, Academic Press (1976).

22. Velick, S.F., Hayes, J.E., Harting, J.: The Binding of Diphosphopyridine Nucleotide by Glyceraldehyde 3-Phosphate Dehydrogenase. J. Biol.Chem. 203, 527-544 (1953).

23. Listowsky, I., Furfine, C.S., Betheil, J.J., Englard, S.: Coenzyme-Induced Changes in the Optical Rotatory Dispersion Properties of Glyceraldehyde 3-Phosphate Dehydrogenase. J.Biol.Chem. 240, 4253-4258 (1965).

Received March 30, 1977

DISCUSSION

Bernhard: Can you distinguish between two exponential decays due to "internal" characteristics of the decay process (is it the same as in aqueous solution?) and the possibility of two types of site environments with comparable ε-NAD affinities?

Gafni: The decay kinetics of ε-NAD bound to GPDH seem to be more complex than in aqueous solution and not really biexponential. Thus from our data one cannot conclude that the ε-NAD molecules bound to different sites on GPDH have different decay kinetics. We feel that the multiexponential decay law originates from excited state interactions of the coenzyme analog molecule with residues in its binding site.

Affinity labeling by alkylating analogues of NAD

J.F. Biellmann, G. Branlant, B.Y. Foucaud and L. Wallén
Institut de Chimie, Université Louis Pasteur, Strasbourg, France

and C. Woenckhaus and R. Jeck
Gustav-Embden Zentrum der Biologischen Chemie, Fachbereich 19 der Universität Frankfurt am Main, West-Germany

Analogues of NAD and NADP with an alkylating function are potentially affinity labels and may be of interest to determine essential residues in the NAD(P) binding site and in the active site. Among these residues, could very well be the base involved in the catalysis of the hydrogen transfer. The relationship of the alkylating agent to NAD(P) can be reflected by the activity of the alkylating agents and closely related analogues as hydrogen acceptors ; on the other hand, the physical properties of the alkylated enzymes and of the alkylated enzymes reduced by dithionites are expected to be similar to those of the complexes of NAD(P) and NAD(P)H.

The adenine ring of NAD can be replaced by other hydrophobic groups, e.g. 5-acetyl-4-methylimidazole, and the properties of these analogues have been shown to be close to those of NAD (1).

[ω-(Acetylpyridine)-alkyl]adenosine diphosphates differ from NAD by two modifications : the nicotinamide is replaced by 2-, 3-, or 4-acetylpyridine and the ribose by an aliphatic linear chain which varies from C-2 to C-6 in order to explore different regions in the vicinity of the active site (2). The loss of the hydrophilic interactions of the ribose is reflected in the kinetic constants of these compounds. The acetyl derivatives can be converted to the bromoacetylated analogues by bromination (3). 3-Chloroacetylpyridine AD (CAPAD), 4-chloroacetylpyridine AD (4-CAPAD) and 3-chloroacetylpyridine ADP (CAPADP) (4) have been prepared from the corresponding diazoketones (5).

The analogues substituted at position 3 of the pyridinium ring with acetyl, bromoacetyl, chloroacetyl and ethylketone groups may be active as hydrogen acceptors. The properties of these analogues are summarised in the Table.

These alkylating analogues can show a different behaviour according to the dehydrogenase studied. In most of the cases they are hydrogen acceptors and affinity labels. Activity and no inactivation have been found for [3-bromoacetylpyridine]-alkyl derivatives and CAPAD with L-lactate dehydrogenase from various sources as well as glutamate dehydrogenase (from beef liver and from yeast for CAPADP) (6). Octopine dehydrogenase from *Pecten maximus* L. is not inactivated by CAPAD nor by 4-CAPAD. These analogues and 3-ethylketonepyridine AD do not act as hydrogen acceptors in enzymic tests with this enzyme and do not show any detectable inhibition, no binary complexes being formed (7). The activity of glutathione reductase from yeast is reduced to 30 % of the initial activity by CAPADP (8). Isocitrate dehydrogenase from beef liver (strictly NADP dependent) is inactivated by CAPADP with a second order kinetic and also by CAPAD. These two analogues are most likely not affinity labels for this enzyme (8).

The inactivation by affinity labeling follows this reaction scheme (9) :

$$E + I \underset{k_{-1}}{\overset{k_1}{\rightleftharpoons}} EI \overset{k_2}{\dashrightarrow} EA$$

E is enzyme, I : inactivator, EI : reversible enzyme-inactivator complex and EA inactivated enzyme.

If $k_2 << k_{-1}$, the kinetic is pseudo-second order. The dissociation constant k_{-1}/k_1, determined according to the classical treatment of the inactivation data should be close to the K_D, K_M or K_I measured for related analogues and in some cases for the alkylating analogues themselves.

Alcohol dehydrogenase from horse liver (LADH)
Strasbourg (10) :

LADH is inactivated by CAPAD with a pseudo-second order kinetic : at pH 7.5, the dissociation constant is 10^{-5} M and the alkylation rate constant 10^{-3} sec^{-1}. NAD protects the enzyme from inactivation. The K_D^{NAD} measured by competition with the inactivator was found to be 26 μM (close to the value determined for the formation of LADH-NAD). Addition of ethanol or pyrazole favours the binding of CAPAD to LADH and increases the alkylation rate constant. Imidazole slows down the reaction by decreasing the binding and rate constants. The specificity of the alkylation by CAPAD is shown by the activity of CAPAD as hydrogen acceptor and the binding of its isostere 3-ethyl-ketone pyridine AD (determined by its activity as hydrogen acceptor and by the quenching of the protein fluorescence). The lack of inactivation by 4-CAPAD and by the mononucleotide : 3-chloroacetylpyridine ribose monophosphate indicates the necessity of binding of the inactivator for the alkylation. A group of pKa 7.2 is found by the pH dependence of the alkylation constants : the binding constant increases with pH and the alkylation rate constant remains close to a constant value. The stoechiometry is one per subunit and there is a linear relationship between the activity and the incorporation of the label. The activity measured by reduction of NAD with ethanol was found to be proportional to that determined by oxidation of NADH by benzaldehyde.

The inactivated enzyme is not reduced by ethanol, nor does it catalyse the ethanol-acetaldehyde exchange. It is reduced by sodium dithionite and the reduced species has an absorption band at 370 nm.

It was found that the label was lost during dialysis (3 to 4 days), but no activity was recovered. No labeled peptide except two with very low labeling (containing Lys-159 and Lys-325) could be found (11). Thus the alkylated group remains unknown (Lys-159 and Lys-325 are far from the active site). Possibly it is the hydroxyl group bound to the catalytic zinc atom which is alkylated.

Frankfurt :

LADH is inactivated by nicotinamide[5-bromoacetyl-4-methylimidazol]dinucleotide, [3-(4-bromoacetylpyridinio)-propyl] adenosine diphosphate and [4-(3-bromoacetylpyridinio)-butyl] adenosine diphosphate (12,13,14) with modification of cysteine side chains. Cys-174 and Cys-46 are ligands of the catalytic zinc in the vicinity of which the nicotinamide of NAD binds (15). [4-(3-Bromoacetylpyridinio)-butyl]adenosine diphosphate preferably alkylates Cys-46 while [3-(4-bromoacetylpyridinio)-propyl]adenosine diphosphate prefers to react with Cys-174. Nicotinamide[5-bromoacetyl-4-methylimidazole] dinucleotide reacts exclusively with Cys-174. This inactivation can only be explained if the coenzyme analogue is bound to the active centre in a reversed position or in a twisted form.

Using labeled compounds, two moles of inactivator are incorporated per mole of enzyme dimer. The relation between the extent of inactivation and enzymic activity is linear in all cases with no hint for half-of-the-sites reactivity.

Alcohol dehydrogenase from yeast (YADH)
Strasbourg (10) :

With YADH, 3-ethylketone pyridine AD is active as hydrogen acceptor and CAPAD gives at 23° a rapid inactivation with a second order kinetic. NAD protects YADH from inactivation. 4-CAPAD and 3-chloroacetylpyridine ribose phosphate do not inactivate YADH. At 4°, CAPAD inactivates YADH with a pseudo-second order kinetic. This is consistent with an affinity labeling where the rate determining step is at 23° the formation of the binary complex YADH-CAPAD whereas at 4° it is the alkylation step. The stoechiometry is one per subunit and a linear relationship activity *vs* incorporation of the label is observed. The label is more stable with YADH than with LADH, but here again no labeled amino acid could be detected (11).

Frankfurt :

Nicotinamide[5-bromoacetyl-4-methylimidazole]dinucleotide, [3-(4-bromoacetylpyridinio)- and [3-(3-bromoacetylpyridinio)-propyl]adenosine diphosphate inactivate YADH modifying Cys-43 (12,13). This cysteine residue is homologous to Cys-46 of the liver enzyme (15). Incorporation of 4 moles per mole of enzyme leads to a complete inactivation. Half-of-the-sites reactivity was not observed. [4-(3-Bromoacetylpyridinio)-butyl]adenosine diphosphate, homologue of the [3-(3-bromoacetylpyridinio)-propyl] derivative also inactivates the enzyme but the labeling is unstable and could not be stabilized by treatment with sodium borohydride. Therefore, it was not possible to identify the modified amino acid residue (16).

Alcohol dehydrogenase from Bacillus stearothermophilus
Frankfurt :

In this enzyme Cys-38 is in a surrounding homologous to that of Cys-43 of YADH and Cys-46 of LADH (17). Reagents specific for sulfhydryl groups do not modify this residue. [4-(3-Bromoacetylpyridinio)-butyl]adenosine diphosphate inactivates the enzyme specifically by alkylation of this residue. Complete inactivation is observed after incorporation of 4 moles per mole tetrameric enzyme with no hint of half-of-the-sites reactivity (18).

L-lactate dehydrogenase (LDH)
Frankfurt and Strasbourg :

All the derivatives with 3-halogenoacetylpyridine ring and 4-CAPAD are not inactivators of lactate dehydrogenase from pig heart and from dogfish (with which only CAPAD and 4-CAPAD were tested) (19) although some are active as hydrogen acceptors.

LDH from pig heart however, is inactivated by [3-(4-bromoacetyl-pyridinio)-propyl]adenosine diphosphate. The number of SH-groups remains unchanged even upon complete inactivation. The pH-dependence of the inactivation reaction reveals a residue with a pKa 6.7 to be involved. This is the pKa of the essential His (20).

Glyceraldehyde-3-phosphate dehydrogenase (GAPDH)
GAPDH from various sources is inactivated by most of the alkylating analogues (cf. Table).

Frankfurt :

In our investigations with [3-(3-bromoacetylpyridinio)-propyl]

Table : Properties of coenzymeanalogues

	ADH			GAPDH			LDH	GDH		ODH	Ster DH
	HL	Y	Bs	r.mus.	Y	st & Bs		Y	bl		
N 5-acetyl-4-methyl-imidazole D	vAD	vAD		vAD			vAD		vAD		
N 5-bromoacetyl-4-methylimidazole D	+	+		+			-		-		
3-(3-AP)-propyl-ADP	D	D	o	o	o		D				
3-(3-BAP)-propyl-ADP	-	+		+	+		-				
4-(3-AP)-butyl-ADP	AD	AD		AD			AD				
4-(3-BAP)-butyl-ADP	+	+	+	+			-				
5-(3-AP)-pentyl-ADP	D	D	o	AD			D				
5-(3-BAP)-pentyl-ADP	+	+		+			-				
6-(3-AP)-hexyl-ADP	o	o	o	o			o				
6-(3-BAP)-hexyl)-ADP				+			-				
3-(2-BAP)-propyl-ADP	-	+		+			-				
3-(4-BAP)-propyl-ADP	+	+		+			+				
CAPAD	vA+	A+		+	+	+	A-		A-	o-	vA+(25)
3-ethylketone pYAD	vA	A		o	o	o	A			o	vA
4-CAPAD	-	-		+	+	+	-				
CAPADP								vA-			vA+(25)

Abbreviations: v: very ; o: no acceptor & donator function ; +: inactivator ; -: not acting as an inactivator ; AP: Acetylpyridinio ; BAP: Bromoacetylpyridinio ; ADP: Adenosine diphosphate ; A: active as hydrogen acceptor ; D: reduced active as hydrogen donor ; HL: horse liver ; bl: bovine liver ; Y: yeast ; Bs: Bacillus stearothermophilus ; st: sturgeon ; Ster DH: 17β-hydroxysteroid DH from human placenta.

adenosine diphosphate, the homologous butyl-compound and nicotinamide [5-bromoacetyl-5-methylimidazole]dinucleotide, GAPDH from rabbit muscle shows a complete loss of activity of hydrogen exchange upon incorporation of 2 moles per mole tetrameric enzyme (21,22). Even at high concentrations of inactivator the amount of incorporation is not altered. The inactivated enzyme can still bind two moles of NAD per tetrameric enzyme. Reactivation of the enzyme was not observed. After complete inactivation, only 2 SH-groups per mole are affected whereas in the native enzyme 4 SH-groups per mole react rapidly with DTNB. In the ultracentrifugation there is no difference with respect to unmodified holoenzyme. Therefore the structure of the protein is not altered by the inactivation reaction (21). GAPDH from yeast is completely inactivated by [3-(3-bromoacetylpyridinio)-propyl]adenosine diphosphate after incorporation of 4 moles per mole tetrameric enzyme. 4 SH-groups as in the native enzyme react with DTNB even after complete inactivation. The rate of reaction however, is greatly reduced.

Yeast and muscle enzyme form after inactivation hybrids with the same electrophoretic mobilities as the hybrids of the native enzymes (21).

Strasbourg (23) :

GAPDH from yeast, rabbit muscle, sturgeon and Bacillus stearothermophilus are rapidly and irreversibly inactivated by CAPAD. No activity of CAPAD and 3-ethylketone pyridine AD as hydrogen acceptors could be detected ; but 3-ethylketone pyridine AD is a competitive inhibitor with respect to NAD. GAPDH from sturgeon and from Bacillus stearothermophilus have been most studied because of the stability of their apoenzyme*. The kinetic of alkylation of the sturgeon apoenzyme is biphasic at low CAPAD concentration and monophasic at concentrations higher than 10^{-5} M where at pH 7.6 the dissociation constant is 2.6 10^{-5} M and the alkylation rate constant 0,016 sec^{-1}. The inactivation rate is decreased by NAD. The pH dependence of the overall inactivation rate reveals a group with a pKa of 8.5. The stoechiometry has been determined with substoechiometric amounts of inactivator : there is an incorporation of 1 per subunit and a linear relationship between activity and labeling.

The alkylated enzyme shows one reactive SH/subunit (DTNB) as does the native enzyme, and a wide absorption band at 345 nm (pH 7.0 ; ε = 2.500/subunit).

This band is sensitive to pH and to the presence of phosphate. The intensity of this band is linearly related to the extent of inactivation.

Does the alkylation of Cys-149 by CAPAD reveal another reactive Cys ? Is the high wavelength absorption due to charge transfer between the pyridinium ring and Tyr-317 ? Or is another residue than Cys-149, His-176 for instance, alkylated ? The absorption could then be attributed to an interaction between Cys-149 and the pyridinium ring. Similar absorption bands are observed for other alkylated GAPDH.

The 4-CAPAD inactivates GAPDH from sturgeon so rapidly that no kinetic study could be achieved. The inactivation is slowed down by NAD. The alkylated enzyme does not show any

---

* With those apoenzymes, the esterolytic activity towards p-nitrophenyl acetate is very low compared to that of the rabbit muscle enzyme (24).

DTNB reactive SH group and in 8 M urea 9 ± 1 SH/tetramer are titrated. There is an absorption at 435 nm (ε = 10.000/ subunit). Its intensity is linearly related to the amount of alkylating agent and this band is also sensitive to the presence of phosphate. From these experiments one could conclude that Cys-149 is alkylated.

The determination of tridimensional structure of some of these derivatives may answer some of these questions.

Conclusion :

The synthesis of NAD alkylating analogues was performed by classical chemical methods on rather complex molecules. The molecular weight of these inactivators is probably the highest among the inactivators prepared until now and their efficient use should encourage further synthesis of alkylating analogues of more complex substrates.

The modifications of NAD described here for most of the alkylating analogues do generally not abolish their activity as hydrogen acceptors and this allows a correlation between the constants determined by the enzymic activity and those measured by the inactivation kinetics. This situation is favourable for the demonstration of affinity labeling.

Furthermore the use of highly specific inactivators of this kind is a powerful tool in the investigation of biochemical and evolutional relations between enzymes from different species.

Acknowledgements :

The Strasbourg group thanks Prof. C.I. Bränden and his group for numerous stimulating discussions. The Centre National de la Recherche Scientifique (Laboratoire associé n° 31), the Délégation Générale à la Recherche Scientifique et Technique (DGRST) (Contracts 73.7.1312, 74.7.0220 and 76.7.0356) and the Fondation pour la Recherche Médicale Française are acknowledged for the financial support of the work done in Strasbourg. The work in Frankfurt was supported by grants from the Deutsche Forschungsgemeinschaft.

References :

(1) Woenckhaus, C.: in "Topics in Current Chemistry" (F.L. Boschke ed.), Vol. 52, p. 209. Springer Verlag, Berlin and New York, 1974.

(2) Woenckhaus, C., Jeck, R.: Methods in Enzymology (1977) in press, Academic Press, New York.

(3) Woenckhaus, C., Zoltobrocki, M., Berghauser, J.: Hoppe Seyler's Z. Physiol. Chem. 351, 1441 (1970).

(4) Biellmann, J.F., Eid, P., Goulas, P.P., Hirth, C.: in preparation.

(5) Biellmann, J.F., Branlant, G., Foucaud, B.Y., Jung, M.J.: FEBS Letters, 40, 29-32 (1974).

(6) Goulas, P.P., Fourcade, A., Venard, R.: unpublished results.

(7) Olomucki, A., Thomé-Beau, F., Biellmann, J.F., Branlant, G.: Europ. J. Biochem., 56, 109-116 (1975).

(8) Goulas, P.P.: unpublished results.

(9) Gold, A.M., Fahrney, D.: Biochemistry, 3, 783 (1964). Kitz, R., Wilson, I.B.: J. Biol. Chem., 237, 3245-3249 (1962).

(10) Foucaud, B.Y.: Thèse de Doctorat-ès-Sciences, Strasbourg, 1976.

(11) Jörnvall, H.: unpublished results.

(12) Jörnvall, H., Woenckhaus, C., Johnscher, G.: Eur. J. Biochem., 53, 71 (1975).

(13) Jörnvall, H., Woenckhaus, C., Schättle, E., Jeck, R.: FEBS Letters, 54, 297 (1975).

(14) Jeck, R., Zumpe, P., Woenckhaus, C.: Justus Liebigs Ann. Chem., 961 (1973).

(15) Brränden, C.I., Jörnvall, H., Eklund, H., Furugren, B.: in "The Enzymes" (P.D. Boyer, ed.) Vol. 11, p. 176, Academic Press, New York, 1975.

(16) Woenckhaus, C., Jeck, R., Jörnvall, H.: unpublished results.

(17) Bridgen, J., Kolb, E., Harris, J.I.: FEBS Letters, 33, 1 (1973).

(18) Woenckhaus, C., Jeck, R., Harris, J.I.: in preparation.

(19) Biellmann, J.F., Marchal-Rosenheimer, N., Samama, J.P.: in preparation.

(20) Woenckhaus, C., Schättle, E., Jeck, R., Berghäuser, J.: Hoppe-Seyler's Z. Physiol. Chem., 353, 559 (1972).

(21) Dietz, G., Woenckhaus, C., Jaenicke, R., Schuster, I.: Z. Naturforsch. (1977) in press.

(22) Klatte, W.: unpublished results.

(23) Branlant, G.: Thèse de Doctorat-ès-Sciences, Strasbourg, 1974 ; Wallén, L.: unpublished results.

(24) Abdallah, M., Wallén, L.: unpublished results.

(25) Biellmann, J.F., Branlant, G., Nicolas, J.C., Pons, M., Descomps, B., Crastes de Paulet, A.: Europ. J. Biochem., 63, 477-481 (1976) ; Goulas, P.P., Nicolas, J.C.: unpublished results.

Received January 31, 1977

DISCUSSION

Hollaway: Although two molecules of your inhibitor per tetramer of GAPDH are sufficient to inhibit the enzyme completely, can you get another two molecules in to make a tetramer containing four inhibitor molecules?

Woenckhaus: No. But after complete inactivation even at high concentration by incorporation of two moles coenzyme analogs [3-(3-b moacetyl pyridinio) -propyl]-adenosine pyrophosphate per tetramer the enzyme binds two more moles of coenzyme analog [3-(3-acetylpyridinio)-propyl]-adenosine pyrophosphate or NAD^+ without reactivation either dehydrogenase activity nor esterade activity.

Werber: Do you think that the biphasic pattern of inactivation of liver ADH with your analog might be due to the labeling of a second site, which has a weaker affinity for the analog? Especially in view of the fact that the biphasic pattern occurs at high concentrations of the analog?

Biellmann: The stoichiometry is always one per subunit. The biphasicity is observed with enzyme of different origin and we have no explanation for the biphasicity. But we know that for NAD^+ there is only one binding site per polypeptide chain and by extension we presume that for the alkylating analog the situation is the same. By fluorescence, we detect only one site per polypeptide chain for the ethylketone analog.

Seydoux: Does inactivated rabbit muscle GAPDH still bind NAD^+ with formation of a racker band?

Woenckhaus: No. The coenzyme analog [3-(3-acetylpyridinio)-propyl]-adenosine pyrophosphate is bound with K_D = 1 mM and NAD^+ with K_D = 12 mM to the inactivated enzyme, but we could not find any change in absorption between 340 and 380 nm.

Dieter: By reaction of NAD^+ and AMP with α-brom-β-ketobutyraldehyde we synthesized ε-analogues of NAD^+ and AMP bearing an acetyl group in the 7-position of the ethylenic ring. The UV-spectroscopic data of the compounds strongly suggest that the carbonyl group is involved in the mesomeric system of the heteroaromatic ring and one would therefore expect that the methyl protons of the acetyl group are rather acidic. However, we were not able to brominate this group under varying conditions. What do you mean is the explanation for this?

Biellmann: Are you certain of the structure of your compound?

Dieter: Based on our NMR-, and UV- and fluorescence evidences, yes. The substituted ε-adenine ring has a pK of 1.9.

Dunn: Your kinetics of inactivation clearly show two nonequivalent classes of sites for liver ADH. What are the relative amplitudes of the two phases of the inactivation time course, and do you think that your biphasicity arises from half-site reactivity?

Biellmann: The relative amplitudes of the two phases seem to depend somewhat on the origin of the enzyme, but the two phases are always present. We have no explanation.

Dunn: What do you find if you prepare enzyme that is only half-active when you assay the protein via isoelectric focusing?

Biellmann: We never could observe any separation by different techniques, but work is not easy in this direction, due to the loss of the label.

IMMOBILIZED ADENINE COENZYMES IN GENERAL LIGAND AFFINITY CHROMATOGRAPHY AND THEIR USE AS ACTIVE COENZYMES

K. Mosbach
Biochemical Division, Chemical Center, University of Lund, S-220 07 Lund 7, SWEDEN

There is considerable advantage in affinity chromatography in a versatile ligand or "general ligand" with a broad biological affinity-binding spectrum in that a new synthetic route leading to suitable affinity matrices does not have to be devised for every putative biospecific purification. Adenine coenzymes meet this requirement and the first examples that appeared in the literature on their application involved model separations of dehydrogenases using immobilized NAD, at which time the term "general ligand" was first coined. (The examples given and coenzyme analogs described here will be primarily taken from our own work; lack of space does not permit to discuss the excellent work carried out in other laboratories in this area; for a more general review on this subject see K. Mosbach: Advances in Enzymology, ed. A. Meister, 1977, in press). By virtue of its generality, however, this approach appears at first sight to sacrifice something of the very specificity which is the keystone of affinity chromatography. However, it has become apparent that bioelution provides adequate compensation in retaining specificity, as will be discussed in some depth later.

Of the about 2000 different enzymes which have been assigned a specific number (Enzyme nomenclature 1973 + supplement), about one third require that one of the following four adenine coenzymes, NAD (170), NAD(P) (50), NADP (115), CoA (free) (80), and ATP (220), participate in the enzymic reactions. In addition, there are enzymes with masked nucleotide binding sites and further there is a number of enzymes with binding sites for nucleotide effector molecules, which further increases the number of enzymes that may be affinity-bound by the above four general ligands.

In many general ligand affinity chromatography studies reported, it has been shown that the entire coenzyme molecule is not essential for binding. Fragments of the coenzyme, usually the "half-molecule" thereof, were found to retain the affinity and specificity of the intact parent moiety. For instance AMP has been found to be an adequate substitute in many situations for the more labile NAD. Major emphasis will be given to so called "preassembled" adenine coenzymes, i.e. ligands modified by substitution with a spacer, carrying a functional group suitable for binding to a matrix, to which it is then coupled as opposed to ligands coupled to a matrix to which spacers have already been attached.

A. GENERAL LIGAND AFFINITY CHROMATOGRAPHY

I. Some Adenine Nucleotides Used as General Ligands

A considerable amount of effort has been expended on synthesising preassembled analogs of 5'-AMP, 2',5'-ADP, and 3',5'-ADP, which constitute the corresponding "half-molecules" of NAD, NADP, and CoA, respectively. The first general ligand thus tailored for immobilization and of unambiguous, defined structure was N^6-(6-aminohexyl)-AMP (Fig. 1). In subsequent studies the corresponding N^6-(6-aminohexyl)-2',5'-ADP and N^6-(6-aminohexyl)-3',5'-ADP were prepared (1). All these analogs are modified at the same site, i.e. position N^6, the exocyclic nitrogen on the adenine moiety, and by the same spacer molecule.

A number of affinity chromatographic studies have also been reported with other AMP analogs, including 8-(6-aminohexyl)-amino-AMP, substituted at position 8 of the adenine moiety. The corresponding 2',5'-ADP substituted in position 8 and ribosyl-linked AMP prepared by reaction of periodate-treated AMP with a dihydrazide, followed by coupling to CNBr-activated agarose (alternatively by reaction of periodate-treated AMP with a spacer-matrix).

ATP (ADP)-analogs. Four different points of attachment have been chosen for ATP or ADP derivatization, including the N^6- and C-8-positions of the adenine moiety, the ribose and in addition the terminal phosphate group. These preparations have found application in particular in the affinity chromatography of kinases.

NAD and NADP-analogs. Considerable efforts have concentrated on the synthesis of NAD-analogs with a spacer molecule and carrying a functional group suitable for covalent binding. Both N^6-, C-8- and ribose-substituted NAD-analogs have been prepared. To obtain a N^6-substituted NAD-analog, the N^1 position of the adenine ring has been quaternized with iodoacetic acid followed by a Dimroth rearrangement leading to a carboxymethyl group at position N^6 (2). The spacer can be extended by carbodiimide condensation of hexamethylenediamine (see structure in Fig. 3).

Analogous to NAD, the same points of attachment for derivatization of NADP have been chosen. The synthetic procedures applied for the synthesis of N^6-, C-8- and ribose-substituted NADP were also equivalent. For attachment to the ribose moieties however, only the NMN-ribose is available.

I. In general it can be said that the preassembled NAD and NADP analogs and their corresponding half molecules 5'-AMP and 2',5'-ADP, derivatized in the adenine moiety either at position N^6 or C-8 have proven to be excellent general affinity chromatography ligands for NAD- and NADP-dependent dehydrogenase. Both types of compounds seem equally well suited. By contrast substitution at the nicotinamide portion, gave, as expected, poor affinity material for this class of enzymes. Ribose substitution, however, yields useful bioaffinity material although with weaker affinity. For the other major class of enzymes utilizing ATP, i.e. kinases and ATPases, substitution at all the positions of ATP (ADP) mentioned gave in general good affinity material.

II. (Bio)adsorption. Despite the non-specific factors often involved in the binding of biological macromolecules to ligands, biospecific adsorption is probably involved and of importance in most of the reported purifications of enzymes using general ligand affinity chromatography. Two examples from our own work displaying "true" biospecificity serve to illustrate how enzymes are able to recognize subtle variations in structures of immobilized ligands. The ligands N^6-AMP and N^6-2',5'-ADP, comprising the half molecules of NAD and NADP, respectively, were immobilized on Sepharose (1) (Fig. 1). The only difference between the two preparations was the additional phosphate residue on the 2'-position of the ribose. When a mixture of NAD and NADP dependent dehydrogenases was applied to the latter, the NAD-dependent enzyme lactate dehydrogenase was washed through unretarded as expected. The two NADP-dependent enzymes present were bound and could subsequently be eluted off. Conversely lactate dehydrogenase would bind to the 5'-AMP Sepharose gel, but the NADP-dependent 6-phosphogluconate dehydrogenase would not. The NADP-dependent glucose-6-phosphate dehydrogenase did, however, have a very weak affinity for the AMP gel, consistent with the observation that this enzyme is inhibited in free solution to some degree by AMP. Nevertheless, these gels are useful not only for purification of individual enzymes but also for quick group separation of NAD- from NADP-dependent enzymes.

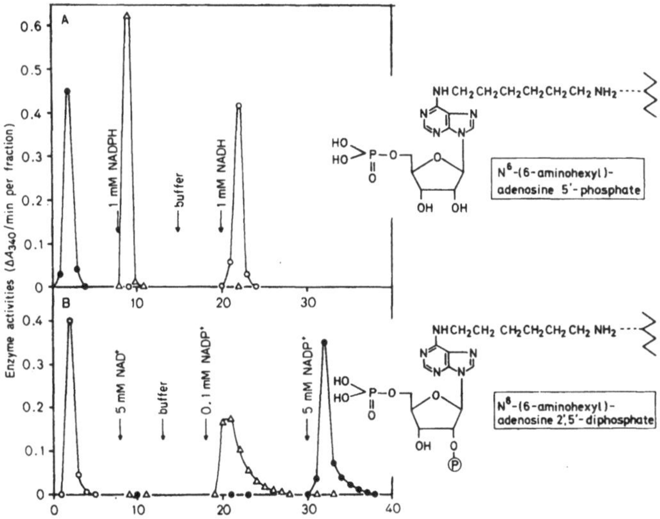

Fig. 1 Affinity chromatography of mixtures of lactate dehydrogenase, glucose-6-phosphate dehydrogenase and phosphogluconate dehydrogenase on Sepharose-bound N^6-(6-aminohexyl)-adenosine 5'-monophosphate (A) and on Sepharose-bound N^6-(6-aminohexyl)-adenosine 2',5'-diphosphate (B). Applications of NAD(H) and NADP(H) dissolved in irrigant buffer are indicated by arrows. (○—○) Lactate dehydrogenase; (△—△) Glucose-6-phosphate dehydrogenase; (●—●) 6-phosphogluconate dehydrogenase.

In the other example to be given here, the enzyme succinate thiokinase, a CoA-dependent enzyme, has affinity for the Sepharose-bound 3',5'-ADP-analog corresponding to the half molecule of coenzyme A, but not for the closely related 2',5'-ADP-analog (3).

Alternative to the binding to catalytic sites, biospecific binding has also been utilized in the purification of enzymes affinity-bound to the effector site. Thus glycogen phosphorylase b would bind to N^6-AMP immobilized to Sepharose, allowing its efficient purification (AMP is known to be an allosteric effector of the enzyme).

Related is the reported purification of transhydrogenases obtained by binding the enzyme to the nucleotide effector site using an immobilized 2',5'-ADP-analog (4) to be discussed later.

III. (Bio)elution. In the two-step process involved in affinity chromatography, i.e. adsorption followed by elution, the importance of the latter's inherent potential for introducing even greater specificity has been increasingly recognized. Using a general ligand with affinity for a large number of enzymes, makes greater demands on the specificity of the elution process. Elution from general ligands has been affected by both specific and non-specific means. To the latter category belong changes of temperature and pH, elution through changes in ionic strength, and elution with hydrophobic agents such as ethylene glycol. Many successful purifications of enzymes from general ligands applying such non-specific elution conditions have been reported. In the author's opinion however, bioelution seems to be the procedure of choice. This implies the use of either a soluble counterligand or other ligands or combinations of ligands for which the enzyme has higher affinity. The eluting nucleotide is simply removed by dialysis or gel chromatography and the cost is rather low since the concentrations of the eluting agent is usually in the millimolar or less range. Normally short pulses suffice, although by applying gradients when necessary, additional resolving power can be gained. The method is more subtle, more additional information of theoretical interest as to the nature of enzyme binding sites can be obtained, and most important the specificity and therefore resolving power is far higher.

To illustrate the potential and power of bioelution, a few examples will be given from work in the author's laboratory on the elution of dehydrogenases affinity-bound to the general ligand N^6-(6-aminohexyl)-AMP.

An example on bioelution utilizing binary complex formation is the separation of the five isozymes of lactate dehydrogenase with a gradient of NADH (Fig. 2) (5). All five isozymes could be separated, illustrating the high resolving power and sensitivity of the system considering that the various K_{diss} constants of the LDH-NADH binary complex are within the narrow range of $0.4-2 \cdot 10^{-6}$M. To obtain completely pure isozymes however, rechromatography of the fractions from each peak area had to be performed. The utilization of ternary complex formation to elute enzymes has also been applied. In the first example, a coenzyme-inhibitor combination was used for the specific elution of the SS isozyme of alcohol dehydrogenase (a steroid-active form) by applying a pulse of NAD and cholic acid, a steroid inhibitor (6). Simultaneous addition of NAD and pyruvate allows formation of a strong ternary dead-end or abortive complex, specifically eluting lactate dehydrogenase (7). Finally, application of specific NAD-adducts led to specific elution of the three enzymes malate dehydrogenase, alcohol dehydrogenase and lactate dehydrogenase (8).

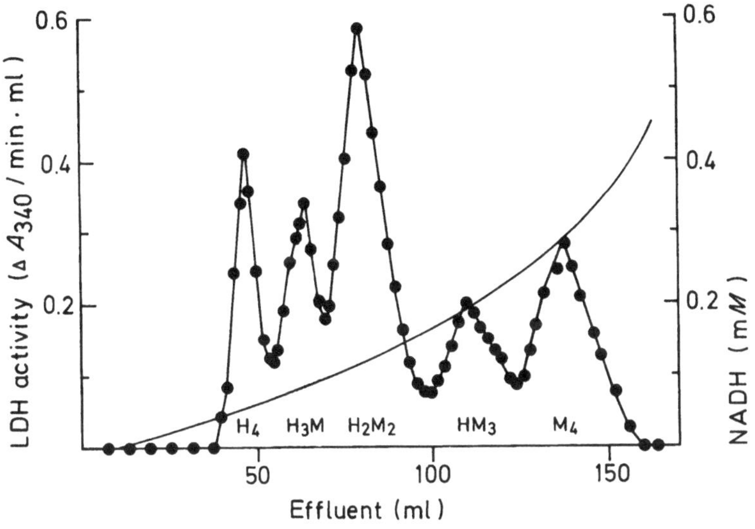

Fig. 2 (Bio)elution of LDH isozymes affinity-bound to N^6-(6-aminohexyl)-adenosine 5'-monophosphate using a concave gradient of NADH.

IV. Applications in enzyme purification and separation. In this paragraph a few examples will be given dealing with the application of general ligand affinity chromatography including (a) purification and enrichment of enzymes from crude samples, (b) purification of enzymes on a preparative scale (to illustrate the resolving power of the procedure), (c) resolution of mixtures of isozymes on a preparative scale, (d) removal of contaminants, and (e) separation of active from non-active species. (a) As an example on purification of enzymes from crude extracts the case of transhydrogenase from Pseudomonas aeruginosa can be given, which has been purified 1.300-fold in one step using N^6-substituted 2',5'-ADP-Sepharose and NADPH as eluant (4). A 17.000-fold purification of human serum lactate dehydrogenase has been obtained in a single step using immobilized N^6-AMP, with an NAD-pyruvate adduct as eluant (8). These examples suffice to illustrate the potential of general ligands not only in "conventional" affinity chromatography but also to their power to allow enrichment of enzymes present at very low levels in fluids or tissues. (b) As an example on enzyme purification on a preparative scale will be mentioned the purification of dogfish lactate dehydrogenase using immobilized N^6-AMP followed by elution with a pulse of 0.12 mM reduced NAD-pyruvate adduct (8). After removal of the pyridin nucleotide on Sephadex G-200, the enzyme crystallized readily. About 30 mg of pure enzyme were obtained in this single-step procedure from 100 grams of dogfish muscle using 30 ml of the gel. It is rather seldom though that the affinity chromatography technique alone yields a homogeneous protein in a single step direct from tissue extracts as in this case, especially if this involves purifications of greater than 1.000-fold. For such cases it is more advisable to apply a gross fractionation procedure prior to the biospecific step. (c) Isozyme purification on preparative scale has been applied in the next example. Following ion-exchange chromatography (6), a liver extract was applied to a N^6-AMP-Sepharose column. Elution with 0.2 mM NAD and 1.5 mM of the steroid inhibitor cholic acid, yielded homogeneous, steroid-active alcohol dehydrogenase SS isozyme with a yield of 30 mg from 260 g of liver. (d) An example on the usefulness of affinity chromatography in separating contaminants is the removal of various dehydrogenases associated with cytochrome oxidase. As a result of their adsorption onto an NAD-matrix, the activity of the oxidase was enhanced substantially when calculated as a function of heme aa_3 content (9).

V. Applications to fundamental studies. Apart from their use in the separation and purification of biological macromolecules, in particular enzymes, adenyl general ligands have recently been applied to a few more fundamental investigations. As will be seen from the examples, it has been of the outmost importance that the ligands have a chemically well-defined "natural" structure.

1. Binding site

Bioaffinity ligands have recently been used to distinguish between the catalytic and effector sites of nicotinamide nucleotide transhydrogenase from Pseudomonas aeruginosa (4). Since NADP acts both as a substrate and an effector, conventional kinetic experiments were difficult to interpret. However, differences in behaviour of this enzyme in presence of N^6-AMP-Sepharose or N^6-2',5'-ADP-Sepharose cast some light on the problem. The enzyme could be eluted from a column of the former with either 1 mM NADPH or NADH but from the 2',5'-ADP column it could be displaced only with NADPH. This was taken as a strong indication that there are at least two different binding sites, one for both NADH and NADPH, (as substrates), the other binding NADPH only (as an allosteric effector).

Rapid screening of unknown enzymes for their binding properties and possible coenzyme participation is also feasible with the various coenzyme (or coenzyme fragment) columns now to hand.

2. Enzyme mechanism

An excellent example of the use of affinity chromatography in exploring enzyme mechanisms, although not involving a general immobilized ligand, is the following: In the absence of an appropriate cofactor, lactate dehydrogenase was shown to have no affinity for a Sepharose linked oxamate derivative. However, this enzyme is known to form a strong ternary complex together with NADH with pyruvate with which oxamate is closely analogous. In the presence of NADH the enzyme was indeed strongly adsorbed to the column. These data strongly imply that NADH induced a binding site of the enzyme in line with a proposed compulsory-ordered mechanism (10).

Bioelution has also been involved in the study of enzyme mechanism. Following from an early preliminary study on elution of lactate dehydrogenase from AMP-columns using a tight dead end complex, NAD and pyruvate, a thorough study on rat heart citrate synthase bound to a N^6-ATP-column was carried out using a number of combinations of substrates and substrate analogs (11). It transpired that although neither oxalacetate nor CoA slone would effect elution, in combination they sharply eluted the enzyme. A closely analogous putative dead end complex mixture, however, citrate and acetyl-CoA, was ineffective in eluting the enzyme. (R)-malate together with CoA eluted the enzyme rapidly whereas the (S)-malate form and CoA led to slow elution.

Affinity chromatography of mammalian phosphofructokinase on N^6-AMP, ribose-linked ATP, and the above N^6-ATP-Sepharose, has been compared (12). It was found that the capacity of the N^6-ATP ligand for this enzyme was 90 times greater than that of the N^6-AMP or the ribose-attached ATP; blue dextran-Sepharose was also less effective. The kinase was biospecifically eluted with ADP plus fructose-6-phosphate, apparently forming a strong dead end complex. These studies demonstrate the usefulness of affinity chromatography in confirming mechanistic evidence adduced from steady state kinetics, on for instance the order in enzyme mechanisms and in allowing convenient screening for dead-end complexes.

3. Kinetic constants

The range of application of general ligand affinity chromatography has recently been extended to quantitative investigations of the interactions between enzymes and specific ligands. Various isozymes of lactate dehydrogenase were adsorbed on a column of immobilized N^6-AMP. On eluting with a gradient of NADH, a linear relation was observed between the concentration of nucleotide in the eluant required to displace a particular isozyme and the reported dissociation constants for the corresponding binary enzyme-nucleotide complexes (13). This relation was then utilized to determine unknown K_{diss}-constants of other isozymes. Using this gradient affinity chromatography technique, K_{diss}-constants of various coenzyme fragments used as eluants for N^6-AMP-affinity bound alcohol dehydrogenase could also be determined.

B. IMMOBILIZED ACTIVE COENZYME PREPARATIONS

1. Retention and Regeneration Methods

Perhaps the most intriguing facet of the subject of immobilized coenzymes is their capacity to retain activity and biospecificity when bound either to a water-soluble or a particulate support. The development of efficient enzyme-catalyzed processes has been recognized as being a prerequisite to the use of enzyme technology on an industrial scale. Many of these enzyme-catalyzed chemical processes require the participation of readily dissociable cofactors. As these are often very expensive, their economical utilization requires techniques both for their retention in the reaction mixture and for their regeneration. In addition, in cases where it is essential not to perturb the medium, for instance in certain medical and analytical applications, coenzyme immobilization is essential. Various retention and regeneration methods have hitherto been applied to coenzymes. Amongst the alternatives, coenzyme analogs immobilized on water soluble supports and used entrapped within semi-permeable membranes together with a recycling enzyme system constitute the most closely studied procedure to date.

2. Application in Enzyme Electrode and Enzyme Reactor

Two examples from our own studies on application of immobilized coenzymes in analysis and enzyme technology serve as illustrations of the work in this field (14). The N^6-substituted NAD analog, N^6-[(6-aminohexyl)-carbamoylmethyl]-NAD, was immobilized on dextran and coentrapped with soluble lactate dehydrogenase and glutamate dehydrogenase within a dialysis bag encasing a cation-sensitive enzyme-electrode. In the presence of glutamate in an assay medium, which also contained pyruvate as substrate for the recycling enzyme lactate dehydrogenase, ammonium ions were generated which were recorded by a NH_4-sensitive electrode. A linear relationship was found between the glutamate present and the current response obtained.

Using the same dextran-NAD preparation, a model enzyme reactor reaction was carried out. Inside an ultrafiltration apparatus the NAD-preparation was kept in proximity with soluble lactate dehydrogenase and alanine dehydrogenase. On continuously pumping substrate through the unit, alanine was produced from pyruvate. A similar two-enzyme system was also applied to alanine production on a small scale, but using a polyethyleneimine-coupled NAD-analog, entrapped together with soluble enzymes in cellulose triacetate fibres with the NAD-preparation being recycled at a rate of 222 cycles/h (15).

A rather different enzyme-coenzyme arrangement has been reported for regeneration purposes, which simultaneously solves the retention problem. Here an immobilized enzyme-coenzyme complex is built up on a matrix so that the coenzyme analog of NAD, N^6- [(6-aminohexyl) carbamoylmethyl]-NAD, is sitting permanently in the active site of the enzyme, liver alcohol dehydrogenase (16). On adding a second substrate, the coenzyme is regenerated in a coupled substrate recycling manner, as outlined in Fig. 3. The recycling rate in this type of immobilized NAD-molecule is high and has been estimated to be about 3.400 cycles/h. However, the relatively small number of such coupled systems known, which may be of practical interest, seems to restrict a wider use of this type of preparation. The development of this approach had been influenced by previous work on phosphorylase b, which was permanently activated through coupling of the N^6-AMP-analog which acted as effector.

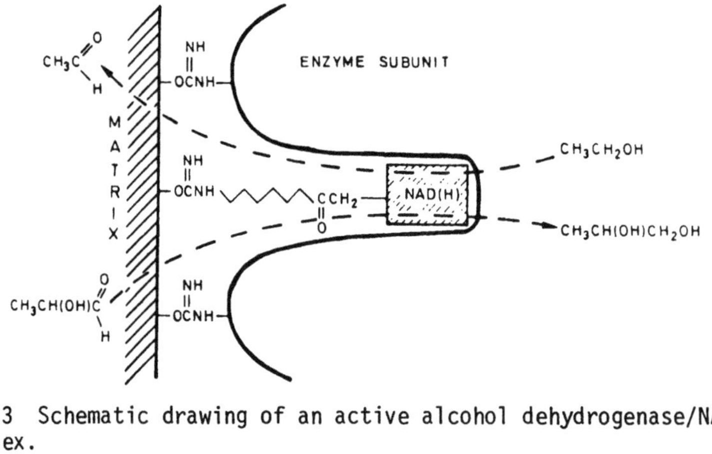

Fig. 3 Schematic drawing of an active alcohol dehydrogenase/NAD/Sepharose complex.

3. Discussion on Coenzymic Activity of Immobilized Coenzyme Preparation

In most of the publications on the use of immobilized coenzymes attention has been focused either on the aspect of affinity chromatography or on the use of coenzymically active preparations, but very seldom have both been considered together. Nevertheless both aspects are intimately related to one another and much useful information can be gained from such joint appraisal. Reports on coenzymically active preparations are scarce and of rather recent date. In most cases preassembled coenzymes, carrying a spacer with a functional group, have been used for immobilization. The coenzymes tested include in particular NAD but also to some extent NADP and ATP. Generally speaking the coenzymic activities of such analogs are somewhat lower than those of the parent coenzyme under comparable conditions and decrease further upon binding due in part to diffusional hindrance exerted by the support, showing the following general relationship: coenzyme>coenzyme analog>coenzyme bound to soluble support>coenzyme bound to particulate support. As example can be given yeast alcohol dehydrogenase showing the following relative "coenzymic activities": NAD = 1, N^6-NAD = 0.61, water soluble dextran-bound = 0.16, and Sepharose 4B-bound = 0.07 (measured at constant nucleotide, substrate, and enzyme concentrations; judged from the few data available on K_m-values of these coenzyme analogs, they appear to be rather close to those of the parent nucleotide).

In closing I wish to point again to the fact that the type of general ligands discussed in the last section, i.e. the various adenine coenzymes, are unique in allowing an answer to the often raised question of whether true biospecific adsorption is involved in an affinity chromatographic system. If a coenzyme analog of e.g. NAD is "coenzymatically active" when immobilized to soluble or particulate supports, thus allowing the enzyme to reach to the nicotinamide position, where the oxidation/reduction takes place, then it is extremely likely that the same biospecific interaction is prevailing when the enzyme is affinity-bound to its immobilized coenzyme. No such definite proof for biospecific binding can be obtained from an enzyme's elution behaviour alone. Even if an enzyme is bioelutable with soluble biospecific ligands, this may simply be the result of conformational changes introduced, or changes in charge obtained near the active site on ligand binding, both of which would lead to an enzyme's elution also from a non-biospecific matrix. Nevertheless in the predominant number of cases, an observed effective bioelution is a very strong indication that the enzyme also has been biospecifically bound either through complete or at least partial biospecificity.

REFERENCES

1. Brodelius, P., Larsson, P.-O., Mosbach, K.: The synthesis of three AMP-analogues: N^6-(6-aminohexyl)-adenosine 5'-monophosphate, N^6-(6-aminohexyl)-adenosine 2',5'-bisphosphate, and N^6-(6-aminohexyl)-adenosine 3',5'-bisphosphate and their application as general ligands in biospecific affinity chromatography. Eur. J. Biochem. 47, 81-89 (1974).

2. Lindberg, M., Larsson, P.-O., Mosbach, K.: A new immobilized NAD^+ analogue, its application in affinity chromatography and as a functioning coenzyme. Eur. J. Biochem. 40, 187-193 (1973).

3. Barry, S., Brodelius, P., Mosbach, K.: General ligand affinity chromatography: N^6-(6-aminohexyl) 3',5'-ADP Sepharose as an affinity adsorbent for the CoA-dependent enzyme, succinate thiokinase. FEBS Lett. 70 No. 1, 261-266 (1976).

4. Höjeberg, B., Brodelius, P., Rydström, J., Mosbach, K.: Affinity chromatography and binding studies on immobilized adenosine 5'-monophosphate and adenosine 2'5'-bisphosphate of nicotinamide nucleotide transhydrogenase from *Pseudomonas aeruginosa*. Eur. J. Biochem. 66, 467-475 (1976).

5. Brodelius, P., Mosbach, K.: Separation of the isoenzymes of lactate dehydrogenase by affinity chromatography using an immobilized AMP-analogue. FEBS Lett. 35 No. 2, 223-226 (1973).

6. Andersson, L., Jörnvall, H., Mosbach, K.: Preparative purification of homogeneous steroid-active isozyme of horse liver alcohol dehydrogenase by affinity chromatography on an immobilized AMP-analog. Analyt. Biochem. 69, 401-409 (1975).

7. Ohlsson, R., Brodelius, P., Mosbach, K.: Affinity chromatography of enzymes on an AMP-analogue: Specific elution of dehydrogenases from a general ligand. FEBS Lett. 25 No. 2, 234-238 (1972).

8. Kaplan, N.O., Everse, J., Dixon, J.E., Stolzenbach, F.E., Lee, C.-Y., Lee, C.-L.T., Taylor, S.S., Mosbach, K.: Purification and separation of pyridine nucleotide-linked dehydrogenases by affinity chromatography techniques. Proc. Nat. Acad. Sci. 71 No. 9, 3450-3454 (1974).

9. Holbrook, J., Bucher, J., Penniall, R.: The binding by an NAD^+-affinity matrix of contaminating dehydrogenases in cytochrome *c* oxidase preparations. Hoppe-Seyler's Z. Physiol. Chem. 357, 623-627 (1976).

10. O'Carra, P., Barry, S.: Affinity chromatography of lactate dehydrogenase. Model studies demonstrating the potential of the technique in the mechanistic investigation as well as in the purification of multisubstrate enzymes. FEBS Lett. 21 No. 3, 281-285 (1972).

11. Mukherjee, A., Srere, P.A.: Purification of and mechanism studies on citrate synthase. Use of biospecific adsorption-elution techniques. J. Biol. Chem. 251, 1476-1480 (1976).

12. Ramadoss, C.S., Luby, L.J., Uyeda, K.: Affinity chromatography of phosphofructokinase. Arch. Biochem. Biophys. 175, 487-494 (1976).

13. Brodelius, P., Mosbach, K.: Determination of dissociation constants for binary dehydrogenase-coenzyme complexes by (bio)affinity chromatography on an immobilized AMP-analogue. Analyt. Biochem. 72, 629-636 (1976).

14. Davies, P., Mosbach, K.: The application of immobilized NAD^+ in an enzyme electrode and in model enzyme reactors. Biochim. Biophys. Acta 370, 329-338 (1974).

15. Marconi, W., Prosperi, S., Giovenco, S., Morisi, F.: Entrapment of coenzymically active NAD^+ polymers in fibres. J. Mol. Catalysis 1 No. 2, 111-120 (1976).

16. Gestrelius, S., Månsson, M.-O., Mosbach, K.: Preparation of an alcohol-dehydrogenase-NAD(H)-Sepharose complex showing no requirement of soluble coenzyme for its activity. Eur. J. Biochem. 57, 529-535 (1975).

Received February 7, 1977

DISCUSSION

Kiltz: The elution diagram, where you have separated the LDH isozymes is very interesting in respect to primary and three-dimensional structure. You found that the M_4 isozyme is stronger bound to the AMP column as the H_4 isozyme. This is in agreement with the isozyme typical amino acid exchanges. The hydrophobic adenine binding pocket seems to be slightly smaller in the M_4 isozyme, so we expected a higher affinity of adenine to this isozyme.

Mosbach: I should stress that the elution pattern obtained on addition of a NADH-gradient seems primarily to be governed by the strength of the binary complex LDH-NADH formed, which is known to vary for the five isozymes in accordance with the elution pattern obtained. Nevertheless, based on your findings. one ought to attempt to obtain separation already in the adsorption step which could be done by addition of excess of the isozymes in relation to the amount of AMP-gel used.

Banaszak: Since you did not show electron patterns of crude extracts, have you ever found active proteins such as heme proteins, as chromatographing with dehydrogenases?

Mosbach: I did show briefly some elution patterns obtained from crude extracts; in my experience I have not noticed cochromatography of heme proteins with dehydrogenases. In fact I know of a report in the literature (Hoppe-Seyler's Z. Physiol. Chem. 357, 623 (1976)) where NAD-dependent dehydrogenases present as contamination in cytochrome oxidase preparations were removed on passing it over an NAD^+-matrix.

Dunn: For the liver ADH regenerating system (Fig. 3), it seems likely to me that only one site per enzyme dimer is occupied by the Sepharose NAD^+. If this is actually so, then your experiment suggests that liver ADH is active when only one site is occupied by substrate and coenzyme. Do you agree?

Mosbach: I agree with your assumption that only one site per enzyme dimer is occupied by the Sepharose-NAD^+. We obtain increased activity on external addition of NAD^+ since the other site seems free to interact; I agree with your conclusion.

In this context I might mention preliminary findings obtained by Dr. Andersson and myself in which liver ADH was strongly bioaffinity bound in the presence of isobutyramide to an immobilized NADH-analogue. After inactivation of the free subunit with iodoacetate the dimer was dissociated from the matrix and shown to be active.

Werber: Why does NAD^+ in some cases cause elution of the enzymes from the affinity columns, and in other cases a pulse of NAD^+ is inefficient, and, therefore, an adduct - or mixture - of NAD^+ and a specific ligand is required to cause elution?

Mosbach: In general, binary complexes, e.g. NAD-enzyme, are weaker than ternary complexes, i.e. complexes of enzymes e.g. with NAD^+-adducts or dead-end complexes such as NAD^+-pyruvate as shown in the slide. Thus, if NAD^+ is not sufficiently strong to cause bioelution from the AMP-column, simultaneous addition of pyruvate will lead to stronger complex formation thus eluting the enzyme (= lactate dehydrogenase). In addition, it should not be forgotten that the efficiency in elution of the various nucleotides or nucleotide-combinations added is concentration-dependent.

Jeck: In his preprint, Dr. Mosbach regrets that affinity chromatographic aspects of immobilized coenzyme analogs seldom have been considered together with the coenzymatic activity of these preparations. In this context I wish to add a few more experimental data, we obtained in Dr. Woenckhaus' laboratory, which show one of the major restrictions towards a wider use of immobilized NAD^+ analogs, especially in recycling processes.

In order to avoid the expensive use of coenzymes in the elution medium of affinity chromatography we synthesized a coenzyme analog of NAD^+ showing low affinity towards dehydrogenases. The analog differs from NAD by one substitution: The adenine linked ribose is substituted by a trimethylene group.

The covalent link to the Sepharose 4 B matrix was performed via hexamethylene group to the exocyclic adenine amino group. The free coenzyme analog (or its trifluoroacetyl derivative) is active as a hydrogen acceptor in enzymic tests with various dehydrogenases. The maximum turnover number ranges from 2 % to 110 % of the turnover number of NAD^+. The immobilized coenzyme analog shows little or no affinity to several dehydrogenases but could be successfully used to separate glyceraldehyde 3-phosphate dehydrogenase from a mixture of several proteins including two more dehydrogenases (LDH and ADH from liver). Glyceraldehyde 3-phosphate dehydrogenase is retained on the column and eluted with a salt gradient.

The major feature of this immobilized coenzyme analog is, that it is completely inactive as a hydrogen acceptor in enzymic tests with dehydrogenases, thus showing the restriction I mentioned at the beginning.

Mosbach: The new compound you describe is very interesting indeed. The lack of "coenzymic activity" in its immobilized state could very well be due to the particulate matrix used. I would not be surprised that on using your analog bound to soluble dextran and using excess of enzyme you will obtain activity. I suggest as reading on this particular subject a forthcoming publication in Adv. Enzym. 1977 (K. Mosbach; in press). The affinity behavior you describe indicates ionic interaction in the binding process.

Blankenhorn: I have to comment on the immobilized "flavin"-material you have been showing. Actually this compound does not belong to the flavin class although the trivial name is acriflavin but it is rather an acridine derivative.

Mosbach: I know, however, I only mentioned it briefly since "acriflavin" can be reduced in the immobilized state by soluble NADH and thus is somewhat related to your bimolecular model systems.

Sund: Is it possible to determine quantitatively the number of nucleotides bound covalently to the matrix and, if this is possible, do you have any information with respect to the distance between two bound nucleotides? I ask this question because I would like to know if LDH or other enzymes can be bound not only to one but also to two nucleotides. And if so is there a difference in the affinity in the binding to one or to two nucleotides?

Mosbach: Yes, the number of nucleotides bound to the matrix can and ought to be determined quantitatively and is frequently carried out based on phosphate analysis. I do not recall at the moment the distances we have calculated for various nucleotide concentrations and it lies certainly within the range of possibility that a situation as you describe can occur in particular considering the existing network of the gel. However, on using gels with low nucleotide substitution level, e.g. 2 μmoles of nucleotide/ml packed wet gel as e.g. commercially available N^6-substutited AMP-Sepharose or 2',5'-ADP-Sepharose gels carry, the risk of perturbation of the elution is negligible.

Holbrook: May I take up your request to interpret affinity chromatography results mechanistically. First a result: Supernatent malate dehydrogenase and H_4 LDH of pig have similar charges, and three-dimensional structures, and bind NADH equally tightly, and in the same way. Yet we, and many others, have consistently observed that soluble malate dehydrogenase will hardly bind to affinity columns based upon Cibachrome blue, AMP or NAD^+, while LDH binds very tightly. Would you agree that this contradiction between solid phase and solution binding is yet another indirect indication that NADH binding to malate dehydrogenase involves a protein rearrangement which is absent in LDH?

Mosbach: Assuming affinity chromatography of soluble malate dehydrogenase has been carried out on well-defined, preferentially pre-assembled NADH-analogs, and that no binding has occurred, then I would assume that the fact that in contrast hereto LDH binds strongly is in part due to additional hydrophobic binding between the spacer and the hydrophobic region present in the enzyme. However, I would not be surprised that bioaffinity binding can be obtained with the proper NAD ligand. If it still doesn't work, then a possible explanation could be the different steric arrangements prevailing on solid-phase effecting the behavior. I suggest the following: use a defined NAD ligand, e.g. N^6-[(6-aminohexyl)carbamoylmethyl]-NADH, test it in free solution, if it is functioning as coenzyme, bit it to a matrix. If it doesn't work then, it could be steric problems not allowing the proper rearrangement of the enzyme to bind to the immobilized ligand or that the terminal NH_2-group of the nucleotide ligand, which is now involved in the binding to the matrix, was required for activity.

Walk: I would like to come back to the remark of Dr. Dunn according to malate dehydrogenase.

In order to study the behavior of malate dehydrogenase during affinity chromatography one has to know, which isozyme is put onto the column. We found that the cytoplasmic form from pig heart does not bind to 5'AMP whereas the mitochondrial malate dehydrogenase does.

In addition the ionic strength during chromatography has to be selected carefully, since the binding of mitochondrial malate dehydrogenase occurs only at low ionic strength. Truly biospecific elution of the adsorbed mitochondrial form is obtained with low concentrations of NADH (36 µM).

Mosbach: Yes, I am aware of that interesting recent paper of yours in Eur. J. Biochem.

The Interaction of Glutamate Dehydrogenase with Ligands.

R. Koberstein, H. Dieter, K. Markau and H. Sund
Universität Konstanz, Fachbereich Biologie, Postfach 7733,
D-7750 Konstanz, West Germany

INTRODUCTION

Beef liver glutamate dehydrogenase (GluDH) is able to bind a variety of nucleotides at its active and regulatory sites. ADP and GTP and related purine nucleotides affect the enzymatic activity of GluDH in a complex manner [1]. In addition, ADP and GTP antagonistically influence the association of GluDH to higher aggregates in the presence of NAD(P)H. In order to elucidate the underlying mechanisms, kinetic and equilibrium binding measurements were performed. In this communication, the binding data will be discussed whereas the kinetic results will be presented in another section of this symposium [2].

Two types of binding experiments will be discussed: tight binding with dissociation constants $K_D < 10^{-3}$ M and weak binding ($K_D > 10^{-3}$ M). These definitions are chosen arbitrarily by the applicability of two different methods. Tight binding is detected by equilibrium dialysis which allows determination of K_D values up to about 10^{-3} M. At higher dissociation constants this method is not feasible since the practicable enzyme concentration normally is limited to about 100 μM. In special cases, optical methods (protein fluorescence or difference spectroscopy) can help, but the only method which is applicable in a wide field is nuclear magnetic resonance (NMR). This method which is not yet so common in biochemistry will be briefly described.

The NMR signals of ligands are broadened when bound to a macromolecular system. In some special cases the signals are shifted in the bound state but the relaxation times (especially the transversal one, T_2) are always shortened as a consequence of the reduced mobility which prevents perfect averaging of the anisotropic dipolar spin-spin interactions. If the exchange rate between the free and the bound state is slow compared to the linewidth only the spectrum of the free ligand is obtained whereas that of the bound ligand is often not detectable due to the very large broadening. In the opposite case only one averaged spectrum is generated and broadening is even induced at high excess of ligand [L] over the binding sites [E]. Therefore, provided the exchange rate is high enough the method is able to detect weak binding sites.

If the shift of the signals can be neglected, as apparent from all our measurements, the equation describing the linewidth $\Delta\nu$ (at half height) [3] reduces to

$$[E]/\Delta T_2^{-1} = (K_D + [L])\,(T_{2b} + {}^{1}/_{k^-}),$$

$$\text{with } \Delta T_2^{-1} = T_2^{-1} - T_{2f}^{-1} = \pi(\Delta\nu - \Delta\nu_f)$$

where T_{2f} and T_{2b} are the transversal relaxation times in the free and the bound state, respectively; K_D is the dissociation constant and k^- dissociation rate constant. If T_{2b} is small compared to $1/k^-$ (intermediate exchange) both rate constants can easily be determined. Unfortunately, this condition is not fulfilled in our experiments but nevertheless we obtain K_D from a plot of $[E]/\Delta T_2^{-1}$ versus [L] as intercept with the [L] axis and we find a minimum value, k_1^-, for the dissociation rate from the reciprocal slope of this plot. In principle T_{2b} can be estimated from dispersion of the relaxation times if the distances of the interacting nuclei are known. In our case, however, this is not possible since we do not know the exact conformation of the bound ligands and we must not assume that the contribution of the protein nuclei is negligible. We can estimate, however, from additional measurements whether the real k^- is much higher than the calculated limit.

In this communication results of tight binding studies of the ternary systems GluDH, ADP, GTP and GluDH, NADPH, GTP will be presented. For the latter system, the correlation between ternary complex formation and dissociation of the enzyme into the unimer is also discussed. From the investigations of weak binding, low affinity sites for ADP and other mononucleotides and their interaction in the bound state with the coenzyme are demonstrated. Furthermore, the rate constants for the substrate binding and the mutual tightening between coenzyme and substrate are given. It is shown that NMR spectroscopy is the method of choice for the detection of fast binding sites, yielding data which, in connection with kinetic and equilibrium measurements, help to complete the model of enzyme action.

RESULTS AND DISCUSSION

Preliminary equilibrium dialysis had indicated very poor binding of GTP to GluDH in the presence of orthophosphate. This was in accordance with earlier observations [4]. In order to clarify the question whether ADP and GTP exclude each other from binding to GluDH or not [5 - 7], we searched for a buffer system in which on one hand nucleotides were strongly bound to the enzyme and on the other hand the enzyme was completely stable for at least 4 hours. We found triethanolamine (50 mM), KCl (0.1 M), pH 7.6 (TRA buffer) to fulfill these requirements, whereas TRIS buffer, pH 7.6 (with or without 0.1 M KCl), caused 50 % inactivation of GluDH within 30 - 180 min depending on the conditions.

First the binary complexes E · NADPH, E · ADP and E · GTP (E = GluDH) were reinvestigated in the new buffer system. The two former complexes showed the same binding behavior as in 0.067 M phosphate, pH 7.6, whereas the E · GTP complex yielded an about tenfold lower dissociation constant ($K_{E,GTP}$ = 35 μM) in TRA buffer than in phosphate ($K_{E,GTP}$ = 325 μM). The ternary system GluDH, ADP, GTP was investigated only in TRA buffer.

Fig. 1A shows the binding of [^{32}P]GTP to GluDH at increasing concentrations of ADP. Fig. 1B is the analogous representation of the binding of [^{32}P]ADP to GluDH at increasing GTP concentrations. At first sight, both figures suggest competitive binding of the two nucleotides since all lines exhibit the same intercept of the abscissa. At very high concentrations of the second ligand, however, finite limiting slopes are obtained for the binding of the first ligand. Moreover, evaluation of the sum of bound ADP plus GTP clearly demonstrates that up to 1.5 sites per polypeptide chain are occupied at high GTP and moderate ADP concentrations (Fig. 2). These results, which exclude competition of the two ligands for the same site, are best

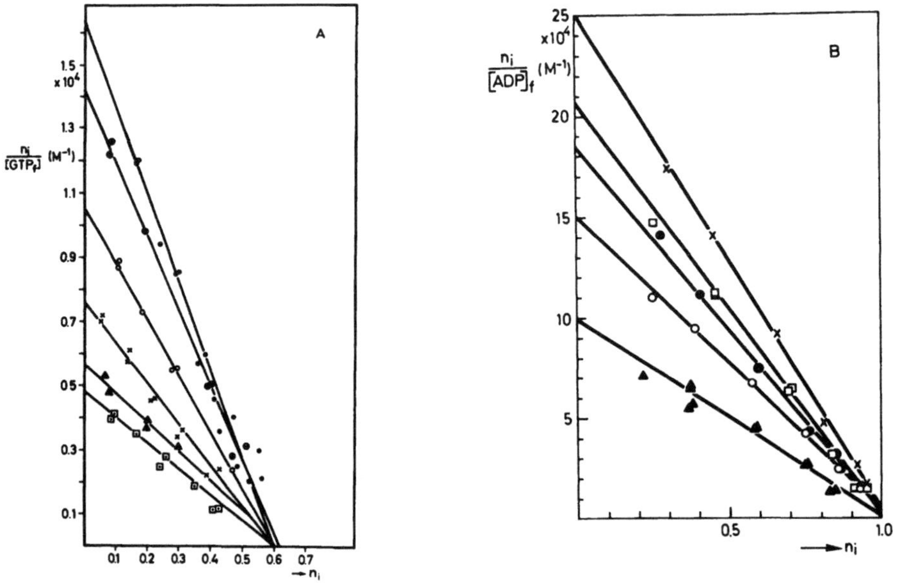

Fig. 1A Binding of [γ-^{32}P]GTP to GluDH in the presence of ADP. Scatchard plot of data obtained by equilibrium dialysis in TRA buffer pH 7.6 at 20° C. $[E_o]$ = 36 μM with respect to the polypeptide chain.

● without ADP		x $[ADP_t]/[E_o]$ =	1.28
◉ $[ADP_t]/[E_o]$ = 0.18		◬ " =	2.90
o " = 0.52		⊡ " =	11.80

Fig. 1B Binding of [β-^{32}P]ADP to GluDH in the presence of GTP. Scatchard plot of data obtained by equilibrium dialysis in TRA buffer pH 7.6 at 20° C. $[E_o]$ = 30 μM

x $[GTP_t]/[E_o]$ = 1.15	o $[GTP_t]/[E_o]$ = 19.35
□ " = 3.33	▲ " = 66.6
● " = 6.45	

explained by the following scheme comprising different but negatively interacting binding sites for the two nucleotides, which are bound more loosely in the ternary complex than in the respective binary complexes.

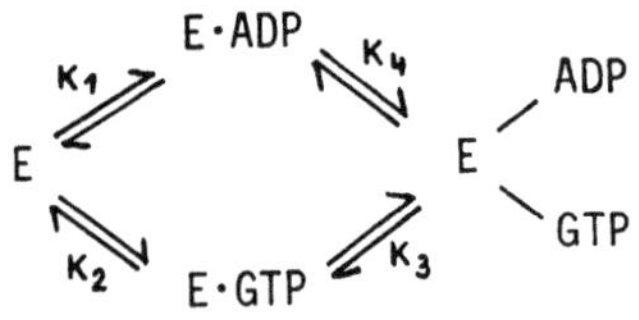

K_1 = 2.7 μM, K_2 = 35 μM,

K_3 = 10 μM, K_4 = 120 μM

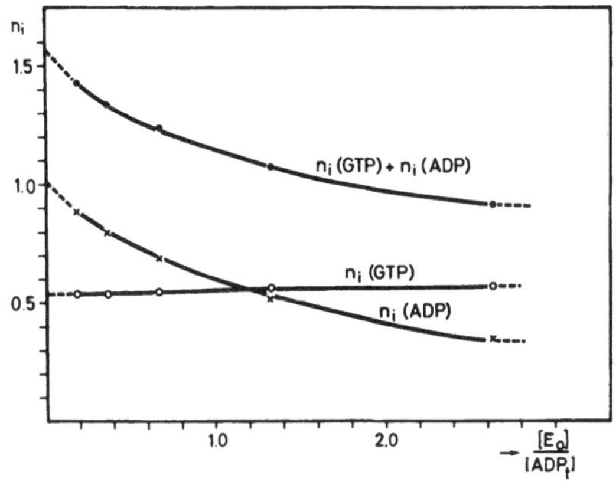

Fig. 2 Simultaneous binding of ADP and GTP to GluDH. n_i, number of occupied binding sites per polypeptide chain. $[E_0]$ = 30 μM, $[GTP_t]$ = 2 mM, other conditions as in Fig. 1.

For this mutual loosening of the nucleotides a coupling energy [8] of $\Delta G_{coup.}$ = + 3.3 kJ/Mole can be calculated. A further support of this mechanism is obtained by computer-fitting of the titration data with the above scheme and dissocition constants.

In the presence of GTP and NAD(P)H GluDH dissociates from a highly associated state (rodlike particles) to the unimer (comprised of 6 polypeptide chains) [1]. In order to correlate the saturation of the enzyme with these ligands and its state of association equilibrium binding measurements of NADPH and GTP were performed and compared with the sedimentation behavior of the enzyme at defined degress of ligand saturation. Since GTP or NADPH alone are not able to dissociate the enzyme even at vast excess concentrations [1], one has to expect cooperativity between the two ligands in order to understand their synergetic stabilization of the dissociated state of the enzyme. Indeed, sigmoidal titration behavior was observed for the binding of GTP in the presence of saturating concentrations of NADPH. This is expressed in the Scatchard plot (Fig. 3), which again shows the strongly reduced affinity for GTP in phosphate buffer (lower curve) as compared to TRA buffer (upper curve). An analogous Scatchard plot indicating positive cooperativity had been obtained earlier [7] for the NADPH binding at saturating GTP concentrations. The Hill coefficients calculated from the titra-

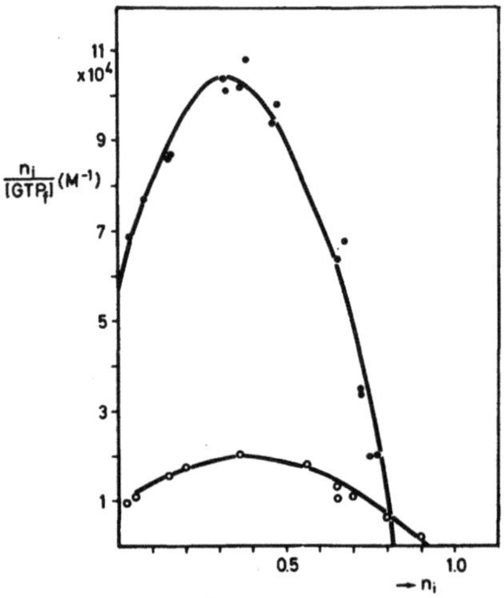

Fig. 3 Binding of $[\gamma\text{-}^{32}P]GTP$ to the GluDH-NADPH complex. Scatchard plot of data obtained by equilibrium dialyses at pH 7.6 and 20^{0} C. • in TRA buffer; o in M/15 phosphate buffer. $[E_0]$ = 36 μM; $[NADPH_t]$ = 600 μM.

tion data are identical for GTP and NADPH (n_H = 1.65 ± 0.05).

Now the sedimentation behavior of the GluDH·GTP·NADPH complex was investigated. In all cases of partial saturation of this ternary complex two separable peaks were observed in the sedimentation pattern: a faster moving peak with about s_{20} = 27 S and a slower one with about s_{20} = 14 S corresponding to the highly associated and fully dissociated particles, respectively. This result was obtained with saturating GTP and low NADPH concentration as well as with saturating NADPH and low GTP concentration. The occurrence of two peaks indicates a very slow equilibration between the associated and the dissociated particles, since otherwise a single peak with an intermediate sedimentation constant would have been expected.

From the area under the slower peak, the relative amount of dissociated GluDH can be determined. As shown in Table 1, there is a very good agreement between the degree of ternary complex formation and the extent of dissociation. This holds true for GTP as well as for NADPH as the "dissociating" ligand.

Therefore, the ternary complex exists in measurable amounts only with the

Table 1 Correlation of ternary complex formation and dissociation into the unimer in the system GluDH, GTP and NADPH in TRA buffer, pH 7.6, at 20° C.

a) GTP at saturating concentration

[NADPH]/[E]	0	0.11	0.43	2.2
relative saturation with NADPH (%)	0	9	39	98
amount of dissociated enzyme (%) *	0	11	36	100
s_{20} of slow peak (S)	-	14.1	14.5	15.0
s_{20} of fast peak (S)	29.0	28.2	26.8	-

b) NADPH at saturating concentration

[GTP]/[E]	0	0.17	0.68	3.4
relative saturation with GTP (%)	0	16	64	99
amount of dissociated enzyme (%) *	0	13	57	100
s_{20} of slow peak (S)	-	13.8	14.4	15.0
s_{20} of fast peak (S)	30.0	29.3	25.5	-

*calculated as area under the slow peak related to the whole area under the sedimentation diagram.

dissociated form of the enzyme. A further consequence of the above mentioned correlation is that saturation of the six NADPH and the six GTP binding sites is necessary to stabilize the dissociated state. At equilibrium complete dissociation into the unimer at half of the sites saturation with the ligand as reported from kinetic measurements [9] can be excluded.

Kinetic studies have shown that ADP not only induces effects characterized by the known low dissociation constant but also competitive inhibition with K_I in the millimolar range [2] suggesting a second weak binding site. The conventional methods must fail for such high K_D values as the practicable enzyme concentrations are in the order of 100 μM. NMR is the only method which is feasible in this case. Indeed, we could find the expected binding site by this method. We also obtained data on the interactions between different ligands.

Fig. 4 shows results for ADP. We obtain K_D = 6 mM in good agreement with the kinetic data and k_1^- = 2000 sec^{-1}. T_{2b} cannot be neglected since 1.) the broadenings of both adenine lines are not equal, 2.) the broadening decreases with increasing temperature (this will be discussed more detailed below) and 3.) $[E]/\Delta T_2^{-1}$ is not independent of the enzyme concentration. The last effect is due to the reversible association of the enzyme which increases the rotational relaxation time τ_R. From the known association constant [10] we calculate that the average molecular weight M_n (which is proportional to the volume and thus roughly proportional to τ_R) increases by 34 % upon increasing the enzyme concentration from 38 to 72 μM whereas the

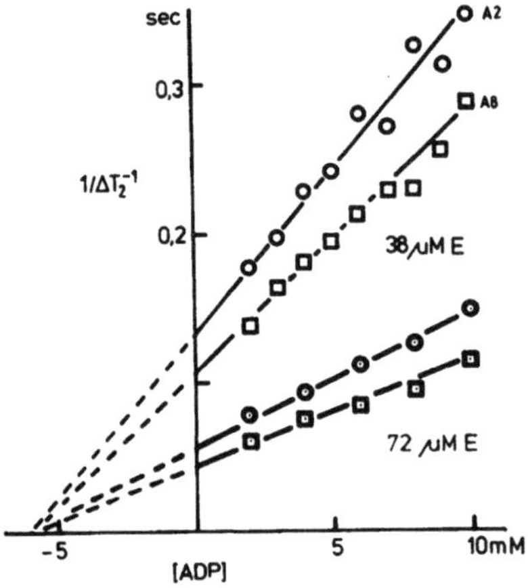

Fig. 4 Relative reciprocal broadening of the adenine signals of ADP versus ligand concentration with two different enzyme concentrations (polypeptide chain with MW = 56 000). M/15 phosphate buffer in D_2O, pD (uncorrected) 7.6, 23° C. A2 and A8 refer to the adenine protons in 2 and 8 position respectively. Measurements are taken with a Bruker HFX 90 spectrometer in the Fourier-transform mode with up to 1024 decays.

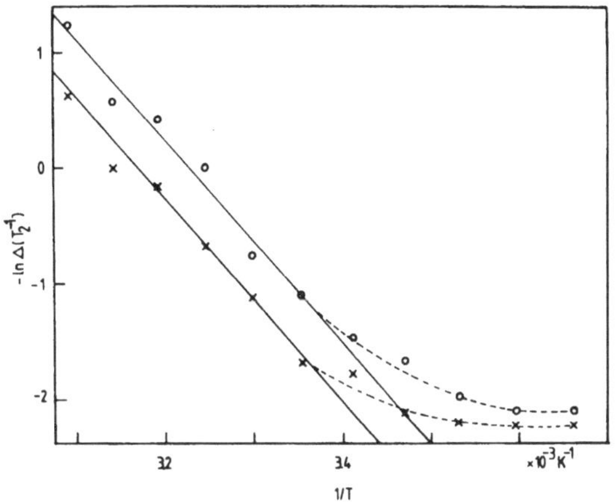

Fig. 5 Arrhenius-plot of the broadening of ADP signals. Conditions as in Fig. 4.

relative broadening only increases by half this amount. We assume therefore that $1/T_{2b}$ has approximately the same value as k^- which is estimated to be in the region of 4000 sec^{-1}. The true value of k^- might be still higher since the temperature dependence shows unexpected anomalies. Fig. 5. shows an Arrhenius representation of the relative broadening from which an activation energy E_a = 73 kJ/mol is calculated.

Even if the binding process itself does not contribute to the broadening ($1/k^- \ll T_{2b}$) (it would induce an opposite temperature effect) the "activation energy" (temperature dependence of τ_R) should not exceed 20 kJ/mol (under the assumption that the change of the viscosity, which gives the main contribution to this value, can be described by the macroscopic data). The much higher experimental activation energy could be explained by residual mobility of ADP in the bound state, E_a describing the hindrance of this motion. The curvature at the lower temperature then indicates that the mobility is frozen and the temperature dependence is described by a superposition of τ_R and the exchange rate. Unfortunately, it is not possible to measure the broadening in a temperature range where E_a is only determined by the exchange rate.

AMP and ATP show the same K_D as ADP (the absolute broadening being somewhat smaller in the case of ATP, which might be even more mobile in the bound state). The additional phosphate groups thus appear to be not bound at this site. K_D of 2',5'-ADP, is higher by a factor of approx. 2, K_D of 2'-AMP is even higher, around 40 mM. The phosphate in 2'-position thus provides positive, that in 5'-position negative contribution to the free energy of binding. ADPR has a similar K_D as the mononucleotides but the broadening is larger indicating reduced mobility in the bound state. (The thermal analysis which suffers from the large broadening has not yet been fully completed but "E_a" is around 12 kJ/mol.)

Fast and weak binding is not only observed with the adenine but also with other nucleotides. GTP is bound with a K_D of 6 mM. Fig. 6 shows the competition between ADP and GTP at the same binding site with K_I = 5 mM (the difference to K_D is not significant in the light of experimental errors).

On the basis of these results we looked for the kinetic effects of high concentrations of GTP. We found inhibition of the reaction with norvaline, described by a K_I in the millimolar range. With the enzyme from Candida utilis, competitive inhibition by GTP with K_I = 10 mM is also observed with glutamate as the substrate [11]. We thus conclude that the guanine nucleotides also have two different binding sites that are both kinetically relevant. The pyrimidine nucleotides are bound to a weak binding site, too. K_D's between 40 to 80 mM for IMP < CMP < UMP are found and the corresponding competitive inhibition is observed. In contrast, no line broadening of

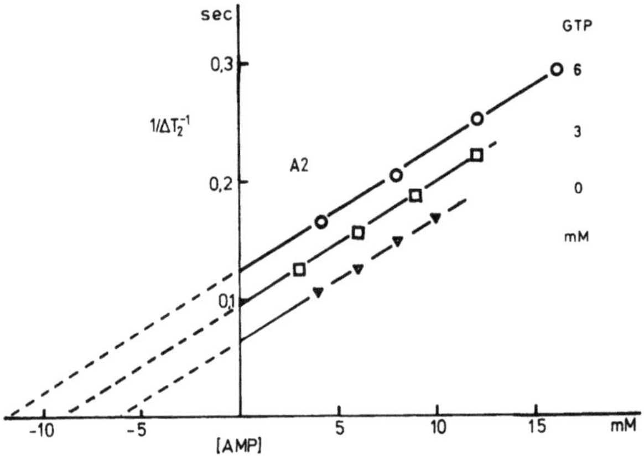

Fig. 6 Competitive replacement of AMP (adenine-2-signal) by GTP at the fast binding site. Conditions as in Fig. 4.

of NMN is observed under these conditions.

The results with the coenzymes are more complicated. There is significant broadening of the signals leading to $k_1^- = 800\ sec^{-1}$, much higher than k^- determined by stopped-flow or steady-state kinetic measurements [2]. The specificity of these sites is demonstrated by their complete sharpening at low GTP concentrations. At concentrations above 2 mM, $K_D = 1.5$ mM is observed for NADH and NADPH but especially for NADH much lower values are obtained at low enzyme and coenzyme concentrations. Most probably we observe the combined effects of at least two binding sites. (The contribution of the active one calculated on the base of stopped-flow data is not negligible at low coenzyme concentrations.) The quantitative analysis of the whole range of concentrations is under progress.

The binding of ADP is strongly influenced by NADH. Fig. 7 shows the primary and secondary plots for ADP in the presence of different amounts of NADH. The nonlinearity of the secondary plot is confirmed by additional measurements with a fixed ADP concentration. The best fit to the data is obtained by approximately 4-fold weakening of ADP binding by $K_I = 80\ \mu M$ which equals K_D of NADH in the presence of ADP at the high affinity site. This partial inhibition, however, does not describe the total interaction between both nucleotides in the bound state, the lines of ADP being further sharpened

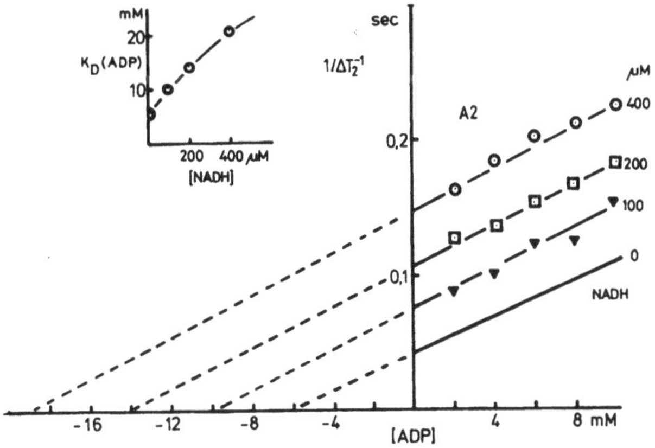

Fig. 7 Partial competitive interaction between NADH at the active site and ADP at the fast binding site. The secondary plot shows the dissociation constants from the main picture depending on NADH. Conditions as in Fig. 4.

upon addition of very large concentrations of NADH. The influence of low ADP concentration on the broadening of NADH is very small. This is in contrast to the known interaction with the active binding site and again shows that the main contribution to the line broadening of the coenzyme comes from a third binding site (the rate constants for the nonactive site [2] are much too low for the observation by NMR). The signals of NADH are significantly sharpened, however, in the presence of high concentrations of ADP indicating strong interference, if not identity between the fast exchanging sites of both nucleotides. The quantitative analysis of this interaction is limited at the moment by the performance of the spectrometer.

For the binding of glutamate, K_D values were obtained by difference spectroscopy [6] but, of course, kinetic measurements were not feasible in this way. The NMR data yield K_D = 50 mM in good agreement with the reported results and k_1^- = 4000 sec^{-1}. Kinetic models thus can be based on rapid equilibrium of the binary enzyme-substrate complex. The temperature dependence and the influence of the enzyme concentrations is much smaller than with the mononucleotides suggesting the true k^- to be only slightly higher than k_1^-. The signals of glutamate (or glutarate) become very sharp in the presence of NADH. This shows, in agreement with the stopped-flow data [2], that the dissociation rate of the ternary complex is much lower and proves, on the other hand, that we indeed observe the active site in this case. The lines of NADH are also sharpened in the presence of glutamate, but this effect is small at high coenzyme and enzyme concentrations. This demonstrates again that the line broadening of the coenzyme is due at least partially to a site which is not identical with the normal active one.

Neither ADPR nor α-NAD^+ shows any influence on the substrate. This demonstrates that the nicotinamide system is responsible for the mutual tightening and that the correct conformational position of this part is indispensable. On the other hand, modification of the adenine system, as for instance, the ethylene bridge of ε-NAD^+ does not prevent the mutual tightening.

The signals of monocarboxylic amino acids such as norvaline or leucine are not significantly broadened under the present conditions. Since the binding of these substances at least at the regulative site is relatively tight ($K_I \approx 2$ mM) we conclude in accordance with stopped-flow data [2] that the exchange is very slow. The broadening is more expressed with the competitive inhibitor acetate [12], and especially with the more hydrophobic acids 4-methylvalerate and benzoate, which are more efficient inhibitors. This indicates much faster exchange than with amino acids. One can thus speculate that these compounds compete for a binding site for the γ-carboxyl group of glutamate which is responsible for the fast binding of the normal substrates.

The preceding results can be summarized in a general model for fast and specific binding of ligands at a functional protein. The ligands are bound very fast but with low specificity at a first site where they are held long enough for passing over to the correct orientation and conformation necessary for the final binding site. The fast prebinding sites thus provide "funnels for entropy" which help to economize the binding process.

REFERENCES

1. Sund, H., Markau, K. and Koberstein, R.: Glutamate Dehydrogenase. In "Subunits in Biological Systems" Part C (Vol. 7 of Biological Macromolecules, a Series of Monographs, ed. by S.N. Timasheff and G.D. Fasman), Marcel Dekker, New York 1975, p. 225-287.

2. Markau, K. and Weber, K.: The Mechanism of Glutamate Dehydrogenase: Some Kinetical Aspects. This Volume, p. 426.

3. Swift, T.J. and Connick, R.E.: NMR Relaxation Mechanism of O^{17} in Aqueous Solution of Paramagnetic Kations and the Lifetime of Water in the First Coordination Sphere. J.Chem. Phys. 37, 307-320 (1962).

4. Price, N.C. and Radda, G.K.: Desensitation of Glutamate Dehydrogenase by Reaction of Tyrosine Residues. Biochem. J. 114, 419-427 (1969).

5. Frieden, C.: Glutamate Dehydrogenase. V. The Relation of Enzyme Structure to the Catalytic Function. J. Biol. Chem. 238, 3286-3299 (1963).

6. Fisher, H.F.: Glutamate Dehydrogenase-Ligand Complexes and Their Relationship to the Mechanism of the Reaction. Edv. Enzymol. 39, 369-417 (1973).

7. Koberstein, R. and Sund, H.: Studies of Glutamate Dehydrogenase. The Influence of ADP, GTP and L-Glutamate on the Binding of the Reduced Coenzyme to Beef-Liver Glutamate Dehydrogenase. Eur. J. Biochem. 36,

545-552 (1973).

8. Weber, G.: Ligand Binding and Internal Equilibria in Proteins. Biochemistry 11, 864-878 (1972).

9. Huang, C.Y. and Frieden, C.: The Mechanism of Ligand Induced Structural Changes in Glutamate Dehydrogenase. Studies of the Rate of Depolymerization and Isomerization Effected by Coenzymes and Guanine Nucleotides. J. Biol.Chem. 247, 3638-3646 (1972).

10. Markau, K., Schneider, J. and Sund, H.: Studies of Glutamate Dehydrogenase. The Mechanism of the Association-Dissociation Equilibrium of Beef Liver Glutamate Dehydrogenase. Eur. J. Biochem. 24, 393-400 (1971).

11. Neumann, P., Markau, K. and Sund, H.: Yeast Glutamate Dehydrogenase from Candida utilis as a Model for the Mammalian Enzyme - the Interaction with Effectors at High Concentrations. 10th Internat. Congr. Biochem. Hamburg 1976, Abstr. No. 07-5-160.

12. Markau, K. and Steinhübel, I.: Kinetic Measurements with Monocarboxylic Acids as Substrates and Effectors of Glutamate Dehydrogenase. FEBS Letters 28, 115-120 (1972).

Acknowledgment: This work was supported by grants from the Deutsche Forschungsgemeinschaft (Sonderforschungsbereich 138: Biologische Grenzflächen und Spezifität) and the Fonds der Chemischen Industrie.

Received March 25, 1977

DISCUSSION

Seydoux: Can you explain the stoichiometry of 0.6 molecules GTP bound per polypeptide chain? Maybe this result is related to the occurrence of two classes of sites in the tight binding of the first six NADH molecules.

Koberstein: GTP always contains GDP and GMP. The value of 0.6 is already corrected for these compounds. Recent enzymatic determination of the GTP purity, however, indicated an additional impurity. This leads to a further reduction to about 0.5 molecules of GTP bound per polypeptide chain. In the binary enzyme-GTP complex no evidence for a second class of GTP sites is obtained from dialysis experiments. Dr. Markau's NMR measurements, however, show a set of weak GTP sites with a K_D of about 6 mM. Their relation to the tight GTP sites is not yet known.

Veeger: I am not happy at all with Dr. Koberstein's remark, that he prefers 0.5 mole/GTP bound per protomer. This result of about 0.65 mole is much more in accord with the data of Robillard (1) with a NAD spinlabel and the data of Brochon with fluorescence decay curves. Robillard et al. showed that about two third of the binding sites have high affinity, about one third much lower affinity. Similarly Brochon et al. (2) showed the presence of two sets of NADPH binding sites in approximately 2:1 ratio. To the latter interpretation one might make some molecular physical objections, as the results of Gafni and Brand (3) show with lifetime studies of NADH bound to ADH, but there are other good arguments to accept the interpretation of Brochon. Thus it could be quite possible taking these data together that glutamate dehydrogenase acts as a two third of the sites enzyme.

(1) A. Zantema, W.E. Trommer, H. Wenzel and G.T. Robillard, Eur. J. Biochem. 72, 175-184 (1977)
(2) J.C. Brochon, Ph. Wahl, J.-M. Jallon and M. Iwatsubo, Biochemistry 15, 3259-3265 (1976)
(3) A.Gafni and L. Brand, Biochemistry 15, 3165-3171 (1976)

Trommer: Dr. Koberstein, you have shown very nicely the effects of phosphate buffer in GTP binding. I think there is real good evidence that phosphate indeed is bound to the GTP binding site.

In collaboration with George Robillard in Groningen (A. Zantema, W.E. Trommer, H. Wenzel and G.T. Robillard, Eur. J. Biochem. 72, 175 (1977)) we have studied the binding of a spin-labeled derivative of NAD^+ to bovine GluDH. The spin-label is attached to the exocyclic aminofunction of the adenine ring. This analogue has nearly the same kinetic constants as natural NAD^+.

Addition of phosphate, pyrophosphate and tripolyphosphate enhance the binding of spin-labeled NAD^+ to the enzyme-2-oxoglutarate complex in phosphate free buffer. In presence of GTP which even more strongly enhances binding no further increase can be observed. However, in presence of adenine nucleotides as ADP, phosphate addition again leads to tighter binding of spin-labeled NAD^+. This shows clearly that there are distinct binding sites for GTP and ADP and that phosphate binds to the GTP site.

Süvdi: In your preprint you are talking of a general model which, perhaps, sounds slightly better than "Philosophy". Now, would you care to re-formulate this general model in terms of (1) encounter complex, (2) rotational diffusion, and (3) diffusion-limited association reaction?

Markau: If the substrate binding would occur in one step, the high specificity would require that it approaches the active center practically within a zero angle of the phase space. There must exist, therefore, intermediates with restricted specificity and lifetimes much above those of normal encounter complexes in which this angle is reduced by interaction with the "lattice". By NMR we just observe the tightest, and perhaps the most effective, of these states. I am convinced that there are other prestates, beginning with a type of simple encounter complex, which are out of range of detection. If these "funneling" mechanisms are very effective (low specificity and low free activation energy), the reaction might appear to be diffusion controlled, at least at low concentrations.

THERMODYNAMICS OF THE LDH REACTION

Hans-Jürgen Hinz and Franz Schmid
Institut für Biophysik und Physikalische Biochemie,
Universität Regensburg, FBB, D-8400 Regensburg, West Germany

INTRODUCTION

Characterization of energetics of protein ligand interactions affords quantitative criteria which constitute a basis for the understanding of the more complex reactions occuring in the living cell. In biology the energetic concept has never played such a fundamental role as in physics or chemistry despite a unanimous agreement, that the same physical principles are operative in matter and living cells. Fascination by results of mechanistic, kinetic, and structural investigations made it easy to forget that the driving force for every reaction, including the realization of a specific macromolecular structure of a protein is the result of energetic and entropic interactions on the molecular level. Only knowledge of the energy parameters provides that "understanding" of the process which enables one to make predictions. Emphasizing the indispensability of the energetic aspect is not to diminish the significance of the other studies but merely intends to draw attention to the mutually complementary function of the various fields of research. The unifying view which in principle may be possible on the basis of energetic parameters is certainly far from being realized. The first steps undertaken into that direction still require a maximum of information on the structure, the kinetics and the chemistry of the system under investigation, in order to render a molecular interpretation of the energy parameters possible.
The reaction catalyzed by lactate dehydrogenase (LDH; EC1.1.1.27) furnishes an appropriate system for thermodynamic studies due to the extensive amount of data on chemical structural and kinetic aspects of the process (1,2,3,4) and the scarcity of thermodynamic parameters with the exception of standard Gibbs

free energy values. The free energy parameters, however, are to a particularly large extent very difficult to interpret in molecular terms. Although the ambiguity with respect to mechanisms is characteristic of thermodynamic quantities the degree of indeterminacy decreases e.g. with increasing order of the temperature derivatives of free energy functions. Thus determination of enthalpy-, entropy-, and heat capacity values imposes itself for the establishment of an energy-structure-function relationship. Model free enthalpies can only be obtained by direct calorimetric measurements, and the accuracy of heat capacity changes especially of reactions involving macromolecular species is in most cases higher than that of indirect approaches. Therefore the thermodynamic studies on the lactate dehydrogenase system centered around direct calorimetric measurements of the heats associated with the various intermediary reactions, supplemented by determinations of the corresponding equilibrium constants. The actual catalytic reaction has been approximated by using the substrate analogous inhibitors oxamate and oxalate (5) instead of pyruvate and lactate, respectively, in the two reactions forming the ternary complexes LDH-NADH-Oxamate and LDH-NAD^+-Oxalate. The sequence of reactions 1,2,3,4 assumed to simulate one turnover of the catalytic process, is presented in Fig.1.

Figure 1 Scheme on which interpretation of the thermodynamic data is based. The reactions 1,2,3,4 are to simulate the intermediary processes occurring in the actual catalytic sequence.

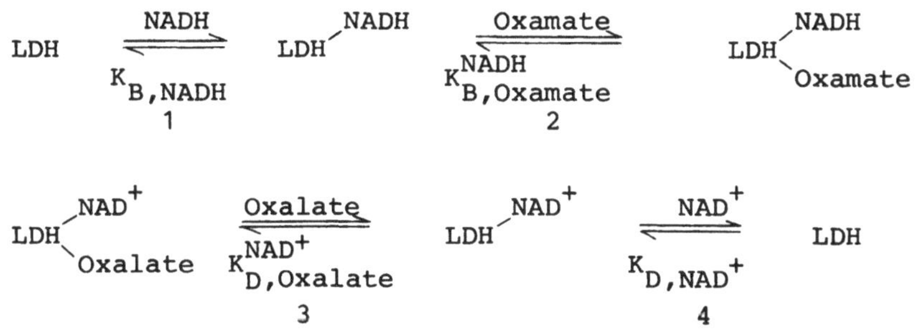

LDH is to denote both the M_4 isozyme, predominating in

skeletal muscle, and the H_4 species, prevalent in heart tissue. These two most divergent isozymes prepared of pig tissue have been subject of the thermodynamic studies.

EXPERIMENTAL SECTION

Pig skeletal muscle and pig heart muscle isozymes, NADH, NAD^+ (both grade I), Adenosine, AMP, ADP and ADP-ribose (ADPR) were obtained from Boehringer, Mannheim, Germany. The enzymes were supplied as 1% crystalline suspensions in ammonium sulfate. Reagent grade potassium phosphate and potassium oxalate were purchased from E.Merck, Darmstadt; sodium oxamate was an Aldrich product. Demineralized and quartz distilled water was used for the buffer solutions. The enzyme crystals were collected from the suspension by centrifugation (40000g, 15min), dialyzed for 24h at 4°C under nitrogen against the buffer solution, with three changes of the buffer thousand times the volume of the enzyme solution. The resulting enzyme solutions had the following activities in the standard test (Biochemica Informationen, Boehringer): heart muscle isozyme: 400-450 units/mg; skeletal muscle isozyme: 600-670 units/mg. Enzyme concentrations were determined using an absorption coefficient $A=1.4cm^2mg^{-1}$ at 280 for either isozyme. The ratio of the absorption of the enzyme at 280nm to that of 260nm was 1.98 for the heart muscle isozyme and 1.83 for the skeletal muscle isozyme. These ratios, the activity, and fluorescence spectra served as criteria of purity when enzyme preparations of different lots were employed. Coenzymes, coenzyme fragments and substrate analogues were dissolved in the last dialysate. Their concentrations were determined spectrophotometrically using the following absorption coefficients: NADH:$6.23\cdot10^3M^{-1}cm^{-1}$ at 340nm; NAD^+:$18.0\cdot10^3M^{-1}cm^{-1}$ at 260nm; Adenosine: $14.9\cdot10^3M^{-1}cm^{-1}$ at 260nm; AMP:$15.3\cdot10^3M^{-1}cm^{-1}$ at 260nm. ADP: $15.4\cdot10^3M^{-1}cm^{-1}$ at 260nm; for ADP-ribose an absorption coefficient of $15.3\cdot10^{-3}M^{-1}cm^{-1}$ at 260nm was assumed (6). Oxamate and oxalate concentrations were based on weight. Methods have been described in detail in references (7,8,9,10,11).

RESULTS

Thermodynamic Parameters for Binary Complex Formation.

The equations for the variation with temperature of the equilibrium constants $K_{B,NADH}$ and K_{B,NAD^+} and the enthalpies $\Delta H_{B,NADH}$ and $\Delta H_{B,NAD^+}$ involved in binding of NADH and NAD^+ to the H_4- and M_4 isozymes in 0.2M potassium phosphate buffer at pH=7.0, are listed below.

LDH H_4:

(1) $\ln K_{B,NADH} = -\dfrac{19673.635}{T} - 83.920 \ln T + 556.582$

(2) $\ln K_{B,NAD^+} = -\dfrac{9460.132}{T} - 42.023 \ln T + 279.192$

(3) $\Delta H_{B,NADH} = -667\,\vartheta - 25575$ cal/(mol of LDH H_4)

(4) $\Delta H_{B,NAD^+} = -334\,\vartheta - 16043$ cal/(mol of LDH H_4)

LDH M_4:

(5) $\ln K_{B,NADH} = -\dfrac{44929.146}{T} - 163.437 \ln T + 1093.557$

(6) $\ln K_{B,NAD^+} = -\dfrac{14949.553}{T} - 61.273 \ln T + 406.852$

(7) $\Delta H_{B,NADH} = -1299\,\vartheta + 2294$ cal/(mol of LDH M_4)

(8) $\Delta H_{B,NAD^+} = -487\,\vartheta - 14206$ cal/(mol of LDH M_4)

ϑ indicates temperature in degrees centigrade, T refers to K. The standard errors of the estimates of the enthalpies are: eq.(3) 1190; eq.(4) 773; eq.(7) 1796 eq. (8) 2314, all values in cal/(mol of enzyme). The number of decimals after the point in equations (1),(2),(5),(6) has been retained with the consideration that the experimental equilibrium constants at 25°C should be reproduced, when applying the equations to T= 298.15 K. This view was also adopted for the analogous equations (9),(10),(13),(14).

Thermodynamic Parameters for Ternary Complex Formation.

A compilation of the formulae pertaining to the dependence on temperature of the binding constants, $K^{NADH}_{B,Oxamate}$ and $K^{NAD^+}_{B,Oxalate}$ and the heats, $\Delta H^{NADH}_{B,Oxamate}$ and $\Delta H^{NAD+}_{B,Oxalate}$, associated with the reactions between the binary complexes LDH-NADH and LDH-NAD^+ of the two isozymes and oxamate and oxalate, respectively, in 0.2M potassium phosphate buffer at pH=7.0 is given in equations (9)-(16).

LDH H_4:

(9) $$\ln K^{NADH}_{B,Oxamate} = -\frac{53164.143}{T} - 203.825 \ln T + 1351.125$$

(10) $$\ln K^{NAD^+}_{B,Oxalate} = -\frac{46927.453}{T} - 170.609 \ln T + 1142.530$$

(11) $$\Delta H^{NADH}_{B,Oxamate} = -1620\,\vartheta - 19814 \text{ cal/(mol of LDH } H_4)$$

(12) $$\Delta H^{NAD^+}_{B,Oxalate} = -1356\,\vartheta + 2588 \text{ cal/(mol of LDH } H_4)$$

LDH M_4:

(13) $$\ln K^{NADH}_{B,Oxamate} = -\frac{64906.159}{T} - 239.180 \ln T + 1590.624$$

(14) $$\ln K^{NAD^+}_{B,Oxalate} = -\frac{52796.880}{T} - 186.714 \ln T + 1251.843$$

(15) $$\Delta H^{NADH}_{B,Oxamate} = -1901\,\vartheta - 3384 \text{ cal/(mol of LDH } M_4)$$

(16) $$\Delta H^{NAD^+}_{B,Oxalate} = -1484\,\vartheta + 14275 \text{ cal/(mol of LDH } M_4)$$

ϑ again relates to temperature in degrees centigrade, T to the absolute temperature scale. The ΔH values comprise the following standard errors in cal/mol of enzyme: eq.(11): 2091; eq.(12): 1346; eq.(15): 1805; eq.(16): 2690.

Thermodynamic Parameters for Binary Complex Formation between LDH H_4 and Coenzyme Fragments.

Apparent thermodynamic values for Gibbs free energies of binding, association enthalpies and entropies, and heat capacity changes concomitant with complex formation between pig heart muscle LDH and various coenzyme fragments have been summerized in Table I. To facilitate a comparison the corresponding quantities for the reactions with the coenzymes have been included.

Table I Apparent thermodynamic parameters for binary complex formation between pig heart muscle LDH and various coenzyme fragments in 0.2M potassium phosphate buffer pH=7.0, 25°C. All values refer to one mol of ligand and are reported with standard errors. *(ADPR=Adenosine 5'diphosphoribose).

Ligand	ΔG_B (kcal/mol)	ΔH_B (kcal/mol)	ΔS_B (cal/K·mol)	ΔC_{p_B} (cal/K·mol)
Adenosine	-2.90 ± 0.05	-6.1 ± 0.3	-10.7 ± 1.2	$+3\pm10$
AMP	-3.98 ± 0.09	-3.2 ± 0.2	$+2.6\pm1.0$	-103 ± 7
ADP	-4.03 ± 0.06	-5.9 ± 0.4	-6.3 ± 1.5	-74 ± 14
ADPR*	-4.96 ± 0.03	-6.4 ± 0.2	-4.9 ± 0.4	-99 ± 8
NAD^+	-4.76 ± 0.04	-6.1 ± 0.2	-4.5 ± 0.8	-84 ± 8
NADH	-7.39 ± 0.08	-10.6 ± 0.3	-10.8 ± 1.3	-167 ± 9

DISCUSSION

Studies on Coenzymes and Ternary Complex Analogues.

Thermodynamic analysis of intermediary steps of the lactate dehydrogenase reaction according to the scheme in Fig.1 has heavily depended on direct calorimetric measurements, since they allow the most unambiguous determination of enthalpy values and their temperature coefficients. Figures 2 and 3 provide a survey of the variation with temperature of the Gibbs free energies of binding, ΔG_B, the association enthalpies ΔH_B, and the product $T\Delta S_B$ for the various complexes formed by the two isozymes. Inspection of the graphs displays distinctly one common feature of all reactions: ΔG varies only very slightly with temperature whereas ΔH and ΔS show a pronounced temperature dependence. This behaviour is typical (as far as investigations of that thermodynamic characteristic exist) of processes involving proteins in aqueous solution (12,13,14), since e.g. gas phase reactions or reactions involving solids generally exhibit rather temperature insensitive enthalpies and entropies but, as a consequence, strongly varying free energies.

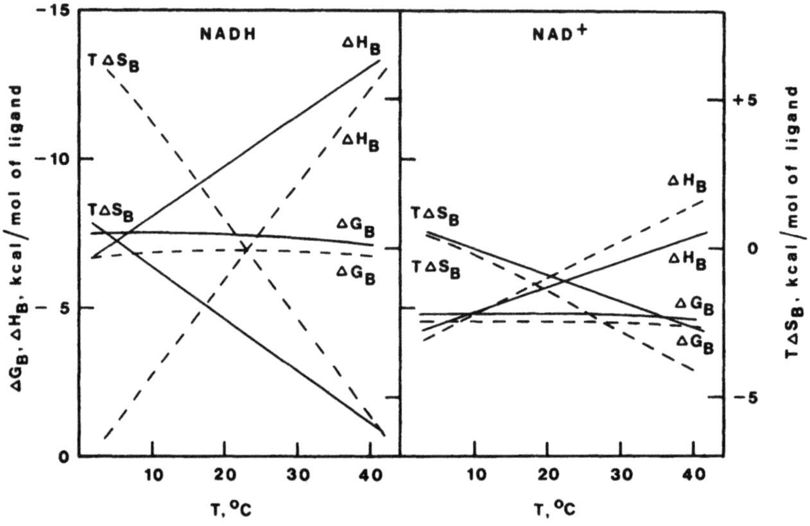

Fig.2 Variation with temperature of the thermodynamic parameters for binary complex formation with the reduced and oxidized coenzyme in 0.2 M potassium phosphate buffer, pH=7.0, solid lines: LDH H_4; dashed lines: LDH M_4.

This disparity implies that large heat capacity changes are frequently associated with reactions in which proteins participate, while in the course of reactions between gases the heat capacity difference of the products relative to the reactants is comparatively minute. To rationalize the heat capacity differences between apoenzyme and liganded species one has to infer structural changes preferentially of the macromolecular system (which comprises the protein itself and the spheres of short range interactions like hydration, preferential binding etc.) that pervade larger portions of the protein than just the immediate area of contact at the binding site. The close connection between heat capacity and molecular vibrations is suggestive of interpreting the heat capacity changes, at least partially, as resulting from changes of easily excitable modes of rotation and vibration. Another contribution probably stems from alterations of hydrophobic hydration.

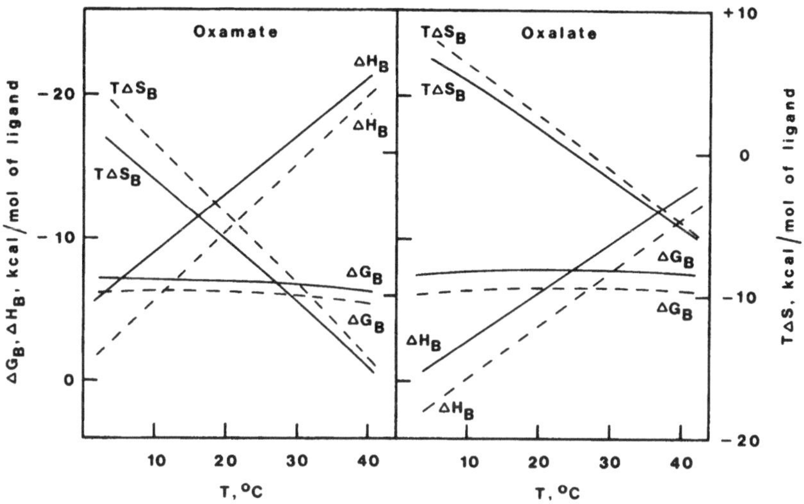

Fig.3 Variation with temperature of the thermodynamic parameters for ternary complex formation between the binary complexes LDH-NADH and LDH-NAD^+ and oxamate and oxalate, respectively, in 0.2 M potassium phosphate buffer, pH=7.0, solid lines: LDH H_4 dashed lines: LDH M_4.

A third contribution may originate from temperature induced changes in the degree of ionization of charged groups. However, although participation of such a reaction is likely to be reflected in the apparent ΔC_p changes detected for ternary complex formation with oxamate, a discussion of this effect requires information on the pH-dependence of the enthalpy and is therefore not undertaken in this context. An examination of the temperature coefficients of the heats of binding in column 5 of Table II shows that without exception the ΔC_{p_B} values are negative. This apparently means that binding of the ligands induces a reduction of the number of vibrational modes of the system (tightening of the complex) and probably also leads to a partial reversal of hydrophobic hydration. This interpretation is in excellent agreement with the results of X-ray studies which demonstrate a contraction to occur on ternary complex formation (4). Conclusions, that analogous reactions are associated with binary complex formation, are only indirectly suggested by the experimental result, that apoenzyme crystals shatter when soaked with coenzyme the better binding NADH being more effective than the less strongly binding NAD^+.

Table II Apparent thermodynamic parameters for the lactate dehydrogenase reaction of the pig heart and skeletal muscle enzymes at 25°C in 0.2 M potassium phosphate buffer pH=7.0. The starred values refer to the heart muscle isozyme. The numbers in the first column refer to the scheme on page 2; all thermodynamic parameters relate to binding reactions and pertain to one mol of binding site.

Reaction	ΔG_B kcal/mol of ligand	ΔH_B kcal/mol of ligand	ΔS_B cal/(K·mol of ligand)	ΔC_{p_B} cal/(K·mol of ligand)
1	-7.39*±0.08	-10.6*±0.3	-10.8*±1.3	-167*± 9
	-6.91 ±0.08	- 7.6 ±0.5	- 2 ±2	-325 ± 10
2	-6.73*±0.06	-15.1*±0.5	-28.0*±1.9	-405*± 21
	-6.03 ±0.05	-12.7 ±0.5	-22.4 ±1.9	-475 ± 12
3	-7.74*±0.03	- 7.8*±0.3	0*±0.1	-339*± 11
	-6.48 ±0.08	- 5.7 ±0.7	+ 2.6 ±2.6	-371 ± 20
4	-4.76*±0.04	- 6.1*±0.2	- 4.5*±0.8	- 84*± 8
	-4.50 ±0.35	- 6.6 ±0.6	- 7.0 ±3.2	-122 ± 11

Comparison of the corresponding ΔC_{p_B} values in Table II shows, that for either isozyme the ΔC_{p_B} values are negative i.e. implying a contraction, and considerably larger for binding of NADH than for association of NAD^+. A collation of all ΔC_p values pertinent to the reactions with the two isozymes demonstrates, that consistently the ΔH values associated with reactions of the skeletal muscle enzyme display a larger temperature coefficient. Following the same line of reasoning as before this imports that analogous reactions involve larger structural alterations in the LDH M_4- than in the LDH H_4 isozyme. In discussions on thermodynamics of biological macromolecules frequently the terms "entropy driven" and "enthalpy driven" are being used. How misleading that terminology can be in case of the occurrence of large ΔC_p-values, is easily verified by inspection of Figures 2 and 3. Depending on the temperature the same reaction would be referred to as "entropy-" or "enthalpy-driven", although it is rather unlikely that any change in the mechanism of interaction occurred as a result of the temperature variation. Thus using these terms without consideration of possible heat capacity changes does not appear enlightening.

Studies on Coenzyme Fragments.

The rationale for the investigations on binding of coenzyme fragments to LDH was to evaluate the contributions of the various groups of the coenzyme to the thermodynamic parameters. Looking at the ΔG_B values in Table I reveals that the ligands can be grouped into four classes, according to their free energies. Addition of the first phosphate group to adenosine makes the binding free energy more negative by approximately 1 kcal/mol. Introduction of the second phosphate practically does not change this value, but extension of the fragment by the ribose lowers the free energy again by 1 kcal/mol. In comparison to ADP-ribose the oxidized coenzyme binds slightly more loosely,however it is still the same level of roughly -5 kcal/mol. A large decrease in the ΔG_B value to -7.4kcal/mol indicates considerably better interaction of the reduced coenzyme. This trend is not reflected in the ΔH values. Adenosine exhibits a surprisingly large negative, within the accuracy of the measurements, temperature invariant enthalpy of -6kcal/mol. Therefore the relatively weak binding is a result of the unfavourable negative entropy change involved in complex formation. Introduction of the first phosphate actually reduces the energetic contribution and leads to a positive ΔS_B value at 25°C. Concomittantly a structural change (in the sense previously outlined) is induced, as judged on the basis of the negative heat capacity change. The second phosphate contributes approximately the same favourable energy, however it does not appear in the ΔG_B term since it is compensated by a negative entropy contribution. Addition of the ribose moiety renders ΔH_B more negative and ΔS_B less negative as compared to

ADP. The oxidized coenzyme shows a slightly more positive ΔH_B, whereas the reduced coenzyme shows an appreciably more negative binding energy by more than 4 kcal/mol which is only partly offset by the unfavourable entropy change. The pronouncedly more negative heat capacity change when NADH binds, indicates that the reduced coenzyme induces larger structural alterations than the other fragments and the oxidized coenzyme. Adenosine does not cause changes in the structure of the macromolecule, as inferred from the heat capacity data, while AMP ADP, ADPR and NAD^+ seem to involve similar structural adjustments of the protein.

It still appears to be premature to speculate in detail on the molecular events which cause these changes of the thermodynamic parameters. What one may state with due reservation is that hydration phenomena will play a significant role and that the experimental ΔH_B- and ΔS_B-values are the result of a delicate balance between enthalpic and entropic contributions involved in forming and destructing bonds between the enzyme, the coenzyme and the surrounding medium. The thermodynamic analysis shows, however, quite clearly the inadequacy of any interpretation of Gibbs free energy data in molecular terms.

REFERENCES

1. Sund, H., (Ed.): Pyridine Nucleotide-Dependent Dehydrogenases, Berlin, Heidelberg, New York, Springer Verlag (1970).
2. Holbrook, J.,J., and Gutfreund, H.: Approaches to the Study of Enzyme Mechanisms. Lactate Dehydrogenase, FEBS Letters, 31, 157-169 (1973).
3. Everse, J., and Kaplan, N.O.: Lactate Dehydrogenases: Structure and Function, Advances Enzymol. 37, 61-133 (1973).
4. Holbrook, J.J., Liljas, A., Steindel, S.J., and Rossmann, M.G.: Lactate Dehydrogenase, The Enzymes, 3rd Ed. 11, 191-291 (1975).

5. Novoa, W.B., Winer, A.D., Glaid, A.J., and Schwert, G.W.: Lactic Dehydrogenase V. Inhibition by Oxamate and Oxalate J.Biol.Chem. 234, 1143-1148 (1959).
6. Dawson, R.M.C., Elliot, D.C., Elliot, W.H., and Jones, K. M., (Eds.): Data for Biochemical Research Oxford, Clarendon Press, 2nd edition (1969).
7. Hinz, H.-J., and Jaenicke, R.,: Calorimetric Investigation of Binding of NADH to Pig Muscle Lactate Dehydrogenase, Biochem.Biophys.Res.Commun. 54, 1432-1436 (1973).
8. Hinz, H.-J., and Jaenicke, R.,: Thermodynamics of Complex Formation between Nicotinamide Adenine Dinucleotide and Pig Skeletal Muscle Lactate Dehydrogenase, Biochemistry, 14, 24-27 (1975).
9. Schmid, F., Hinz, H.-J., and Jaenicke, R.,: Detection of Absorption Changes of Pig Heart Muscle Lactate Dehydrogenase in the Wavelength Range between 260nm and 320nm on Ternary Complex Formation with NAD^+ and Oxalate, Hoppe-Seyler's Z.Physiol.Chem. 357, 241-245 (1976).
10. Schmid, F., Hinz, H.-J., and Jaenicke, R.: Thermodynamic Studies of Binary and Ternary Complexes of Pig Heart Lactate Dehydrogenase, Biochemistry, 15, 3052-3059 (1976).
11. Hinz, H.-J., Schmidt, R., Scheurmann, W., and Jaenicke, R.: Thermodynamic Studies of Binary and Ternary Complexes of Pig Skeletal Muscle Lactate Dehydrogenase, Biochemistry, to be published 1977.
12. Wadsö, J.,: Biochemical Thermochemistry, in Thermochemistry and Thermodynamics, Skinner, H.A., Ed. MTP International Review of Science, 10, 1-56, (1972) Butterworths, London, University Park Press, Baltimore.
13. Bode, W., Hinz, H.-J., Jaenicke, R., and Blume, A.,: Calorimetric Studies on the in vitro Polymerization of Pr.mirabilis Flagellin, Biophys.Struct.Mechanism, 1, 55-64 (1974).
14. Hinz, H.-J., Weber, K., Floßdorf, J., and Kula, M.-R.,: Thermodynamic Studies on the Specificity of L-Isoleucine·tRNA Ligase of Escherichia coli MRE 600, Eur.J.Biochem.

71, 437-442 (1976).

Received January 31, 1977

DISCUSSION

Trommer: Dr. Hinz, I have to point out some contradictory results with respect to the binding of the NAD^+ fragments to pig heart LDH. We have studied the binding of spin-labeled (SL) derivatives of NAD^+ and of all of its structural components, i.e. SL-adenine, SL-adenosine, SL-AMP, SL-ADP, SL-ADPR and SL-NAD^+ in which the label is attached to the exocyclic amino group of the adenine ring (H.R. Wenzel, G. Pfleiderer, W.E. Trommer, K. Paschenda and A. Redhardt, BBA 452, 292 (1976)). The maximum velocity of SL-NAD^+ with LDH reaches 70 % of the value for natural NAD^+, so it should be a rather good model.

In accordance with your results we observe a strong increase in binding when going from adenosine to AMP and also a decrease from ADPR to NAD^+. However, in contrast to your results, ADP binds considerably more strongly than AMP and there is no further increase upon attachment of the ribose. If the spin-label should change the binding behavior this effect should be similar for all the analogues and at least the relative order should be correct. Moreover, the actual binding constants we have found for SL-AMP and SL-NAD^+ correspond well with published data.

Hinz: As you mention in your comment, SL-NAD^+ is a rather good model since it reaches 70 % of the maximum velocity of NAD^+ in the catalytic reaction, but it still is a model which obviously does not exhibit quite the same properties as the natural coenzyme. Thus there appears to be a possibility that also the series of spin-labeled fragments of NAD^+ may react in a way slightly different from that of the unlabeled species. I cannot quite follow your argument that, if the spin-label would change the binding characteristics it should do it consistently in the same manner throughout all the reactions with the different coenzyme fragments.

Your ESR signal depends critically on the electronic environment of the paramagnetic label. There is no a priori reason why a potential perturbation by the label of the binding equilibrium of the fragment should always go into one direction in view of the various conformational readjustments of the macromolecule. The trend of the Gibbs free energy values reported for pig heart muscle LDH in our communication, that AMP and ADP binding is very similar while there occurs a further improvement for ADPR, is in complete agreement with the reports by H. Geyer (Hoppe-Seyler's Z. Physiol. Chem. (1967) 348, 823-832) and A. McPherson (J. Mol. Biol. (1970) 51, 39-46) on the H_4 and M_4 isozymes of human LDH and dogfish LDH, respectively. Though, admittedly, this is no direct proof, since LDH of different species was employed, it appears to be reasonable circumstantial evidence, since interspecies similiarity of similar LDH isozymes has been well established.

Holbrook: In the results shown in Fig. 2 for NADH binding to both pig H_4 and M_4 LDH the value of ΔG_B varies by less than 0.3 kcal/mole of ligand between 3° C and 40° C. How can you reconcile this with the observations of Stinson & Holbrook (Biochem. J. 131, 719-728 (1973))) that the equilibrium dissociation constant for NADH binding increases by more than 10-fold (> 1.4 kcal/mole) in the same temperature range?

Hinz: Inspection of Fig. 7 in the publication cited above shows that in the temperature range between 3° C and 40° C the dissociation constant for the reaction between pig skeletal muscle LDH and NADH varies by a factor of 4.4 while that for the analogous reaction involving pig heart muscle LDH changes by a factor of approximately 14 on the basis of a linear extrapolation of the van't Hoff plot. The linear van't Hoff plot is principally not justified as the strongly varying ΔH values associated with the reactions of both isozymes demonstrate and as actually is clearly seen in Fig. 7 of the paper by Stinson and Holbrook for the temperature dependence of the equilibrium constant of the pig skeletal muscle enzyme. Thus the factor of 14 for the variation with temperature of the dissociation constant of the reaction between pig heart muscle LDH and NADH certainly provides an overestimate. Dr. Holbrook's comment is based on Fig. 2 of our communication which shows the calculated dependence on temperature of the corresponding thermodynamic parameters as derived from the equilibrium constant obtained at 25° C (which is in close agreement with the one reported by Stinson and Holbrook) and the calorimetrically measured enthalpies and heat capacities. That the agreement between calculated and experimental results is good, has been demonstrated in Fig. 1 of the paper by Schmid, Hinz & Jaenicke (Biochemistry 15 (1976) 3052-3059) for the pig heart muscle enzyme. However, the variation with temperature of the equilibrium constant for binary complex formation between pig heart LDH and NADH, as reported by Stinson and Holbrook, appears to be somewhat larger, even if one takes the approximate nature of the van't Hoff plot into account. It should be stressed in this context, however, that the minor divergence in the temperature dependence of the equilibrium constant does not at all interfere with the most important conclusion drawn on the basis of the calorimetric results, that the variation with temperature of ΔH and ΔS is large in comparison to that of ΔG.

Veeger: Could it be considered that in addition to changes in vibrational modes and hydrophobic interactions, also changes in water structure ice-like structures and higher viscosity could give an additional contribution?

Hinz: Besides the three possible sources mentioned in the lecture (1) alterations in the number of excitable modes of energy absorption (vibration and rotation) of the macromolecule, (2) change of hydrophobic hydration and (3) temperature induced changes in the degree of ionization of charged groups, which have to be considered to account for the large negative heat capacity changes observed on binary and ternary complex formation with the coenzymes and substrate analogues, respectively, there are, of course, other molecular interactions of potential importance, like the creation of charged pairs or breaking of hydrogen bonds. (J.M. Sturtevant, Proc. Natl. Acad. Sci. US (1977), in press).

However, the three contributions mentioned first might be considered to be of major significance. Changes in water structure and ice-like structures with concomitant higher viscosity are to my knowledge synonymous with the term "hydrophobic interactions", at least they have been introduced by

Frank and Evans (J. Chem. Phys. 13 (1945) 507-532) to rationalize the macroscopic thermodynamic parameters ΔG, ΔH and ΔS of non-polar groups in water. Similar molecular interpretations have been given by Kauzmann (Adv. Prot. Chem. 14 (1959) 1-63) and Tanford (The Hydrophobic Effect,J. Wiley, New York (1973)). However, the rather pictorial aspect of this interpretation should be emphasized (see e.g. A. Holtzer and M.F. Emerson, J. Phys. Chem. 73 (1969) 26-33), particularly in view of the actual correlation times of water molecules in the vicinity of hydrophobic groups (E. v. Goldammer and H.G. Hertz, J. Phys. Chem. 74 (1970) 3734-3755).

Südi: You have given us an intuitively very pleasing explanation of heat capacity changes in molecular terms. Now I wonder whether you would care to attempt a similar interpretation of the entropy changes in molecular terms?

Hinz: As I mentioned in my lecture, the thermodynamic parameter which is relatively most open for an interpretation in molecular terms is the temperature coefficient of the reaction enthalpy, Δc_p. Thus, I had rather abstain from any further interpretation of entropy changes involved in protein ligand interactions as long as there is no additional non-thermodynamic evidence on specific molecular events.

THE EQUILIBRIUM NADH + $NADP^+ \rightleftharpoons NAD^+$ + NADPH AS STUDIED BY TRANSHYDROGENASE FROM AZOTOBACTER VINELANDII

C. Veeger and J. Krul

Department of Biochemistry, Agricultural University, Wageningen, The Netherlands

INTRODUCTION

The equilibrium constant of the reaction NADH + $NADP^+ \rightleftharpoons NAD^+$ + NADPH, has been measured indirectly by Olsen and Anfinsen (1) and Engel and Dalziel (2) with the enzyme glutamate dehydrogenase. The measurements by Engel and Dalziel (2) indicate that the equilibrium constants of the NAD(H) and NADP(H) dependent reactions catalyzed by glutamate dehydrogenase show dependence on the ionic strength. The constant was measured directly by Kaplan et al. (3) with the aid of the non-energy linked transhydrogenase from mammalian tissue. The K-value was found to be near unity in agreement with the general idea that the two couples differ only slightly in standard redox potentials. The non-energy linked transhydrogenase from mammalian mitochondria catalyzes this reaction at a higher maximum velocity from the right to the left than in the reverse direction. In order to explain the value of the equilibrium constant near unity under these conditions Rydström (4) proposed a mechanism in which the transhydrogenase protein could be activated and deactivated. Since it had been reported by Lee and Ernster (5) that addition of ATP to the submitochondrial transhydrogenase causes a drastic shift in the K_{obs} from near unity to a value of about 500, Rydström et al. (4) explained this by assuming activation of inactivated transhydrogenase at the expense of an energy-rich bond. Fisher and Kaplan (6) suggested that the activation of the energy-dependent transhydrogenase is a function separate from the increase in the apparent equilibrium constant of the reaction. These observations prompted us to reinvestigate the equilibrium constant of the pyridine nucleotide couples as catalyzed by pure preparations of the non-energy dependent transhydrogenase from Azotobacter vinelandii.

RESULTS AND DISCUSSION

Effect of Ionic Strength and pH

The value of the equilibrium constant of the pyridine nucleotide couples as can be calculated from the data of Olsen and Anfinsen (1), Engel and Dalziel (2) and as measured by Kaplan et al. (3) was measured in the presence of pure transhydrogenase. It became clear that K_{obs} measured at a constant temperature depends on the ionic strength and the pH. At higher ionic strength (I) and lower ambient pH the K_{obs}-values increase.

Table I *Dependence of the equilibrium constant of the equilibrium NADH + NADP$^+$ $\rightleftharpoons$ NAD$^+$ + NADPH on the ionic strength.* K_{obs}-values were obtained in 2 mM Tris-HCl buffer pH 7.6 to which KCl was added, at 25°C. The reaction was started by the addition of 150 µM NADPH + 150 µM NAD$^+$ and a catalytic amount of transhydrogenase. Samples were taken at different time intervals and the concentrations of oxidized and reduced pyridine nucleotides determined.

KCl	I	$\sqrt{I}$	K_{obs}	log K
0	0.0015	0.039	0.33	-0.481
5 mM	0.0061	0.078	0.35	-0.456
10 mM	0.0115	0.107	0.38	-0.420
20 mM	0.0216	0.146	0.45	-0.366
50 mM	0.0516	0.227	0.48	-0.318
100 mM	0.1014	0.318	0.54	-0.267

The effect of the ionic strength cannot be attributed to a specific ion effect, since it is independent of the kind of ion used to adjust to the desired ionic strength. The effect of pH on the equilibrium constant was studied at constant ionic strength (I = 0.01) in potassium phosphate buffer of variable pH (Table II).

Table II *Effect of the pH on K_{obs} at a constant ionic strength.* The reaction was started by the addition of 200 µM NADPH + 200 µM NAD^+ and a catalytic amount of enzyme at 25°C. Samples were taken and the concentration oxidized and reduced pyridine nucleotide determined.

Buffer	pH	K_{obs}
potassium phosphate		
ionic strength 0.01	6.11	0.52
	6.53	0.44
	6.99	0.41
	7.45	0.38

In order to exclude a specific enzyme-effect the same equilibrium was studied using glutamate dehydrogenase as a catalyst. The equilibrium was established by starting the reaction with both reduced pyridine nucleotides and the substrates present in the same mixture. After the equilibrium had been reached, samples were taken and heated to destroy the glutamate dehydrogenase. To the cooled (25°C) mixture transhydrogenase was added and by taking samples the new equilibrium constant was measured again (Table III). The sum of the concentrations of the pyridine nucleotides is not the same before and after heating the mixture, but this must be attributed to heat destruction of the cofactors. The K_{obs}-values in

Table III *Equilibrium constant as established by glutamate dehydrogenase and subsequently by transhydrogenase in the same mixture.* The reaction was started with glutamate dehydrogenase in the presence of 0.75 mM ammonium chloride, 0.27 mM α-oxoglutarate, 27 mM glutamate in 2 mM potassium phosphate buffer pH 7.6.

	Equilibrium concentrations (µM)				
Enzyme	NADH	NADPH	NAD^+	$NADP^+$	K_{obs}
+ glutamate dehydrogenase	45.5	27.0	84.0	96.0	0.52
heating (6 min. 80°C)	44.0	19.0	77.0	90.0	-
+ transhydrogenase	40.5	23.0	80.0	86.0	0.53

the reaction catalyzed by glutamate dehydrogenase and subsequently in the same medium by transhydrogenase are identical thus excluding a specific enzyme-effect. Since the ionic strength influences the equilibrium constant, the reaction between the pyridine nucleotides can be regarded as more or less ionic. For charged reactants the rate constant k of a reaction shows a dependence on the charges of the reactants and on the ionic strength, according to the equation derived by Brönsted and Bjerrum (7):

$$\log k = \log k_o + sZ_AZ_B\alpha\sqrt{I} = 2Z_AZ_B\alpha\sqrt{I} \qquad (1),$$

in which A and B are ions with charge Z_A and Z_B respectively; k and k_o are the rate constants in the medium and at I = 0 respectively; the numerical value of 2α in water at 25°C is 1.02. Application of this equation is restricted to point charges and low ionic strength. For an equilibrium situation as defined in the introduction, this equation changes into

$$\log K_{obs} = \log K_o + A\sqrt{I} \qquad (2),$$

in which

$$K_{obs} = [NADPH]\,[NAD^+] \,/\, [NADH]\,[NADP^+] \qquad (3),$$

K_o is the equilibrium constant at I = 0 and

$$A = 2\alpha\,(Z_{NADP^+} \cdot Z_{NADH} - Z_{NADPH} \cdot Z_{NAD^+}). \qquad (4)$$

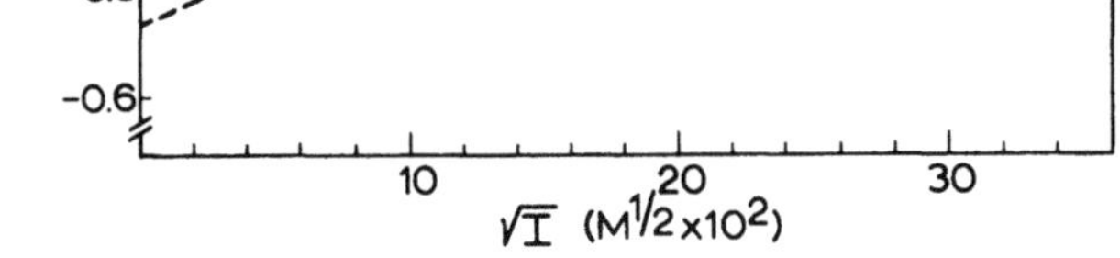

Fig. 1 *Dependence of the equilibrium [NADPH].[NAD⁺]/[NADH].[NADP⁺] on the ionic strength.* Data from Table I.

A linear behaviour of log K_{obs} vs $\sqrt{I}$ is observed at $I < 0.02$ using the data of Table I as plotted in Fig. 1. Extrapolation to zero ionic strength gives $K_{obs} = 0.30$. This corresponds to a difference in standard redox potential of the two pyridine nucleotide couples of 15 mV, the $NADP^+$/NADPH couple being the more electronegative. The slope of the curve of log K vs $\sqrt{I}$ is about one. These results agree very well with the values obtained by Engel and Dalziel (2), by studying the oxidative deamination of glutamate with each of both nicotinamide nucleotides separately, as catalyzed by glutamate dehydrogenase at pH 7.0 in phosphate buffers of varying ionic strength. It can be derived from the results of Engel and Dalziel (2) that the differences in slope of the linear parts of the two ionic-strength dependent curves is about one. This difference is slope reflects the dependence of the equilibrium constant of the two couples NAD^+/NADH and $NADP^+$/NADPH on the ionic strength; similarly the change of log K_{obs} vs $\sqrt{I}$ reflects this effect.

Dependence of K_{obs} on the Initial Pyridine Nucleotide Concentrations.

The equilibrium as reported above is reached upon starting the reaction from both directions with about equimolar concentrations of donor and acceptor. However, with a higher initial $[NAD^+]/[NADPH]$ ratio the thermodynamic equilibrium, as obtained with an initial 1:1 ratio, is not reached. In the reverse reaction the initial $[NADP^+]/[NADH]$ ratio does not influence the K_{obs} (Table IV). Upon addition of either NADH or NADPH the reaction restarts and the new values of K_{obs} approach those as given in Table I.

Table IV shows that NADPH is more effective than NADH in establishing conditions, which lead to normal thermodynamic equilibrium. The values calculated for K_{obs} under the extreme conditions of these experiments are only approximate because small variations in the actual NADPH concentration cause a relatively large shift in the value of this constant. In case of adding NADH only small changes in the concentrations of NADPH and $NADP^+$ are induced; however, the $[NADP^+]/[NADPH]$ ratio declines considerably. In order to explain the effects of changing the initial $[NAD^+]/[NADPH]$ ratio

Table IV *Dependence of K_{obs} of the pyridine nucleotide couples on starting conditions.* K-values as obtained with different initial ratios of nucleotides at 25°C. The reaction was started with the substrates at the concentrations indicated and a catalytic amount of enzyme. Samples were taken at different time intervals and assayed for the concentration of oxidized and reduced pyridine nucleotides; as final concentrations are taken values which were constant over a 15 minute period.

Buffer	Approximate starting concentrations (μM)				Final concentrations (μM)				K_{obs}
	NADH	NADPH	NAD^+	$NADP^+$	NADH	NADPH	NAD^+	$NADP^+$	
0.1 M Tris-HCl pH 7.6		50	100		32.0	15.0	69.5	40.0	0.81
		50	150		37.1	11.7	114	41.5	0.89
		50	250		40.0	10.1	213	44.0	1.2
		50	500		44.5	8.5	461	45.0	1.8
		100	500		72.0	13.5	468	87.0	1.1
		500	500		290	210	232	302	0.55
		450	100		94.7	363	17.0	111	0.58
	500			100	427	61.6	89.0	23.1	0.56
	100			400	29.0	69.5	79.5	315	0.60
0.02 M Tris-HCl pH 7.6		50	250		44.0	10.5	188	43.0	1.0
		250	50		47.0	191	6.0	52.0	0.47

upon K_{obs} it is important to summarize briefly some of the kinetic data of this enzyme. From studies of Van den Broek et al. (8,9) it is known that $NADP^+$ is a very strong inhibitor of the reactions catalyzed by this enzyme. The oxidized enzyme forms a dead-end complex with $NADP^+$, a suggestion also made for the Pseudomonas enzyme (10). In addition it was shown that NADPH activates the reaction and enhances the rate of its own formation in the reaction of NADH + $NADP^+$ → NADPH + NAD^+. This reaction shows a lag, which can be abolished by $MgCl_2$ and NADPH (8). The highest rates of NADPH formation are obtained at about equal concentrations of NADPH and $NADP^+$. It was thus suggested that the $[NADP^+]/[NADPH]$ ratio is of major importance in the regulation of the enzyme activity. From anaerobic titration experiments it is clear that in the presence of high concentrations of $NADP^+$ the

Table V *Effect of reduced substrates on the K_{obs} of the pyridine nucleotide couples.* The reaction was started with an initial ratio of [NAD$^+$]/[NADPH] of about 10 in 0.1 M Tris-HCl buffer pH 7.6 at 25°C. Samples were taken at different time intervals and the concentrations of pyridines nucleotides determined, and NAD(P)H added.

Addition after equilibrium		Final concentrations (μM)				K_{obs}
NADH μM	NADPH μM	NADH μM	NADPH μM	NAD$^+$ μM	NADP$^+$ μM	
-	-	44	10	500	44	2.6
-	21	68	13	482	62	1.5
-	46	91	16	459	85	0.95
-	75	109	20	437	108	0.74
-	100	129	25	414	130	0.62
-	150	164	40	376	159	0.58
-	-	43	7	478	48	1.6
92	-	128	10	483	44	0.86
196	-	218	14	484	39	0.80
400	-	386	18	488	34	0.67

reduction of the enzyme by either NADH of NADPH is strongly retarded, (11). It is nevertheless clear that under unfavourable initial conditions the reaction can proceed, although slowly, towards equilibrium in those cases where enough NADPH is present to activate the enzyme. Equilibrium is reached even at a high final concentration of the inhibitor NADP$^+$ by starting the reaction with NADP$^+$ + NADH in a 4:1 ratio. At equilibrium the [NADP$^+$]/[NADPH] ratio becomes about 4-5 (Table IV).

When the opposite reaction NAD$^+$ + NADPH → NADH + NADP$^+$ is started at a high [NAD$^+$]/[NADPH] ratio, equilibrium is not reached although the [NADP$^+$]/[NADPH] ratio at apparent equilibrium is about 4-5. Thus it could be concluded that the reaction does not stop because of an unfavourable ratio [NADP$^+$]/[NADPH], however, the data in Table IV indicate that the [NADP$^+$]/[NADPH] ratio is of less importance than the actual final NADPH concentration. A survey of all experiments performed leads to the interesting observation that independent of the NADP$^+$-concentration the reaction reaches thermodynamic equilibrium as soon as the NADPH concentration exceeds a concentration of about 25 μM. An estimate shows that under conditions of the experiment of Table V, a 50% decline of the value of K_{obs} is obtained at about 13 μM NADPH, a value in reasonable agreement with the reported K_m-value of NADPH of 15 μM.

Mechanistic aspects

It is obvious that the dependence of the equilibrium constant on the ionic strength and the pH must be due to the main difference between the di- and triphosphopyridine nucleotides, thus to the 2'-phospho group. The question how a slope of one can be obtained in the plot of log K vs $\sqrt{I}$ remains to be answered. Taking all charges into account (actually not permitted in the Brönsted approach used since the charges should be very close to each other) at pH 7.5 the diphosphate backbone carries two negative charges and the 2'-phosphate group also has two negative charges; in addition the oxidized nucleotides carry a delocalized positive charge. Substitution in equation (2) of the net charges gives a slope of 2. This value does not change upon omitting the diphosphate backbone charges. However, in case only one of the two charges of the 2'-phospho-group is taken into account, the value of the theoretical slope becomes 1 in accordance with the experimental value.

In the light of the data discussed previously this does not seem unreasonable. The effect of pH on K_{obs} shows that a lowering of the charge, arising from the second ionization of the 2'-phospho group makes the couples NAD^+/NADH and $NADP^+$/NADPH more equivalent since K_{obs} increases. In favor with this view is the observation (12) that NADPH is a better coenzyme for lactate dehydrogenase at low pH (6.0) than at higher pH values. The ratio of the activities with NADH and NADPH increases from about 40 at pH 6.0 to about 1200 at pH 7.6. In addition the activity with NADPH declines sharply beyond pH 6.5, while that with NADH increases. When the same numerical values for the charges of the nucleotides are substituted in the equations for the slopes of the glutamate dehydrogenase-catalyzed reactions with NADH or NADPH, the theoretical values of 2 and 3 respectively agree fairly well with the experimental values of 1.4 and 2.4 respectively, obtained by Engel and Dalziel (2). All other numerical values for the charges give larger deviations of the theoretical values from the experimental ones. The validity of the Brönsted-Bjerrum equation in two independent studies with different enzymes points to the fact that for thermodynamic equilibrium conditions, the negative charge of the second 2'-phosphoryl-O^- must be very close to the positive pyridine ring in $NADP^+$ to give charge neutralization and to influence the N-C4 of NADPH. This conclusion is necessary to satisfy the requirement of point charges. It follows that the charge of the first 2'-phosphoryl-O^- must be at a sufficiently large distance from the reaction centre (N-C4) to be without effect.

Biellman and Samana (13) have provided evidence that an anti-conformation of the nicotinamide-ribose glycosidic bond is present in NAD^+ bound to horse liver alcohol dehydrogenase and dogfish lactate dehydrogenase. Examination with space-filling models shows that in an anti-conformation the 2'-phospho group of NADPH in a M-helical form can closely approach the 4A proton. Hydrogen bonding is not possible between the amide group and the 2'-phospho group, as proposed by Sarma and Kaplan (14) for the synconformation. Blumenstein and Raftery (15) postulated from ^{31}P N.M.R.-studies electrostatic interaction between the positively charged nitrogen of the oxidized pyridine ring and a negatively charged oxygen of the diphosphate backbone. No differences in the ^{31}P-chemical shifts of NAD^+ and $NADP^+$ were found. As the proton shifts (16) and the carbon shifts (17) as induced by changing the pH have the same direction, it was postulated that the interaction of the phosphate backbone oxygen and the nitrogen causes redistribution of charge in the pyridine ring. Thus not only stacking but also electrostatic interactions induce the shifts in the oxidized nucleotides. Examination of the data of Blumenstein and Raftery (17) on ^{13}C-natural abundance shifts of the nicotinamide carbons of NAD^+ and $NADP^+$ induced by changing the pH, reveals some interesting facts not discussed by the authors. At pH 2.5 the ^{13}C-shifts of NAD^+ and $NADP^+$ are similar. The upfield shifts of the nicotinamide ring carbons are for NAD^+ at pH 7.5 and for $NADP^+$ at pH 5.1 still comparable. However, when the pH of the $NADP^+$ solution is raised to 8.0 further shifts of all nicotinamide carbons are induced. The second ionization of the secondary phosphate group of $NADP^+$ has a pK around 6.1 which indicates that double and not single ionization of this group causes the nucleotides to behave differently in the ^{13}C-resonances of the pyridine ring. It can be derived that the ^{13}C-shifts of the adenine moieties of NAD^+ at pH 7.5 and of $NADP^+$ at pH 8.0 are comparable.

Table VI *Proton shifts of reduced and oxidized pyridine nucleotides induced by 1 M NaCl.* All resonances were taken with a Varian XL-100-15D with Fourier transform. The samples were lypholyzed with D_2O to remove exchangeable protons. The pH of the oxidized nucleotides was adjusted with NaOD to pH 7.0. The reduced nucleotides had a pH of 7.9. All concentrations are about 10 mM. + reflects upfield shifts and - reflects downfield shifts in the presence of salt compared to absence of salt. The proton shifts of nucleotides in the presence relative to those in the absence of 1 M NaCl are given.

atom	NAD^+	$NADP^+$	NADH	NADPH
N-H2	+3.3	+1.2	+2.4	+1.3
N-H6	+4.7	+4.6	-	-
N-H4	-0.3	+0.2	+2.0	+2.0
N-H5	+0.8	+1.3	-	-
A-H8	+3.1	+0.5	+2.9	+1.8
A-H2	-1	+1	+0.2	-0.2
N-H1'	-1.1	+0.8	+4.1	+3.3
A-H1'	+1.6	-2.9	+0.4	-0.5

Some N.M.R. measurements were performed to study the influence of 1 M NaCl upon the chemical shifts of the pyridine nucleotides in 10 mM concentration in D_2O. The results are summarized in Table VI. The shifts are all relatively small (below 0.05 ppm) compared with those caused by overall stacking effects (18, 19). Major differences in the proton shifts are observed at N-H4: hardly any shifts in the oxidized nucleotides, upfield shifts in both reduced forms. Although the N-H5 and N-H6 peaks are hidden by the HDO peak, it can be concluded that increased stacking cannot explain the differences in shifts observed. A similar conclusion can be drawn from the proton shifts induced in the adenine ring. The shifts at

A-H8 have the same direction but the magnitude of the effects is less for $NADP^+$ and NADPH. In addition small but nevertheless totally different shifts are observed at A-H2 of the different pyridine nucleotide forms.

In the A-H1' of both $NADP^+$ and NADPH downfield shifts are observed upon addition of salt, which can be attributed to a reversal of the shielding effect of the adjacent negative charge of the 2'-phospho group, since in NAD^+ and NADH the A-H1' shows a small upfield shift. In the reduced forms upfield shifts of N-H1' are observed upon increasing the salt concentration; in the oxidized nucleotides, however, shifts in opposite direction are observed.

The combination of changes in stacking, changes in electrostatic interactions and changes in shielding make prediction and interpretation rather difficult. It is, however, clear that salt induces proton shifts which are different in NAD^+ as compared with $NADP^+$ and in NADH as compared with NADPH. Nevertheless one conclusion seems allowed: the ionization of the second 2'-phosphoryl-OH has a marked influence on the properties of the NADP(H) couple. The results of Table VI suggest that ionization of the first 2'-phosphoryl-OH has a shielding effect on the proton of the ribose connected to the adenine ring (A-H1'). Such a condition can only be achieved in the folded form.

Table VII summarizes ^{31}P shifts observed upon the addition of 1 M NaCl to the oxidized and reduced pyridine nucleotides. It is clear from the data that the ^{31}P shifts of the backbone phosphates which are induced by 1 M NaCl are downfield and about equal for both the oxidized and reduced pyridine nucleotides. In addition the 2'-phospho group of $NADP^+$ shows a downfield shift of about equal magnitude. The major difference is the ^{31}P shift of the 2'-phospho group of NADPH induced by NaCl, downfield but smaller. Thus the decline of charge in the 2'-phospho group of NADPH by the addition of NaCl (probably by interaction of the phospho-O^- with Na^+) is much less than the decline of charge in the backbone phosphates and in the 2'-phospho group of $NADP^+$. This can only be achieved in a folded form of NADPH, in which the 2'-phospho group is H-bonded. This does not exclude that $NADP^+$ is in a folded form, the positive pyridine ring is main-

Table VII *^{31}P shifts of oxidized and reduced pyridine nucleotides induced by 1 M NaCl.* The spectra were measured with a Varian XL-100-15D spectrometer at 40.48 Mcs with Fourier transform. The samples were recorded in D_2O at pH 7.9 (not corrected for D_2O) at a concentration of about 10 mM. Shifts relative to the internal standard $(CH_3O)_3$ P = O.

compound	^{31}P at bound to position					
	A-2'	Δ	A-5'	Δ	N-5'	Δ
$NADP^+$ without NaCl	15.2	13.5	-580.8	14.0	-592.2	14.0
$NADP^+$ plus 1 M NaCl	28.7		-566.8		-578.2	
NADPH without NaCl	25.0	4.9	-579.0	12.4	-579.0	12.4
NADPH plus 1 M NaCl	29.0		-566.6		-566.6	
NAD^+ without NaCl	-	-	-581.9	16.6	-593.3	16.6
NAD^+ plus 1 M NaCl	-		-565.3		-576.7	
NADH without NaCl	-	-	-578.9	12.0	-578.9	12.0
NADH plus 1 M NaCl	-		-566.9		-566.9	

ly in interaction with the backbone phosphates, as in NAD^+ (cf. ref. 15,20) The magnitude of the shift in the 2'-phospho group of $NADP^+$ induced by 1 M NaCl points to the fact that the positive charge of the pyridine ring is localized in that case mainly around the N-atom and not around C-4. It is, however, clear that after addition of 1 M NaCl the relative shifts of the 2'-phospho group of NADPH and $NADP^+$ are equal, which points to identical character of this group in the reduced and oxidized forms of the triphosphopyridine nucleotide at high ionic strength.

In conclusion: the difference between the two pyridine nucleotide couples disappears at high ionic strength; at low ionic strength the thermodynamic equilibrium is established by a reaction in which at least the oxidized and the reduced triphosphopyridine nucleotide participate in their folded form; in no other configuration the charges can be counted as has been done here, while in the open form the ionic strength will not influence the equilibrium constant. This conclusion, however, is only valid for the free coenzymes in thermodynamic equilibrium. It does not exclude the possibility that the coenzymes in either form bind to the enzyme in their open conformations.

ACKNOWLEDGEMENTS

We are grateful to Dr. R. Gast and Dr. C.J. van Schagen for cooperation and discussions with the N.M.R. experiments. This study has been carried out under auspices of the Netherlands Foundation for Chemical Research (S.O.N.) with financial aid from the Netherlands organization for the Advancement of Pure Research (Z.W.O.).

REFERENCES

1. Olson, J.A. and Anfinsen, C., J. Biol. Chem. 202, 841-856 (1953).
2. Engel, P.C. and Dalziel, K., Biochem. J. 105, 691-695 (1967).
3. Kaplan, N.O., Colowick, S.P., Zatman, L.J. and Ciotti, M.M., J. Biol. Chem. 205, 31-43 (1953).
4. Rydström, J., Teixeira da Cruz, A. and Ernster, L., Eur. J. Biochem. 17, 56-62 (1970).
5. Lee, C.P. and Ernster, L., Biochim. Biophys. Acta. 81, 187-190 (1964).
6. Fisher, R.R. and Kaplan, N.O., Biochemistry 12, 1182-1188 (1973).
7. Brönsted, J.N., Z. Physik. Chem. 102, 169-207 (1922).
8. Van den Broek, H.W.J. and Veeger, C., in (H. Sund ed) Pyridine Nucleotide-dependent Dehydrogenases, Springer Verlag Heidelberg, 335-349 (1970).
9. Van den Broek, H.W.J., Santema, J.S. and Veeger, C., Eur. J. Biochem. 24, 55-62 (1971).
10. Cohen, P.T. and Kaplan, N.O., J. Biol. Chem. 245, 2825-2836 (1970).

11. Van den Broek, H.W.J. and Veeger, C., Eur. J. Biochem. 24, 63-71 (1971).
12. Navazio, F., Ernster, B.B. and Ernster, L., Biochim. Biophys. Acta. 26, 416-421 (1957).
13. Biellmann, J.F. and Samana, J.P., FEBS Lett. 38, 175-178 (1973).
14. Sarma, R.H. and Kaplan, N.O., Biochemistry 9, 539-548 (1970).
15. Blumenstein, M. and Raftery, M.A., Biochemistry 11, 1643-1648 (1972).
16. Sarma, R.H. and Kaplan, N.O., Biochem. Biophys. Res. Comm. 36, 780-788 (1969).
17. Blumenstein, M. and Raftery, M.A., Biochemistry 12, 3585-3590 (1973).
18. Jardetzky, O. and Wade-Jardetzky, N.G., J. Biol. Chem. 241, 85-91 (1966).
19. Catterill, W.A., Hollis, D.P. and Walter, C.F., Biochemistry 8, 4032-4036 (1969).
20. Czeisler, J.L. and Hollis, D.P., Biochemistry 14, 2781-2785 (1975).

Received March 8, 1977

DISCUSSION

Engel: Can you be sure that the pH effects that you see are not due to the instability of reduced pyridine nucleotide coenzymes in phosphate buffer at acid pH's?

Veeger: The answer is 'yes', we are sure. We checked the sum of the concentrations of oxidized and reduced forms of the two pyridine nucleotides before and after the reaction. Equilibrium was established within one minute and only followed for about five minutes to establish whether equilibrium had been achieved. In this time period the sums of the concentrations were constant.

Engel: I am a little uneasy about the interpretation of the slope in your Fig. 1. This slope includes the contributions of four unknowns, making the numerical interpretation dependent on several assumptions and correspondingly uncertain. I wonder whether it would not be worth measuring the ionic strength dependence of log K_{obs} for a pyridine nucleotide dependent dehydrogenase with uncharged substrates - e.g. alcohol dehydrogenase with isopropanol and aceton. The slope in that case would contain only two unknowns, the effective charges on NAD^+ and NADH, and ought to make possible a more confident interpretation of the slopes seen in your experiments and ours.

Veeger: We shared your uneasiness. But after measuring it several times we became confident about the results, because they were quite reproducible and very consistent with your glutamate dehydrogenase data. You certainly make a good suggestion for further investigation. But I am afraid one can do it only with a diphosphopyridine nucleotide enzyme. It is more difficult for the triphosphopyridine nucleotide, and it is particularly the charge on this form which is responsible for the slope observed. We tried all combinations of possible charges on the different pyridine nucleotide forms but only in the case discussed a slope of one could be obtained.

Section IV. Structure Function Relationship

Chairman: A.C.T. North

FUNCTIONAL SIGNIFICANCE OF THE STRUCTURE OF LIVER ALCOHOL DEHYDROGENASE.

Carl-Ivar Brändén
Department of Chemistry
Agricultural College of Sweden
S-750 07 UPPSALA, Sweden

The crystallographic studies of horse liver alcohol dehydrogenase have revealed certain basic features of the molecular architecture of the coenzyme and substrate binding areas. These features provide some stringent restrictions on possible catalytic mechanisms of action of this enzyme. In addition the structure and its correlation with solution studies have provided some ideas about the factors which might be important for coenzyme and substrate specificity. In particular, evidence will be presented which demonstrate the importance of the carboxamide group of the coenzyme for proper positioning of the nicotinamide ring. Furthermore the geometry of the substrate-binding pocket has been correlated to the observed stereospecificity of alkyl-substituted cyclohexanol and steroid molecules as substrates for the liver alcohol dehydrogenases.

The results to be described here are based on the following X-ray studies; independent structure determinations of the orthorhombic apoenzyme to 2.4 A resolution (1) and a triclinic enzyme-NADH-dimethyl-sulfoxide complex to 4.5 A resolution (H. Eklund and C.-I. Brändén to be published), ligand binding studies by difference Fourier techniques of the orthorhombic complexes enzyme-ADPribose (2) enzyme-imidazole (3), enzyme-NADH-imidazole and enzyme-tetrahydroNAD to 2.9 A resolution (E. Zeppezauer to be published) as well as enzyme-1,10-phenanthroline (3), enzyme-imidazole-pyridine adenine dinucleotide and enzyme-imidazole-3iodo pyridine adenine dinucleotide to 4.5 A resolution (4).

A. SOME STRUCTURAL FEATURES OF ALCOHOL DEHYDROGENASE.

A detailed description of the apoenzyme structure has recently been published (1, 5). Here we show in Fig. 1 a schematic diagram of one molecule with the observed position of bound coenzyme and an assumed position of substrate. Each subunit of the molecule is divided into two domains, one of which binds the coenzyme whereas the other binds the two zinc atoms of the subunit.

The catalytic zinc atom is situated in the middle of the subunit close to the junction between the two domains. The

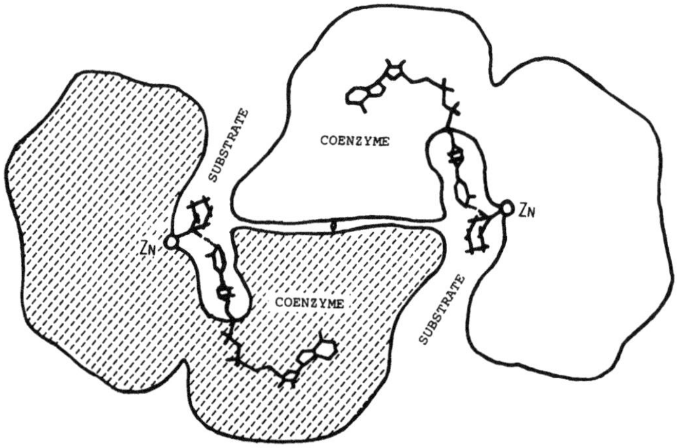

Fig. 1 Schematic diagram of a section through the dimeric LADH molecule with bound coenzyme projected onto this section. The substrate molecule, cyclohexanol, has been positioned by modelbuilding assuming direct binding to zinc.

domains are oriented relative to each other in such a way that the catalytic zinc atom is accessible to the solvent in some regions of the interface. These accessible regions can be described as one crevice in which the nicotinamide moiety of the coenzyme is bound and one deep pocket where the substrate is assumed to bind.

The assumption that the substrate binds in this pocket is based on the following evidence. The substrate must bind in the vicinity of the nicotinamide group and there are no other possibility available than this pocket. The inhibitor molecules imidazole and 1,10-phenanthroline which have been shown to compete with substrate binding are bound in this pocket (3). The geometry of this pocket can be correlated to the results of extensive kinetic mapping of the substrate binding area as will be discussed later.

Three protein ligands anchor the catalytic zinc atom firmly to the bottom of the substratebinding pocket, two sulphur atoms from Cys 46 and Cys 174 and one nitrogen atom from His 67. A water molecule or hydroxyl ion, depending on pH, completes a tetrahedral coordination in the apoenzyme structure.

The substratebinding pocket can be divided into three parts (see figure 2), a hydrophilic bottom where the catalytic action occurs, a large hydrophobic barrel through which the substrate must enter and a rim where both polar and nonpolar residues are present. The bottom part contains the following hydrophilic groups: 1. The zinc atom and its coordination

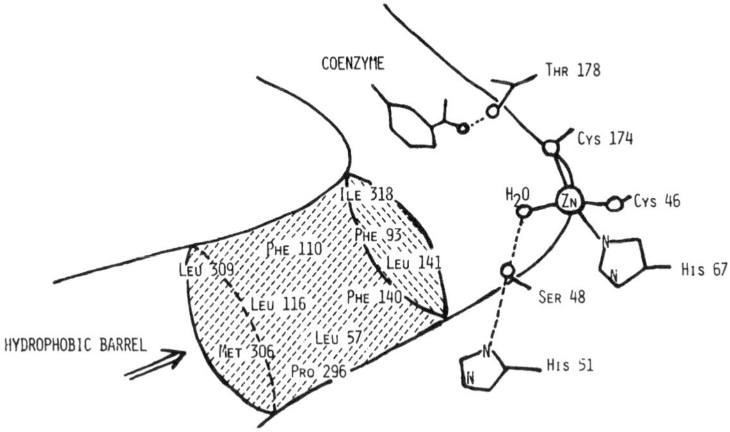

Fig. 2 Schematic diagram of the active site of LADH as observed from X-ray studies.

sphere including the water molecule which projects into the pocket. 2. The nicotinamide moiety of the coenzyme and the hydroxyl group of Thr 178. 3. A general acid-base system consisting of the hydroxyl group of Ser 48 and the imidazole group of His 51. The oxygen atom of the zinc bound water molecule forms a hydrogen bridge to the serine oxygen which in turn is hydrogen bonded to the imidazole nitrogen.

The hydrophobic barrel provides a completely non-polar lining of the pocket by the side chains of residues Leu 57, Phe 93, Phe 110, Leu 116, Phe 140, Leu 141, Pro 296 and Ile 318 from one subunit and Met 306 and Leu 309 from the other subunit. Hydrophobic regions of large substrates such as cyclohexanol, steroids and ω-hydroxy fatty acids would be expected to bind to some of these residues.

B. COENZYME BINDING.

The coenzyme molecule is bound in an extended conformation within a crevice of the NAD^+ binding domain. It is positioned by a number of contacts which extend over the whole coenzyme molecule. These contacts are more frequent in the AMP-part which is thus firmly anchored to the enzyme in agreement with binding studies of coenzyme fragments (5). Some of the important contacts are; a hydrophobic pocket for the adenine part, a hydrogen bond between the 2' oxygen of the adenosine ribose and Asp 223 and a salt bridge between the pyrophosphate group and Arg 47.

The nicotinamide moiety is positioned in the active site with the A-side of the ring facing the zinc atom and the B-side facing the hydrophobic wall of the parallell pleated sheet of the NAD-binding domain. The distance from the zinc atom to the centre of the nicotinamide ring is 4.5 A and approximately the same or somewhat shorter to the C4 atom of this ring. The only polar contact between the nicotinamide moiety and the protein is a hydrogen bond between the side chain of Thr 178 and the carbonyl oxygen of the carboxamide group. A similar hydrogen bond is formed to a serine residue in lactate dehydrogenase (M. Rossmann, private communication).

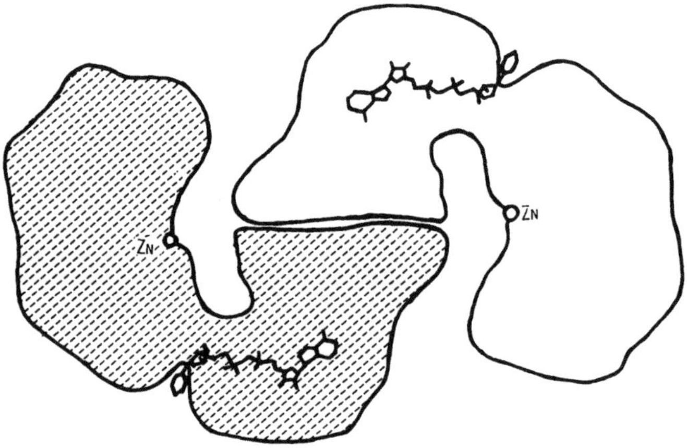

Fig. 3 Schematic diagram of the binding of pyridine adenine dinucleotide to LADH.

In order to deduce the possible importance of this hydrogen bond for proper positioning of the nicotinamide moiety we have studied the binding of coenzyme analogues where this group is either removed, pyridine-adenine-dinucleotide, $PyAD^+$, or substituted with an iodine atom, 3-I $PyAD^+$. In both these complexes the binding of the pyridine mononucleotide part is the same, but it is very different from that of the NMN-part of the coenzyme (see Fig. 3). The pyridine ring is positioned on the surface of the molecule quite far from the active site, the distance from zinc to the centre of the pyridine ring being 13 A. The AMP-parts of these analogues, on the other hand, bind in a similar manner as the corresponding part of the coenzyme. The hydrogen bond between the carboxamide group and Thr 178 is thus quite essential for positioning the nicotinamide moiety of the coenzyme in the active site.

C. SUBSTRATE BINDING.

By model building experiments we have been able, in collaboration with Dr. H. Dutler, ETH, Zürich to correlate the geometry of the substratebinding pocket described here with kinetic results in solution of systematically varied alkylsubstituted cyclohexanone derivatives and bridged carbocyclic ketones of stable conformations as substrate molecules (Dutler, H. and Brändén, C.-I. to be published). The kinetic studies by Dutler (6) were designed to obtain a map of the free space available for substrate molecules at the active site using the "diamond lattice" concept (7). In our modelbuilding we assumed that the oxygen atom of the alcohol binds directly to zinc and that the H-atom to be transferred is 2.5 A away from and points towards the C4 atom of the nicotinamide ring. Using these assumptions we built a cyclohexanol molecule with chair conformation and the hydroxyl group in axial position into the active site and studied the steric effects of all possible substitutions of methyl groups. We found that in all compounds where the activity was decreased or abolished as measured kinetically (6), the substituted methyl group on the cyclohexanol skeleton overlapped sterically with the enzyme or coenzyme. This is schematically illustrated in Fig. 4. For the remaining substitutions which did not affect the kinetic parameters of the derivatives as substrates we found that space was available for the corresponding substitution in the substrate binding pocket.

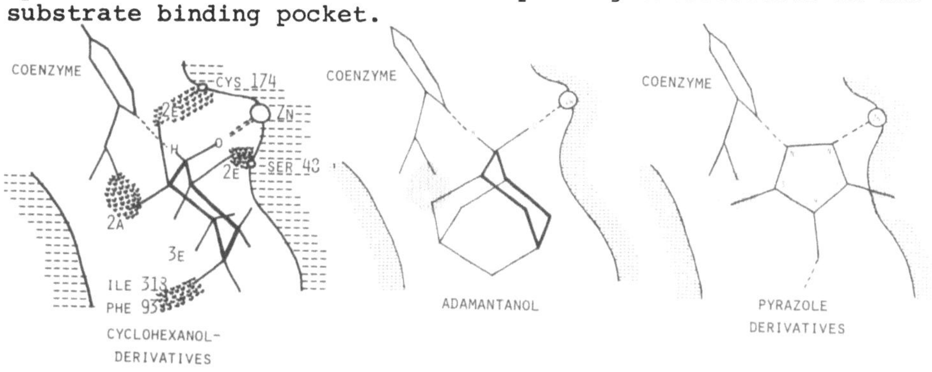

Fig. 4 Schematic diagram of different molecules built into the substrate binding pocket of LADH. Light shades represent protein, darker shades represent steric overlap between atoms of the protein or coenzyme and atoms from the added molecule.
(a) Methylsubstituted cyclohexanol. Substitution in position 2e overlaps Cys 174, in 2'e Ser 48, in 2a the carboxamide group of the coenzyme and in 3e Ile 318 and Phe 93.
(b) Adamantanol. The CH_2group corresponding to substitution in position 2a of cyclohexanol overlaps the carboxamide group of the coenzyme.
(c) Methylsubstituted pyrazole. Substitution in position 3 overlaps the carboxamide group, in position 5 Ser 48 whereas substitution in position 4 is directed into the hydrophobic barrel of the pocket.

Using this position of the cyclohexanol ring we could deduce that adamantanol is inactive as a substrate (8) because this rigid molecule has a CH_2 group in one of the sterically hindered substituted positions of cyclohexanol, 2a (see figure 4b). Bicyclo(3.2.1)-octan-2-ol which lacks this CH_2 group but otherwise is identical to adamantanol is active as a substrate (6) in complete agreement with our X-ray model.

We have also used these results to study possible binding of 3-hydroxy steroid molecules in the active site. The A-ring of these molecules can be considered as a partly substituted cyclohexanol ring. It has been shown kinetically that the enzymes from horse (isozyme SS and ES but not EE), rat and human liver catalyze the oxidation of 3β-hydroxy groups of both 5α and 5β steroids (9, 10). Corresponding 3α-hydroxy-steroids were not active as substrates. The X-ray model is based on the horse EE isozyme. However, since only six amino acid differences have been observed between the E and the S chain (11) it can be safely assumed that the structure of the S-subunit is similar to our model except in the vicinity of these differences.

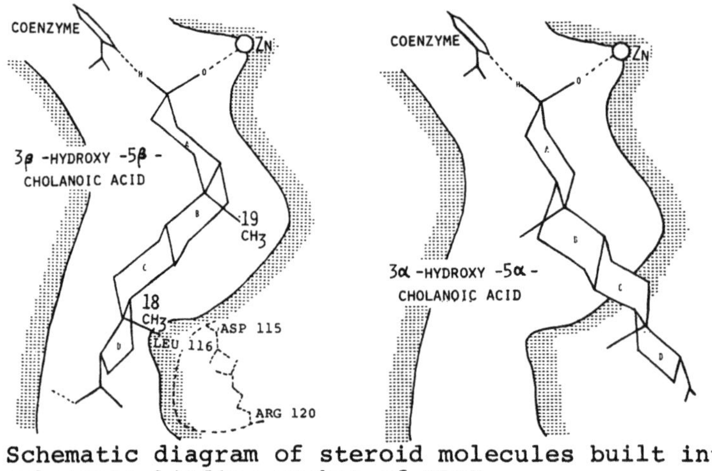

Fig. 5 Schematic diagram of steroid molecules built into the substrate binding pocket of LADH.
(a) 3β-hydroxyl-5β-cholanoic acid.
(b) 3α-hydroxyl-5β-cholanoic acid.

Our model building studies show that 3β-hydroxy-5β-cholanoic acid fits very nicely in the substrate binding pocket except at one particular point (see Fig. 5a). The methyl group at position 18 of the steroid overlaps sterically with the side chain of Leu 116 close to the outer rim of the hydrophobic barrel. This interaction would prevent the steroid to bind in the assumed productive orientation in the E-subunit. However, in the S-subunit the situation could be quite different. Leu 116 is part of a loop on the outside of the molecule with no organized secondary structure. There is a saltbridge in the E-subunit from Asp 115 to Arg 120 which might contribute significantly to the conformation of this loop and thus to the observed position of Leu 116. This salt bridge cannot occur either in the S-subunit or in the rat enzyme since Asp 115 is substituted by serine (or a deletion) and asparagine, respectively. It is thus quite possible that the residues in the vicinity of 115 have a somewhat different structure in these enzymes and thus no steric overlap for our assumed orientation of the steroid molecule. We could thus predict, on the basis of these modelbuilding studies, that 3β-hydroxy-5β-steroid analogues which lack the methylgroup at position 18 might be substrates for the EE isozyme.

On the other hand 3α-hydroxy-5β-steroids cannot bind in a productive mode as is evident from Fig. 5b. The pocket is sufficiently narrow to create impossible overlaps both at the junction between rings A and B on one side of the pocket and for the whole of ring D on the other side. Similar results were obtained for modelbuilding of the 5α-steroids.

Using this model we have also examined possible binding of different pyrazole derivatives which are powerful inhibitors in the presence of NAD^+. It has been suggested (12) that one nitrogen atom of the pyrazole ring binds to zinc and the other to C4 of the nicotinamide ring. Figure 4c shows schematically the result of positioning a pyrazole ring in this manner in our model. One nitrogen atom superimposes the oxygen atom of the cyclohexanol molecule and the other nitrogen atom superimposes the hydrogen to be transferred.

It is obvious that substitutions in 3- and 5-positions give steric overlap with enzyme and coenzyme which is consistent with the much larger dissociation constants for these derivatives (13). K_i is 0.22 μM for pyrazole compared to 7100 μM for 3,5-dimethyl pyrazole. Alkyl substitutions in the 4-position, on the other hand, are not only sterically possible but would be expected to contribute to tighter binding since they are positioned in the hydrophobic barrel. This is also the case. K_i for 4-methylpyrazole is 0.08 μM and 0.001 μM for 4-pentylpyrazole (14). The introduction of charged groups on these alkyl chains would decrease the binding strength when they are inside the barrel but would lead to further enhancement of binding and specificity if they are suitably introduced to form salt bridges with charged groups at the rim of the pocket.

This excellent agreement between modelbuilding in the X-ray structure and kinetic data in solution for the cyclohexanol derivatives, steroids and pyrazole derivatives strongly indicates firstly that the geometry of the active site described here is relevant for deductions about the catalytic action of the enzyme and secondly that the substrate in all probability binds directly to the zinc atom. Furthermore it is obvious that the X-ray structure can be used not only to obtain correlations with existing kinetic data but also to design substrate or inhibitor molecules of desired properties. Work in this direction is now in progress.

D. MECHANISM OF ACTION.

The X-ray structure determination shows that the only groups in the enzyme that may be responsible for binding and polarization of the reactive part of the substrate are the zinc atom, the zinc bound water molecule, the hydroxyl group of Ser 48 and, indirectly, the imidazole group of His 51 through its hydrogen bond to Ser 48.

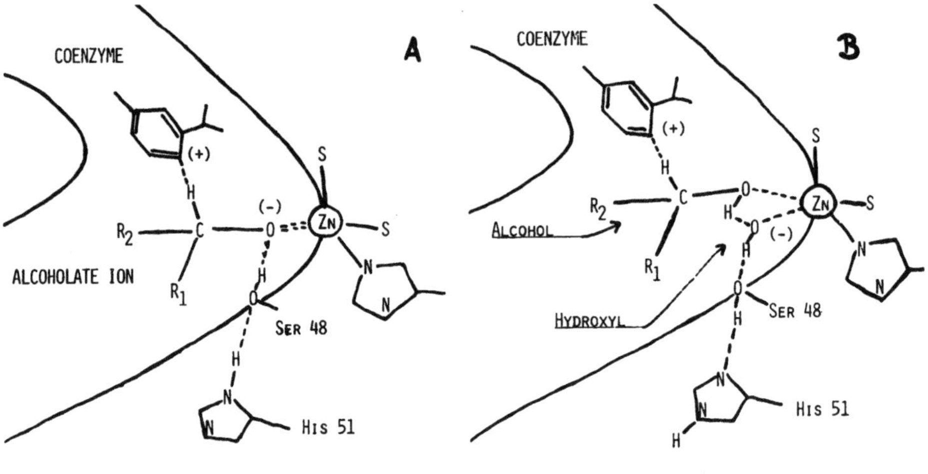

Fig. 6 Schematic diagram of assumed productive LADH-NAD$^+$-alcohol complex suggested by
a) Brändén et al. (5).
b) Dworschack and Plapp (15).

We have suggested a mechanism (5) based mainly on electrophilic catalysis mediated by the active site zinc atom. In this mechanism we assume that alcohol binds directly to zinc as the negatively charged alcoholate ion (see figure 6a). Formation of this ion is facilitated by the previous step in the reaction, NAD$^+$ binding, which perturbs the pK of the zinc bound water from about 9.0 to about 7.5. The resultant hydroxyl ion bound to zinc can combine with the hydrogen atom of the hydroxyl group of alcohol to produce water and alcoholate ion. The hydrogen bond system of Ser 48 and His 51 might be involved both in the proton release upon NAD$^+$ binding and in substrate binding and polarization. The water molecule is in the hydrophobic substrate binding pocket, whereas the εN atom of His 51 points into the solution at the surface of the molecule. It might thus be energetically more favourable for proton release to occur through this system of hydrogen bonds to the polar surface of the molecule than directly into the pocket.

A modification of the mechanism suggested by us has recently been proposed (15) in order to explain some recent structure-reactivity studies (15, 16, 17). These studies indicate that no charge develops on the substrate during catalysis and hence that release of the proton of the alcohol and the hydride transfer occur simultaneously. A general acid-base catalysis would be consistent with these results. In the modified mechanism (14) the hydroxyl ion is assumed to remain bound to zinc while the neutral alcohol molecule adds as a fifth ligand to zinc. The suggested productive LADH-NAD^+-alcohol complex is schematically illustrated in figure (6b). The zinc bound hydroxide ion is postulated to be the general base catalyst for the oxidation of alcohols, and the zinc bound water molecule as the general acid catalyst for the reduction of aldehydes. Hopefully a decisive answer to this point will be obtained by X-ray studies now in progress of suitably chosen triclinic enzyme-coenzyme-substrate complexes which might be crystallized because of favourable equilibrium constants.

References:

1. Eklund, H., Nordström, B., Zeppezauer, E., Söderlund, G., Ohlsson, I., Boiwe, T., Söderberg, B.-O., Tapia, O., Brändén, C.-I., Åkeson, Å.: Three-dimensional Structure of Horse Liver Alcohol Dehydrogenase at 2.4 Å Resolution. J. Mol. Biol. 102, 27-59 (1976).

2. Abdallah, A., Biellmann, J.-F., Nordström, B., Brändén, C.-I.: The Conformation of Adenosine Diphosphoribose and 8-BrAdenosine Diphosphoribose when Bound to Liver Alcohol Dehydrogenase. Eur. J. Biochem. 50, 475-481 (1975).

3. Boiwe, T., Brändén, C.-I.: X-ray investigation of the binding of 1,10-phenanthroline and imidazole to horse liver alcohol dehydrogenase. Eur. J. Biochem. in press (1977).

4. Samama, J.-P., Biellmann, J.-F., Zeppezauer, E., Brändén, C.-I.: The Crystal Structure of Complexes between Horse Liver Alcohol Dehydrogenase and the Coenzyme Analogues 3-Iodopyridine- and Pyridine Adenine Dinucleotide. Eur. J. Biochem. in press.

5. Brändén, C.-I., Jörnvall, H., Eklund, H., Furugren, B.: Alcohol Dehydrogenases. The Enzymes (P.D. Boyer, ed.) 3rd edn. Vol 11 pp 93-190 (1975) Academic Press, New York.

6. Dutler, H.: Structure-activity relationship as studied with an enzyme of known structure: The liver alcohol

dehydrogenase-catalyzed reduction of alicyclic ketones in "Structure-Activity Relationship in Chemoreception" (G. Benz ed.) pp 65-73 (1976). Information Retrieval, London.

7. Prelog, V.: Specification of the stereospecificity of some oxido-reductases by diamond lattice sections. Pure and Applied Chemistry 9, 119-126 (1964).

8. Graves, J.M.H., Clark, A., Ringold, H.J.: Stereochemical Aspects of the Substrate Specificity of Horse Liver Alcohol Dehydrogenase. Biochemistry 4, 2655-2671 (1965).

9. Reynier, M., Theorell, H., Sjövall, J.: Studies on the Stereospecificity of Liver Alcohol Dehydrogenase for 3β-Hydroxy-5β-steroids. Acta Chem. Scand. 23. 1130-1136 (1969).

10. Cronholm, T., Larsén, C., Sjövall, J., Theorell, H., Åkeson, Å.: Steroid Oxidoreductase Activity of Alcohol Dehydrogenases from Horse, Rat and Human Liver. Acta Chem. Scand. B29, 571-576 (1975).

11. Jörnvall, H.: Horse Liver Alcohol Dehydrogenase. On the Primary Structures of the Isoenzymes. Eur. J. Biochem. 16, 41-49 (1970).

12. Theorell, H., Yonetani, T.: Liver Alcohol Dehydrogenase-DPN-Pyrazole Complex. Biochem. Z. 338, 537-553 (1963).

13. Theorell, H., Yonetani, T., Sjöberg, B.: On the Effects of Some Heterocyclic Compounds on the Enzymic Activity of Liver Alcohol Dehydrogenase. Acta Chem. Scand. 23, 255-260 (1969).

14. Dahlbom, R., Tolf, B.R., Åkeson, Å., Lundqvist, G., Theorell, H.: On the inhibitory power of some further pyrazole derivatives of horse liver alcohol dehydrogenase. Biochem. Biophys. Res. Comm. 57, 549-553 (1974).

15. Dworschack, R.T., Plapp, B.V.: Mechanism of alcohol dehydrogenase. Biochemistry. in press (1977).

16. Klinman, J.P.: Yeast alcohol dehydrogenase reaction. Biochemistry 15, 2018-2026 (1976).

17. Blackwell, L.F., Hardman, M.J.: Effect of Substrate Structure on the Pre-Steady-State Kinetics of Oxidation by Liver Alcohol Dehydrogenase. Eur. J. Biochem. 55, 611-615 (1975).

Received February 19, 1977

DISCUSSION

Theorell: It is remarkable how fast time goes - it is now 30 years since the liver ADH was crystallized in my lab in Stockholm. I am no admirer of Adolf Hitler, but without knowing it he rendered a service to biochemistry when he chased away a young Dane from Denmark over to Sweden. His name was Roger Bonnichsen; he came to my lab and had substantial part in producing the first liver ADH crystals, today so beautifully described by Brändén. Let me pick up one little detail from the lecture we just heard. The enzyme molecule, ADH, has the active groups buried in a pocket, the dimensions of which must be decisive for which substrates and competitive inhibitors can be operative together with the coenzyme.

Already in 1963 we found that pyrazole is a fairly good competitive inhibitor (K_I = 220 nM). Later we found that the alkyl group in the 4-position of pyrazole increases the inhibitory power. 4-methyl-pyrazole has a K_I value of 13 nM, and each additional $-CH_2$ group doubles the inhibitory power; the last one we obtained before I came here was that for 4-octyl-pyrazole (K_I = 0.03 nM). It will be interesting to see if the inhibition disappears when the pyrazole molecule becomes too large to find place in the substrate pocket. C_{11}- and C_{14}-acid are next on the schedule.

We have carried out experiments on rats with methyl-pyrazole and alcohol together with Rolf Blomstrand. The alkyl-pyrazoles give an 80 % inhibition of the combustion. This indicates that liver ADH is responsible for 80 % of the combustion at least in rats.

Trommer: I should like to present some of our results showing interesting differences in the stereochemistry of the substrate arrangement in LDH and ADH. In addition to a coenzyme-substrate analogue composed of NADH and pyruvic acid which gives intramolecular hydride transfer in the presence of LDH and alanine dehydrogenase (H. Kapmeyer, G. Pfleiderer and W.E. Trommer, Biochemistry 15 (1976) 5024-5028), we have now prepared the stereochemically pure L- and D-Lac-NAD^+ (I) in which again the substrate is attached via a methylene spacer to the complete coenzyme (H. Hüth, U. Grau and W.E. Trommer, unpublished results).

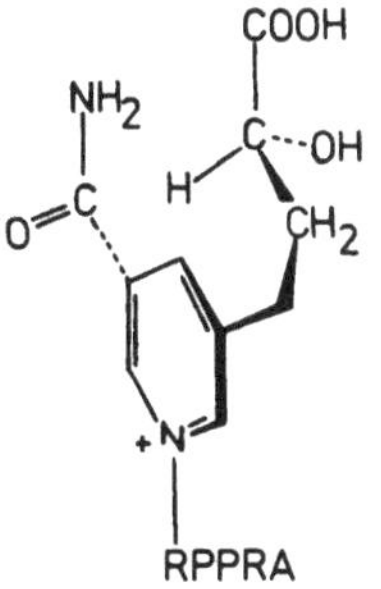

L-LAC-NAD^+ (I)

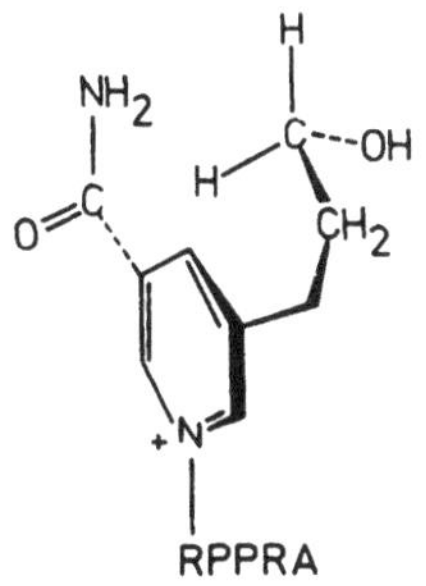

(II)

The L-diastereomer undergoes an intramolecular redox reaction in the presence of LDH under the formation of the corresponding NADH derivative with a pyruvic acid side chain. In addition, Dr. Rossmann could show that the structure of L-Lac-NAD^+ could be fitted very nicely to the electron density of ternary LDH complexes.

In contrast to this the D-diastereomer is neither converted by a D-lactate specific LDH nor by the L-specific enzyme. Since D-lactate LDH is A side specific as well, this implies that the substrate is pointing into the opposite direction relatively to the nicotinamide ring. An analogue in which lactate is attached via a spacer to the carboxamide function is presently being synthesized in our laboratory.

The second analogue which is shown in the figure (II) differs from Lac-NAD^+ in that it does not contain the carboxyl group and it should therefore be a appropriate transition state analogue of ADH. However, it is neither converted by ADH from yeast or horse liver. As a matter of fact, it is even a rather poor inhibitor. From these results one might speculate that in ADH the substrate is in the opposite arrangement as in LDH, that ADH is a "D"-specific enzyme. This fits very well the structural data just presented by Dr. Brändén and will be further shown by the elegant experiments of Dr. Dutler in his talk.

Brändén: The main reason why your analogue is not a substrate to ADH and even a poor inhibitor is that a substitution in the 5-position of the nicotinamide ring is sterically interfering with the protein (zinc and Cys 174), your analogue can thus not enter the active site of liver ADH and would also be expected to bind poorly. The different behavior of your analogue towards LDH and liver ADH reflects the fact that the catalytic domains of LDH and liver ADH are differently oriented with respect to the common structural features of the coenzyme-binding domain.

Kempfle: The dye tetraiodofluorescein is bound to several NAD-dependent dehydrogenases and it was shown by X-ray investigations (P.M. Wassarman and P.J. Leutz (1971) J. Mol. Biol. 60, 509-522) that the dye is bound to the adenine binding moiety of NAD(H). Now we take this as a tool to determine enzyme-coenzyme-binding constants since it is possible to displace the dye from its binding site on the enzyme by the interesting coenzymes (M. Kempfle et al. (1977) Z. Anal. Chem., in press). It is obviously a competition phenomenon.

If we relate the ability of some interesting effectors to displace the dye (D) or not with their enzymatic activity (A) or their influence on it we find the following - only qualitative - results:

	NMN	ADP	GDP	NAD(H)	NADP(H)	ε-NAD	α-NAD	NAD(H)*	NADP(H)*
	D A	D A	D A	D A	D A	D A	D A	D A	D A
liver ADH	- -	+ o	- o	+ +	- -	- +	+ -	+ -	- -
M_4 LDH	- -	+ o	- o	+ +	- -	- +	+ -		

D: - no displacement
   + displacement

A: + activity
   o no influence on activity
   - no activity

* PLP enzyme

Now my question: Have you any explanation for this behavior especially for the NADP(H), ε-NAD, α-NAD and the PLP-treated liver ADH? Does that mean that the adenine moiety is not necessary for activity, because ε-NAD is not able to fit into the coenzyme-binding cleft?

Brändén: Your interesting data fits very nicely to the structures. The extra phosphate group of NADPH prevents the adenosine moiety from being bound properly due to repulsion of Asp 223 and thus shows no displacement. α-NAD would be expected to bind in the adenosine pocket in agreement with your displacement data, but shows no activity due to wrong stereochemistry for positioning of the nicotinamide part. Your data indicate that both the dye and α-NAD bind at the same time. The extra goup in the adenine part of α-NAD might allow that part to bind at different positions depending on whether the dye is bound or not. Thus it does not displace the dye but binds in an incorrect orientation. My prediction would be that the dye is an inhibitor, not because of competition but because it forces the α-NAD to bind incorrectly.

Jeck: I want to make a remark concerning the coenzyme binding cleft of liver ADH. NAD analogs in which the adenine linked ribose is substituted by n-butyl or n-pentyl residues show considerably higher Michaelis constants than NAD^+ but they also show much higher maximum turnover numbers. The pentyl derivative is reduced by the enzyme and ethanol at a rate 12 times faster as compared to NAD. Can you explain this feature on the basis of liver ADH conformation without having further kinetic data.

Brändén: The rate limiting step of liver ADH is NADH dissociation. Since your analogue binds weaker than NADH one would expect that the dissociation rate is faster and thus the turnover number higher. The same situation has been shown by Plapp to occur when Lys 228 is modified.

Werber: Isn't one of the differences between the tetracoordinated and pentacoordinated Zn^{++} mechanism in liver ADH reflected also in the state of ionization of His 51? Does one have an imidazolium ion in the tetracoordinated - and a neutral imidazole group in the pentacoordinated mechanism?

Brändén: No. The difference is only in positioning the hydrogen atom on the Nδ atom of the imidazole ring of His 51 in the case of tetracoordinated zinc and on the Nε atom in the other case.

Gutfreund: One of the important differences between the reactions of pig heart lactate dehydrogenase and horse liver alcohol dehydrogenase is the point in time when the proton, which is a product, is released.

In LDH we have shown that the proton abstracted from lactate by histidine 195 is released after hydride transfer, while in LADH it is released during binary and ternary complex formation. How do you correlate proton release during enzyme-substrate complex formation with the structure changes you have described.

Brändén: One of the major structural differences between apoenzyme and ternary complexes with coenzyme and substrate involves helix α1 of the catalytic domain comprising residues 46-55. This helix is fixed at one end in approximately the same position in both structures by the zinc ligand

Cys 46. The other end which is in the region between the two domains has moved closer to the catalytic domain in the ternary complex. This movement involves Ser 48 and His 51 which form a hydrogen bond system with the zinc bound water molecule in the apoenzyme structure. We have previously suggested that the proton is released from the zinc bound water through this hydrogen bond system. At the present resolution we can only say that the conformational change should effect the pK of this system but can give no molecular details. It is also apparent that the introduction of the positive charge of NAD^+ effects this pK since ternary complexes of NADH and aldehyde exhibit very similar structural changes as those of NAD^+ and alcohol.

Veeger: Is there enough space near the pyridine ring positions 2 or 6 to make σ-bond formation possible, as Peter Hemmerich proposes, prior to H^+ transfer?

Brändén: The orientation of substrate and coenzyme is such that a covalent bond between the substrate and C2 or C6 of the nicotinamide ring cannot be formed. The distance is far too large.

Luisi: On the basis of the structural differences between apo and holoenzyme, can it now be considered as established that the two subunits in the holoenzyme are conformationally not equivalent?

Brändén: The distance between the structural zinc and the catalytic zinc is 20.2 Å in both subunits in the apoenzyme. In the holoenzyme these distances are 19.7 Å in subunit 1 and 18.3 Å in subunit 2. We believe that this difference of 1.4 Å reflects structural non-equivalence of the two subunits in the holoenzyme structure.

SUBSTRATE ORIENTATION IN THE ACTIVE SITE OF LIVER ALCOHOL DEHYDROGENASE

Hans Dutler
Laboratorium für Organische Chemie, ETH Zürich
CH-8092 Zürich, Switzerland

Studies on the specificity are of particular interest in the case of enzymes whose three-dimensional structure is known in detail. Under these circumstances it is possible to correlate the results obtained by the kinetic experiments with appropriate results available from the structural investigations. By such a correlation method we expect to be able to deduce the structure of the enzyme-substrate and enzyme-product complexes, and to explore the molecular motions taking place during reaction. This type of information is hardly obtained from structural investigations alone, since these complexes represent transient species which cannot normally be isolated and crystallized.

The specificity of the investigated enzyme must be broad enough to allow comparison of the reactivity of many different substrates. Although an enzyme normally is highly specific with respect to metabolites with which it is contacted in the living cell, it may show a low specificity with respect to synthetic substrates. Liver alcohol dehydrogenase represents a favourable case, since it catalyzes the reduction and oxidation of a large variety of non-physiological substrates. Among these substrates, alkylcyclohexanones are of particular interest, since their structure can be varied systematically and within very wide limits. By changing the size and position of the alkyl groups it is possible to search for interactions between enzyme and substrates. In this sense such a substrate in the active site can be compared with an imprisoned squid extending its tentacles in all directions around its body and thereby sensing the topography of the cage. In view of the interpretation of the kinetic results in terms of interactions it is very important that the substrate groups coming into contact with groups of the active site are all of the same kind. This is of course ideally realized when alkylcyclohexanones are used at the substrates.

The available structural information about liver alcohol dehydrogenase concerns the apoenzyme and the enzyme-NAD^+ complex (1-3). For this reason the correlation of kinetic and structural results is done on the basis of rates of product formation and of the structure of enzyme-NAD^+-product com-

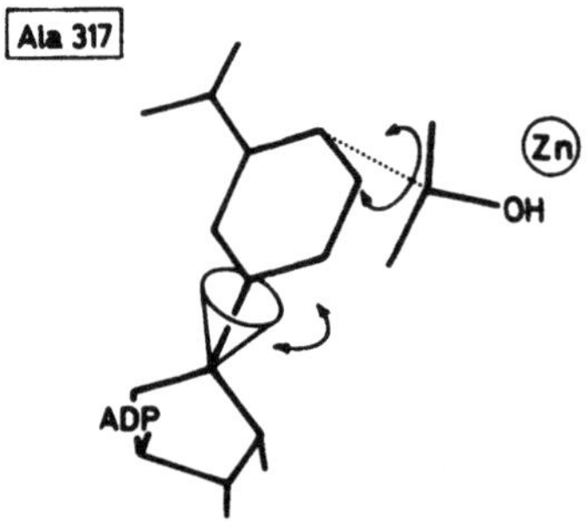

Fig. 1 Restricted changes in orientation of ribose, nicotinamide, and reacting carbon and hydroxyl of cyclohexanol.

plexes. Before carrying out these studies it is important, however, to characterize the interactions occuring between the active-site groups and coenzyme, and between coenzyme and substrate. Additional constraints which need to be considered are created by the conformational rigidity of the cyclohexane ring and the steric requirements of the transfer of the hydride anion from coenzyme to substrate. The information needed to characterize these constraints is readily available from structural and theoretical investigations.

The ADP-ribose part of the coenzyme is quite rigidly fixed in the active site and hence acts as an anchor for the nicotinamide moiety, which can change its orientation within the freedom of motion provided by the glycosyl linkage. This movement is intramolecularly restricted due to interactions between the pyridine ring and pyrophosphate, and depends on the puckering modes (4) of the ribose (Fig. 1). The structural analysis clearly shows that the nicotinamide ring appears in this orientation with the A-side facing the zinc atom and the carboxyamide group pointing towards Ala-317. This is in accordance with the known stereochemistry of the hydride transfer with liver alcohol dehydrogenase.

The steric requirements of the hydride transfer restrict the orientation of the reacting carbon and hydroxyl of the cyclohexanol relative to the pyridine ring. It is very likely that at the final stage of transfer the hydride anion leaves the pyridine ring perpendicularly and approaches the tetrahedraly coordinated reacting carbon of the cyclohexanol along a straight line. The resulting steric arrangement of the pyridine and the cyclohexanol ring allows for rotation about the axis determined by the hydride transfer (Fig. 1). Since the hydroxyl oxygen with great probability is liganded to the

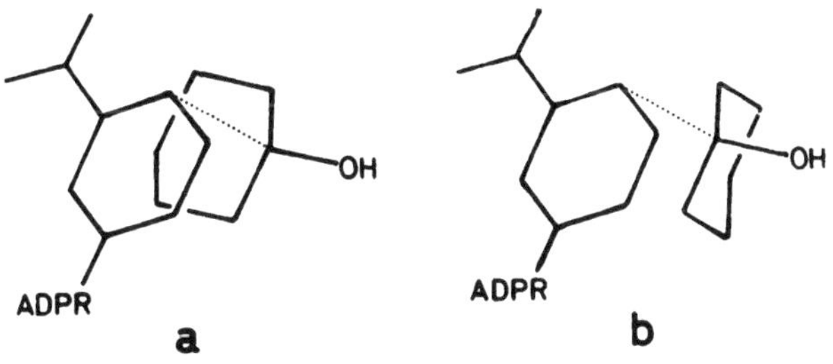

Fig. 2 Orientation of the cyclohexanol relative to the pyridine ring. (a) chair conformation for axial approach, (b) chair conformation for equatorial approach.

active-site zinc atom this rotation is considerably limited.

The orientation of the cyclohexanol ring is further limited by inherent constraints of the cyclohexanol ring, which is known to exist in a number of different conformations. Among these the two chair conformations are of lowest energy and therefore are expected to be the preferred ones; in one case the hydride anion approaches equatorially, in the other axially (Fig. 2). Inspection of models clearly shows that the interactions occuring between the cyclohexanol and nicotinamide ring can appreciably disfavour the conformation for axial approach. This steric repulsion depends on the torsion angle about the axis determined by the hydride transfer and is of considerable importance for orientations which allow the hydroxyl oxygen to be liganded to the zinc atom. Since the described constraints are little dependent on the position and size of the alkyl groups at the cyclohexanol ring, the orientation of the cyclohexanol relative to the nicotinamide ring and the orientation of the latter within the active site are expected to be the same for all products considered. Therefore, in order to deduce these geometric parameters and to collect information about the interactions occuring between the alkyl groups and groups of the active site, the cyclohexanol rings of all products can be superimposed.

The method applied to determine at which position of the cyclohexanol ring an alkyl group leads to favourable, unfavourable or no interactions with groups of the active site and with the nicotinamide ring consists in comparing the measured rates of product formation with those expected if no such interactions occured. When the rates are comparable it is concluded that no interactions occur. A significantly smaller observed rate is taken as an indication of unfavourable and a significantly larger observed rate as an indication of favourable interactions.

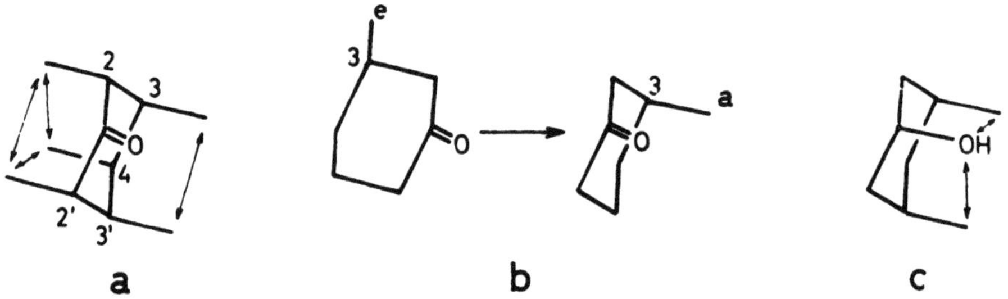

Fig. 3 1,3-Diaxial interactions. (a) interactions in cyclohexanone, (b) change from the more stable to the less stable conformation in the case of 3-alkylcyclohexanone, (c) additional interactions in cyclohexanol.

The following criteria were used to define the expected rates of product formation: The cyclohexanone substrates in which the alkyl group is axial when they are reduced in the required conformation are less stable than those in which it is equatorial (5). This relative instability is caused by 1,3-diaxial interaction involving alkyl and hydrogen (Fig. 3). Hence it can be argued that reduction of the cyclohexanones with axial alkyl groups requires a conformational change of the cyclohexane ring from the more stable to the less stable conformation (Fig. 3) and that this conformational change represents one of the factors responsible for the lower rate of product formation as compared to that observed for formation of the unsubstituted cyclohexanol. Since this slowing down of the rate of reduction depends on the energy difference between the two conformers, the same arguments imply that the rate decreases with increasing size of the alkyl group. If this conformational change occurs prior to substrate binding, the low rate is due to a low concentration of the substrate in the required conformation and, if it occurs after substrate binding, the conformational change represents a slow step added to the catalytic process. A second factor, which lowers the rate of reduction of the conformer with axial alkyl, constitutes the formation of the axial hydroxyl during hydride transfer. This process is accompanied by the formation of additional 1,3-diaxial interactions with the newly generated axial hydroxyl (Fig. 3) and thereby contributes appreciably to the slowing down of the hydride transfer. From these considerations it is evident that cyclohexanol products with equatorial alkyl are formed at rates comparable to that of unsubstituted cyclohexanol.

The kinetic experiments were done using racemic 2-, 3-, and 4-alkylcyclohexanones with the alkyl groups methyl, ethyl, isopropyl, and t-butyl. Reduction of these substrates was studied on a preparative scale and under conditions where the coenzyme NADH was continously regenerated by auxiliary enzymes (glucose-6-phosphate dehydrogenase and aldolase). Product formation was followed by gas chromatography of the α-methoxy-α-trifluoromethylphenylacetyl esters of the products (6,7) (Taylor, K.E. and Good, P., unpublished work).

Table Rate constants for the reduction of racemic alkylcyclohexanones at pH 8 and 25° C. Cyclohexanone is reduced at a rate of 24.4 s^{-1}. Decreasing rates are framed by thin lines and constant rates by heavy lines.

Position	a				e			
	Me	Et	i-P	t-B	Me	Et	i-P	t-B
				s^{-1}				
2	0	0.01	0	0	0	0	0	0
2'	1.3	0.8	0	0	0	0	0	0
3	0.18	0	0	0	0.1	0	0	0
3'	0.18	<0.1	<0.05	<0.05	12.9	13.3	12.0	7.0
4	10.6	10.6	10.7	-	12.8	13.0	12.0	-

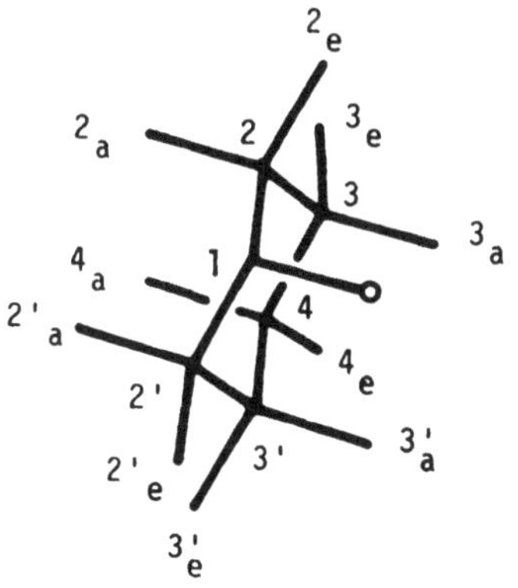

Fig. 4 Positions of alkyl groups in cyclohexanol.

The results which are summarized in the Table, show that the rate for the products with alkyl groups in the positions $2_a'$, $3_a'$, $3_e'$, 3_a and 4_e (for the nomenclature compare Fig. 4) obey the rules outlined for product formation with equatorial approach of the hydride anion. The finding that no alkylcyclohexanols were formed in which alkyl groups appear in positions 2_a, 2_e, $2_e'$, and 3_e allows to conclude that alkyl groups in these 4 positions create severe unfavourable interactions with groups of the active site or with the nicotinamide moiety of the coenzyme. In one case only, in position 4_a, alkyl groups seem to be strongly stabilized by favourable interactions (8).

a b c

Fig. 5 Products of the bridged carbocyclic ketones. (a) Adamantanol, (b) bicyclo[3.3.1]nonan-2-ol, (c) bicyclo[3.2.1]octan-2-ol.

These kinetic results are very convincingly confirmed by the behaviour of the completely rigid ketones adamantanone, bicyclo[3.3.1]nonan-2-one, and bicyclo[3.2.1]octan-2-one (Fig. 5). Adamantanone is very slowly reduced as is expected for a substrate, which can be considered as a highly rigid cyclohexanone derivative in which hydride transfer must occur axially. In strong contrast, the two bridged bicyclic ketones are reduced at rates about a third of that of cyclohexanone. Both these substrates are reduced because they avoid the unfavourable interaction at the position corresponding to 2_a of the cyclohexanol ring in the conformation for equatorial hydride transfer.

The kinetically obtained information was then used in close co-operation with Dr. C.I. Bränden, Agricultural College of Sweden, Uppsala, to identify the groups involved on the side of the active-site pocket (9) (Dutler, H. and Bränden, C.-I. to be published). For this purpose the active-site pocket including the nicotinamide moiety of the coenzyme was built with Kendrew-type models on the basis of the information on the three-dimensional structure of the enzyme-NAD^+ complex. Models of the alkylcyclohexanols formed as products were placed into the active-site pocket such that their hydroxyl oxygen atom appears in the position where the water oxygen atom was observed in the apoenzyme. Within the freedom of motion allowed by the steric constraints discussed above the alkylcyclohexanol models were allowed to change their orien-

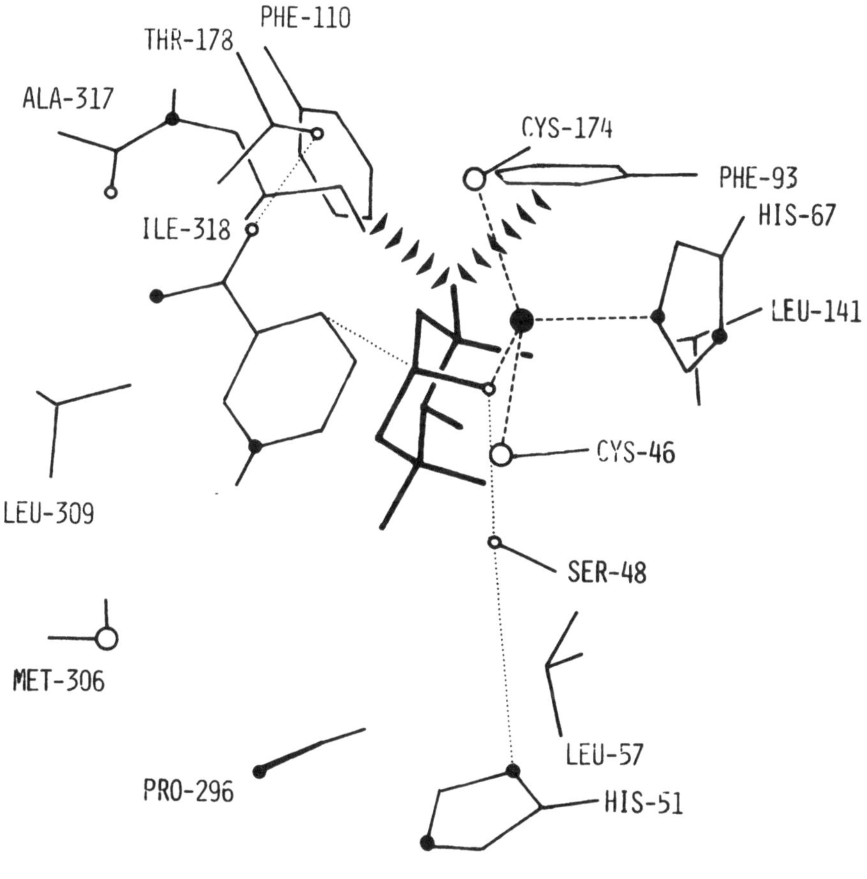

Fig. 6 Orientation of the 3-alkylcyclohexanols in the active site. The unfavourable interations are indicated by arrows. (●) Zinc, (○) sulfur, (•) nitrogen, (o) oxygen, (------) bond to the ligands of zinc, (········) hydrogen bridges and path of hydride transfer.

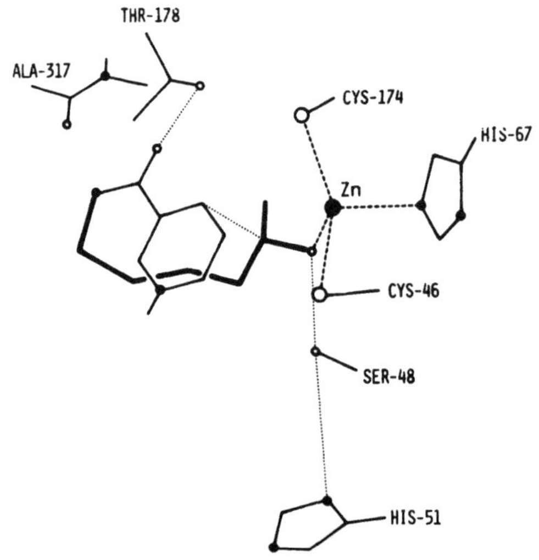

Fig. 7 Orientation of the reactive NAD^+-substrate covalent compound in the reduced form.

tation. This fitting process was carried on until full agreement was achieved between the interactions observed in the model and those determined kinetically. The resulting orientation and the interactions are shown in Fig. 6. The deduced orientation of the products is in excellent agreement with the finding that the NAD^+-substrate covalent compound shown in Fig. 7 is reactive (10) and looses its reactivity when the aliphatic chain is shortened.

On the basis of the deduced orientation of the considered products in the active site it is possible to predict the molecular motions taking place during reaction. According to the well documented view on the steric arrangement of the reactive centers for nucleophilic addition to a carbonyl group (11), the hydride anion, which is leaving the tetrahedraly coordinated reacting carbon of the dihydropyridine, approaches the carbonyl carbon at an angle of ca. 107° with the C-O bond. During reaction the cyclohexane ring bends away until its reacting carbon becomes tetrahedraly coordinated. When it is assumed that the oxygen as well as the cyclohexane ring should undergo a minimum change in position, the approaching hydride anion together with the (dihydro)pyridine ring has to give away. The molecular motions involved in this process are visualized in Fig. 8.

There exists convincing evidence that during reaction the cyclohexane ring is very little changing its orientation, if

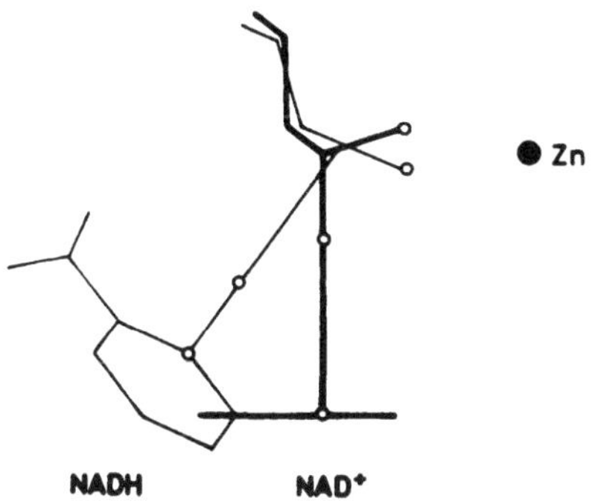

Fig. 8 Schematic drawing of the molecular motions taking place during reaction. View along the C_2-C_2' axis of cyclohexanol.

at all. The kinetic results, which are concerned with the enzyme-substrate complexes or the transition states, led to an excellent agreement with the results from model building, which is concerned with the enzyme-product complexes. This indicates that during reaction the critical interactions between the alkyl groups and the active site are very little changed. Furthermore the good reactivity for highly rigid steroid substrates of the steroid-active isozyme, which differs only by 6 aminoacids from the normal ethanol-active isozyme (12,13), demonstrates that the reaction can definitively proceed without appreciable movement on the side of the substrate.

From considerations concerned with the steric constraints it is clear that the glycosyl linkage provides enough freedom of motion for the (dihydro)pyridine ring to change its position. The movements, which this ring must undergo in order to satisfy the steric requirements are: rotation about the C'_4-C'_1 axis of ribose and about the glycosyl linkage, and change of the tilt of the (dihydro)pyridine ring relative to the cyclohexane ring. In this sense the (dihydro)pyridine ring acts like a windshield wiper. Also the carboxyamide group may take part in these movements. According to the model it seems possible that in the dihydropyridine it is out of plane and forms a hydrogen bond to the main chain carbonyl oxygen of Ala-317 and in the pyridine ring it is in plane and forms a hydrogen bond to the side chain of Thr-178.

References

1. Eklund, H., Nordström, B., Zeppezauer, E., Söderlund, G., Ohlsson, I., Boiwe, T., Döderberg, B.-O., Tapia, O., Bränden, C.-I., Akeson, A.: Three-dimensional Structure of Horse Liver Alcohol Dehydrogenase at 2.4 A Resolution. J. Mol. Biol. 102, 27-59 (1976).
2. Bränden, C.-I., Jörnvall, H., Eklund, H., Furugren, B.: Alcohol Dehydrogenases. The Enzymes (P.D. Boyer, ed.) 3rd edn. Vol 11 pp 93-190 (1975) Academic Press, New York.
3. Abdallah, D., Biellmann, J.-F., Nordström, B., Bränden, C.-I.: The Conformation of Adenosine Diphosphoribose and 8-BrAdenosine Diphosphoribose when Bound to Liver Alcohol Dehydrogenase. Eur. J. Biochem. 50, 475-481 (1975).
4. Sundaralingem, M.: Structure and Conformation of Nucleosides and Nucleotides and Their Analogs as Determined by X-Ray Diffraction, in "Chemistry, Biology and Clinical Uses of Nucleoside Analogs", Annals of the New York Academy of Sciences, edit. Block, A., 255, 1-42 (1975).
5. Eliel, E.L. (1962) Stereochemistry of Carbon Compounds, McGraw-Hill, New York.
6. Helmchen-Zeier, R.E.: Ueber die enzymatische Reduktion von 2-, 3- und 4-Alkylcyclohexanonen. Dissertation ETH 4991 (1973).
7. Dale, J.A., Dull, D.L., and Mosher, H.S.: α-Methoxy-α-trifluoro-methylphenylacetic Acid, a Versatile Reagent to the Determination of Enantiomeric Composition of Alcohols and Amines. J. Org. Chem. 34, 2543-2549 (1969).
8. Dutler, H.: Structure-Activity Relationship as Studied with an Enzyme of Known Structure. The Liver-Alcohol-Dehydrogenase-Catalyzed Reduction of Alicyclic Ketones, in "Structure-Activity Relationships in Chemoreception" (Benz, G. Ed.) p. 65-73, Information Retrieval, London (1976).
9. Dutler, H.: Probing the Active Site of Liver Alcohol Dehydrogenase. Biochem. Soc. Transactions. In Press (1977)
10. Karabatsos, G.J., Miedema, J.A., Nunez, V.C., Stamoudis, V.C., May, M., and Kroon, D.: Synthesis of NAD^+-substrate Covalent Compounds and their Reactivity with Alcohol Dehydrogenase. Abstracts, 173rd ACS Meeting, New Orleans (1977).
11. Bürgi, H.-B.: Zur Beziehung zwischen Struktur und Energie. Bestimmung der Stereochemie von Reaktionswegen aus Kristallstrukturdaten. Angew. Chem. 87, 461-475 (1975).

12. Reynier, M., Theorell, H., Sjövall, J.: Studies on the Stereospecificity of Liver Alcohol Dehydrogenase for 3 - Hydroxy-5 -steroids. Acta Chem. Scand. 23, 1130-1136 (1969).

13. Cronholm, T., Larsen, C., Sjövall, J., Theorell, H., Akeson, A.: Steroid Oxidoreductase Activity of Alcohol Dehydrogenases from Horse, Rat and Human Liver. Acta Chem. Scand. B29, 571-576 (1975).

Received May 23, 1977

DISCUSSION

Rossmann: It would appear, judging by the beautiful stereo slides we have seen and making some assumptions as to assignment of ligands, that the "hand" of the alcohol in LADH and L-lactate in LDH are opposite:

		substrate	
		L	D
nicotinamide	A	LDH, MDH	LADH
	B	GPD	

Does this mean that there are only four different types of dehydrogenases?

Dutler: If one considers only the stereospecificity with respect to the configuration at the reacting carbon of coenzyme and substrate, there exist indeed four possibilities, for all of which examples can be cited. This means that the stereospecificity is primarily determined by coenzyme-protein and substrate-protein interactions and not by coenzyme-substrate interactions. I must, however, stress the point that the stereospecificity with respect to the reactive carbon of the substrate of one enzyme is only comparable to that of another when the substrates are structurally similar. The criteria used to describe the chirality of a substrate (by L,D; S,R; Si,Re) are in no way related to the steric requirements of the corresponding active site. For this reason the stereospecificity of LDH is not comparable to that of LADH. It can only be stated that stereochemically the same hydrogen (H_{Re}) is transferred:

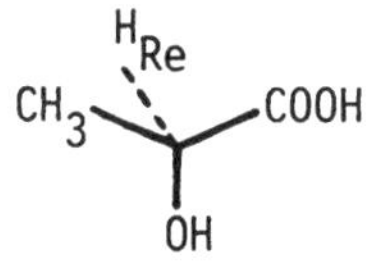

L-lactic acid

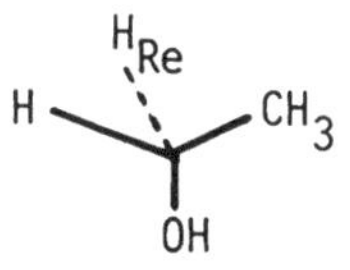

ethanol

Engel: Could you tell us whether the rates you showed are extrapolated maximum rates for each substrate or whether they are measured at constant sub-saturating concentrations?

Dutler: A maximum rate of substrate reduction can of course be determined by the usual methods. However, since two products are formed from the ketones under consideration, these maximum rates represent sums of the rates of formation of these two products. What we really want to determine are the maximum rates of formation of the individual products. Due to the competition arising from the substrate forming the two different enzyme-substrate complexes these values are much more difficult to obtain. This kind of extensive analysis has not yet been carried out.

Südi: I just don't see it quite clearly what role "Transition State Structure" plays in your considerations. As far as I could see all your detailed considerations referred to
(1) The structure of two stable ternary complexes,
(2) The "Dynamic Stereochemistry" of Dunitz & Bürgi.

Do you need anything else in addition, except a frequency for the interconversion of these two relatively stable structures?

Dutler: The measured rate can either be an expression of the rate of the catalytic step or of the dissociation constant of the enzyme-substrate complex. In the first case it is the structure of the transition state and in the second case it is the structure of the enzyme-substrate complex which plays the most important role in determining the overall rate of the reaction. Preliminary experiments indicate that for the products which require the alkyl group to be equatorial the rate is controlled by the catalytic step and for those products which require the alkyl group to be axial it is an expression of the dissociation constant of the enzyme-substrate complex. Therefore we are dealing with a comparison of information about the structure of the enzyme-product complex, as studied in the model, with that of the transition state or the enzyme-substrate complex.

EQUILIBRIUM STUDIES AND KINETICS OF REACTIVATION, REFOLDING AND REASSOCIATION OF LACTIC DEHYDROGENASE AND GLYCERALDEHYDE-3-PHOSPHATE DEHYDROGENASE

Rainer Jaenicke and Rainer Rudolph

Institut für Biophysik und Physikalische Biochemie , FBB , Universität Regensburg, D 8400 Regensburg , West Germany

The folding of the polypeptide chain as it is synthesized on the ribosome occurs spontaneously without the need of a matrix or any additional information beyond the specific amino acid sequence of the given protein and its aqueous environment. This has been verified by detailed experimental studies on the refolding and reactivation of enzymes after denaturation in various solvent media (1-3). The native structure of a given protein in its natural environment is determined by a kinetically accessible minimum of the free energy (2). In the case of oligomeric proteins energetically favourable interchain, in addition to the intrachain interactions, provide the minimum on the energy surface leading to a certain well-defined state of association. The particular biological significance of the assembly may depend on whether or not the subunits of the specific enzyme contain sufficient information to provide a catalytically active entity. There exists no general answer to this question which has been the subject of equilibrium studies reviewed previously (4). The two enzymes, lactic dehydrogenase (LDH) and glyceraldehyde-3-phosphate dehydrogenase (GAPDH) are of special interest in the given context: Both show a close structural relationship; however, LDH represents a non-cooperative tetramer consisting of kinetically independent subunits, while GAPDH is characterized by strong positive or negative cooperativity and half-site activity. Indirect evidence from a variety of experimental techniques indicates that the monomeric enzymes are catalytically inactive (4-8). For LDH there exist experimental results which contradict this finding (9,10).

The following equilibrium and kinetic studies deal with

I. Characterization of the initial and final states in the dissociation-reassociation cycle ;

II. Correlation of dissociation-denaturation-deactivation, and reassociation-renaturation-reactivation ;

III. Conditions providing maximum reactivation ;

IV. Kinetics of reactivation, renaturation, and reassociation ;

V. Kinetic constraints in the thermodynamically determined process of reconstitution.

EXPERIMENTAL

Pure homologous isoenzymes of LDH from pig heart and skeletal muscle (LDH-H_4 and LDH-M_4), NADH, NAD^+, and substrates were obtained from Boehringer (Mannheim). GAPDH from yeast (Y-GAPDH) was a gift of Dr.H.Durchschlag. The enzymes represent coenzyme free apoenzymes. Dithiothreitol was supplied by Calbiochem (Luzern), ultra-pure guanidine.HCl and urea by Schwarz-Mann (New York), bovine serum albumin and trypsin by Serva (Heidelberg). All other reagents were A-grade substances from Merck (Darmstadt). Quartz bidistilled water was used throughout.
Stock solutions of the enzymes were prepared by dialysis (4°C) using 0.2M phosphate buffer pH 7.6. All solvents contained 1-10mM EDTA plus 0.1-10mM dithiothreitol.
Equilibrium measurements were performed in 0.1M phosphate (20°C). Dissociation (deactivation) of the native enzymes was achieved by dilution with 1M glycine/H_3PO_4 pH 2.3, 0.1M citrate pH 2.3, or 0.2M phosphate pH 2.3 or 7.6 in the presence of 6M guanidine·HCl or urea. The time of denaturation at 20°C varied from 5 to 120 min. For reassociation (reactivation) the aforementioned denaturation mixtures were diluted with 0.2M phosphate or 0.05M triethanolamine, final pH 7.6.
The kinetics of reactivation or renaturation were analyzed by taking aliquots at defined times or by direct measurement of protein fluorescence (F_{rel}; λ_{exc}=275nm, λ_{em}=340nm) (Hitachi-Perkin-Elmer). Further characterization made use of circular dichroism (CD) (Roussel-Jouan), sedimentation analysis (Beckman), and determinations of the heat stability of the enzymes. For details see (5-7,11).

RESULTS AND DISCUSSION

The application of reactivation and refolding kinetics to the problem of subunit activity in oligomeric enzymes is based on the assumption that dissociation, unfolding, and deactivation are reversible processes under the given experimental conditions. Reversibility was demonstrated by denaturation and renaturation under varying concentrations of the denaturants (H^+, guanidine·HCl, urea). The final product of renaturation and the enzyme in its initial state were compared directly using available physicochemical or enzymological criteria. A comparison of the kinetics of reactivation after dissociation in various denaturants showed that no significant differences can be detected regarding the rates and final products of renaturation.

I. Characterization of the Initial and Final States in the Dissociation-Reassociation Cycle

Data from equilibrium measurements characterizing the enzymes in their initial and final tetrameric state and under dissociating conditions are summarized in Table 1.

Table 1 Characterization of lactic dehydrogenase and glyceraldehyde-3-phosphate dehydrogenase in their native, dissociated, and refolded states

Native (refolded) at pH 7.6	LDH-H_4	LDH-M_4	Y-GAPDH
$s_{20,w}$ (S)	7.38 (7.38)	7.60 (7.60)	7.50 (7.50)
Molecular weight $x10^{-3}$	142 (142)	140 (140)	143 (144)
Specific activity (IU/mg)	405 (388)	639 (655)	176 (177)
F_{rel} (%)	100 (94)	100 (104)	100 (99)
Fluorescence maximum (nm)	339 (339)	339 (339)	337 (337)
CD , $\Delta\epsilon_{282}$ ($M^{-1}.cm^{-1}$)	38.2 (37.6)	57.3 (58.2)	-62.3 (-63.3)
Yield of reactivation (%)	57	83	73
Dissociated at pH 2.3 , A_{sp}=0			
$s_{20,w}$ (S)	2	3	2
Molecular weight $x10^{-3}$	37.5	36.5	37.0
F_{rel} (%)	39	34	39
Fluorescence maximum (nm)	336	338	336
CD , $\Delta\epsilon_{282}$ ($M^{-1}\cdot cm^{-1}$)	3.1	7.0	0

All data in Table 1 belong to homogeneous systems, i.e. monodisperse tetramers in their native and renatured state, and monomers under dissociating conditions. To separate out inactive high aggregates, and to characterize the reactivated enzymes, solutions were concentrated and fractionated by ultrafiltration, centrifugation, millipore filtration, and gelfiltration. Comparison of the reactivated tetramers with the native enzyme made use of (a) repeated denaturation-renaturation, and (b) direct determination of the enzymological and physico-chemical properties of the products of reactivation and the native enzyme. The tetrameric species shows 100 ± 4 % and the higher aggregates 0 % of the activity of the native enzyme. Similarly the spectral and hydrodynamic properties (Table 1), and the heat stability coincide for both the renatured (N^*) and the native enzyme (N) within the limits of error (< 5%) ; cf. (7,11,12). On the other hand, the spectral characteristics of the inactive aggregates (e.g. after acid dissociation) resemble those of the monomeric fraction in the same denaturant. Obviously the aggregation process causes the fixation of the enzymes in a denatured state. The different denaturants generate different final states of unfolding. At acid pH appreciable residual "helicity" is preserved while in guanidine the enzymes under consideration, like other proteins (1), may be considered to be unfolded to a high degree. Maximum unfolding seems to be provided by superimposing the effects of guanidine·HCl and low pH (13,14). The characteristics of the different states may be summarized as follows:

N Native LDH-H_4 and LDH-M_4 as well as Y-GAPDH in the absence of denaturants represent homogeneous tetramers even in the range of low enzyme concentrations applied in the enzymatic test and in reactivation experiments .

D Acid pH (pH 2.3), urea (6M), guanidine·HCl (6M), and urea or guanidine·HCl (6M) at pH$\approx$2 cause complete dissociation to inactive monomers characterized by an increasing degree of denaturation .

N^* Removal of the denaturants leads to recovery of enzymatic activity with parallel reconstitution of the native quaternary structure. After separation of inactive higher aggregates the final product of reactivation and the native tetramers turn out to be indistinguishable regarding all available biochemical and physico-chemical criteria ($N=N^*$) .

II. Correlation of Dissociation-Denaturation-Deactivation, and Reassociation-Renaturation-Reactivation : Equilibrium Measurements

The initial and final states during dissociation and reassociation are the active tetramers (N,N^*), characterized by specific spectral and hydrodynamic properties for the different enzymes, and the deactivated and (partially) denatured monomers (D_i), characterized by a lower fluorescence emission and CD. In the following experiments only the pH-dependent dissociation, denaturation, and deactivation will be considered. Other denaturants lead to comparable results (11,14).

As shown by ultracentrifugation, spectroscopy, and optical tests, dissociation ($\sim\Delta s$), conformational changes ($\sim\Delta F_{rel}$), and deactivation ($\sim\Delta A_{sp}$) show a close parallelism under equal conditions of the experiments (Fig.1). Varying the solvent, e.g. by applying chloride instead of phosphate (I=0.2), in the case of LDH-M_4 leads to a shift of the limiting pH of dissociation

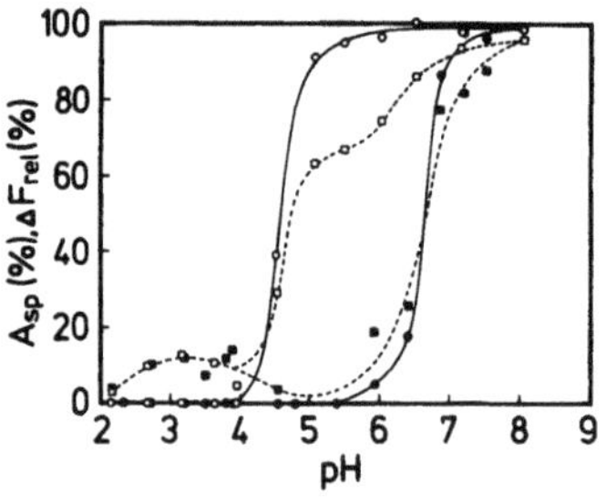

Fig.1 pH dependent deactivation-reactivation and denaturation-renaturation: LDH-H_4 in 0.1M phosphate. Incubation at c=0.21 µM, 20°C.
o deactivation, ● reactivation (A_{sp}),
□ denaturation, ■ renaturation (ΔF_{rel})

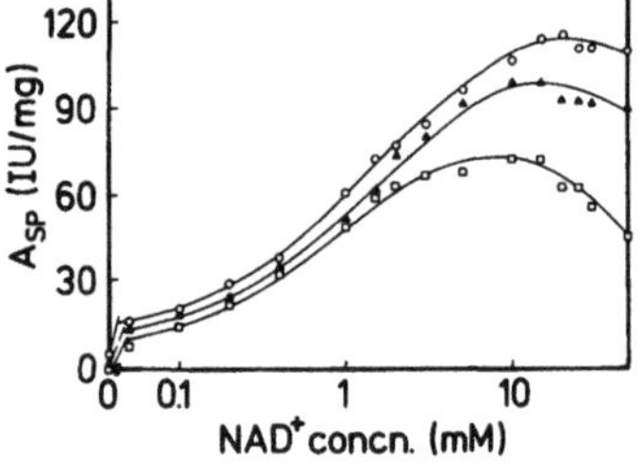

Fig.2 Effect of NAD^+ on the reactivation and stability of N^*,25°C: Y-GAPDH after denaturation in 0.1M citrate pH 2.3 + 6M guanidine·HCl. Reactivation at c=0.31µM in 0.05 M triethanolamine pH 7.6 after 24h (o), 48h (▲), 142h (□).

and deactivation of one pH unit (7). In the case of Y-GAPDH, A_{sp} and ΔF_{rel} run closely parallel while $s_{20,w}$ is anomalously increased due to the formation of enzymatically inactive high aggregates. The sharp transition within 0.5 pH units points to a high cooperativity of the three processes. Fluorescence emission clearly indicates that local conformational changes cause characteristic denaturation profiles which in part precede the deactivation and the gross changes regarding the secondary and quaternary structure of the enzyme.More pronounced discrepancies reported previously (6) are mainly caused by differences in the buffers applied (12). As taken from Fig.1 (LDH H_4) deactivation-reactivation or denaturation-renaturation as monitored by fluorescence may show hysteresis effects which corresponds to the observation that in a certain pH range reassociation irreversibly leads to inactive aggregates.

In summary, LDH and GAPDH show coincident pH dependent profiles of dissociation, denaturation, and deactivation. The reconstitution of the enzyme to its native structure shows the same parallelism of the three processes at a higher pH value, as proven for LDH-H_4. In the intermediate pH range polymerization to irreversibly denatured higher aggregates takes place; the tetrameric enzyme is metastable in this pH range, while the dissociated monomers show rapid aggregation.

III. Conditions Providing Maximum Reactivation

The extent of reactivation depends in a critical way on the conditions of the solution in the deactivation and reactivation processes. As pointed out (Fig.1) formation of inactive aggregates competes with the process of reconstitution. Therefore, kinetic experiments should be performed under conditions characterized by equilibrium constants of deactivation and reactivation far from 1, i.e. under essentially irreversible conditions . For optimum conditions of deactivation and reactivation see Experimental. Incubation in guanidine·HCl or urea at pH 7.6 leads to lower reactivation as compared to guanidine or urea at pH 2.3. For GAPDH,citric acid pH 2.3 was used to stabilize the enzyme in guanidine·HCl over longer periods of incubation; for LDH, glycine/H_3PO_4 served the same purpose.
As expected from earlier hybridization experiments with LDH-H_4 and LDH-M_4 at high dilution (15) or from dissociation studies with Y-GAPDH (16), reactivation is promoted by raising the temperature. Since reactivation of GAPDH under optimum conditions proceeds fast, 15°C were applied to slow down the rate for the determination of the reaction order (cf.Fig.3).

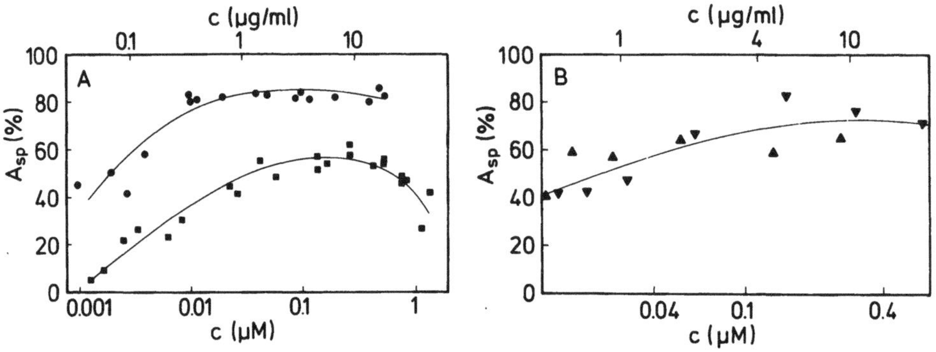

Fig.3. Effect of enzyme concentration on the extent of reactivation.
A. Reactivation of LDH-H_4 (■) and LDH-M_4 (●) at 20°C after deactivation in 1M glycine/H_3PO_4 pH 2.3. Reactivation in 0.2M phosphate pH 7.6.
B. Reactivation of Y-GAPDH at 15°C (▲) and 25°C (▼) after deactivation in 0.1M citrate pH 2.3 + 6M guanidine·HCl. Reactivation in 0.05M triethanolamine + 10mM NAD^+.

Addition of the coenzyme (NADH or NAD^+) in saturating concentrations leads to a significant increase of the yield of reactivation in the case of LDH-H_4 and Y-GAPDH (Fig.2), while LDH-M_4 does not show this effect. Regarding the concentration of the enzyme optimum yields of reactivation are observed at c=0.1-0.5µM (Fig.3). The decrease at low concentrations may be caused by c-dependent deactivation (5,7) and/or insufficient time to reactivate the enzyme due to the higher than first order rate equation (see IV). At high concentrations aggregation obviously competes with refolding and tetramer formation. The maximum yield does not exceed 85% for the following possible reasons: (a) instability at low c, and kinetically determined competition of reactivation and aggregation at high c, or (b) instability of the denatured and dissociated enzymes.

In summary, there is no general set of conditions available which would guarantee high yields of reactivation. Occasionally the coenzyme or substrates lead to an increase of the yield; however, side reactions caused by some kind of chemical modification may compete with this optimizing effect. Conditions far from equilibrium generally lead to higher yields as compared to equilibrium conditions. Apparently kinetic constraints become important in this connection (see V).

IV. Kinetics of Reactivation, Renaturation, and Reassociation

Dissociation and deactivation are first order processes which are strongly enhanced with increasing concentrations of the denaturant (17,18). The reconstitution of the tetramer must consist of first order transconformation

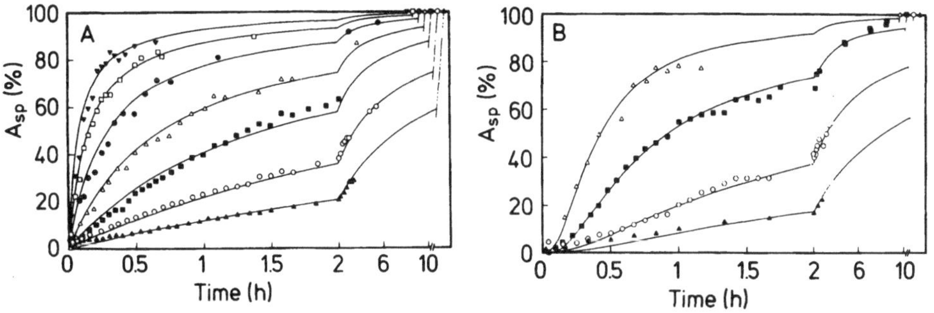

Fig.4 Kinetics of reactivation of LDH after deactivation in 1M glycine/ H_3PO_4 pH 2.3. Reactivation in 0.2M phosphate pH 7.6 (20°C).

A. LDH-M_4 . c in nM: 149 (▼), 79 (□), 41 (●), 17 (△), 8.1 (■), 2.2 (○), 0.96 (▲). Solid lines are calculated according to an irreversible bimolecular mechanism with k_2=23.4 ($mM^{-1}\cdot sec^{-1}$).

B. LDH-H_4 . c in nM: 367 (△), 91 (■), 19 (○), 7.0 (▲). Solid lines are calculated according to an irreversible mono-bimolecular mechanism with k_1=1.45·10^{-3} (sec^{-1}) and k_2= 5 ($mM^{-1}\cdot sec^{-1}$). The same rate constants hold for reactivation kinetics after deactivation in 0.2M phosphate pH 7.6 or pH 2.3 containing 6M urea or 6M guanidine·HCl.

plus second order reassociation processes. Analyses of the reaction order of the reactivation, therefore, allows one to discriminate between monomeric and oligomeric active species: If reactivation is of higher than first order the isolated subunit cannot be an active entity. Furthermore, it is important to compare the time and concentration dependence of the renaturation with the recovery of enzymatic activity in order to find out whether the formation of both the gross structure and the active center is determined by the same rate limiting steps.

Reactivation. The time course of reactivation of LDH-M_4 is characterized by a second order reaction (7). LDH-H_4 and Y-GAPDH show sigmoidal reactivation kinetics (11,12). At low enzyme concentrations the reaction order equals $n \approx 2$, while at high concentrations n approaches 1. This is easily interpretable by the fact that the rate of the second order association reaction is increased at high enzyme concentrations so that a first order refolding process as a precursor or consecutive reaction of dimer formation becomes rate determining (Fig.4) . From the foregoing equilibrium data it is evident that under the present conditions reactivation is essentially irreversible since readjusting the pH from pH 2.3 to pH 7.6 will shift the $D \rightarrow N^*$ equilibrium towards the tetramer. As a consequence some of the single elementary steps in the $D \rightarrow N^*$ transition must proceed in an irreversible fashion. Obviously the reassociation of LDH-M_4 can be described by a simple association process

$$nD \xrightarrow{k_2} N^{\star} \qquad (1)$$

For the sigmoidal traces the minimum scheme describing the data is an uni-bimolecular reaction sequence according to

$$2D \xrightarrow{k_1} 2D^{\star} \xrightarrow{k_2} N^{\star} \qquad (2)$$

with $D^{\star}$ as an hypothetical intermediate (11). Comparing this result with the reactivation kinetics after dissociation in guanidine·HCl or various other denaturants identical rate constants are obtained. Since the various states of denaturation (D_i) differ widely regarding their residual structure there must exist a common intermediate (D) in the specific reactivation reactions according to

$$nD_i \longrightarrow nD \xrightarrow{k_1} nD^{\star} \xrightarrow{k_2} N^{\star} \qquad (3)$$

Obviously the various extent of residual structure in the denatured state has no significant influence on the kinetics of reactivation in terms of rate determining nucleation steps.

Renaturation. As shown in Table 1 the intrinsic fluorescence of the (acid) denatured enzymes is decreased significantly compared to the native state. The kinetics of renaturation after readjusting to neutral pH are characterized by biphasic or triphasic profiles which include fast processes belonging to either transconformation reactions in the isolated subunits, or to the titration of groups in the vicinity of specific fluorophores.

The analysis of the slower portion of the recovery of native fluorescence

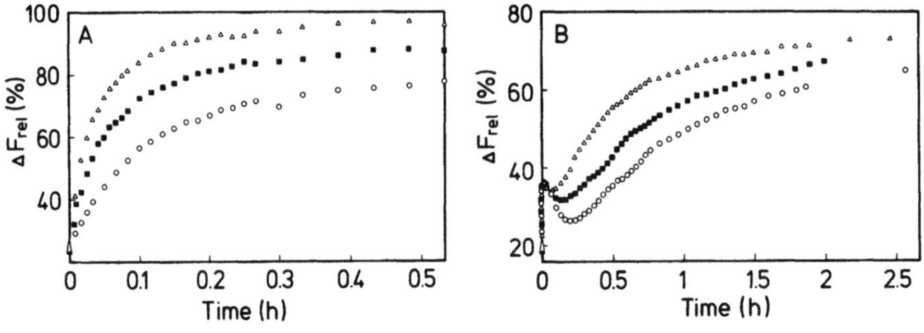

Fig.5 Fluorescence kinetics of LDH after deactivation in 1M glycine/H_3PO_4 pH 2.3. Renaturation in 0.2M phosphate pH 7.6 (20°C).
A. LDH-M_4 . c in nM : 305 (▲), 155 (■), 63 (o) .
B. LDH-H_4 . c in nM : 320 (▲), 138 (■), 66 (o) .

yields a reaction order of renaturation comparable to that of reactivation at comparable enzyme concentration (Fig.5). Both renaturation and reactivation run parallel within the limits of experimental error; both exhibit higher than first order kinetics, i.e. in both processes association is rate limiting under the given conditions of concentration and temperature. Obviously the changes in fluorescence in the present renaturation experiments reflect both isomerization and association.

Since reactivation and renaturation are determined by association processes isolated monomers of LDH and GAPDH cannot be active. Whether dimers or tetramers are the first to show enzymatic activity cannot be answered on the basis of the present experiments. As shown by hybridization experiments with LDH (4,10,15) the trimer can be excluded as an additional intermediate.

Regarding the kinetics of reassociation direct experimental evidence is still missing ; indirect evidence comes from the aforementioned fluorescence data.

In summarizing the kinetics of reactivation and renaturation three facts are of importance:

1. The denatured states (D_i) of the given enzymes refer to fully dissociated and deactivated molecules which are unfolded to a different extent.
2. In order to reduce the formation of inactive high aggregates the kinetics $D \rightarrow N^*$ refer to essentially irreversible processes.
3. In order to guarantee refolding and reassociation to occur independently low enzyme concentrations have to be used.

Under the given conditions the rate limiting step for reactivation and renaturation is an irreversible second order process which in certain cases contains an unimolecular (first order) precursor reaction reflecting some kind of transconformation. The fitting of all reactivation data with at maximum two constants k_1 and k_2 is possible assuming the monomeric subunits to be inactive. The latter result is not at all trivial, especially because within the tetramer the individual subunits of LDH are kinetically independent (19). For aldolase which (like LDH) is not allosterically regulated and tetrameric under non-denaturing conditions previous experiments have shown that artificially generated isolated monomers have at least partial activity (20-22).
The two rate constants k_1 and k_2 observed for the reactivation of LDH-H_4 hold for experiments using a variety of denaturants, suggesting the transformation of a common intermediate to a final state of renaturation to be rate limiting in the different pathways of reactivation.

V. Kinetic Constraints in the Thermodynamically Determined Process of Reconstitution

In the foregoing experiments there are three observations which may be considered to indicate that the product of refolding or reactivation represents the *kinetically accessible state of minimum free energy* .

1. Reactivation far from the tetramer ⇌ monomer equilibrium generates maximum yields due to the fact that in the transition range inactive aggregates are formed. Obviously strong medium-range interactions (especially interchain ones) compete with the nearest-neighbor interactions dominating the folding of the polypeptide chain *in vivo* which is based on specific nucleation sites (3). These sites may be combined in a different manner if the complete chain is present. On the other hand, conditions far from conditions *in vivo* may generate new nucleation centers or wrong interactions between the specific backbone nucleation sites predetermined by the amino acid sequence. The effects become increasingly important if the enzymes are incubated under destabilizing conditions for longer periods of time.

2. Although the effect of NAD^+, NADH, or substrates has not unequivocally been determined there seems to be an increase of the rate and yield of reactivation for $LDH-H_4$ and Y-GAPDH. This suggests nucleation to play a role in the process of reactivation, in accordance with earlier results (23).

3. As mentioned (Fig.5), the extent of reactivation depends in a characteristic way on the enzyme concentration: At $c > 0.1$ mg/ml an increasing amount of an inactive high molecular weight fraction is observed. Again kinetic constraints are responsible for this effect, since above a critical concentration the rate of the formation of inactive aggregates exceeds that of tetramerization.

It is evident that the aforementioned kinetic constraints depend on experimental conditions which are far from conditions *in vivo* . Therefore, it is doubtful whether they have biological implications.

Denaturation in various solvent media leads to distinct final states of structural disintegration. Reactivation from these different states of denaturation proceeds with identical kinetics. This implies that the residual structures in the denatured states have no bearing as nucleation centers in the process of reactivation to products indistinguishable from the native enzymes. Thermodynamic control seems to be sufficient for the attainment of the well-defined energy minimum of the native quaternary structure.

CONCLUSIONS

Whether or not subunits of enzymes which are oligomeric under non-denaturing conditions are active on their own seems to be an academic question because the isolated subunits might be in a different conformational state due to the solvation of "internal surfaces". Obviously the amino acid sequence provides the correct folding pathway to generate a specific tertiary structure with inter-subunit contact surfaces which then lead to association as a necessary requirement for the formation of the *active quaternary structure*. However, there is experimental evidence for oligomeric enzymes with intrinsic subunit activity.

Whether subunit interfaces serve as nucleation centers in the reshuffling of the active conformation of the subunit cannot be decided on the present level of accuracy in the kinetics of reactivation. These kinetics *in toto* may be described by a consecutive mechanism

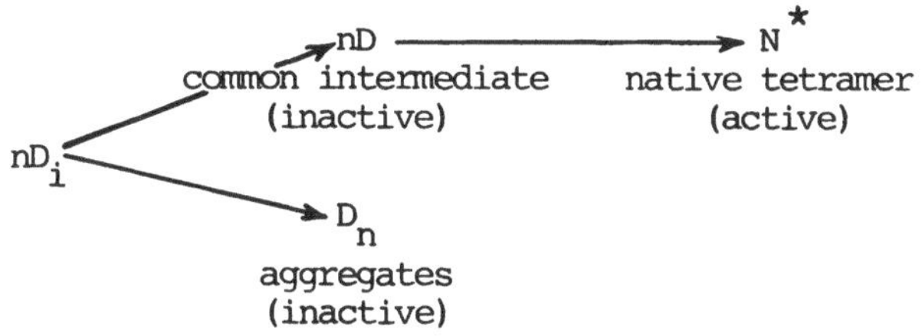

Under the given experimental conditions all transitions belong to systems far from their equilibrium state; all but N^* represent catalytically inactive species. Since the enzymes under concern are intrinsically tetrameric the equilibrium

$$\underset{\text{(active)}}{P_n} \rightleftharpoons \underset{\text{(inactive)}}{nP}$$

cannot be assumed to provide regulatory capacity.

Acknowledgements. This study was supported by grants of the Deutsche Forschungsgemeinschaft and the Fonds der Chemischen Industrie. Thanks are due to Dr.H.Durchschlag and Mrs. G.Durchschlag for their generous gift of Y-GAPDH, and to Dr.E.Westhof for critical discussions and help with computer programming. Excellent technical assistance of Mrs. Eugenie Zech and Miss Ingrid Heider is gratefully acknowledged.

REFERENCES

1. Tanford,C.: Protein denaturation. Advan.Prot.Chem. 23,121-282 (1968)
2. Wetlaufer,D.B., Ristow,S.: The acquisition of three-dimensional structure of proteins. Annu.Rev.Biochem. 42, 135-158 (1973)
3. Anfinsen,C.B., Scheraga,H.A.: Experimental and theoretical aspects of protein folding. Advan.Prot.Chem. 29, 205-300 (1975)
4. Jaenicke,R.: Quaternary structure and conformation of LDH and GAPDH. in Sund,H.(Ed.) "Pyridine Nucleotide-Dependent Dehydrogenases",Springer Verlag, Berlin,Heidelberg,New York 1970, p.71-90
5. Bartholmes,P., Durchschlag,H.,Jaenicke,R.: Molecular properties of LDH under conditions of the enzymatic test. Eur.J.Bioch.43,101-108 (1973)
6. Jaenicke,R.: Reassociation and reactivation of LDH from the unfolded subunits. Eur.J.Biochem. 46, 149-155 (1974)
7. Rudolph,R., Jaenicke,R.: Kinetics of reassociation and reactivation of pig-muscle LDH after acid dissociation. Eur.J.Biochem. 63,409-417(1976)
8. Chan,W.W.-C., Mosbach,K.: Effects of subunit interactions on the activity of LDH studied in immobilized enzyme systems.Biochemistry,15,4215-4222 (1976)
9. Cho,I.C., Swaisgood,H.:Surface-bound LDH.Biochim.Biophys.Acta 334,243-256 (1974)
10. Tenenbaum-Bayer,H., Levitzki,A.: Refolding of LDH subunits and their assembly to the functional tetramer.Biochim.Biophys.Acta 445,261 (1976)
11. Rudolph,R.,Heider,I.,Westhof,E.,Jaenicke,R.: Mechanism of refolding and reactivation of LDH-H_4 after dissociation in various solvents. In preparation
12. Jaenicke,R.,Engelhard,M.,Kraus,E.,Rudolph,R.: Reversible dissociation of glycolytic enzymes.Biochem.Soc.Trans.3, 1051-1054 (1975)
13. Tsong,T.Y.:An acid induced conformational transition of denatured cyt c in urea and guanidine·HCl. Biochemistry 14, 1542-1547 (1975)
14. Gerschitz,J.,Rudolph,R.,Jaenicke,R.: Kinetics of reactivation of aldolase after denaturation and dissociation. Biophys.Struct.Mechanism, in press
15. Jaenicke,R.,Koberstein,R.,Teuscher,B.:Molecular properties of LDH at low protein and high salt concentrations. Eur.J.Biochem.23,150-9 (1971)
16. Bartholmes,P.,Jaenicke,R.: Molecular properties of GAPDH in the presence of ATP and KCl. Biochem.Biophys.Res.Commun. 64, 485-492 (1975)
17. Anderson,S., Weber,G.: Reversible acid dissociation and hybridization of LDH. Arch.Biochem.Biophys. 116, 207-223 (1966)
18. Wassarman,P.M., Burgner,J.W.: Kinetics of unfolding of LDH in the presence of guanidine·HCl. J.Mol.Biol. 67, 537-542 (1972)
19. Holbrook,J.J., Liljas,A., Steindel,S.J., Rossmann,M.G.: Lactate dehydrogenase.In Boyer,P.D.(Ed.) "The Enzymes", Academic Press,New York(3) 1975. Vol.XI A. p.191-292.

20. Chan,W.W.-C., Mort,J.S., Chong, D.K.K., Macdonald,P.D.M.: Studies on protein subunits III. Kinetics. J.Biol.Chem. 248,2778-2784 (1973)

21. Rudolph,R., Engelhard,M., Jaenicke,R.: Kinetics of refolding and reactivation of aldolase. Eur.J.Biochem. 67, 455-462 (1976)

22. Rudolph,R., Westhof,E., Jaenicke,R.: Kinetic analysis of the reactivation of aldolase after denaturation and dissociation with guanidine·HCl. FEBS-Letters in press

23. Teipel,J.W., Koshland, D.E.: Kinetic aspects of conformational changes in proteins. Biochemistry 10, 792-805 (1971)

Received January 31, 1977

DISCUSSION

Gutfreund: I should like to make two comments which confirm Professor Jaenicke's conclusions. We have studied the temperature dependence of the renaturation of pig heart LDH. The percentage of recovery of active enzyme drops from about 80 % at 10° C to 10 % at 7° C. We concluded that refolding was controlled by kinetics rather than thermodynamics. To obtain active enzyme the conditions (environment) have to be the right ones for the correct pathway to be the fastest one. There is no reason to believe that the active enzyme must be in the lowest energy conformation.

We have carried out renaturation expriments in the presence of NADH and have followed reactivation and NADH binding (by nucleotide fluorescence) simultaneously. The rates of reactivation with and without NADH and the rates of formation of the NADH binding site appeared to be identical.

Werber: Hydrophobic interactions have been suggested to be involved in the refolding process. In connection with this, did you see any influence of the type of salt and its concentration on this process?

Jaenicke: Indications for hydrophobic bonds may be taken from the anomalous temperature dependence in the reactivation of LDH H_4 and M_4 (Jaenicke, 1974; Rudolph and Jaenicke, 1976). Examples for the expected influence of high salt concentrations have been reported for staphylococcal nuclease (H.F. Epstein et al., J.Mol.Biol. 60 (1971) 499-508) and TMV-protein (M.A. Lauffer and C. Stevens, Adv. Virus Research 13 (1968) 45). Effects of specific ions which are clearly demonstrated by the different stability of LDH-M_4 towards pH dependent dissociation and deactivation in the presence of phosphate and chloride (Rudolph and Jaenicke, 1976) may be caused by preferential solvation rather than hydrophobic effects. There do exist systematic studies regarding the influence of specific ions on the process of refolding: S.W. Schaffer et al., J. Biol. Chem. 250 (1975) 8483-8486.

Werber: Does the destruction of active sites precede dissociation of the oligomeric enzymes?

Jaenicke: There is a number of hints in this direction from various experiments, including our own data. Especially, there is good evidence that the fully deactivated state is reached under conditions where neither denaturation nor dissociation are complete. However, it is important in this context to compare the system under identical experimental conditions; the better these conditions correspond to each other the closer coincide deactivation and dissociation (cf. Rudolph and Jaenicke, 1976).

Werber: Could you comment on the reasons for having a lower degree of refolding at "high" protein concentration. Should not the refolding be faster the higher the subunit concentration is?

Jaenicke: "Refolding" represents an isomerization process. Therefore, one would expect it to be independent of subunit concentration. What we observe, and what has been quoted as an example of kinetic constraints of structure formation is the optimum curve of reactivation as a function of concentration. As mentioned this profile may be explained by the formation of "wrong aggregates". At high protein concentration this side reaction becomes fast compared to the process of refolding and reassociation to active tetramers.

Kiltz: Are you sure that the "aggregates" do not simply represent enzyme molecules characterized by a higher degree of denaturation? This could block the refolding process due to the fact that nucleation centres are lacking.

Jaenicke: Conformational analysis under conditions favoring the formation of N^* shows that the aggregates are fixed in the D state. No significant deviations or specific properties could be detected within the limits of error. Further evidence that the degree of denaturation has no detectable influence on the refolding process comes from the fact that e.g. in the case of Y-GAPDH maximum yields are obtained after maximum destruction of residual structure in the process of denaturation.

Kiltz: Do you have any idea with respect to the amount of residual tertiary structure in your denatured M_4 and H_4 LDH? Residual structure could be very important in the initiation of the refolding process.

Jaenicke: According to the operational criteria from ORD, CD and other spectroscopic techniques the residual secondary and tertiary structure varies in a wide range depending on the denaturing conditions. As I pointed out, however, this has no influence on the kinetics of reactivation and renaturation, even in case of the most powerful denaturants (cf. Gerschitz et al., Biophys. Struct. Mechanism 3 (1977)). To my knowledge there is no convincing evidence that essentially all residual structural elements indeed are broken down even in concentrated guanidine·HCl at low pH or high temperature. Therefore, we cannot exclude that nearest neighbor nucleation centres which were formed during biosynthesis remain unperturbed in the denaturation-renaturation cycle.

Trommer: The general disadvantage of dehydrogenases investigated so far is that dissociation to the single polypeptide chains can only be achieved under denaturating conditions. Therefore, the discrimination between isomerization and reassociation steps during the regain of enzyme activity becomes very difficult.

Tetrameric glucose dehydrogenase which has been studied in our laboratory by Dr. Pauly (H.E. Pauly & G. Pfleiderer (1975) Hoppe-Seyler's Z. Physiol. Chem. 356, 1613-1623) spontaneously and reversibly dissociates to monomers at pH 8.5. Dissociation is induced by the deprotonation of one lysyl ε-amino group per subunit and leads to inactive monomers showing unchanged optical rotatory dispersion and immunological reactivity (H.E. Pauly & G. Pfleiderer (1977) Hoppe-Seyler's Z. Physiol. Chem. 358, 287). This well-defined monomeric state may be obtained as an intermediate in the renaturation of glucose dehydrogenase from an unfolded state, too.

It would be worthwile to study the kinetics of renaturation and reassociation of this enzyme with regard to the correlation of subunit interactions and catalytic functions.

Jaenicke: This is a beautiful system indeed. However, I do not believe that there exist oligomeric enzymes which can be dissociated without structural changes. So far no protein-protein interaction has been shown to occur according to the hard sphere model. Therefore, I doubt whether anybody can escape the problem of discriminating isomerization and reassociation which seems to me even a useful tool in answering relevant biochemical questions.

Hollaway: You showed that NAD^+ is very effective in increasing the yield of GAPDH in renaturation experiments. We know that NAD^+ "tightens" the structure of the enzyme molecule, and other nucleotides, including NADH, ATP, AMP and ADP-ribose, loosen the structure. How does the latter class of ligands affect the yield?

Jaenicke: Since we have found that the effect of ATP on Y-GAPDH is rather complex (P. Bartholmes and R. Jaenicke, Biochem. Biophys. Res. Commun. 64 (1975) 485-492) we did not perform refolding experiments in the presence of ATP. With NADH a problem arises from a side reaction: At the low enzyme concentrations required for our kinetic experiments slow irreversible inactivation occurs which does not allow a proper determination of the final value of reactivation.

Mannervik: If we want to address ourselves to the problem of biosynthesis of proteins, I believe that although the information for folding of the polypeptide chain is certainly implicit in the sequence of amino acids, the conditions for folding in vivo may be very difficult to reproduce. For example ribosomes attached to membranes, such as the endoplasmic reticulum, may afford a hydrophobic and anisotropic environment which is very different from the conditions in an aqueous solution. It is also known that many proteins get carbohydrate substitutions, and I believe that the folding pathway may be steered by successive substitutions. Also addition of other prosthetic groups may serve the same purpose. Finally, some proteins may (like insulin) by synthesized in the form of a precursor, which has the proper structure for folding, and after acquisition of the correct three-dimensional structure be hydrolyzed by specific proteolysis.

My question is: Is anything known about the enzymes you have discussed which relates to the structure of the nascent polypeptide chains formed in the biosynthesis and the problems of possible posttranscriptional modifications?

Jaenicke: Folding of the polypeptide chain as it is synthesized and still attached to the ribosome or to membranes is an attractive but so far unproven idea. The same holds for a post-translational control of folding. In principle we cannot answer these questions on the basis of our in vitro experiments. Fortunately, the isozymes of LDH or Y-GAPDH (as far as we know today) are no glycoproteins nor do they originate from precursors. On the other hand W.W.-C. Chan and K. Mosbach have shown that refolding on matrices leads to an enzymatically active structure providing the appropriate intersubunit contacts to form native oligomers.

From this and the fact that N* equals N we believe that the specific intracellular conditions do not play an important role in the acquisition of the native three-dimensional structure for the enzymes under consideration.

Perham: I should like to make three points, one of which will include a question for Dr. Jaenicke. First, to take up Dr. Mannervik's comment, it seems that carbohydrate is enzymically added to certain proteins at asparagine residues in specific amino acid sequences. These self-same specific sequences are present in other proteins but no carbohydrate is added. It is likely,therefore,that carbohydrate will be attached only when the susceptible sequence presents itself to the enzyme in a suitable three-dimensional way. This implies that the carbohydrate is added after folding.

Secondly, it is important not to forget the biological implications. In vivo, proteins are synthesized from the N-terminus and sequential folding from the N-terminus has been envisaged as a possible mechanism. Most studies of reversible denaturation in vitro have been carried out with intact proteins and this is an important difference. However, although the published evidence is limited and it may be dangerous to generalize, it does appear that the bulk of the amino acid sequence is required to ensure formation of stable refolded structures. I hope Dr. Jaenicke will comment on this.

Thirdly, with regard to kinetic constraints in refolding, Cyrus Levinthal some 10 years ago pointed out that protein folding has to take place in a biologically useful time. It takes only seconds for catalytic activity to appear in biosynthesis. If a protein was free to search all possible conformations in a purely thermodynamic folding process it would take far too long. For that reason, Levinthal postulated the need for a kinetically determined folding pathway. As Dr. Jaenicke has clearly described, all the evidence points that way.

Jaenicke: With respect to the second point there is a wealth of indirect evidence which suggests "domains" to represent structural elements as intermediates in the process of refolding (cf. multiphasic profiles of denaturation and renaturation, N-terminal "arm" in LDH, refolding and reactivation e.g. of tryptophane synthetase after proteolytic cleavage and denaturation of the domains etc.). The best approach to check the sequential folding mechanism would be denaturation-renaturation studies on a highly specific enzyme synthesized from the C-terminal end. The case of ribonuclease is only a first step in this direction. Future improvements of the Merrifield technique or classical methods may provide us with a definite answer to this question.

Veeger: In studying the reconstitution of the dimeric flavoproteins glutathione reductase and lipoamide dehydrogenase from the monomeric apoenzymes, we observed similar behavior as you described, e.g. hardly any reactivation below 10° C, parallel regain of activity (> 90 %) and molecular physical parameters. However, in contrast to the enzyme as isolated, the FAD-prosthetic groups could be replaced from the active reconstituted enzymes by adding FMN. With time, however, this property of replacing FAD by FMN disappeared, but it took about 36 hrs. Thus it seems to me that these reconstituted proteins might contain domains, which show partial damage which is slowly repaired. My question to you in this connection is: Do the reconstituted enzymes show identical binding affinities for their coenzymes and/or substrates? Small differences do not need to show up in the enzyme-kinetic parameters.

Jaenicke: Experiments performed so far prove all physico-chemical and enzymological properties of N and N* to be identical. Binding constants for the coenzymes or substrates have not been determined yet.

Engel: My question follows on in a way from a previous one regarding residual structure in unfolded enzymes. I wonder whether you can hold out any hope to those of us who are struggling with enzymes like bovine glutamate dehydrogenase which do not seem to want to refold. You, and also Dr. Gutfreund, have given us evidence that coenzymes do not nucleate refolding of unfolded dehydrogenase subunits. On the other hand, one knows that, in the reverse direction, ligands often slow down unfolding. I wonder whether there is any evidence that the presence of coenzymes or other ligands in the denaturing medium may promote the retention of productive nucleation centres, in unfolded enzymes.

Jaenicke: To my knowledge there is no experimental proof that the unfolded enzymes still bind coenzymes or substrates in a specific way. On the contrary, denaturation can be measured by the release of these ligands. On the other hand, there is clear evidence that the enzymes pick up free ligands during renaturation. E.g. Y-GAPDH shows a significant decrease of its A_{280}/A_{260} ratio for N* compared to N, because N* takes up all trace amounts of NAD^+ which are released from the irreversibly denatured fraction of the enzyme.

One type of experiments which we are just about to start with, is the refolding of dehydrogenases after covalent attachment of coenzyme analogs like the bromoacetyl derivative of Dr. Woenckhaus.

With this approach we hope to be able to protect essential binding domains or nucleation centres in the enzymes.

STUDIES ON DEHYDROGENASES FROM HALOBACTERIUM OF THE DEAD SEA

H. Eisenberg, W. Leicht, M. Mevarech and M.M. Werber
Polymer Department, The Weizmann Institute of Science,
Rehovot, Israel

INTRODUCTION

Bacteria of the genus Halobacterium are obligate halophiles, i.e. they require for their growth an extracellular salt concentration higher than 2 M NaCl (1). In order to balance the osmotic pressure of their environment, these organisms have developed the ability to maintain a high intracellular salt concentration, which can reach up to 4 M KCl and 2 M NaCl (2). Such high salt concentrations can, in some cases, lead to the disruption of the native structure of proteins (3). Thus, survival of halophilic microorganisms was made possible by adaptation of the whole biochemical machinery of the cell to the extreme environmental conditions. Indeed, the proteins and enzymes from Halobacteria are active at multimolar concentrations of salt, which are also required to maintain their stability (4). Upon lowering the salt concentration, most of the halophilic enzymes are inactivated. The identification of the structural properties that confer salt-dependent stability and the study of structure-function relationships in halophilic proteins are the main objectives of our work.

We report here on some functional and structural properties of two dehydrogenases of Halobacterium of the Dead Sea: malate dehydrogenase (MDH) and glutamate dehydrogenase (GDH). At low salt concentrations, the abovementioned inactivation process is accompanied by large conformational changes and by dissociation of the enzymes into subunits. In the case of MDH, dissociation could be reversed and activity restored merely by increasing the salt concentration (5).

PURIFICATION AND MOLECULAR PROPERTIES OF HALOPHILIC DEHYDROGENASES

The lability of halophilic enzymes at low salt concentration imposes restrictions on the choice of purification techniques. We have therefore developed a purification procedure in which, at all steps, the salt concentration is kept high enough to prevent inactivation. This procedure,

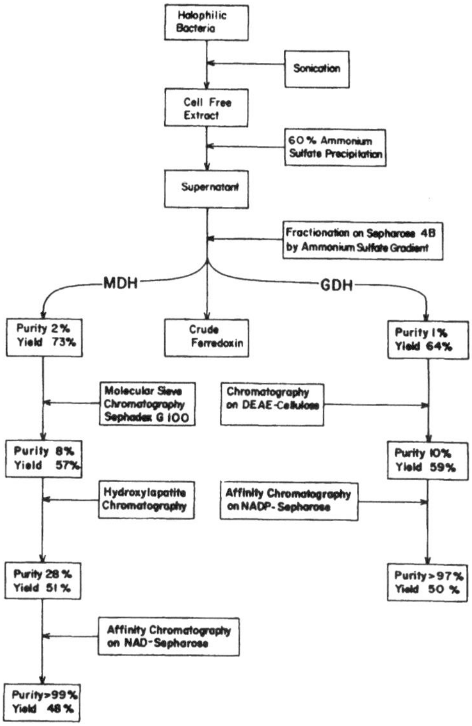

Fig. 1: General purification scheme of halophilic proteins.

which can be applied in part to the purification of other halophilic enzymes, is schematically presented in Figure 1; a 2 Fe-ferredoxin was also obtained in high yield and purity.

The first chromatographic step is based on the observation that halophilic proteins are adsorbed to Sepharose gels at ammonium sulfate concentrations higher than 60% saturation. The proteins can then be eluted by decreasing the concentration of the same salt (6). The second step - chromatography on DEAE-cellulose - involves adsorption as in the case of Sepharose and fractionation according to charge: the enzymes are adsorbed to DEAE-cellulose at high ammonium sulfate concentration prior to elution by a conventional increasing concentration gradient of sodium chloride. By this modification of ion-exchange chromatography inactivation of the enzymes, due to exposure

Table 1 Amino Acid Composition of MDH and GDH from *Halobacterium* of the Dead Sea

Amino Acid	Mole Percent		Amino Acid	Mole Percent	
	MDH	GDH		MDH	GDH
Lys	2.6	3.0	Val	8.9	8.6
His	2.3	0.7	Met	1.3	1.0
Arg	4.7	4.6	Ile	4.7	4.7
Asx	14.1	13.9	Leu	6.5	6.9
Glx	13.1	14.2	Tyr	2.6	3.7
Thr	5.0	6.0	Phe	2.6	3.0
Ser	5.5	5.3	Cys	0	*
Pro	4.4	4.1	Trp	0.8	*
Gly	10.7	10.0	Amide	8.4	10.4
Ala	10.2	11.2			

*Not determined

to low ionic strength, can he prevented. The final step in the purification procedure is affinity chromatography on immobilized NAD (for MDH) or NADP (for GDH). In this affinity gel the pyridine nucleotide is coupled to Sepharose through an hexamethylenediamine spacer which is attached to the C-8 position of the adenine ring (7).

The molecular weight of MDH was found to be 84,000 daltons, from sedimentation equilibrium, and that of its subunits about 39,000 daltons, from SDS gel electrophoresis (8). Accordingly, the enzyme is believed to be a dimer. The molecular weight of the GDH subunit was found to be approximately 52,000 daltons, from SDS gel electrophoresis. Both MDH and GDH displayed an absolute specificity for one type of coenzyme, NADH and NADPH respectively.

The amino acid composition as well as the amide content of halophilic MDH and GDH are shown in Table 1. In both enzymes, the most remarkable feature - which will be discussed below - is the high molar excess of acidic residues (Asp and Glu) over basic ones (Lys and Arg).

KINETIC AND STABILITY PROPERTIES

The dependence of the activity of MDH and GDH on the concentration of NaCl is presented in Figure 2. The optimal salt concentration for the activity

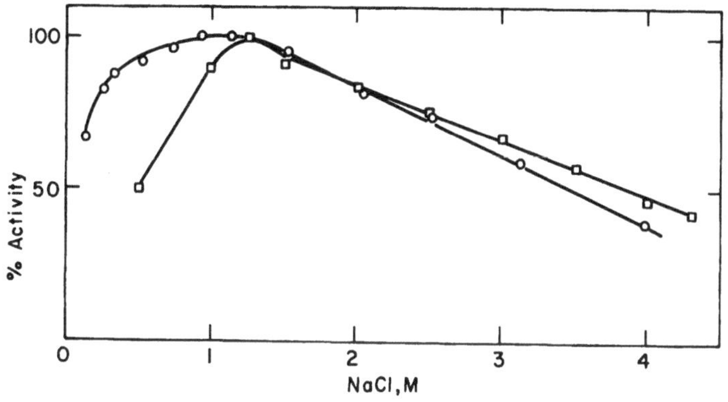

Fig. 2 NaCl concentration-activity profile of halophilic dehydrogenases. □-□, MDH. Assay conditions: 0.15 mM oxalo-acetate, 0.1 mM NADH, 0.01 M Na-phosphate, pH 7.3, T=23°C. ○——○, GDH. Assay conditions: 30 mM α-ketoglutarate, 0.12 mM NADPH, 0.1M NH_4Cl, 0.1M Tris, pH 7.8, T=23°C

of both enzymes was 1-1.2 M, but they maintained a high activity at higher salt concentration as well. The Na_2SO_4 concentration-activity profile of GDH has its optimum around 0.45 M, i.e. at an ionic strength of 1.35 M. The K_m values of GDH for NADPH were measured at the optimal activity concentrations of NaCl and Na_2SO_4 (Table 2). For comparison, the K_m value for bovine GDH is 25 μM and in GDH from other species it is in the range 25-200 μM (9). There seems thus to be very little, if any, dependence of the K_m value upon the nature of salt and its concentration. Neither is

Table 2 V_{max} and K_m Values of Halophilic GDH for NADPH at Various Salt Concentrations at pH 7.8, 30°

Type of Salt	NaCl	Na_2SO_4	
Salt concentration (M)	1.0	0.47	1.74
K_m (μM)	24	14	18
V_{max}	23.5	25.0	7.5

the K_m value of halophilic GDH different from that obtaining with GDH from

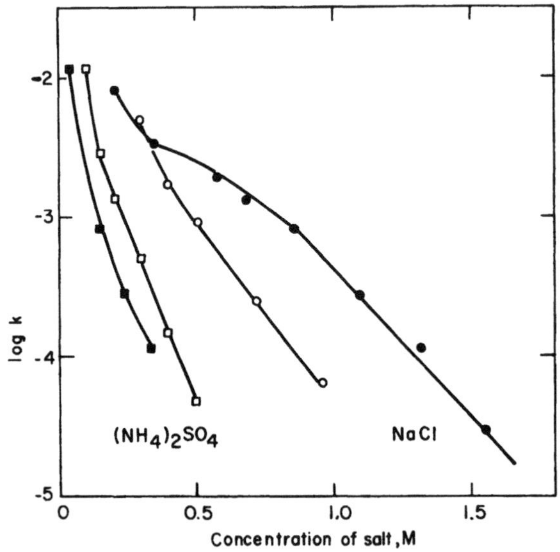

Fig. 3 Salt concentration dependence of the inactivation rate constants k(sec^{-1}) of halophilic dehydrogenases; ●——●, ■-■ , MDH at 30°C; ○——○, □-□ , GDH at 45°C.

other species. Therefore, most of the effect of changing the salt concentration (see Fig.2) on the activity is due to changes in the V_{max} parameter (Table 2).

The remarkable stability of halophilic enzymes at high salt concentrations has been previously reviewed (4). Thus, the "shelf-life" of MDH and GDH in 4-4.3 NaCl, at room temperature (20-25°), is almost infinite, since samples of both enzymes aged 3 years are still fully active. However, the stability is rapidly lost upon lowering the salt concentration. MDH was found to be more labile than GDH in this respect. Inactivation rates were shown to be first-order with respect to enzyme concentration. The rate constants of inactivation of the two enzymes at various NaCl and $(NH_4)_2SO_4$ concentrations are shown in Figure 3. (For the determination of k, enzymes were incubated at various salt concentrations and aliquots assayed at various times under standard conditions (cf. Fig.2).)

In contrast with the activity profiles, which display an optimum salt concentration, the stability monotonously decreases with salt concentration. However, the inactivation rate constant k is a complex function of salt concentration and type. Below an ionic strength of 0.2 M the salt dependence of the inactivation rate constants may be treated as a simple

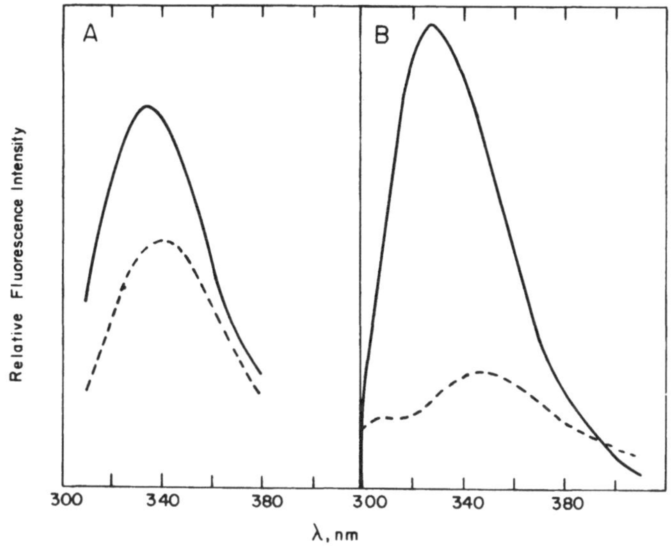

Fig. 4 Intrinsic fluorescence emission spectra of halophilic dehydrogenases at high and low salt concentrations. A. GDH (——) in 4 M NaCl-10mM Hepes, pH 7.2, (----) in 0.125 M NaCl-10 mM Hepes, pH 7.2. Excitation wavelength - 295 nm. B. MDH (——) in 4.3 M NaCl-10 mM Na-phosphate, pH 7.3, (----) in 10 mM Na-phosphate pH 7.3. Excitation wavelength-280 nm.

ionic strength effect, independent of salt type. At sodium chloride concentrations higher than 0.85 M (MDH) or 0.5 M (GDH) and at ammonium sulfate concentrations higher than 0.2 M for both enzymes, a linear dependence of the logarithm of the rate constants of inactivation upon salt concentration is observed. A comparison of the slopes of log k (of inactivation) vs. salt concentration shows that ammonium sulfate is much more effective than sodium chloride in preventing inactivation. Both the linearity of the dependence of log k on salt concentration and the fact that the order of effectiveness follows the Hofmeister series (10), indicate that, in this salt concentration range of the profile, salting out effects (11) are prominent.

STRUCTURAL PROPERTIES: FLUORESCENCE AND CIRCULAR DICHROISM (CD)

The effect of lowering the salt concentration on the intrinsic fluorescence emission spectra of GDH and MDH is shown in Figure 4. In both cases, a considerable quenching of the fluorescence intensity as well as a red shift of the maximum are observed. The far-UV CD spectra of GDH at high and low salt concentrations are shown in Figure 5. At the high salt

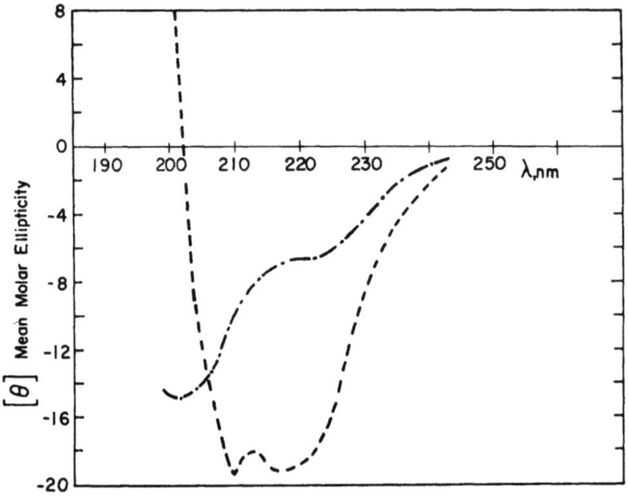

Fig. 5 Circular dichroism spectra of halophilic GDH at high and low salt concentrations. The spectra were recorded in 4 M NaCl, 50 mM phosphate, pH 6.6 (- - - - -) and 50 mM phosphate, pH 6.6 (-.-.-.-.-).

concentration, the CD spectrum in the range of 205-230 nm is consistent (12) with about 50% α-helical structure, whereas upon lowering the salt concentration, a major conformational change occurs; the secondary structure is, however, not completely abolished. Thus, low salt inactivation of halophilic MDH and GDH is accompanied by considerable structural changes, as reflected both in the intrinsic fluorescence and CD spectra. The rates of decrease of the fluorescence intensity at various salt concentrations parallel in both enzymes the rates of inactivation, thus establishing a correlation between the maintenance of the integrity of the active conformation and the overall ordered structure of the halophilic dehydrogenases. From the nature of the structural changes occurring upon inactivation, exposure of tryptophanyl residues to a more polar environment and loss of α-helical structure, one can deduce that the native conformation is more compact, and more ordered than the inactive one; nevertheless, it was not established that these processes occur at the same rate. One possible mechanism of inactivation is that the loss of the native structure is accompanied by dissociation of the oligomeric dehydrogenases into subunits. This hypothesis is strongly supported by the observation that, in the case of MDH, the rate-determining step of reactivation is second-order with respect to enzyme concentration (5).

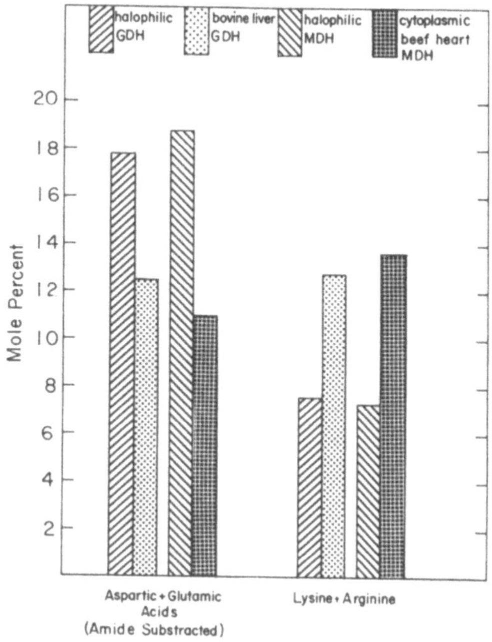

Fig. 6 Comparison of acidic and basic amino acids in dehydrogenases from non-halophilic (16,17) and halophilic sources (this work).

NATURE OF THE STABILIZING FORCES OF HALOPHILIC DEHYDROGENASES

The nature of the forces involved in conferring the outstanding stability of halophilic enzymes has been a much debated issue (4). The high salt concentrations required by halophilic enzymes in order to achieve maximum stability have been suggested to play a major role in shielding of negative charges (13) and supporting of hydrophobic interactions (14). As already mentioned, salting-out effects are important in maintaining the native structure of halophilic GDH and MDH. These effects are operative mainly by promoting hydrophobic interactions at the contact area between subunits or in the interior of the protein.

A striking difference between halophilic dehydrogenases and non-halophilic ones is expressed in their respective amino acid compositions (Figure 6). Thus, the excess of acidic over basic residues, in the two dehydrogenases discussed, is about 17% (11% when corrected for amide content). These values are similar to those obtained by Reistad for the total of the cytoplasmic proteins in several halophilic organisms (15). This might constitute the structural basis of halophilism, since in mesophilic as well as thermophilic (18) malate dehydrogenases, for example, this excess - uncorrected for amide content - does not exceed 10%. In thermophilic dehydrogenases (18,19) only subtle structural changes - as compared with the mesophilic enzymes - could be demonstrated in order to account for their stability at high temperatures.

Thus, a characteristic feature of the halophilic dehydrogenases is their dependence on multimolar concentrations of salts for stability. At low salt concentration, presumably, the electrostatic repulsion between the negative charges on the enzyme surface and/or at the contact area between its subunits is the main cause for disruption of the native structure, leading to inactivation. On the other hand, it may be that the role of the large excess of negative charges in halophilic enzymes is to maintain an appropriate hydration layer around the protein surface in the presence, in the cell, of a multimolar concentration of salt. The latter would cause dehydration, aggregation and eventually denaturation. This deleterious effect of salt is prevented in halophilic proteins by the high charge density on their surface. Therefore, halophilic dehydrogenases are successfully adapted to their extreme environment, without essentially altered functional properties.

ACKNOWLEDGEMENT

We are grateful to the Stiftung Volkswagenwerk for generous support of this work.

REFERENCES

1. Breed, R.S., Murray, E.G.D., Smith, N.R., Eds.: "Bergey's manual of determinative bacteriology", 7th edn., William and Wilkins, Baltimore, p.207 (1957).
2. Ginzburg, M., Sachs, L. and Ginzburg, B.Z.: Ion metabolism in a Halobacterium. I. Influence of age of culture on intracellular concentrations. J.Gen.Physiol. 55, 187-207 (1970).

3. Jaenicke, R., Koberstein, R. and Teüscher, B.: The enzymatically active unit of lactic dehydrogenase. Molecular properties of lactic dehydrogenase at low-protein and high salt concentrations. Eur.J.Biochem. 23, 150-159 (1971).
4. Lanyi, J.K.: Salt-dependent properties of proteins from extremely halophilic bacteria. Bacteriol.Rev. 38, 272-290 (1974).
5. Mevarech, M. and Neumann, E.: Malate dehydrogenase isolated from extremely halophilic bacteria of the Dead Sea. II. Effect of salt on the catalytic activity and structure. Biochemistry, submitted (1977).
6. Mevarech, M., Leicht, W. and Werber, M.M. Hydrophobic chromatography and fractionation of enzymes from extremely halophilic bacteria using decreasing concentration gradients ammonium sulfate. Biochemistry, 15, 2383-2387 (1976).
7. Lee, C.-Y. and Kaplan, N.O.: Characteristics of 8-substituted adenine nucleotide derivatives utilized in affinity chromatography. Arch.Bioch.Bioph. 168, 665-676 (1975).
8. Mevarech, M., Eisenberg, H. and Neumann, E.: Malate dehydrogenase isolated from extremely halophilic bacteria of the Dead Sea. I Purification and molecular characterization. Biochemistry, submitted (1977).
9. Smith, E.L., Austen, B.M., Blumenthal, K.M. and Nyc, J.F.: Glutamic dehydrogenases, in "The Enzymes", 3rd edn., Vol. XIA (Boyer, P.D., Ed.), Academic Press, New York, pp.293-367 (1975).
10. Robinson, D.W. and Jencks, W.P.: The effect of concentrated salt solutions on the activity coefficient of acetyltetraglycine ethyl ester. J.Amer.Chem.Soc. 87, 2470-2479 (1965).
11. von Hippel, P.H. and Schleich, T.: The effects of neutral salts on the structure and conformational stability of macromolecules in solution, in "Biological Macromolecules", Vol. II (Timasheff, S. and Fasman, G.D., Eds.), Marcel Dekker, New York, pp.417-574 (1969).
12. Chen, Y.-H. and Yang, J.T.: A new approach to the calculation of secondary structures of globular proteins by optical rotatory dispersion and circular dichroism. Biochem.Biophys.Res.Commun. 44, 1285-1291 (1971).

13. Baxter, R.M.: An interpretation of the effect of salts on the lactic dehydrogenase of Halobacterium salinarium. Can.J.Microbiol. 5, 47-57 (1959).
14. Lanyi, J.K. and Stevenson, J.: Studies on the electron transport chain of extremely halophilic bacteria. IV. Role of hydrophobic forces in the structure of menadione reductase. J.Biol.Chem. 245, 4074-4080 (1970).
15. Reistad, R.: On the composition and nature of the bulk protein of the extremely halophilic bacteria. Arch.Mikrobiol. 71, 353-360 (1970).
16. Siegel, L. and Engelard,S.: Beef heart malic dehydrogenase. III. Comparative studies of some properties of m-malic dehydrogenase and s-malic dehydrogenase. Biochim.Biophys.Acta, 64, 101-110 (1962).
17. Moon, K. and Smith, E.L.: Sequence of bovine liver glutamate dehydrogenase. VIII. Peptides produced by specific chemical cleavages; the complete sequence of the protein. J.Biol.Chem. 248, 3082-3088 (1973).
18. Biffen, J.H.F. and Williams, R.A.D.: Purification and properties of malate dehydrogenase from Thermus aquaticus, in "Enzymes and proteins from thermophilic microorganisms" (Zuber, H., Ed.), Birkhäuser Verlag, Basel, pp.157-167 (1976).
19. Hocking, J.D. and Harris, J.I.: Glyceraldehyde 3-phosphate dehydrogenase from an extreme thermophile, Thermus aquaticus in, "Enzymes and proteins from thermophilic microorganisms" (Zuber, H.,Ed.), Birkhäuser Verlag, Basel, pp.121-133 (1976).

Received February 7, 1977

DISCUSSION

Sund: The molecular weight of the polypeptide chain of glutamate dehydrogenase is 52.000 according to the SDS gel electrophoresis. What is the molecular weight of the native enzyme and what is the number of polypeptide chains from which the native enzyme is built?

Eisenberg: We have not yet determined in reliable fashion the molecular weight of the halophilic glutamate dehydrogenase and the number of peptide chains forming the native enzyme. These determinations are somewhat more difficult than for the non-halophilic enzymes and work on this problem is currently in progress.

Kempfle: Can you use the method of density gradient ultracentrifugation with high CsCl densities for molecular weight determination? What are the problems arising? The advantage of this method will be that you can use very low concentrations of the protein.

Eisenberg: The method of density gradient ultracentrifugation in CsCl solutions can be used for molecular weight determinations in protein solutions. J.B. Ifft (Biophys. Chem. 5, 137 (1976); Methods in Enzymology, Vol. 27, Part D, Academic Press, New York, 1973, p. 128) has described details and difficulties associated with this procedure. The method has proven most useful for the determination of buoyant densities and "net" hydrations under various experimental conditions. Molecular weight results have been very scant because of the technical difficulties associated with the method. We have not yet attempted to determine the molecular weight of the dehydrogenases from the halophilic bacteria in this way.

Jaenicke: Are there any specific properties with respect to the partial specific volume in terms of marked deviations from the sum of the amino acid increments?

There exist pronounced stabilizing effects of high salt concentrations in non-halophiles which are related to activity as well as stability (e.g. R. Jaenicke, R. Koberstein and Brigitte Teuscher, Eur. J. Biochem. 23 (1971) 150-159). Obviously "normal" enzymes are adapted to a wider range of variation of salt concentration. Could you comment on this fact in terms of physical enzymology or evolution?

Eisenberg: The calculated value (from the sum of amino acid increments) of $\overline{v}_2$ of halophilic malate dehydrogenase is 0.72. We did not experimentally determine $\overline{v}_2$, as was done for other systems (E. Reisler, Y. Haik and H. Eisenberg, Biochemistry 16, 197 (1977)). The apparent quantity ϕ' equals 0.68 ± 0.02; it is defined by

$$1 - \phi'\rho \equiv (1 - \overline{v}_2\rho^o) + \xi_3\,(1 - \overline{v}_3\rho^o)$$

where (loc. cit.) subscript 3 refers to salt and ξ_3 is a preferential interaction parameter. Further work on the study of interactions with salt and with water in solutions of halophilic enzymes in in progress.

It seems that the activity of the non-halophilic lactate dehydrogenase, referred to by Dr. Jaenicke, decreases with time and aggregation occurs at 37° C at 4.6 M NaCl. The halophilic enzymes so far studied by us are active and highly soluble at high ionic strength, they are unstable at low ionic strength. We cannot, with the limited amount of information on hand, offer generalized statements. Thus, whereas the halophilic malate dehydrogenase is unstable already below 2 M ionic strength, the halophilic glutamate dehydrogenase is significantly more stable at low ionic strengths, down to 1 M of NaCl.

Sund: Can you speculate a little more about the result that (1) in the halophilic dehydrogenases the content of Arg and Lys is low and that of Asx and Glx is high compared to other dehydrogenases and (2) that the amino acid composition of the halophilic dehydrogenases is rather similar. What is the helix content and do they exhibit a broad substrate specificity?

Eisenberg: We believe that the high net charge residing on the halophilic dehydrogenases, as a result of the significant excess of acidic over basic amino acid residues, prevents aggregation and precipitation at the high ionic strength under physiological conditions. (This composional feature also leads to the relatively high solubility in ammonium sulfate.) We have speculated on this in our text. The excess of acidic over basic residues reflects the overall composition of cytoplasmic proteins from halophilic organisms (Reistad, Arch. Mikrobiol. 71, 353 (1970)). The α-helix content of native halophilic glutamate dehydrogenase was found to be high (cf. text). We have not, so far, investigated other than the natural substrates for the two enzymes we have purified.

Hinz: Is the specific activity of halophilic dehydrogenases and "normal" dehydrogenases similar at comparable salt conditions?

Eisenberg: If comparable salt conditions refer to the enzyme at its activity maximum then the specific activity of halophilic glutamate dehydrogenase is roughly the same as glutamate dehydrogenase from bovine liver. On the other hand, the maximum activity of halophilic malate dehydrogenase is only about 7 % of that of pig malate dehydrogenase at comparable conditions.

Veeger: It seems to me that the fluorescence emission of malate dehydrogenase is mainly caused by tyrosine which is mainly quenched upon lowering the NaCl concentration. In order to discriminate between the different contributions you should use the difference method of Weber and Young.

Eisenberg: Our belief that the fluorescence is mainly due to the tryptophans is based on the fact that excitation at 280 and 295 nm results in emission spectra of apparently the same shape, with a maximum at 328 nm.

Harris: Did you assay for glyceraldehyde 3-phosphate dehydrogenase in your halophile extracts?

Eisenberg: Most species of *Halobacterium* apparently do not metabolize sugars. However, in some species (*H. sacchavorum* (Tomlinson, Koch and Hochstein, Can. J. Microbiol. 20, 805 (1974)), and *H. marismortui* (Volcani, Ph.D. Thesis, Hebrew University, Jerusalem) sugar metabolism could be observed and at least in one of them (*H. sacchavorum*) glyceraldehyde 3-phosphate dehydrogenase activity was present. We did not yet assay for glyceraldehyde 3-phosphate dehydrogenase activity in the extracts of our halophilic bacteria.

ORGANIZATION OF A BIFUNCTIONAL ENZYME : *Escherichia coli* ASPARTOKINASE I-HOMOSERINE DEHYDROGENASE I. Relationships between the catalytic and regulatory functions

Paolo TRUFFA-BACHI and Elisabeth FONTAN

Département de Biochimie et Génétique Microbienne - INSTITUT PASTEUR
75724 PARIS - Cedex 15 (France)

The threonine-sensitive aspartokinase and homoserine dehydrogenase (AK I-HDH I) activities of *Escherichia coli* K12 are carried by a tetrameric enzyme of molecular weight 344,000 (1). Each polypeptide chain possesses the sites for the two catalytic functions (1,2). Two domains have been described : one, carrying the aspartokinase activity, is located towards the amino-terminal region of the polypeptide chain (3,4), the second, located at the carboxy-terminal end, possesses the homoserine dehydrogenase activity (4).

The enzyme is subject to an allosteric equilibrium : L-threonine stabilizes an inactive T-form while potassium ions and L-aspartate shift the equilibrium to an active R conformation (5).

Sulfhydryl groups titration by a variety of reagents results in the inactivation of the aspartokinase activity and in the desensitization of the homoserine dehydrogenase towards L-threonine inhibition (6,7), suggesting that cysteinyl residues may play an essential role in the catalytic properties of this enzyme.

We shall report here some recent observations about the effects of carboxymethylation of AK I-HDH I by iodoacetic acid (8,9). The correlation between the extent of carboxymethylation and the inactivation and/or desensitization of the dehydrogenase will be discussed in term of conformational changes and oligomeric structure.

R E S U L T S

The reaction of iodoacetic acid (2 mM) with the AK I-HDH I (2.8 μm) leads to the inactivation of the aspartokinase activity ; the inactivation results from a pseudo-first order process with an half-life of 75 minutes (fig. 1). The inactivation of the dehydrogenase results from two pseudo-first events having different rate constants ; the rate constant for the first inactivation step (obtained after substraction of the contribution of the second phenomenon) is 15 minutes (fig. 1). The half-life of the second inactivation step is of 16 hours. While the aspartokinase remains sensitive to threonine during the inactivation process, the dehydrogenase becomes insensitive to the feed-back inhibitor with a kinetic of desensitization similar to the one observed for its inactivation.

The number of cysteines titrated was determined by utilization of ^{14}C-iodoacetic acid (5.6 mCi/mmol). The kinetic of radioactive material incor-

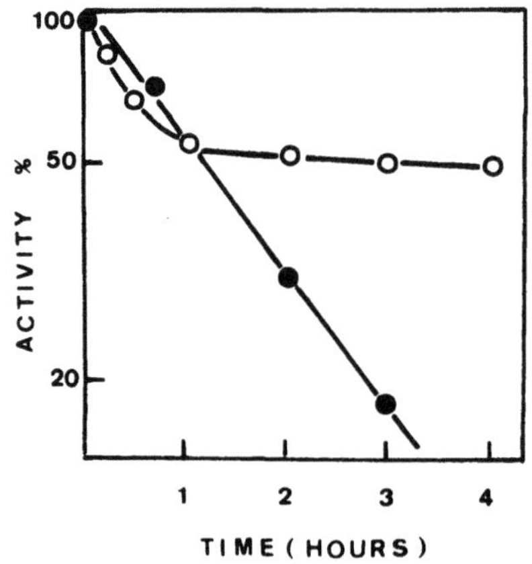

Figure 1

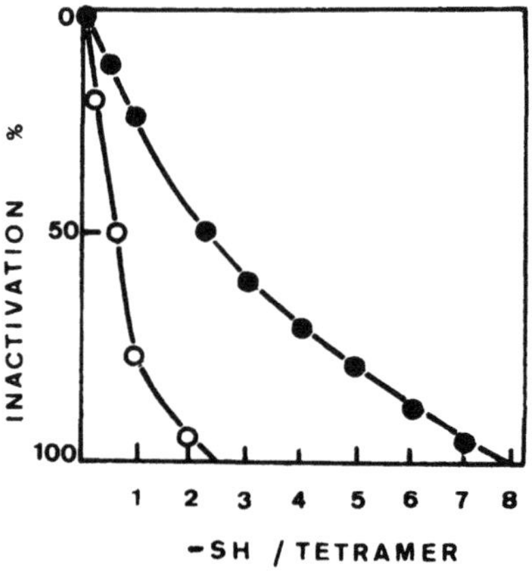

Figure 2

Figure 1 Kinetic of inactivation : aspartokinase (●-●-●) ; homoserine dehydrogenase (O-O). Incubation buffer : Tris-HCl 0.2 M, pH 8.1 containing 2 mM Mg-titriplex and 2 mM iodoacetic acid.

Figure 2 : Correlation between the number of alkylated cysteinyl residues and aspartokinase (●-●-●) or homoserine dehydrogenase (O-O) inactivation.

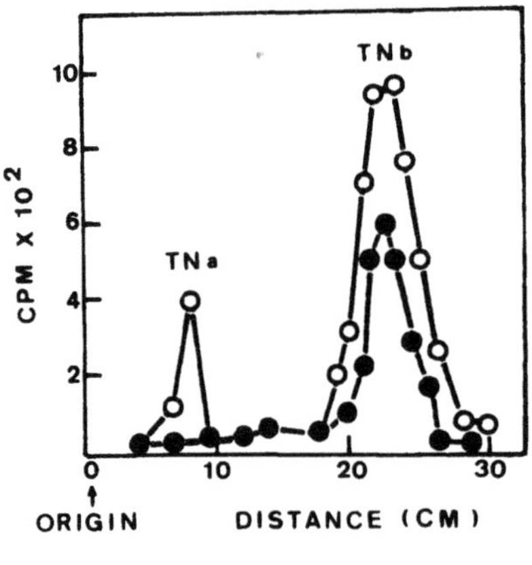

Figure 3

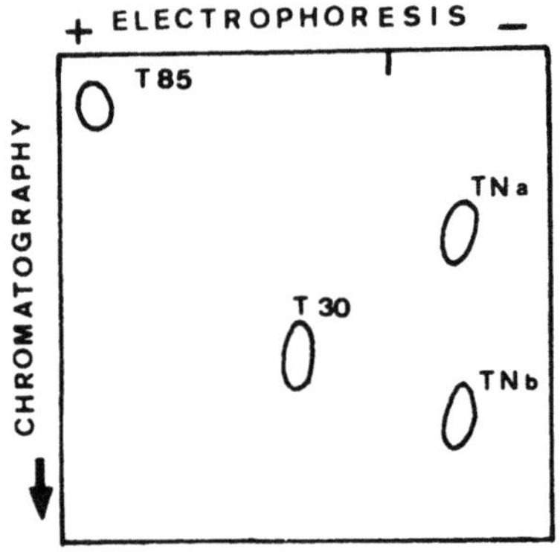

Figure 4

Figure 3 Chromatography of tryptic digests after electrophoresis of ^{14}C-carboxymethylated AK I-HDH I taken when the enzyme has lost 25% (●-●) or 50% (O-O) of the aspartokinase activity.

Figure 4 Autoradiography of the tryptic map of AK I-HDH I alkylated by 2 mM ^{14}C-iodoacetic acid.

poration is a pseudo-first-order process having an half-life of 105 minutes. The number of cysteinyl residues titrated in function of the extent of the inactivation of the aspartokinase and of the dehydrogenase is shown in figure 2. At the end of the carboxymethylation 8 molecules of iodoacetic acid are incorporated per molecule of enzyme (the native enzyme possesses 44 cysteinyl residues). From figure 2 it can be seen that the alkylation of a single -SH per tetramer leads to the complete desensitization of the homoserine dehydrogenase and also to almost all of the effect observed on the dehydrogenase activity (it should be noted that the inactivation of the dehydrogenase reaches only 40 to 50 per cent).

The analysis of the tryptic maps of the enzyme carried out at different times during the alkylation shows that the radioactivity is found only in the TNb tryptic peptide (1). The amount of radioactivity found in this peptide increases in a stoichiometric way until two residues are titrated per tetramer (fig. 3). After the titration of these two equivalents of -SH, other peptides are labelled ; the amount of radioactivity of each peptide at the end of the carboxymethylation is similar. The peptides that are alkylated are TNb, TNa, T30 and T85 (fig. 4).

The discrepancy between the number of radioactive peptides found, *i.e.* four and the expected number, *i.e.* two, can be explained if only half of the 4 protomers of the native enzyme are carboxymethylated. This is indeed the situation : the gel filtration analysis of the carboxymethylated protein reveals that the protein travels as two species equivalent in protein amount, one being an aggregate of molecular weight in excess of 10^6 and the second possessing a molecular weight of 180,000 and representing a dimer (fig. 5). The repartition of the radioactivity is asymmetric : only the aggregate is radioactive, the dimer being barely labelled.

The alkylation of the enzyme results then in a dissociation and an aggregation of the alkylated dimers, while the non alkylated dimers are stable under this form. The study of the dissociation of the enzyme as a function of the number of cysteinyl groups titrated shows that the dissociation occurs only when more than 2 -SH per tetramer are titrated (results not shown).

Various ligands of the enzyme as threonine, homoserine, aspartate, ATP or NADPH protect the protein against the effects of the alkylation on the two activities or on the threonine sensitivity. The analysis of the incor-

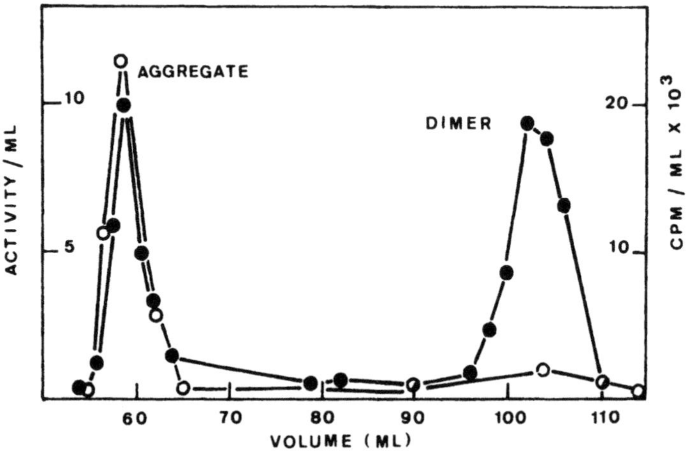

Figure 5 Gel chromatography analysis (on A-1.5 M Agarose gel) of the alkylated AK I-HDH I ; cpm (O-O) ; homoserine dehydrogenase activity (●-●).

poration of radioactive iodoacetic acid shows that in the presence of threonine there is no alkylation, whereas with the other ligands 4 cysteines per tetramer are titrated. The tryptic map of the enzyme carboxymethylated in the presence of homoserine reveals that two tryptic peptides are radioactive : TNb and T60 (fig. 6).

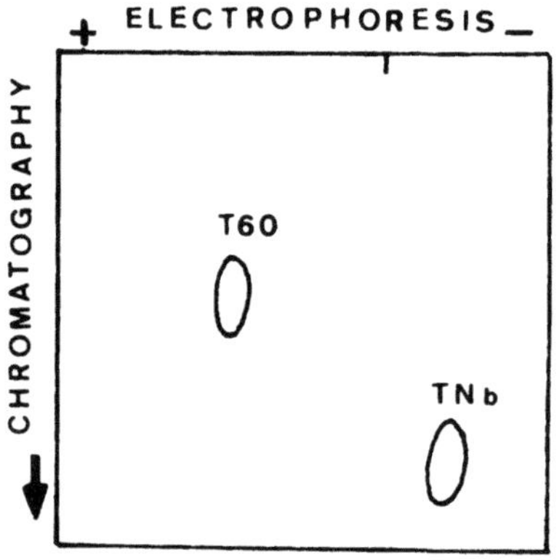

In the absence of ligands there is a stoichiometric relation between the titration of the cysteinyl residue of TNb (located in the aspartokinase region) and the aspartokinase inactivation at least until 50% of the kinase is lost, suggesting that this cysteine could play an essential

Figure 6 Autoradiography of the tryptic map of AK I-HDH I alkylated by 2 mM ^{14}C-iodoacetic acid in the presence of 20 mM L-homoserine.

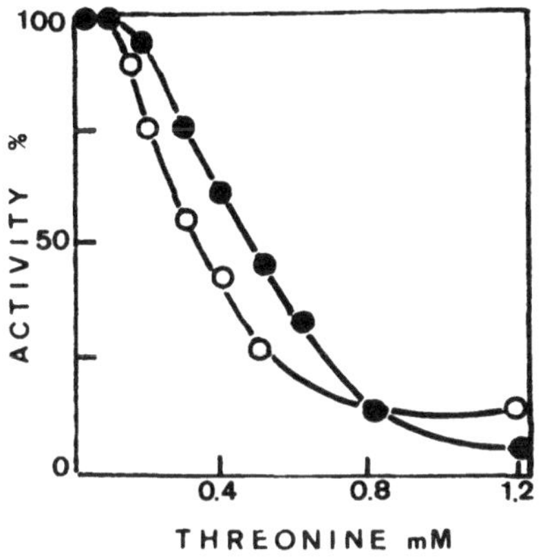

Figure 7

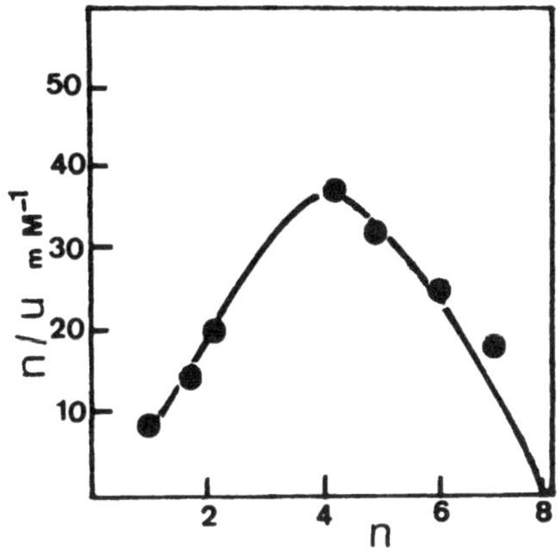

Figure 8

Figure 7 Inhibition curve of aspartokinase by L-threonine ; native enzyme (O-O) ; enzyme after alkylation of two -SH (●-●).

Figure 8 Scatchard plot of ^{14}C-L-threonine binding to the enzyme after alkylation of two -SH. n is the number of bound molar equivalent at the cocentration : u of free threonine.

role in the kinase catalytic site ; however this cysteine is titrated in the presence of homoserine without loss of aspartokinase activity. When homoserine (as well as iodoacetic acid) is removed after the alkylation of the 4 cysteinyl residues, the aspartokinase and the homoserine dehydrogenase loose 50 % of their activities and the homoserine dehydrogenase became desensitized towards threonine.

As mentioned before, during the inactivation process, the aspartokinase remains sensitive to L-threonine inhibition. When 2 -SH per tetramer are titrated the inhibition of the aspartokinase by L-threonine shows cooperative effects identical to the ones obtained with the native enzyme (fig. 7). Threonine binding studies on this alkylated enzyme show that it possesses 8 threonine binding sites as the native enzyme (fig. 8).

DISCUSSION

The *thrA* gene of the threonine operon codes for a polypeptide chain of molecular weight 86,000 (10,1) ; this chain, organized as a tetramer in the native protein, carries the threonine-sensitive homoserine dehydrogenase activities (1). The mild proteolysis of the native enzyme leads to the obtention of a dimer of molecular weight 110,000 possessing the homoserine dehydrogenase activity, desensitized towards L-threonine (4). Another fragment of the native enzyme is obtained by purification of the protein of an *ochre* mutant of the AK I-HDH I ; this fragment carries only the aspartokinase activity, sensitive to threonine inhibition and is a tetramer of molecular weight 180,000 (3,4).

A model of the organization of the AK I-HDH I has been proposed in which each polypeptide chain is organized into two globules one carrying the homoserine dehydrogenase activity and the other the aspartokinase activity (7). The tetrameric structure is maintained by the interactions of four kinase globules surrounded by the dehydrogenase dimers (17).

Affinity labeling of the AK I-HDH I with an analogue of ATP (11), an analogue of aspartate semi-aldehyde (12) or with cobalt complexes of threonine or ATP, as well as studies of the nuclear magnetic resonance of ATP-Mn(II) complexes (13) have been reported. From the affinity labeling experiments using ATP-Co or an ATP analogue, 6-mercapto-9-β-D-ribofuranosyl purine 5'-triphosphate (SHTP), it was concluded that cysteinyl residues should play an important role in the catalytic properties of the aspartokinase (11,14).

In the present studies we show that the inactivation of the kinase by alkylation of the enzyme with iodoacetic acid proceeds with a single pseudo-first-order process with an half-life of 75 minutes. In contrast to the previous reports using different reagents, the kinase inactivation and the homoserine dehydrogenase desensitization proceed with different rates : the desensitization occurs with an half life of 15 minutes. The desensitization and the inactivation of the homoserine dehydrogenase show biphasic kinetics : the first rates, obtained after substraction of the second rates, are identical for the two parameters under study : it should be pointed out that, while the desensitization is complete, the dehydrogenase looses only about half of its activity. From the inactivation rates of the kinase and the desensitization of the homoserine dehydrogenase we can conclude that the two phenomena do not result from a same and unique event.

From previous studies it was shown that incubation of AK I-HDH I with -SH reagents leads to the dissociation of the native enzyme into dimers (15). It was not possible however to decide whether the inactivation of the kinase was related to the dissociation of the enzyme. However, during the inactivation of the kinase by SHTP, dissociation of the enzyme does not occur (11) ; on the other hand, dimeric species found upon incubation of the enzyme in alkaline conditions, in the presence of reducing agent, possesses an aspartokinase activity (16).

In the present studies we have shown that the alkylation of the enzyme and the dissociation of the protein are not directly correlated. From the incorporation studies of the radioactive iodoacetic acid and from the determination of the molecular weight of the alkylated enzyme (on aliquot taken at different time during the reaction) it was found that the enzyme remains a tetramer until more than two -SH are alkylated. From our experiments it is difficult to ascertain whether the dissociation is caused by the alkylation of a third cysteinyl residue on the alkylated tetramer, this causing the dissociation or from the weakening of the interaction forces between aspartokinase globules leading to the dissociation and to the exposition of a cysteinyl residue.

The alkylation of the AK I-HDH I by iodoacetic acid is not a random event and occurs firstly at the cysteinyl residue of the tryptic peptide TNb on one of the four subunits ; furthermore the preferential titration of this cysteine is followed by the alkylation of a same residue on one of the three other subunits. After the titration of this second -SH it has been difficult to distinguish between a sequential titration of the other residues (TNa, T30, T85) or a random alkylation.

From the kinetics of incorporation of radioactive iodoacetic acid and of the inactivation (or of the desensitization of homoserine dehydrogenase) it is clear that *all the molecules of the enzyme are alkylated.* When one equivalent of iodoacetic acid per tetramer is incorporated, the complete desensitization of all the remaining homoserine dehydrogenase activity is observed ; we are not thus dealing with a mixture of molecules of enzyme some of them being alkylated, the other remaining in the native state.

The alkylation of the AK I-HDH I is then a sequential titration of a definite cysteinyl residue ; however this titration is also selective : the only -SH residues that can be titrated are the ones that are carried by the same protomer in which the TNb peptide is already alkylated. After the carboxymethylation it was found that the radioactivity is present only in the aggregated species, the dimers being hardly labelled. The non-alkylated dimer can be similar to the one obtained upon incubation of the native AK I-HDH I under alkaline conditions which leads to the formation of three disulfide bridges, obviously inacessible to iodoacetic acid (14).

It is difficult to distinguish between a preexistent asymmetry of the native molecule or an asymmetry induced by the carboxymethylation of the first -SH residue ; our results can be explained in two ways :

1) Only half of the protomers exposes the cysteinyl residue of the tryptic peptide TNb ; in such a situation, as the titration of this -SH is a prerequisite for the titration of the others -SH, two of the four protomers will never be alkylated.

2) The titration of the TNb cysteinyl residue leads to a conformational change of half of the protomers of the native enzyme in a way in which, not only the -SH of TNb is buried, but all the other cysteinyl residues became inaccessible in two protomers to iodoacetic acid.

This last hypothesis is favoured by the experimental data showing that a conformational change takes place, indeed, after the alkylation of a single -SH. Moreover the affinity labeling experiments with SHTP show that four equivalents are incorporated in the native protein, suggesting that the asymmetry is not preexistent but is induced by the carboxymethylation and not by the formation of a mixed disulfide bridge between an -SH of the protein and the SHTP.

There is a good correlation between the number of cysteinyl residues titrated and the aspartokinase inactivation at least until 50 % of the aspartokinase is lost. Since the TNb tryptic peptide is located on the aspartokinase fragment it was tempting to speculate about the possible role of this cysteine in the kinase catalysis.

The inactivation of the aspartokinase is not however a result of the alkylation of this cysteinyl residue but is related to the conformational change induced by carboxymethylation. This conclusion is reached from the following experimental observations :

1) The alkylation of the enzyme in the presence of L-homoserine does not elicit a modification of the catalytic properties of the enzyme and yet the tryptic map reveals that the cysteinyl residue of the TNb peptide is alkylated.

2) When homoserine, that should stabilize the above mentioned species is removed (as well as iodoacetic acid) the aspartokinase and the homoserine dehydrogenase inactivate and the latter becomes insensitive to L-threonine inhibition. We are then dealing with an enzyme that has similar kinetic properties than the one obtained in the absence of ligand when two TNb are alkylated. In the presence of L-homoserine the enzyme should also present a different conformation in the dehydrogenase region : a cysteinyl residue located in the dehydrogenase globules of the enzyme is also alkylated, T60, that is

never alkylated in the native enzyme in the absence of ligand. The conformational change induced by L-homoserine is probably only local since no differences are found between the spectroscopic properties of the homoserine liganded species and the unliganded enzyme (17).

The conformational change induced by the alkylation of a single -SH results in the complete desensitization of the remaining homoserine dehydrogenase. From our data, it is clear that after carboxymethylation of TNb the dehydrogenase domains are not anymore linked to the conformational changes induced by different ligands binding to the aspartokinase domains (see below). Supplementary information on the loss of the coordinated conformational change of the two globules is brought by the following experiments :

It was reported that ATP-Mg^{++} can replace K^{+} ions for maximum activity of the AK I-HDH I (18,19). There are two types of ATP binding sites, one on the homoserine dehydrogenase fragment and the other on the aspartokinase fragment (20). ATP binding in the presence of homoserine induces a conformational change of the native protein molecule (17) but not of the alkylated species. This suggests that the ATP-Mg^{++} binding site responsible for the activation of homoserine dehydrogenase is the one located in the kinase domain and that in the alkylated enzyme the information received at the level of the binding site of aspartokinase cannot be transmitted to the homoserine dehydrogenase domain. This argument is strenghthened by the one discussed below for threonine binding and for the desensitization of the homoserine dehydrogenase.

During the alkylation process, iodoacetic acid can be removed and the enzyme alkylated to different extent can thus be studied. If iodoacetic acid is removed after the titration of the first 2 -SH, the enzyme is stable for at least one hour. The characteristics of this alkylated AK I-HDH I are remarkable.

1) Threonine inhibition of the aspartokinase shows cooperative effects that are similar to the ones obtained with the native enzyme.

2) Threonine binding studies show that eight molecules of threonine bind to the alkylated enzyme ; the binding is highly cooperative in contrast to the binding to the native protein that follows, in this condition, a Langmuir isotherm (7). This implies that the protein molecule is now in a R-like conformation different from the R configuration of the native AK I-HDH I, nevertheless subject to an allosteric transition.

From the studies of Takahashi *et al.* (21) it is clear that threonine and the aspartate binding sites overlap : aspartate-Co(III) and threonine

Co(III) complexes are mutually exclusive (21). The fact that eight binding sites exist on the alkylated species reinforce the conclusion that the loss of the kinase is not due to the alkylation of a cysteine in the active site but to a conformational change. In this respect it would be very interesting to know whether this species can still bind four molecules of the aspartate-Co(III) complexes. This alkylated enzyme is capable of an allosteric transition as shown by the inhibition curves and by the threonine binding. Eight threonine binding sites, capable of a cooperative interaction, are still present in the alkylated protein suggesting that the organization of the protein of the polypeptide chain carrying the corresponding sites are unaffected by the alkylation.

Four threonine binding sites are localized on the aspartokinase fragment : none is found in the homoserine dehydrogenase proteolytic dimer (4). Where are then localized the other threonine binding sites ? The above mentioned fragments possess in common a polypeptide chain of molecular weight 17,000 ; this shared sequence is organized in both fragments as shown by its insensitivity to the action of proteolytic enzymes as reported by Véron *et al.* (4), for the homoserine dehydrogenase fragment and by Costrejean and Truffa-Bachi (22) for the aspartokinase fragment.

However they should be organized in a different way on each fragment as suggested by the experiments performed with antibodies directed towards one or the other fragment (23). We suggest that in the native enzyme this region can be organized in a way such as it carries the other threonine binding site. This hypothesis is based on the following observation : in the alkylated species (with two -SH residues titrated) the homoserine dehydrogenase is insensitive to L-threonine inhibition, yet eight threonine binding sites are found. As the conformational changes observed seem to be restrained to the kinase region, we postulate that in the native enzyme this region can be organized as two globules, one being the real kinase domain, the other carrying the second subset of threonine sites. The latter must be organized differently when it is present either in the aspartokinase fragment or the homoserine dehydrogenase fragment.

ACKNOWLEDGEMENT

We wish to thank Dr. G.N. Cohen for many helpful discussions.

REFERENCES

1. Falcoz-Kelly, F., Janin, J., Saari, J.C., Véron, M., Truffa-Bachi, P., and Cohen, G.N., *European J. Biochem.*, 28, 507-519 (1972).

2. Truffa-Bachi, P., van Rapenbusch, R., Janin, J., Gros, C., and Cohen, G.N., *European J. Biochem.*, 7, 401-407 (1969).

3. Janin, J., Truffa-Bachi, P., and Cohen, G.N., *Biochem. Biophys. Res. Comm.*, 26, 429-434 (1967).

4. Véron, M., Falcoz-Kelly, F., and Cohen, G.N., *European J. Biochem.*, 28, 520-527 (1972).

5. Janin, J., and Cohen, G.N., *European J. Biochem.*, 11, 520-529 (1969).

6. Truffa-Bachi, P., Le Bras, G., and Cohen, G.N., *Biochim. Biophys. Acta*, 128, 441-449 (1966).

7. Véron, M., Saari, J.C., Villar-Palasi, C., and Cohen, G.N., *European J. Biochem.*, 38, 325-335 (1973).

8. Fontan, E., and Truffa-Bachi, P., manuscript submitted for publication.

9. Fontan, E., and Truffa-Bachi, P., manuscript submitted for publication.

10. Thèze, J., and Saint-Girons, I., *J. Bacteriol.*, 118, 990-998 (1974).

11. Truffa-Bachi, P., and Heck, J. d'A., *Biochemistry*, 10, 2700-2706 (1971).

12. Hirth, C.G., Véron, M., Villar-Palasi, C., Hurion, N., and Cohen, G.N., *European J. Biochem.*, 50, 425-430 (1975).

13. Tilak, A., Wright, K., Damle, S., and Takahashi, M., *European J. Biochem.* 69, 249-255 (1976).

14. Ryzewski, C., and Takahashi, M.T., *Biochemistry*, 14, 4482-4486 (1975).

15. Truffa-Bachi, P., Thèse de Doctorat d'Etat, Paris 1968.

16. Jacques, Y., and Truffa-Bachi, P., *European J. Biochem.*, 62, 485-490 (1976).

17. Costrejean, J.M., Thèse de Doctorat d'Etat ès-Sciences, Paris 1977.

18. Broussard, L. Cunningham, G.N., Starnes, N.L., and Shive, W., *Biochem. Biophys. Res. Comm.*, 46, 1181-1186 (1972).

19. Truffa-Bachi, P., Costrejean, J.M., Py, MC., and Cohen, G.N., *Biochimie*, 56, 215-219 (1974).

20. Wright, J.K., and Takahashi, M., Personal communication.

21. Wright, J.K., Feldman, J., and Takahashi, P., *Biochemistry*, 15, 3704-3710 (1976).

22. Costrejean, J.M., and Truffa-Bachi, P., manuscript in preparation.

23. Costrejean, J.M., Guiso, N., Cowie, D.B., Cohen, G.N., and Truffa-Bachi, P., *European J. Biochem.*, 50, 431-435 (1975).

Received February 2, 1977

DISCUSSION

Harris: Have you evidence for internal sequence homology between the two "domains" of the bifunctional enzyme?

Truffa-Bachi: Dr. Cohen's group in Pasteur have sequenced about half of the polypeptide chain and there is no evidence for sequence homologies between the two domains. Using immunological tools we were able to show that common antigenic determinants are shared between the three isofunctional aspartokinases and the homoserine kinase.

Werber: Does the labeling of aspartokinase sites with Co(III)-complexes occur on the same cysteinyl residues that you alkylate with iodoacetate?

Truffa-Bachi: The amino acid composition of the tryptic peptide isolated by Takahashi and coworker does not contain cysteinyl residues. The complex contains 0.8 ATP and 0.9 Co/Arginine. The peptide should represent at least part of the kinase binding site for the coenzyme.

Harris: In the nonsense mutant where you get the tetrameric enzyme how does the lack of homoserine dehydrogenase function affect the well being of the organism?

Truffa-Bachi: In E. coli K12 the reduction of aspartate-semialdehyde to homoserine is carried out by two isofunctional enzymes. When by mutation, the homoserine dehydrogenase is lost, the other bifunctional enzyme, aspartokinase II-homoserine dehydrogenase II becomes derepressed and allows a normal growth of the bacteria.

Veeger: Have these enzymes been isolated from other organisms and if so, also in the bifunctional form?

Truffa-Bachi: In all the coliform bacteria tested so far, with the exception of Serratia marscens, the two activities are carried by a single polypeptide chain.

In genera as Pseudomonaceae, Azotobacter, S. cerevisiae, Rhodospirullum rubrum and Rhodospirullum tenue, the two activities are catalyzed by two different enzymes, each inhibited by L-threonine.

The fact that in E. coli the two activities are carried by a single polypeptide chain raise the problem of their evolution. The actual aspartokinase I-homoserine dehydrogenase I arises probably by fusion of two genes codying one of the aspartokinase and the other for the homoserine dehydrogenase. However, why in coliforms the two activities are carried by the same polypeptide chain and in other genera they are carried by distinct polypeptide is not clear. What type of selection advantages are generated by the fusion? We do not know; we can only speculate.

Veeger: Since on the pathway to homoserine you need another enzyme to form aspartate semialdehyde, could it be possible that this bifunctional enzyme is actually part of a multi-enzyme complex, of which one or more enzymes have been removed by your purification procedure?

Truffa-Bachi: The aspartate semialdehyde dehydrogenase, that catalyzes the second step of the biosynthetic pathway does not seem to be associated with the aspartokinase I-homoserine dehydrogenase I. Such an association could result indeed in a better functionality in the homoserine synthesis but would also create problems for the lysine biosynthesis.

CHEMICAL PROBES OF TOPOGRAPHY AND SUBUNIT INTERACTIONS IN A SIMPLE DEHYDROGENASE AND A MULTIENZYME COMPLEX

Richard N. Perham
Department of Biochemistry, University of Cambridge, Tennis Court Road, Cambridge CB2 1QW, England

To understand the connexion between structure and activity in oligomeric enzymes many problems require solution. Prominent among them are the folding of the polypeptide chains and the geometrical arrangement of the protein subunits. These problems are usually best solved by X-ray diffraction techniques but the lack of suitable crystals has sometimes caused insuperable difficulty. Analysis of the chemical reactivity of functional groups has often proved valuable in such circumstances, protein topography being defined in terms of groups that are 'accessible' or 'buried' when probed by treatment with suitable reagents (1-3).

A danger in this approach is that chemical modification might disturb the native conformation of the protein. This danger can be circumvented in two ways. In the technique of competitive labelling, chemical modification of target groups is restricted to a trivial extent (4) which makes consequential conformational changes rather unlikely. Alternatively, modifications can be chosen that favour the maintenance of local side-chain interactions and, with them, the preservation of three-dimensional structure. Amidination (5) is such a modification. The reaction appears to be specific for amino groups, which undergo no change of charge in the process, and many modified proteins have been found to retain their biological activity (6-8).

The advantages conferred by choice of the amidination reaction for this purpose find similar expression in the analysis of subunit geometry by means of chemical cross-linking procedures. Among the most successful of these is the use of bis(imidoesters) (9,10), which can provide information about the number of subunits in a particle and the symmetry with which they are arranged.

In this paper I should like to describe some recent work from our laboratory that shows how topography can be probed in rather intimate detail by treatment of the protein with methyl acetimidate and how subunit arrangements can be examined by the use of a novel series of cleavable bis(imido esters). For these experiments we chose a 'simple' enzyme, glyceraldehyde 3-phosphate dehydrogenase (EC 1.2.1.12) from various sources, and a multienzyme complex, the pyruvate dehydrogenase complex of E.coli. The former was selected because of its intrinsic interest (11) and because interpretation of the experiments would be aided by the existence of high-resolution X-ray crystallographic structures (12,13). The latter was chosen because its size and complexity has thus far placed it beyond detailed analysis by X-ray diffraction methods (14). In both instances, important conclusions about function can be drawn. Full accounts of the experimental methods can be found elsewhere (15,16).

GLYCERALDEHYDE 3-PHOSPHATE DEHYDROGENASE

Rabbit muscle glyceraldehyde 3-phosphate dehydrogenase is a tetramer of identical polypeptide chains. When the holoenzyme was treated at pH 8.0 with 0.1 M-methyl acetimidate, the number of amino groups fell rapidly from the starting value of 27 per subunit (molecular weight 36 000) to a value of approximately 5 per subunit. A second treatment with the reagent lowered this number to values between one and two but the enzyme retained 50% of its catalytic activity at the end of the experiment and was still tetrameric (15). Identification of the residues remaining unreacted was achieved by treating the modified enzyme with methyl (1-^{14}C)acetimidate under denaturing conditions. Chemical analysis of the modified enzyme then showed that only one amino group, that of lysine-306, had completely failed to react with the imido ester in the native enzyme. Several other amino groups, those of lysine residues at positions 159, 183, 248, 251, 256/257, 260 and 331, were found to react rather slowly compared with the other lysine residues in the subunit. The location of these poorly reacting residues within the primary structure was achieved by reference to the sequence of the highly homologous pig enzyme (17), as shown in Fig 1.

									10										20					
	Val.	Lys.	Val.	Gly.	Val.	Asp.	Gly.	Phe.	Gly.	Arg.	Ile.	Gly.	Arg.	Leu.	Val.	Thr.	Arg.	Ala.	Ala.	Phe.	Asn,	Ser.	Gly.	
Ac.	Ser.		Ile.		Ile.											Leu.				Leu.	Ser.	Cys		
					30										40									
Lys.	Val.	Asp.	Ile.	Val.	Ala.	Asn.	Asp.	Pro.	Phe.	Ile.	Asp.	Leu.	His.	Tyr.	Met.	Val.	Tyr.	Met.	Phe.	Gln.	Tyr.	Asp.		
	Ala.	Gln.	Val.		Val.						Ala.		Gln.							Lys.				
		50										60										70		
Ser.	Thr.	His.	Gly.	Lys.	Phe.	His.	Gly.	Thr.	Val.	Lys.	Ala.	Glu.	Asp.	Gly.	Lys.	Leu.	Val.	Ile.	Asp.	Gly.	Lys.	Ala.	Ile.	
				Val.		Lys.		Glu.			Met.				Ala.			Val.				Lys.		
								80										90						
Thr.	Ile.	Phe.	Gln.	Glu.	Arg.	Asp.	Pro.	Ala.	Asn.	Ile.	Lys.	Trp.	Gly.	Asp.	Ala.	Gly.	Thr.	Ala.	Tyr.	Val.	Val.	Glu.	Ser.	
	Val.		Asn.		Met.	Lys.		Glu.			Pro.		Ser.	Lys.			Ala.	Glu.		Ile.				
				100										110										
Thr.	Gly.	Val.	Phe.	Thr.	Thr.	Met.	Glu.	Lys.	Ala.	Gly.	Ala.	His.	Leu.	Lys.	Gly.	Gly.	Ala.	Lys.	Arg.	Val.	Ile.	Ile.	Ser.	
						Ile.				Ser.			Phe.						Lys.		Val.			
120										130									140					
Ala.	Pro.	Ser.	Ala.	Asp.	Ala.	Pro.	Met.	Phe.	Val.	Met.	Gly.	Val.	Asn.	His.	Glu.	Lys.	Tyr.	Asp.	Asn.	Ser.	Leu.	Lys.	Ile.	
										Cys				Leu				Ser.	Lys.	Asp.	Met.	Thr.	Val.	
						150										160								
Val.	Ser.	Asn.	Ala.	Ser.	Cys.	Thr.	Thr.	Asn.	Cys.	Leu.	Ala.	Pro.	Leu.	Ala.	[Lys.]	Val.	Ile.	His.	Asp.	His.	Phe.	Gly.	Ile.	
													Val.				Leu.		Glu.	Asn.		Glu.		
		170										180										190		
Val.	Glu.	Gly.	Leu.	Met.	Thr.	Thr.	Val.	His.	Ala.	Ile.	Thr.	Ala.	Thr.	Gly.	[Lys.]	Thr.	Val.	Asp.	Gly.	Pro.	Ser.	Gly.	Lys.	
										Val.												Ala.		
								200										210						
Leu.	Trp.	Arg.	Asp.	Gly.	Arg.	Gly.	Ala.	Ala.	Gln.	Asn.	Ile.	Ile.	Pro.	Ala.	Ser.	Thr.	Gly.	Ala.	Ala.	Lys.	Ala.	Val.	Gly	
Asp.			Gly.											Ser.										
				220										230										
Lys.	Val.	Ile.	Pro.	Glu.	Leu.	Asp.	Gly.	Lys.	Leu.	Thr.	Gly.	Met.	Ala.	Phe.	Arg.	Val.	Pro.	Thr.	Pro.	Asn.	Val.	Ser.	Val.	
																				Asp.				
240																				260				
Val.	Asp.	Leu.	Thr.	Cys.	Arg.	Leu.	Glu.	[Lys.]	Pro.	Ala.	[Lys.]	Tyr.	Asp.	Asp.	Ile.	[Lys.]	[Lys.]	Val.	Val.	[Lys.]	Gln.	Ala.	Ser.	
				Val.			Gly.		Glu.	Cys.	[Ser.]						[Ala.]	Ala.	Met.		Thr.			
						270										280								
Glu.	Gly.	Pro.	Leu.	Lys.	Gly.	Ile.	Leu.	Gly.	Tyr.	Thr.	Glu.	Asp.	Gln.	Val.	Val.	Ser.	Cys.	Asp.	Phe.	Asn.	Asp.	Ser.	Thr.	His.
				Gln.		Phe.							Asp.				Ser.			Ile.	Gly.	Asp.	Asn.	Arg.
290										300										310				
Ser.	Thr.	Phe.	Asp.	Ala.	Gly.	Ala.	Gly.	Ile.	Ala.	Leu.	Asn.	Asp.	His.	Phe.	Val.	[Lys.]	Leu.	Ile.	Ser.	Trp.	Tyr			
	Ile.				Lys.				Gln.		Ser.	Lys.	Thr.				Val.	Val.						

Fig 1 The amino acid sequence of glyceraldehyde 3-phosphate dehydrogenase from pig muscle. The sequence of the lobster muscle enzyme is identical except where indicated below the continuous line of the pig sequence (17). Poorly-reacting lysine residues are boxed.

Location of Abnormally Reacting Residues in a Folding Domain

The poorly reacting lysine residues were all found to lie within the C-terminal half of the polypeptide chain (Fig 1). This result proved capable of an interesting interpretation in terms of the X-ray crystallographic structure of the highly homologous lobster muscle enzyme (15). The N-terminal half of each chain is folded into a domain that binds NAD^+ whereas the C-terminal half forms a domain that is primarily associated with catalytic activity and subunit interactions (12). If a functional group that bears a charge is buried in the hydrophobic interior of a protein, a counter-ion of opposite charge should normally be buried with it (6). Lysine-306 manifests the chemical properties of a buried side chain and inspection of the model of the 'catalytic' domain reveals the reason (Fig 2). This residue forms an intrachain ion-pair with aspartic acid-241 in the hydrophobic environment of the P-axis subunit interface, which is provided by a large region of twisted anti-parallel pleated sheet.

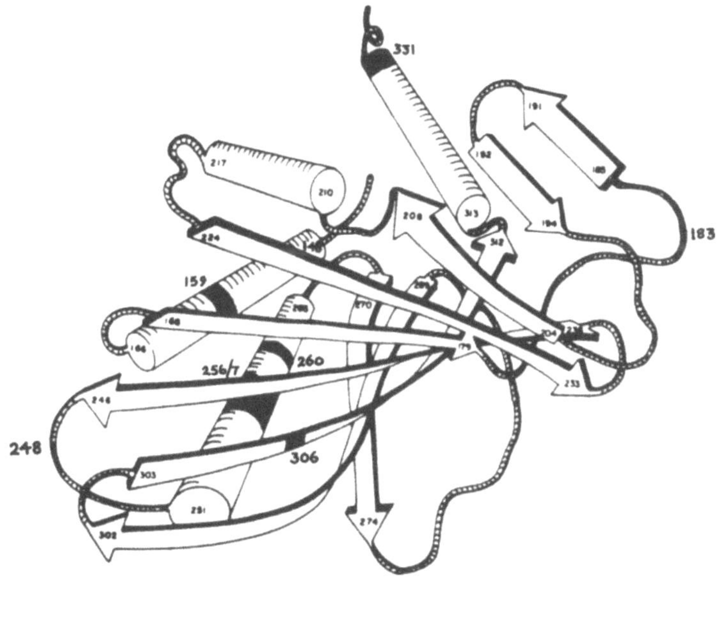

Fig 2 The catalytic (subunit interaction) domain of glyceraldehyde 3-phosphate dehydrogenase from lobster muscle (12)

Intramolecular Ionic Interactions at the Protein Surface

A simple but satisfying explanation of this kind was not immediately apparent for the slowness to react of those other lysine residues picked out by the treatment with methyl acetimidate. Close inspection of the structure of the catalytic domain uncovered no serious restriction of access of the reagent to these amino groups. However, in almost all cases, the poorly reacting amino groups, although from 'surface' lysine residues, could be brought into proximity with negatively-charged side chains that came from residues similarly regarded as being at the enzyme surface. For example, lysine-331 is in a helix on the surface of the subunit and is sterically unhindered, but glutamic acid-332, the C-terminus of the polypeptide chain (Fig 1), carries two negative charges. We propose that this adjacent negative charge will raise the pK of the amino group of lysine-331, which will depress its reactivity towards methyl acetimidate at pH 8.0.

Further inspection of the model (Fig 2) showed that other poorly reacting lysine residues also inhabit environments rich in negative charge. Thus, lysine-159 is close to aspartic acid-163, lysine-251 is close to aspartic acid residues at positions 253 and 254 and lysine-256 is close to glutamic acid-275 of the same subunit (if we make reasonable assumptions about the sequence homology between the rabbit and lobster enzymes) (18). Similarly, lysine-260 is thought to interact with tyrosine-273 and lysine-183 is a subunit side-chain contact (12). The conclusion that intramolecular ionic interactions at the protein surface lead to the depressed reactivity of selected amino groups was supported by the observation that modification of these amino groups, with one exception, proceeded more readily under conditions of high ionic strength (15). The exception was lysine-306, the residue buried as an ion-pair, to modify which would require the dissociation of the enzyme into monomers.

Comparative Studies with Other Glyceraldehyde 3-Phosphate Dehydrogenases

Very similar results were obtained with the enzymes from B.stearothermophilus and from ox muscle and ox liver, supporting the view that the ion-pair involving lysine-306 and aspartic acid-241 is a common feature of the glyceraldehyde 3-phosphate dehydrogenase subunit. No differences in topography could be detected between the enzymes from ox muscle and ox liver, in accord with other evidence that points to the identity of these enzymes (19). The pattern of slowly reacting amino groups in the enzyme from B. stearothermophilus was similar to that of the mammalian enzymes but showed one or two significant differences. For example, lysine-107 in the N-terminal half (NAD^+-binding domain) of the thermophile protein was very slow to be modified but this also correlates nicely with the presence of negatively charged side chains in its vicinity (13).

An Assessment of the Method

The amidination technique has proved to be an informative chemical probe of protein topography. The interpretation was greatly strengthened by reference to the detailed three-dimensional models from X-ray crystallography but the close agreement between the chemical predictions and the structural facts means that the method can now be applied elsewhere with some confidence e.g. to fructose diphosphate aldolase (20). The amino groups of proteins are commonly thought to reside at the molecular surface. Our work complements X-ray crystallographic study in that a side chain has to be reasonably well-localized before it can be seen much beyond the β-carbon atom by X-ray diffraction. The specific patterns of abnormally reacting lysine residues we have detected at the enzyme 'surface' imply that weak intramolecular interactions of this kind, quite apart from the more obvious ion-pairs deeply buried in hydrophobic environments, will be a general feature of protein structure. The potential value of surface ionic interactions in stabilizing proteins from thermophiles is already apparent (21,22). We would like to suggest that they may have a wider significance. If the free energy of a protein molecule is lowered by 6 $kJ.mol^{-1}$, the rate of thermal denaturation should fall 10-fold at 30°C. Such free energies are well within the range expected for ionic interactions of the type we describe (15,20). In the next section there follows a preliminary account of comparable experiments with a structurally complex system.

THE PYRUVATE DEHYDROGENASE COMPLEX OF E.COLI

The pyruvate dehydrogenase multienzyme complex of E.coli contains multiple copies of three different types of polypeptide chain, responsible for the three component enzyme activities. These are: E1, pyruvate decarboxylase; E2, lipoate acetyltransferase and E3, lipoamide dehydrogenase (23,24). The resolved E1 and E3 components show no affinity for each other but assemble round the E2 component. The intact enzyme and the E2 'core' probably have octahedral symmetry (14,23) which implies that the core comprises 24 E2 subunits. The polypeptide chain stoicheiometry (E1:E2:E3) in the fully assembled complex appears to be 2:1:1 (24,25) and the whole particle is significantly bigger than a ribosome (23). We have begun to attack the twin problems of complex topography (26) and subunit arrangement (16).

Treatment of the Complex with Methyl Acetimidate

When the enzyme complex was treated with methyl acetimidate at pH 8.0, pyruvate dehydrogenase activity was rapidly lost although the enzyme complex remained intact. On the other hand, the complex retained about 50% of its E3 activity, measured with exogenous dihydrolipoamide as substrate. The loss in pyruvate dehydrogenase activity was attributed largely to inactivation of the E1 component since substantial activity could be recovered by partly replacing the amidinated E1 with unmodified E1 in reconstitution experiments (26). The ability of the E2-E3 subcomplex to interact productively with active E1 component is therefore retained.

In further experiments, no striking difference in the rates of reaction of methyl acetimidate with the three component enzymes could be detected and most, if not all, of the amino groups in each subunit became modified in an exhaustive treatment. An analysis of the reaction as detailed as that for glyceraldehyde 3-phosphate dehydrogenase is at present beyond us owing to the lack of amino acid sequence information. But it is clear that the E2 component is not greatly masked by the binding of E1 and E3 and that few, if any, lysine residues are involved in the contact regions between subunits (26).

Cross-Linking Experiments with Bis(imido esters)

The treatment of the enzyme complex with dimethyl suberimidate was monitored by sodium dodecyl sulphate-polyacrylamide gel electrophoresis (16). We were surprised to find that the E2 and E3 components were rapidly cross-linked but that cross-links failed to form with the E1 polypeptide chains. Similar cross-linking experiments with the isolated E3 component were consistent with the accepted dimeric structure of this enzyme in free solution but the isolated E1 component, also thought to be a dimer, again failed to yield cross-links. This result cannot be due to any general lack of reactivity of the amino groups in the complex in view of the topographical study with methyl acetimidate described above. It must reflect a rather unusual surface distribution of amino groups that keeps them from the neighbourhood of the E1 subunit interfaces, both in the isolated E1 dimer and in the assembled complex. This may explain why raising the pH to 10 causes specific loss of the E1 component from the complex (23).

The cross-linking experiments with the intact molecule also provided evidence for the persistence of the E3 dimer as an entity in the assembled enzyme complex but the interpretation of the cross-linking patterns was difficult owing to the large number of polypeptide chains involved. To help overcome this difficulty, we have introduced a series of novel periodate-cleavable bis(imido esters). After an initial electrophoretic separation of cross-linked species, treatment with periodate to cleave the cross-links permits the regenerated monomers to be characterized again in a second dimension of gel electrophoresis (16), as shown in the following scheme:

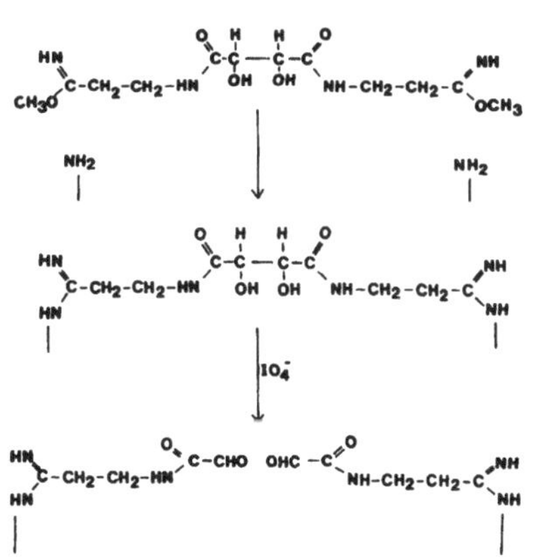

The cross-linked species obtained when the enzyme was treated with this reagent closely resembled those with dimethyl suberimidate and their identification was reinforced by the periodate-cleavage step. That E1 cannot be cross-linked to the E2-E3 subcomplex despite their physical association emphasizes the need to interpret cross-linking experiments with care.

Conclusions

The pyruvate dehydrogenase multienzyme complex of E.coli is a worthy challenge to the protein chemist in terms of its topography, its subunit arrangement and its mechanism of action. The substrate is borne between the contributing active sites by lipoic acid 'swinging arms' covalently attached to the E2 subunits (23,27) and these arms behave as though effectively in free solution on the surface of the enzyme (28). More about the assembly and symmetry may yet be learned by studies of chemical modification. A better knowledge of the arrangement of the constituent enzymes must be a first priority if subunit interactions are properly to be understood.

Acknowledgements

The skilled collaboration of Dr D.L.Bates, Dr J.Coggins, Dr J.M.Lambert and Miss E.A.Hooper and the support of the Science Research Council and the Wellcome Trust are gratefully acknowledged.

REFERENCES

1. Cohen, L.A.: Chemical modification as a probe of structure and function. The Enzymes 3rd Ed. 1, 147-211 (1970).

2. Perham,R.N., Anderson,P.J.: The reactivity of thiol groups in aldolase. Biochem.Soc.Symp. 31, 49-58 (1970).

3. Thomas, J.O.: Chemical modification of proteins. In Companion to Biochemistry (Bull, A.T., Lagnado, J.R., Thomas, J.O. and Tipton, K.F. eds.), pp. 87-138, Longman, London (1974).

4. Kaplan, H., Stevenson, K.J., Hartley, B.S.: Competitive labelling, a

method for determining the reactivity of individual groups in proteins. Biochem.J. 124, 289-299 (1971).

5. Hunter, M.J., Ludwig, M.L.: Amidination. Methods Enzymol.25, 585-596 (1972).

6. Perham, R.N., Richards, F.M.: Reactivity and structural role of protein amino groups in tobacco mosaic virus. J.Mol.Biol. 33, 795-807 (1968).

7. Reynolds, J.H.: Acetimidation of bovine pancreatic ribonuclease A. Biochemistry 7, 3131-3135 (1968).

8. Dworschack, R., Tarr, G., Plapp, B.V.: Identification of the lysine residue modified during the activation by acetimidylation of horse liver alcohol dehydrogenase. Biochemistry 14, 200-203 (1975).

9. Davies, G.E., Stark, G.R.: Use of dimethyl suberimidate, a cross-linking reagent in studying the subunit structure of oligomeric enzymes. Proc.Natl.Acad.Sci.U.S.A. 66, 651-656 (1970).

10. Hucho, F., Müllner, H., Sund, H.: Investigation of the symmetry of oligomeric enzymes with bifunctional reagents. Eur.J.Biochem. 59, 79-87 (1975).

11. Harris, J.I., Waters, M.: Glyceraldehyde 3-phosphate dehydrogenase. In The Enzymes 3rd Ed. 13, 1-49 (1976).

12. Moras, D., Olsen, K.W., Sabesan, M.N., Buehner, M., Ford,G.C., Rossman, M.G.: Studies of asymmetry in the three-dimensional structure of lobster D-glyceraldehyde 3-phosphate dehydrogenase. J.Biol.Chem. 250, 9137-9162 (1975).

13. Wonacott, A.J.: personal communication. See also the article by Dr Wonacott in this volume.

14. DeRosier, D.J., Oliver, R.M.: A low resolution electron-density map of lipoyl transsuccinylase, the core of the α-ketoglutarate dehydrogenase complex. Cold Spring Harb.Symp.Quant.Biol. 36, 199-203 (1972).

15. Lambert, J.M., Perham, R.N.: Folding domains and intramolecular ionic interactions of lysine residues in glyceraldehyde 3-phosphate dehydrogenase. Biochem.J. 161, 49-62 (1977).

16. Coggins, J.R., Hooper, E.A., Perham, R.N.: Use of dimethyl suberimidate and novel periodate-cleavable bis(imido esters) to study the quaternary structure of the pyruvate dehydrogenase multienzyme complex of E.coli. Biochemistry 15, 2527-2533 (1976).

17. Harris, J.I., Perham, R.N.: Glyceraldehyde 3-phosphate dehydrogenase from pig muscle. Nature(London) 219, 1025-1028 (1968).

18. Olsen, K.W., Moras, D., Rossmann, M.G., Harris, J.I.: Sequence variability and structure of D-glyceraldehyde 3-phosphate dehydrogenase. J.Biol. Chem. 250, 9313-9321 (1975).

19. Lambert, J.M., Perham, R.N.: A comparison of the glyceraldehyde 3-phosphate dehydrogenase from ox muscle and liver. FEBS Lett. 40, 305-308 (1974).

20. Lambert, J.M., Perham, R.N., Coggins, J.R.: Intramolecular ionic interactions of lysine residues and a possible folding domain in fructose diphosphate aldolase. Biochem.J. 161, 63-71 (1977).

21. Perutz, M.F., Raidt, H.: Stereochemical basis of heat stability in bacterial ferredoxins and in haemoglobin A_2. Nature(London) 255, 256-259 (1975).

22. Hocking,J.D., Harris,J.I.: Glyceraldehyde 3-phosphate dehydrogenase from an extreme thermophile, Thermus aquaticus. In Symposium on Enzymes and Proteins from Thermophilic Microorganisms (Zuber, H. ed.), pp. 121-133. Birkhäuser-Verlag, Basel (1976).

23. Reed, L.J.:Multienzyme complexes. Acc. Chem. Res. 7, 40-46 (1974).

24. Perham, R.N.: Self-assembly of biological macromolecules. Philos.Trans. R.Soc.London Ser.B 272, 123-136 (1975).

25. Bates, D.L., Harrison, R.A., Perham, R.N.: The stoichiometry of polypeptide chains in the pyruvate dehydrogenase multienzyme complex of E.coli determined by a simple novel method. FEBS Lett. 60, 427-430 (1975).

26. Bates, D.L., Coggins,J.R., Perham, R.N.: unpublished work.

27. Danson, M.J., Perham, R.N.: Evidence for two lipoic acid residues per lipoate acetyltransferase chain in the pyruvate dehydrogenase multienzyme complex of E.coli. Biochem.J. 159, 677-682 (1976).

28. Ambrose, M.C., Perham, R.N.: Spin-label study of the mobility of enzyme-bound lipoic acid in the pyruvate dehydrogenase multienzyme complex of Escherichia coli. Biochem.J. 155, 429-432 (1976).

Received February 7, 1977

DISCUSSION

Werber: Why do you assign the low reactivity of some lysyl residues in glyceraldehyde 3-phosphate dehydrogenase to an increase of their pK value? A higher pK value would also lead to a higher reactivity according to linear free-energy relationships of log k versus pK and the rate difference due to the pK increase should thus not be that large. Did you also try to modify the lysyl residues with a negatively charged modifying reagent, such as trinitrobenzene sulfonate, which be repelled by a negatively charged environment? An alternative interpretation for the low reactivity of the lysyl residues might be that they are involved in ion pairs with the negatively charged residues. Is there any evidence to that effect?

Perham: I think there is no real difference in our explanations. Our own interpretation of the lowered reactivity of certain lysine residues is indeed that they form ionic interactions with ω-carboxyl groups on the surface of the enzyme. This would be an apparent increase in pK. In answer to your second question, we have not carried out modification with negatively charged reagents. Unless we use competitive labeling methods, we must use a modification that causes no charge change at the modified residue.

Kiltz: I wonder if you ever have looked into the temperature stability of your acetimidated glyceraldehyde 3-phosphate dehydrogenase. We recently have modified porcine LDH H_4 with various imidoesters (W. Kapmeyer, W. Keil, H.H. Kiltz, H. Meyer, G. Pfleiderer (1977) Hoppe-Seyler's Z. Physiol. Chem. 358, 39-46; P. Tuengler, G. Pfleiderer (1977), Biochim.Biophys. Acta, in press). All surface lysine have been modified. However, no modification could be found of any of the lysine buried in the quaternary structure. These modified enzymes show a drastically increased stability against thermal and alkaline denaturation. This could be of interest for your enzyme, as this is normally quite labile.

Perham: We have not examined the thermal stability of the amidinated enzyme. I heard your comment earlier in the week about the increased stability of H_4 LDH after amidination with some interest. I can tell you that I have found that tobacco mosaic virus is more stable towards alkaline degradation after amidination, as you find for LDH. The pK of amidinated lysine residue is higher, so that is reasonable.

Brändén: You have correlated the weakly protected Lys residues with the presence of negatively charged residues in the vicinity. Most lysine residues in protein structures are close to negative charges. Do you see any steric or chemical reason why these should be different?

Perham: I wouldn't want to claim too much for our method but I think what we see is something that may not always be obvious from the crystal structure, essential as that was to our interpretation of the present experiments. What happens to side-chains beyond the β-carbon atom may be hard to see unless the residue is well-localized, although the chemical probe can still show an abnormal reactivity to exist. In the light of your question, however, I shall go back and have another look at the models!

North: Considerations of symmetry suggest that an E2 core of 24 subunits is much more probable than one of 16. 24 subunits can be arranged as an isodimensional body with all subunits in identical environments. 16 subunits can be positioned in identical environments only as a ring of 16 or two rings of 8. These would of course be clearly non-iso-dimensional in electron microscopy. Any other arrangement of 16 would require two different environments - and would of course be likely to complicate the interpretation of cross-linking experiments.

Perham: I agree. An E2 core of 16 subunits in isometric arrangement would require D_4 symmetry and two different subunit environments, in other words quasi-equivalence. The only X-ray crystallographic evidence comes from Dr. Derosier's work on the E2 core of the similar 2-oxoglutarate dehydrogenase complex. That is clearly octahedral. It is therefore likely that the pyruvate dehydrogenase complex will also have an octahedral (24-subunit) core. Unpublished studies of the molecular weight of the complex by Dr. A. Rowe and Dr. M.J. Danson support that view.

Veeger: I want to make the brief comment that the pyruvate dehydrogenase complex from *A. vinelandii* can be cross-linked to a high degree without markedly affecting the overall activity.

Section V. Kinetics and Regulation

Chairmen: K. Mosbach and C. Veeger

PRESSURE RELAXATION OF THE EQUILIBRIUM OF THE REACTION CATALYSED BY PIG HEART LACTATE DEHYDROGENASE: A TEST OF THE KINETIC MECHANISM

J. H. Coates, M. J. Hardman and H. Gutfreund
Molecular Enzymology Laboratory, Department of Biochemistry, University of Bristol, Bristol BS8 1TD, U.K.

The information which can be deduced from transient kinetic studies on the mechanism of NAD^+ linked dehydrogenases has recently been reviewed in some detail (1). It is of interest to summarise the differences observed between the behaviour of the tetrameric metal free pig heart lactate dehydrogenase (LDH) and that of the dimeric Zn enzyme horse liver alcohol dehydrogenase (LADH).

With LDH only reactions with the specific substrates lactate and pyruvate were analysed in any detail. In this system the equilibration of the hydride transfer process is much faster than product dissociation. In the direction of lactate oxidation proton release from the enzyme occurs after formation of enzyme bound NADH. The rate constants for the individual steps are set out below.

LADH was studied with different alcohols and aldehydes. With ethanol the hydride transfer is considerably slower than aldehyde dissociation and the "on enzyme equilibrium" of the chemical catalysis is almost completely uncoupled by product dissociation. With benzyl alcohol and other substrates, which result in the formation of products which dissociate more slowly, the hydride transfer process can equilibrate partially. This results in the phenomenon which has been mistaken for "half-site reactivity" (1).

In the reaction of LADH with ethanol the proton release occurs prior to hydride transfer and appears to be associated with a substrate induced conformation change.

In both enzymes there is kinetic evidence for substrate induced conformation changes, associated with pK changes, during the formation of the ternary complex in the direction of NADH oxidation. There is also some evidence of a conformation change during NADH dissociation, but more refined kinetic analysis is necessary to define this process satisfactorily. The investigation reported here was intended to examine whether relaxation studies can add to our knowledge of the kinetic mechanism of LDH.

In a recent paper on the scope of the pressure relaxation technique Davis and Gutfreund (2) have reported some preliminary observations on the application of this technique to the study of the reactions of lactate dehydrogenase. Perturbation of equilibrium can be attempted both through rapid pressure changes at constant pH and through combination of such effects with pH changes due to the presence of a buffer with a large ΔV^{o} of ionisation such as phosphate.

An interesting feature of the lactate dehydrogenase system is that extinction at 340 nm, nucleotide fluorescence and protein fluorescence permit one to observe the rate of change in concentration of the different forms of enzyme bound and free NADH (3,4). Detailed investigations by transient kinetic analysis of stopped-flow experiments in this laboratory (4-6) lead us to propose the following mechanism:

$$E \underset{k_{-1}}{\overset{k_1(NAD^+)}{\rightleftharpoons}} \begin{matrix}NAD^+\\E\end{matrix} \underset{k_{-2}}{\overset{k_2(Lact)}{\rightleftharpoons}} \begin{matrix}NAD^+\\E\\Lact\end{matrix} \underset{k_{-3}}{\overset{k_3}{\rightleftharpoons}} \begin{matrix}NADH\\EH^+\\Pyr\end{matrix} \underset{k_{-4}(H^+)}{\overset{k_4}{\rightleftharpoons}} \begin{matrix}NADH\\E\\Pyr\end{matrix} \underset{k_{-5}(Pyr)}{\overset{k_5}{\rightleftharpoons}} \begin{matrix}NADH\\E\end{matrix} \underset{k_{-6}(NADH)}{\overset{k_6}{\rightleftharpoons}} E$$

The analytical solution of the relaxation equations describing this complex mechanism, potentially in terms of six relaxation times, can be provided in principle but would not be of significant practical use. On the other hand computer simulations of this mechanism can be used to compare experimentally observed relaxations with those predicted from the model.

In the experiments described here the concentrations of enzyme (in sites), NAD^+ and lactate recorded are the initial conditions in solutions left to reach equilibrium at atmospheric pressure prior to the pressure perturbation experiments. The experimental procedure was that previously described (2).

Experiments carried out in TRIS buffer 0.1 M pH8.0 at pressures up to 200 atm. involve a pH change of less than 0.005 and are interpreted here as pressure effects at constant pH. The overall equilibrium of (NADH)(Pyr)/(NAD^+)(Lact) examined at relatively low enzyme concentration (0.07 μM) proved to be essentially unaffected by pressures up to 200 atm. This shows that the product of all the intermediate equilibrium constants remains constant. At enzyme concentrations above 1 μM significant amounts of enzyme bound intermediates are present at equilibrium and pressure relaxations are observable. Observations of protein fluorescence changes, which monitor the concentrations of all NADH containing enzyme complexes, at enzyme concentrations above 1 μM indicated two relaxations. From observation of extinction at 340 nm, which monitors total NADH, free and bound, a single relaxation was recorded corresponding to the slower relaxation observed by protein fluorescence. Since the fast relaxation τ_1 involves only bound NADH and not total NADH it must characterize the movement between bound and free NADH. The slow relaxation τ_2 which involves both bound and total NADH must represent the reduction of NAD^+. The slight dependence of τ_1 on enzyme concentration results in τ_1 = 5 ms at 2.2 μM LDH and 4 ms at 9.9 μM. The dependence on NAD^+ concentration (20 to 400 μM) is even less than that. The slow relaxation, τ_2 = 50 ms at NAD^+ 20 μM and LDH 10 μM, has a small amplitude at constant pH and was studied more extensively in the "pH jump" experiments.

Experiments carried out in phosphate buffer (pH8.0, I = 0.1) result in an increase in pH of about 0.05 for a decrease in pressure of 150 atm. This will lead to a perturbation of the overall equilibrium of the reaction towards increasing NADH concentration. Under these conditions we observed two relaxations with the same relaxation times as those observed at constant pH. The amplitudes however, are different. In phosphate the total observed change in NADH concentration monitored either by protein fluorescence or extinction, is about twice as large as in TRIS. The faster relaxation contributes about 20% of the observed change in protein fluorescence in phosphate buffer and about 70% in TRIS.

The value of $1/\tau_2$ is linearly dependent on enzyme concentration in the range 1 to 10 μM at NAD^+ 20 μM and lactate 2 mM. The reciprocal relaxation time $1/\tau_2$ decreases by a factor of 2 over a 100 fold increase in NAD^+ concentration. This small effect is only noteworthy because it is in agreement with the behaviour of the simulated model.

Simulations were carried out by programming the above kinetic model in "FOCAL" on a PDP8/E computer, using the following rate constants (from ref. 4-7):

$k_1 = 3.0 \times 10^6\ M^{-1}\ s^{-1}$	$k_4 = 1200\ s^{-1}$
$k_{-1} = 340\ s^{-1}$	$k_{-4} = 2 \times 10^{10}\ M^{-1}\ s^{-1}$
$k_2 = 9 \times 10^5\ M^{-1}\ s^{-1}$	$k_5 = 150\ s^{-1}$
$k_{-2} = 4000\ s^{-1}$	$k_{-5} = 3.5 \times 10^6\ M^{-1}\ s^{-1}$
$k_3 = 600\ s^{-1}$	$k_6 = 50\ s^{-1}$
$k_{-3} = 2000\ s^{-1}$	$k_{-6} = 3.0 \times 10^7\ M^{-1}\ s^{-1}$

The initial concentrations of NAD^+, lactate and enzyme were set, the simulation was run to equilibrium and the equilibrium concentrations of the intermediates were noted. To simulate relaxations one of two procedures was used. To simulate the effect of release of pressure at constant pH, k_{-5} and k_6 were decreased by 10%, the overall equilibrium constant remaining unchanged ($K = 2.6 \times 10^{-12}$M). This particular approach was chosen because we concluded from our experimental data that the fast relaxation involved transfer of free to bound NADH. Changing k_{-4} and k_5 produces a simulation which is inconsistent with the experimental observations. The simulation of the combined effects of increase in pH and direct pressure perturbation was obtained by simultaneously changing k_{-5}, k_6 and the hydrogen ion concentration.

The simulation of different conditions also showed only two relaxations of significant amplitude. Only the slower of these appeared when total NADH was recorded; this corresponds to the relaxation observed by extinction at 340 nm in the experiments. The amplitudes of the relaxation recorded for total NADH and for bound NADH are consistent with those calculated from the observed changes in extinction and protein fluorescence respectively.

Reasonable agreement was obtained between observed and simulated relaxation times, both with respect to a particular set of initial concentrations and dependence on enzyme and NAD^+ concentrations. The poorest agreement was obtained for the slow relaxation, where the simulated values of $1/\tau_2$ are about 60% of the experimental ones.

The present investigation is the first detailed examination of the relaxations of the complete reaction of a dehydrogenase. The information available from other transient kinetic and spectrophotometric studies of the reactions of lactate dehydrogenase provided a good background and the promise of complementary information from relaxation studies of such a complex coupled system. It is difficult to say whether any useful information would have been deduced from the relaxation studies without the existing background of kinetic information. In the event the most useful outcome of the studies reported is that they provide a rigorous test of the proposed mechanism.

The two relaxation times studied in detail are interpreted as the following properties of the model. The relaxation time τ_1 is consistent with the relaxation expected for the association of NADH with enzyme. As protein fluorescence decreases, nucleotide fluorescence is enhanced with the same relaxation time, which confirms that protein fluorescence and nucleotide fluorescence describe the same process during NADH binding.

The slow relaxation described by τ_2 can be interpreted in terms of an extension of the treatment of a minimal enzyme mechanism by Bernasconi (8). To interpret our six step model in terms of Bernasconi's three step model

$$E + S \rightleftharpoons ES \rightleftharpoons EP \rightleftharpoons E + P$$

we made the following assumption. We have taken the slowest step for the formation of ENADH (k_5, k_{-5}) as our equivalent of Bernasconi's slow step k_2, k_{-2}. Substituting our rate constants and the concentrations obtained from our simulations into Bernasconi's equations (4.53-4.59) for the slowest relaxation, we obtain the following comparisons. For the condition enzyme 8 μM, NAD^+ 20 μM, Lactate 2 mM, the above calculation gave a reciprocal relaxation time of 26.4 s^{-1}, simulation gave 11.7 s^{-1} and $1/\tau_2$ obtained experimentally is 21 s^{-1}. This treatment also produces agreement with the observed and simulated enzyme concentration dependence of $1/\tau_2$.

The model chosen on the basis of steady state and stopped-flow kinetic experiments is capable of explaining the observed behaviour on pressure relaxation. This investigation provides a clear example of the complementary nature of transient and relaxation kinetics in the study of enzyme mechanism.

J. H. C. and M. J. H. are on sabbatical leave from the University of Adelaide, South Australia and Massey University, Palmerston North, New Zealand, respectively.

REFERENCES

1. Gutfreund, H. Prog. Biophys. Mol. Biol., 29, 161-195 (1975).
2. Davis, J. S. and Gutfreund, H. FEBS Letters, 72, 199-207 (1976).
3. Holbrook, J. J. and Gutfreund, H. FEBS Letters, 31, 157-169 (1973).
4. Whitaker, J. R., Yates, D. W., Bennett, N. G., Holbrook, J. J. and Gutfreund, H. Biochem. J. 139, 677-697 (1974).
5. Boland, M. J. and Gutfreund, H. Biochem. J. 151, 715-727 (1975)
6. Holbrook, J. J., Gutfreund, H. and Südi, J. Biochem. J. 157, 287-288 (1976).
7. Schwert, G. W., Miller, B. R. and Peanasky, R. J. J. Biol. Chem., 242, 3245-3252 (1967).
8. Bernasconi, C. F.: "Relaxation Kinetics". Academic Press, New York p.49-50 (1976).

Received February 19, 1977

DISCUSSION

Hinz: The fact that you can perturb binary complex formation equilibrium between LDH and NADH reconfirms my interpretation of the negative Δc_p involved in that reaction. Is the effect you detect in accordance with a volume decrease?

Gutfreund: Yes, pressure relaxation causes the dissociation of the nucleotide from the complex. We have yet to carry out a detailed analysis of the amplitudes to obtain values for the volume changes during binary and ternary complex formation.

Jaenicke: Professor Gutfreund has already mentioned a couple of experiments we have been doing at high hydrostatic pressures (< 2000 atm). The aim of our equilibrium and kinetic studies is different from the pressure relaxation technique he has been pursuing.

High hydrostatic pressure turns out to be a parameter which may strongly influence equilibria of association (hybridization of isozymes of LDH, tobacco mosaic virus protein A-protein $\rightleftharpoons$ double disk $\rightleftharpoons$ helical aggregates, *E. coli* ribosomes 70S $\rightleftharpoons$ 50S + 30S etc.).

Corresponding to the pressure dependence of specific steps in reaction sequences, high pressure may strongly affect e.g. the enzymatic turnover as soon as the volume of activation or the reaction deviate significantly from zero (yeast glyceraldehyde 3-phosphate dehydrogenase etc.). A careful analysis allows to differentiate clearly between effects on the enzymatic mechanism and/or the structure of the enzyme. In many cases reversible reactions are superimposed by pressure induced (irreversible) denaturation.

In the given context one should keep in mind that more than 70 % of the oceans covering the earth are "high pressure systems"; in the Philippinean Trench there is life under conditions of p > 1000 atm. Obviously "baryphilic" organisms like thermophiles and halophiles represent another example of adaptability in nature which may deserve a closer look.

THE ROLE OF CONFORMATIONAL CHANGES IN THE LIVER ALCOHOL DEHYDROGENASE REACTION MECHANISM

Joseph D. Shore, Herbert R. Halvorson and Karen D. Lucast
Edsel B. Ford Institute, Henry Ford Hospital, Detroit, Mich. 48202, U.S.A.

Past studies from our laboratory (1) demonstrated that the binding of NAD^+ to liver alcohol dehydrogenase perturbs the pK_a of an enzyme functional group from a value greater than 9 to 7.6, and that alcohol binds to the unprotonated form of the perturbed functional group. More recently (2), we have investigated conformational states related to the binding of NAD^+ and subsequent perturbation of the pK_a of the enzyme group. The protein fluorescence of the enzyme is quenched at alkaline pH, with a pK_a of 9.8. The binding of TNS, a coenzyme competitive fluorescent probe, showed the same pK_a. In the presence of NAD^+, the pK_a of the protein fluorescence quenching was shifted toward 7.6, and formation of a ternary complex with NAD^+ and trifluoroethanol resulted in a pH-independent maximal quench. These results were interpreted to indicate that free enzyme exists in two pH dependent conformations with a pK_a of 9.8, with NAD^+ selectively binding to the conformation in which the functional group is protonated. Subsequent to binding NAD^+, a further conformational change occurs related to the perturbation of the pK_a to 7.6, with binding of alcohol to the unprotonated form of the perturbed functional group. Thus, there are two conformational changes in going from free enzyme to ternary complex at neutral pH.

The present study was undertaken in an effort to correlate the binding of oxidized and reduced coenzyme and ADPribose (ADPR), elucidate the conformational isomerizations involved, and investigate the energetics of some of the discrete interactions involved.

MATERIALS AND METHODS

Liver alcohol dehydrogenase was prepared, assayed and titrated by methods previously reported (1). Fluorescence titrations were performed on a Farrand spectrofluorimeter and were analyzed by a least squares fit using a program developed in this laboratory for the NOVA 2/10 computer. Stopped-flow kinetics were performed on a Durrum-Gibson instrument with a fluorescent attachment, interfaced to the computer by a system purchased from On Line Instruments, Inc. This system enabled signal averaging of up to 16 runs, and facilitated calculation of pseudo first order rate constants.

A fluorescent probe technique developed in our laboratory was used to determine the equilibrium binding constants for ADPR and NAD^+, and for the kinetics of NAD^+ binding. This technique is particularly useful for studying ligand binding in which there is no convenient spectrophotometric or fluorescence signal. The dissociation constant of the probe from free enzyme must be determined, and a competitive relationship with the ligand to be studied must be established. The concentrations of enzyme and probe are then set such that less than 5% of the total enzyme and probe concentrations are bound. TNS was used in all studies above pH 9, while ANS, which is five-fold less tightly bound, was used at lower pH values. It was possible to use this technique to study the equilibrium binding of ADPR and NAD^+. For the NAD^+ binding studies, 20 µM pyruvate and a catalytic concentration of lactate dehydrogenase was added to maintain coenzyme in the oxidized state, obviating the problem of slight alcohol contaminants. The probes were also used for kinetic studies of NAD^+ binding. At pH values below 10, where NAD^+ is relatively weakly bound, 50 mM ethanol was used to shift the binding equilibrium by ternary complex formation. Since displacement of the probe is the signal being monitored, the fact that the ternary complex is catalytically active presented no limitations. Independent studies demonstrated that the rate constants for TNS and ANS dissociation were 300-400 sec^{-1}, so that binding reactions as fast as 60 sec^{-1} could be studied. In some cases, the signal to noise ratio of the Durrum instrument was too low for accurate measurements, necessitating computer averaging of 6-10 runs. In order to perform studies at pH values above 10, where the enzyme is unstable, the enzyme syringe contained 1 mM phosphate at pH 7 while the coenzyme was in 0.1 M buffer at the higher pH. Phosphate buffer was used at pH 7 thru 8, pyrophosphate at pH 8.5 through 9.5, and bicarbonate at pH 9.5 to 11.5. All buffer concentrations were .05 M after mixing.

RESULTS

Prior to a consideration of conformational states in the LADH mechanism, the proposed (3) half of the sites reaction mechanism should be evaluated. We have recently completed a study of the reaction of the benzyl alcohol-benzaldehyde system at pH 8.75 and concluded that there is no evidence which mandates proposing site non-equivalence during turnover of the enzyme (4).

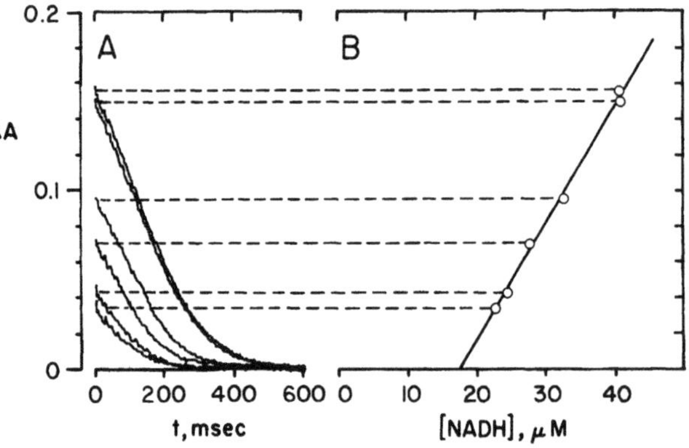

Figure 1 Steady state oxidation of free NADH by benzaldehyde in .05 M pyrophosphate buffer at pH 8.75. Syringe 1, 36 μN enzyme and NADH; Syringe 2, 24 mM benzaldehyde; 360 nm, 25° C, 0.1 m. sec time constant.

Figure 1 demonstrates a typical result of this study, in which enzyme with concentrations of NADH in excess over enzyme were mixed with saturating concentrations of benzaldehyde in a stopped-flow apparatus. An extrapolation of the observed amplitudes of the steady-state turnover of the free NADH (figure 1B) will intercept the abscissa at the concentration of NADH oxidized in the rapid pre-steady state phase. The intercept was 17.75 ± .27 μM, as contrasted to an enzyme concentration after mixing of 18 μN. This experiment required no assumptions regarding extinction coefficients, or elaborate curve fitting procedures, and indicated that two equivalents of NADH per mole of enzyme are rapidly oxidized prior to turnover.

Our previous studies (1,2) indicated that NAD^+ should bind to the enzyme conformation in which a functional group with an alkaline pK_a is protonated. Consequently, the variation with pH of the bimolecular rate constant for NAD^+ binding was determined using the fluorescent probe technique. A typical experiment is demonstrated in figure 2. The results at pH 8 through 11.5 are shown in figure 3, and indicate a pK_a of 10 ± 0.2, with NAD^+ selectively binding to the protonated form of the enzyme.

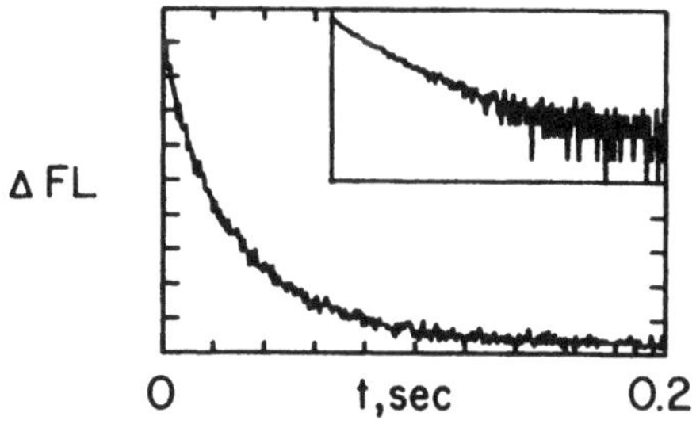

Figure 2 Binding of NAD^+ to LADH at pH 8.5. Concentrations after mixing were: LADH 11.7 μN; ANS, 10 μM; NAD^+, 63 μM and 50 mM ethanol. Excitation, 370 nm, emission cut on filter, 420 nm.

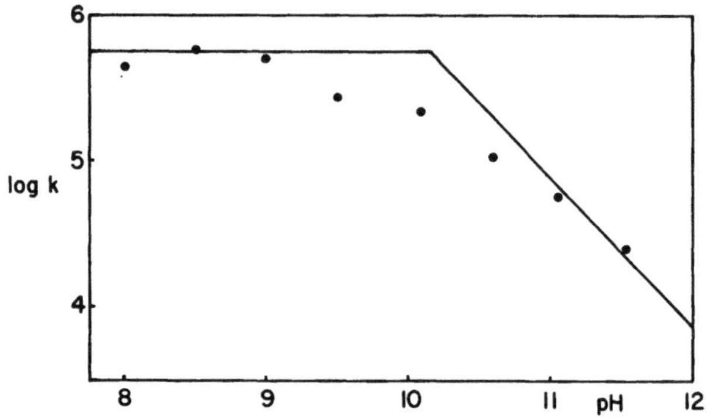

Figure 3 Plot of the pH dependence of the bimolecular rate of NAD^+ binding.

To discern the relationship of ADPR binding to NAD^+ binding, the pK_a for the equilibrium binding of the coenzyme fragment was determined using fluorescence probes. The results are presented in figure 4, and indicate a pK_a of 8.9 ± 0.2.

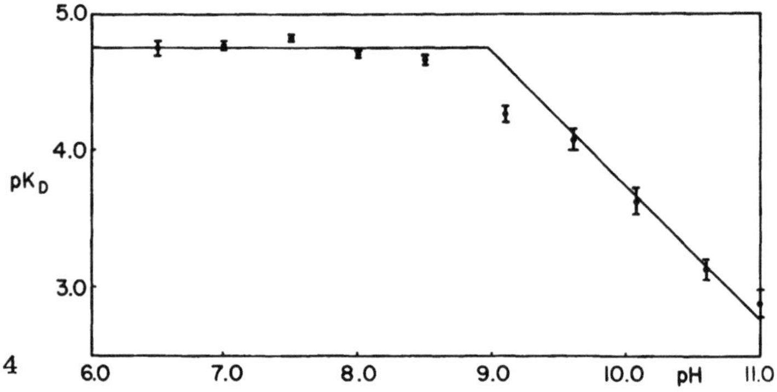

Figure 4

Plot of the pH dependence of the

The bimolecular rate constants for NADH binding at various pH values were also determined in the same range, by mixing coenzyme and isobutramide with enzyme in the stopped-flow fluorimeter. The pK_a for this reaction was 9.0 ± 0.2, indistinguishable from the value for the equilibrium binding of ADPR.

To further delineate the nature of the pH-dependent interactions observed, the temperature variation of the equilibrium binding of TNS, ADP Ribose and NAD^+ were determined.

The results are expressed in Table 1.

TABLE I

Thermodynamic values for binding, pH 7, 20 C

Ligand	ΔG	ΔH	ΔS
ADPR	-6.6	-6.6	0
NAD^+	-5.4	+2.0	28
NAD^+-ADPR	+1.1	+8.6	28

DISCUSSION

The results presented in figure 1 place considerable limitations on any proposed site non-equivalence for LADH. It is unequivocal that two moles of NADH per mole of enzyme are rapidly oxidized in the pre-steady state phase, in agreement with the results of Hadorn et al (5). Since it is also established that benzyl alcohol dissociation primarily limits turnover (4), these results rule out any mechanism in which product dissociation is required for the adjacent subunit to catalyze hydrogen transfer. Although more complicated mechanisms involving site non-equivalence can be proposed, there does not seem to be a necessity to invoke them based on currently available data.

The pH dependence of the binding of ADPR and the rate constants for NADH and NAD^+ binding provide several interesting insights. It is probable from the pK_a values that the rate constant for NADH binding is controlled by the same interaction as ADPR. The NAD^+ binding rate, however, has a pK_a one pH unit higher although all species bind to protonated enzyme forms. If the pK_a of ADPR and NADH binding were due to interaction with ARG-47, the group responsible for binding the pyrophosphate of ADPR (6), one would expect to see the same pK_a in the rate constants for either association or dissociation of NAD^+. However, the pK_a for NAD^+ association is 10 ± 0.2, while for dissociation it is 7.6 ± 0.2. Since the guanidine moiety of arginine has a pK_a of 12.5 it is likely that, in the enzyme, ARG-47 has a similar pK_a and is always protonated in the pH range being studied. Consequently, the pK_a of 9 ± 0.2 observed for the binding of ADPR and NADH must be due to another factor. The simplest explanation would be that due to a conformational isomerization the pK_a values observed for both ADPR and NAD^+ binding are apparent, with the same functional group involved. Without proposing this, it is extremely difficult to reconcile the disparity between the pK_a values for binding NAD^+ and ADPR, a fragment of the NAD^+. The following scheme can adequately describe the results obtained:

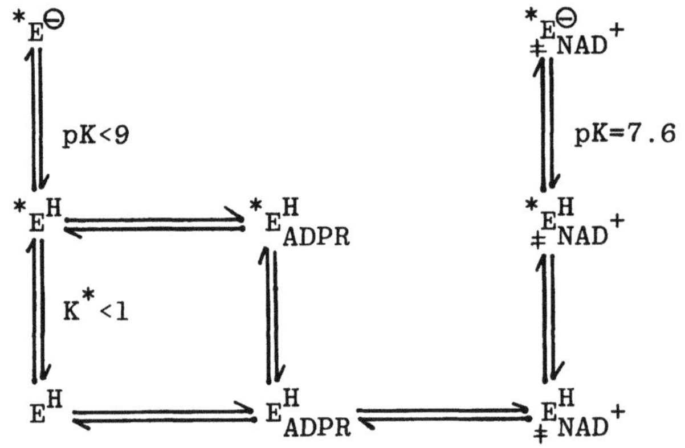

At neutral pH, the enzyme exists in two conformational states, $^*E^H$ and E^H, with both able to bind ADPR and NADH. Oxidized coenzyme, however, can only bind to one conformation. The other conformation, $^*E^H$, is linked to the ionization of a group with a pK_a less than 9.0 with the equilibrium constant for the conformational change, K^*, less than 1. The observed pK_a of 9 for ADPR binding arises from the effect of K^* on the observed equilibrium binding; the observed pK_a of 10 ± 0.2 for protein fluorescence quenching is due to both $^*E^H$ and $^*E^{\ominus}$ being quenched; and the pK_a of 10 ± 0.2 for NAD^+ binding results from E^H being the only conformation able to bind to it. Subsequent to binding of the nicotinamide ring an additional conformational change occurs to form $_{\ddagger}E^{H}_{NAD^+}$, related to the perturbation of the pK_a to 7.6. This conformational change is a necessary corollary since the binary complex is able to dissociate a proton but the unprotonated enzyme form is kinetically unable to bind NAD^+. The unprotonated form of the enzyme-NAD^+ complex selectively binds alcohol and is the catalytically active species.

The thermodynamic results obtained at neutral pH are compatible with the scheme we have presented. The ΔH and ΔS values for ADPR binding indicate a simple binding interaction, while the values for NAD^+ are characteristic of an entropically driven process, the conversion of E^H to $_{\ddagger}E^H$. The ΔG and ΔH values of 1.1 and 8.6 kcal/mole respectively for the difference between NAD^+ and ADPR binding indicate that the nicotinamide interaction (including the conformational change) is unfavorable, and that the driving force for coenzyme binding derives from the ADPR moiety.

There are many unknown quantities in the scheme presented above, such as the values for the pK_a and K^*, as well as the relative affinities of ADPR for E^H and $^*E^H$. Although the present scheme is compatible with currently available data, it is likely that future studies will result in considerable modification and extension of it.

REFERENCES

1. Shore, J. D., Gutfreund, H., Brooks, R. L., Santiago, D. and Santiago, P.: Proton equilibria and kinetics in the liver alcohol dehydrogenase reaction mechanism. Biochem. 13, 4185-4191 (1974).

2. Wolfe, J. K., Weidig, C. F., Halvorson, H. R., Shore, J. D., Parker, D. M. and Holbrook, J. J.: pH-dependent conformational states of horse liver alcohol dehydrogenase. J. Biol. Chem - In Press (1977).

3. Bernhard, S. A., Dunn, M. F., Luisi, P. L., and Schack, P.: Mechanistic studies on equine liver alcohol dehydrogenase. Biochem. 9, 185-191 (1970).

4. Weidig, C. F., Halvorson, H. R. and Shore, J. D.,: Evidence for site equivalence in the reaction mechanism of horse liver alcohol dehydrogenase with aromatic substrates at alkaline pH. Biochem. In-Press (1977).

5. Hadorn, M., John, V., Meir, F. K. and Dutler, H.: Kinetic equivalence of the active sites of alcohol dehydrogenase from horse liver. Eur J. Bioch. 54, 65-73 (1975).

6. Eklund, H., Nordstrom, B., Zeppezauer, E., Soderlund, G., Ohlsson, I., Boiwe, T., and Branden, C.-I.: The structure of horse liver alcohol dehydrogenase. FEBS Letters 44, 200-204 (1974).

Received February 7, 1977

DISCUSSION

Hollaway: How can a ligand, that lowers the pK_a of a group on a protein, bind preferentially to the acid form of the enzyme which contains the group in the acid form. The opposite is expected and this seems to be confirmed in your case by the fact that H^+ is released above pH 7.0. There seems to be some confusion (perhaps my own) between thermodynamically and kinetically determined pK_a values. Could you please explain?

Shore: The pK_a perturbation indicates that, thermodynamically, the ionized form of the enzyme has a much greater affinity for NAD^+. However, the reaction pathway is kinetically ordered so that only the protonated enzyme forms a complex with NAD^+ initially. This is demonstrated by the pH dependence of the rate of association of NAD^+. Structural features of the enzyme at alkaline pH apparently preclude coenzyme binding.

Zeppezauer: We have been studying the binding of ANS to liver alcohol dehydrogenase and found that probably more than one molecule of ANS is bound per subunit (R. Einarsson, Thesis, Uppsala, 1974). This has been confirmed by crystallographic analysis.

Still, the change of fluorescence of the complex liver ADH-ANS upon addition of different competitive ligands e.g. AMP, ADPR and other anions can be described as the change in binding (or quantum yield) of one molecule ANS per subunit. Would you agree with these observations?

Shore: We have also found a response of the fluorescence of bound ANS and TNS probes upon addition of ADPR which is linear with the fractional saturation.

Hinz: There appears to be a problem in rationalizing the different pK shifts induced by binding of NAD^+ and ADPR. From a purely thermodynamic point of view the apparent pK values correspond to apparent Gibbs free energies of ionization. Calorimetric measurements of the temperature coefficient of the reaction enthalpy involved in binding of NAD^+ and ADPR to horse liver ADH (F. Schmid, Thesis, Regensburg 1977) indicate that probably different "conformational changes", in the sense outlined in my lecture, are associated with the two reactions. Thus, it is at least conceivable, that a different environment of the ionizable group leads to an alteration of the free energy of ionization.

Shore: A pK_a shift occurs only when NAD^+ is bound, not ADPR. We have not obtained evidence for a conformational change due to ADPR binding at neutral pH, which is in agreement with crystallographic evidence. Accordingly, the Δc_p for ADPR binding to liver ADH should reflect primarily the differences in hydration, as you pointed out in your talk. We do not choose to call this a "conformational change".

In contrast to the results with ADPR, NAD^+ binding is associated with a pK_a shift and a conformational change. The existence of this conformational change is supported by the positive ΔH at 25^0 C. Subramanian & Ross (Fed. Proc. 36, Abstract 2074, 1977) have pointed out that liver ADH stands alone among dehydrogenases in the sign of ΔH for NAD^+ binding (determined calorimetrically). This we choose to call a "conformational change". We would expect Δc_p for NAD^+ binding to differ from that for ADPR binding by the Δc_p for the conformational change, although Δc_p (conf) = 0 is not ruled out as a possibility.

The Mechanism of Glutamate Dehydrogenase: Some Kinetic Aspects

K. Markau and K. Weber
Universität Konstanz, Fachbereich Biologie
Postfach 7733, D-7750 Konstanz (GFR)

This paper consists of two parts which deal with different approaches to the kinetics of glutamate dehydrogenase. In the first studies of the effectors ADP and leucine are presented under conditions when they act as competitive inhibitors. The second part concerns the binding kinetics of the reduced coenzyme in phosphate buffer that differ even in some qualitative aspects from the results reported for NADPH in Tris buffer.

ADP AND LEUCINE AS INHIBITORS OF GLUTAMATE DEHYDROGENASE

ADP and leucine are well known as activators of glutamate dehydrogenase by increasing V_{max} and suppressing abortive complexes (for a review see [1]). We have shown previously [2, 3] that the inhibitory effects of ADP observed under some conditions are due in pricinple to the same mechanism as the activation and are always prevalent under conditions of low turnover of the enzyme. Similar results were obtained for leucine.

Both effectors showed a first inhibition step which was not clearly separated from the activation under the usual conditions and a second, purely competitive step with relatively high inhibition constants K_{Ih}. Since at that time only one binding site for each effector was known we tried to construct models capable of explaining two-step inhibition by only one binding site.

In order to test these models we sought conditions under which the disturbing activation is minimal. Three approaches were used 1.) very low concentrations of substrate or coenzyme, 2.) saturation of the activation by one effector while observing the other one and 3.) examining the burst reaction using stopped-flow techniques. In addition, the inhibition was studied with the anomalous substrate norvaline. Under all these conditions linear

Lineweaver-Burk plots are obtained. Self-activation and self-inhibition by substrate or coenzyme thus do not play a significant part.

Low Concentration of Substrate or Coenzyme

Fig. 1 shows the results with [glu] < 500 µM, a value defined by a discontinuity in Lineweaver-Burk plot [3]. The inhibition is competitive and the secondary plots (A, B) show the combination of a partial inhibition and a normal one with K_{Ih} = 6 mM. A value for the low inhibition constants K_{I1} is obtained by plotting $K_{MI} = 1/(1/K_M - 1/K_{ML})$ versus [ADP] (C). The value for K_{ML}, the hypothetical Michaelis constant at saturation of the first step, is chosen so that the plot becomes linear. (The same value is obtained by extrapolation of the high concentration part to [ADP] → 0.) We find K_{I1} = 3 µM in excellent agreement with the reported K_D [4]. The strong mutual strengthening of binding between substrate and coenzyme prevents the specification, by these experiments alone, with which of either ADP really competes. Experiments with fixed [NAD^+] and variable [glu] (< 500 µM) yield the same type of inhibition and the same constants.

The measurements with leucine were performed at the beginning under reciprocal conditions: [NAD^+] below the first discontinutiy in the respective Lineweaver-Burk plot [5] (approx. 8 µM) and variable high [glu]. In agree-

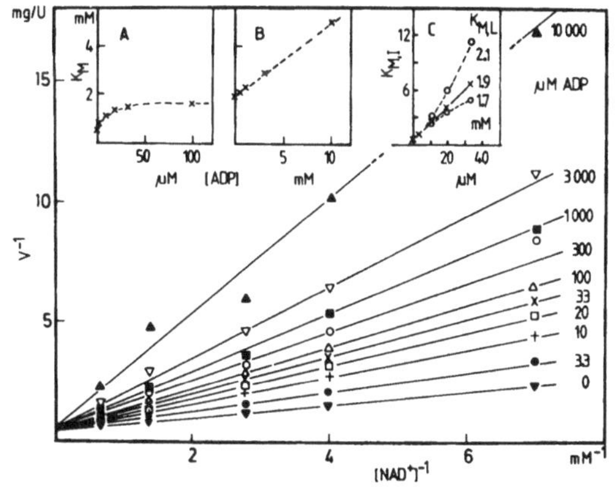

Fig. 1 Inhibition by ADP, A and B: Secondary plots of K_M from the main picture in two ranges of [ADP]. C: Secondary plot of K_{MI} versus [ADP] (see text). [glu] = 170 µM, M/15 phosphate buffer, pH 7.6, t = 20°C.

ment with our expectations we found two-step competitive inhibition also with this effector. K_{I1} = 1.8 mM is 5-fold higher than K_D derived from difference spectroscopy data [6]. The reason for this difference remains unclear since the conventional methods for determining K_D are not sufficiently precise in this region. K_{Ih} = 400 mM is also higher than the value we reported previously. This is mainly due to the limited solubility of leucine which allows determination of this high K_I only by extrapolation. The new value is secured by a larger number of measurements.

Mixed inhibition

With 100 µM ADP or 10 mM leucine, respectively, it was possible to saturate the activation and to remove all nonlinearities of the Lineweaver-Burk plots. Two-step inhibition by the respective other effector then is observed in the whole range of coenzyme and substrate concentrations. The K_{I1} values are both lower by a factor of 1.8 indicating mutual tightening of both effectors and supporting the conclusions from combined activation by both effectors [7]. The activation is saturated by one effector alone (in contrast to reported experiments[8] when obviously nonsaturating concentrations were used) whereas the inhibition is (at least qualitatively) additive. This shows that the dissociation of the product complexes is accelerated by each effector to such an extent that another process, unsensitive to the effectors, becomes rate limiting. Regarding the time constant of this process (approx. 10 sec^{-1}) it could be identified with the release of the proton [9] (probably as NH_4^+).

Two binding sites for each effector

Inconsistent with our first model [2], a significant inhibition amplitude by low concentrations of ADP is also observed with low [NAD^+] and relative high [glu]. Reciprocal results were obtained with leucine. In addition, two-step inhibition is also observed with norvaline as substrate that does not strengthen the binding of the coenzyme. (We found K_{I1} = 6 µM and K_{Ih} = 9 mM; the difference to the preceding values is due to the higher pH (8.0) which was necessary for sufficient reaction velocities.)

These results cannot be explained by only one binding site for each effector, especially since we have found [10] that a microbial enzyme (NADPH-specific from Candida utilis) shows only the inhibition by high effector concentrations. In fact such models are not required, following the detection of a second binding site for ADP with K_D = 6 mM [11, 12]. Furthermore we have found that K_M of norvaline in the limit of vanishing substrate concentration is of the order of 10^{-1} M so that K_{Ih} of leucine can be identified with the active binding site for monocarboxylic acids.

We thus assume a defined weakening of the coenzyme binding by a first ADP binding site and further interaction with a second one. For NAD^+ no residual binding was detected in the presence of intermediate [ADP] [12], but this is most probably due to the limits of the method used in those experiments. NADH on the other hand is loosened by a factor 4-5 upon addition of 100 μM ADP and K_D = 80 μM seemed to remain constant at higher [ADP] [14]. We have now found (by determination of the supernatant [NADH] after ultracentrifugation of a mixture with the enzyme) that K_D increases further in the presence of very high [ADP] (Table 1) and we calculated K_I = 6 mM for this displacement. (After saturating the activation we find competitive inhibition with K_I = 6 mM also in the reverse reaction.)

First we assumed that leucine interacts mainly with glutamate since for NADH we could not find the negative interaction with leucine which has been concluded on the basis of spectroscopic values for NADPH. Stopped-flow data revealed, however, (see below) that leucine simultaneously weakens the first, active binding site and strengthens a second one so that the net effect in the equilibrium studies is normally small. On the other hand, further weakening of the coenzyme binding was not observed with very high [leu] so that K_{Ih} is obviously due to the competition with glutamate as we supposed above. It must thus be assumed that at pH 7.6 the monocarboxylic

Table 1. Competition of high ADP with NADH. Measurements by ultracentrifuge under the assumption of one binding site per polypeptide chain.

[ADP]	1	5	10	15	20	mM
K_D (NADH)	75	106	153	200	245	μM

acids are not only bound much worse than glutamate but that also V_{max} is lower by at least an order of magnitude. Together with the strong increase of this value at high pH this shows the importance of sufficiently high negative charge around the active center for the enzymic reaction.

With norvaline as the substrate no inhibition by leucine is observed. This can, however, not be used as an argument for lacking interaction between leucine and coenzyme since norvaline is an effector of the same type as leucine. The high K_{Il} ($\approx$ 4 mM) shows the importance of the hydrophobic part for this binding site (in fact, glycine and alanine have no effect). On the other hand, K_{Ih} of norvaline ($\approx$100 mM) is lower than that of leucine in agreement with the better substrate properties of the first.

Quantitative determination of the coupling and kinetic models

The linearity of the Lineweaver-Burk plots which intersect at one point and of the secondary plots allows the determination of the parameters in the bilinear rate equation in the case of low [glu]. We compare these values for the free enzyme with those in the presence of 100 μM ADP (a defined state, since the first step is highly saturated whereas the second is practically not yet occupied). Fig. 2 shows the results which are summarized in Table 2 using the nomenclature of Dalziel [15] and the rate equation

$$[E]/v = \emptyset_0 + \emptyset_N/[N] + \emptyset_S/[S] + \emptyset_{NS}/[N][S] .$$

(N refers to the coenzyme, S to the substrate)

$\emptyset_{NS}/\emptyset_S$, the Michaelis constant of NAD^+ for vanishing [glu], equals K_D of NAD^+ for all mechanisms discussed (ternary complex, glutamate not binding compulsorily first) and our value of 750 μM is in reasonable agreement with the reported K_D (400 μM [13]). The remaining difference, if it does not only reflect the experimental errors, might be due to negative coupling between one active center binding NAD^+ and a second still loaded with NADH. The corresponding value with 100 μM ADP thus has also to be considered as K_D of NAD^+ and we obtain a weakening factor of 4 (free energy of coupling 3.3 kJ/mol)similar to that with NADH. The main part of the interaction thus

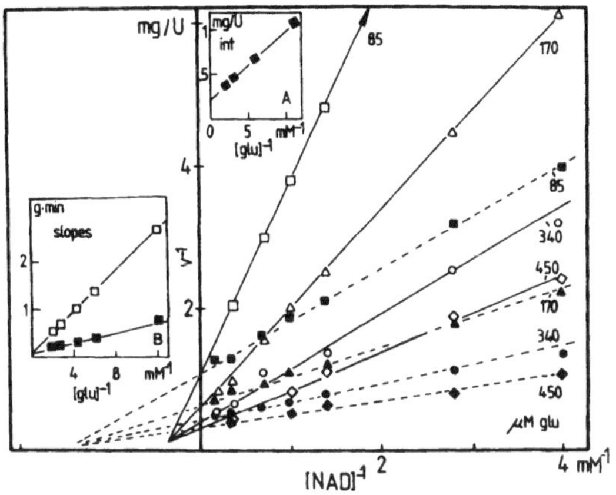

Fig. 2 Determination of the kinetic parameters with low [glu]. A: Secondary plot of the intercepts with the v^{-1} axis in the main picture (with and without effector). B: Secondary plot of the slopes. Open symbols: with effector, closed symbols: without effector. Conditions as in Fig. 1.

Table 2 The kinetic parameters at low [glu], values taken from Fig. 2. * with an equivalent weight of 56 000 for the functional unit, the values are twice as high under the assumption of half-of-the-sites reactivity, see below.

	ϕ_0(mg/U)	ϕ_N(mg·min)	ϕ_S(mg·min)	ϕ_{NS}(mM·mg·min)
control	0.25	90	74	55
100 μM ADP	0.25	90	74	220

	ϕ_N/ϕ_0(μM)	ϕ_S/ϕ_0(μM)	ϕ_{NS}/ϕ_S(μM)	ϕ_{NS}/ϕ_N(μM)
control	360	310	740	630
100 μM ADP	360	310	3,000	2,500

	$1/\phi_N$ ($M^{-1}sec^{-1}$)*	$\phi_{NS}/(\phi_N \cdot \phi_S)$ (sec^{-1})*
control	0.012	8.8
100 μM ADP	0.012	35

is located on the ADPR moiety of the coenzyme. The Michaelis constant of NAD^+ for infinite [glu], $\emptyset_N/\emptyset_0$, should equal K_D of NAD^+ with respect to the E-glu complex if the mechanism would be of the rapid equilibrium type. Our value of 360 μM, however, is not only much higher than the reported one in the presence of glutarate which mimics glu with respect to the binding properties (6 μM [12]) but, in contrast to all known data, also insensitive to ADP. Under very low saturation with substrate the reaction thus has to be considered as practically ordered with the coenzyme as the leading ligand. Then $1/\emptyset_N$ equals k^+, the binding rate constant of NAD^+ to the free enzyme, and with the known K_D also k^- can be calculated. The values (k^+ = 0.012 μM^{-1} sec^{-1}, k^- = 8.8 sec^{-1}) are surprisingly low but they are apparently supported by stopped-flow data on the replacement of NADH by NAD^+ (work under progress). Similar results are obtained for leu with the difference that the energy of coupling is lower (Δ G = 1.7 kJ/mol).

Other adenine nucleotides

ATP is an inhibitor of the same type as ADP with approx. 10-fold higher K_{Il} and the same K_{Ih} (in accordance with the NMR results). The difference in K_{Il}, which governs also the activation, leads occasionally to effects opposite to that of ADP which were sometimes overinterpreted [16]. AMP exhibits the same K_{Ih} but no effect at lower concentrations. Nevertheless it appears to be bound to a site with higher affinity since it induces weak activation under appropriate conditions and reduces the binding of NADH by a small factor (approx. 1.5 following our ultracentrifuge data). Most conclusive are the experiments with mixtures of AMP and ADP. Fig. 3 shows that AMP suppresses completely the inhibition by ADP (the extrapolation of the high concentration part to [AMP] $\rightarrow$ 0 yields 1, the activity of the free enzyme) and then shows its normal inhibition. With the assumption that the equilibrium between ADP and AMP is not disturbed by the coenzyme ($[NAD^+] \ll K_M$ and the binding rate constants of both effectors being very low, see below) we calculate K_D of AMP from the inserted secondary plot by the equation

$$\text{intercept}\ ([AMP]) = 1/\text{slope} = [ADP] \cdot K_D(AMP)/K_D(ADP)$$

and obtain 12 μM in good agreement with direct measurements [17]. Since the

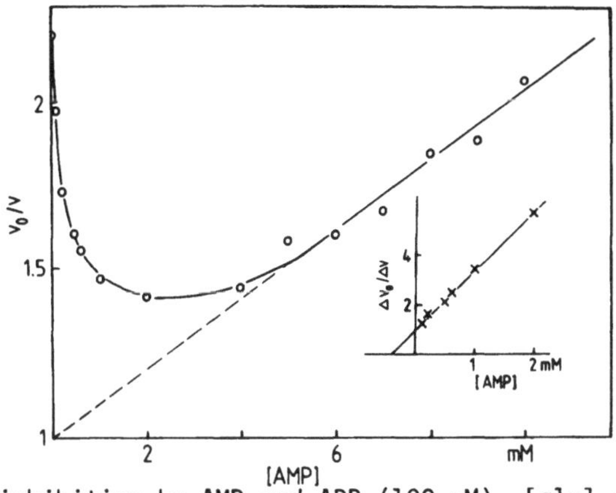

Fig. 3 Mixed inhibition by AMP and ADP (100 μM). [glu] = 170 μM, [NAD^+] = 500 μM. v_0 = activity without effector. Insert: Δv is the difference between the measured activities and the extrapolated line in the main plot, Δv_0 is the value for [AMP] = 0. Conditions as in Fig. 1.

difference in the free binding energy between AMP and ADP is the same as the free energy of coupling between ADP and NAD^+ the following model is strongly suggested: AMP (at the high affinity site) and the coenzymes do not use overlapping binding subsites. (The small interaction with NADH and the ternary substrate complexes can be attributed to electrostatic repulsion which for NAD^+ is compensated by the positive charge.) ADP (and ATP), however, are bound to an additional subsite for phosphate (e.g. a basic residue) which is also used by the coenzymes and is characterized by a free energy increment of 3.3 kJ/mol. ADPR was also tested and shows unlimited competitive inhibition with K_I = 260 μM. (Slight deviations at the very highest concentrations might indicate a second inhibition step also for this nucleotide.) Apparently ADPR uses the normal coenzyme binding site and we conclude that the oxidized nicotinamide ring does not supply a negative contribution to the free binding energy. The small residual binding of NADH in the presence of ADPR which has to be expected on the base of its K_D (15 μM) is much too weak (60 mM) to be detected. The inhibition constant of ADPR increases also by a factor of 4 in the presence of 100 μM ADP. This gives further support to the proposed model. Binding of ADPR to the high affinity site is suggested by some stopped-flow data but effects on the steady state kinetics could not be detected.

The burst reaction, comprising all reaction steps up to the formation of the

E-NADH-oxoglutarate-(NH_4^+) complex, is also competitively inhibited by ADP and leucine (Fig. 4). The first inhibition step is saturated if the enzyme is preincubated with 50 μM ADP (the high concentration of the enzyme prevents the determination of K_I). Without preincubation the first inhibition step is characterized by $K_I \approx$ 120 μM. The second inhibition step again shows K_I = 6 mM independent of preincubation. In agreement with the NMR data the binding to the second site thus is very fast. For the high affinity site the hypothesis has been formulated [18] that very fast binding with high K_D and no influence on the coenzymes is followed by a slow adaptation step which causes strong tightening of the ADP and its interaction with the coenzyme. If, in the frame of this model, our value of K_I is interpreted as K_D of the first step, it must be assumed, in contrast, that already this fast prebinding causes interaction with the coenzyme. On the other hand the first order plots of the reaction are practically linear also in the presence of ADP and also some stopped-flow data indicate that

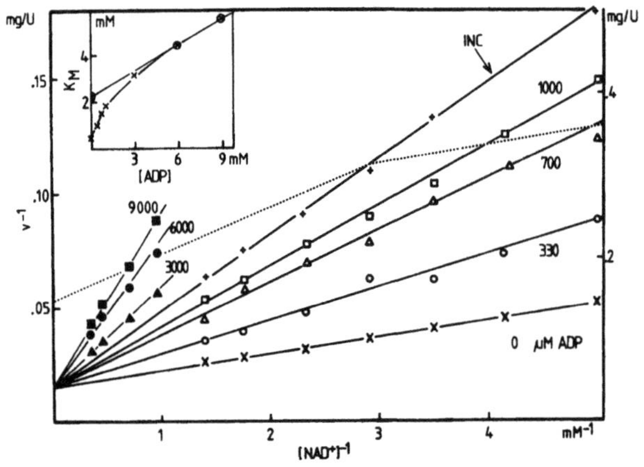

Fig. 4 Inhibition of the burst reaction. The extrapolated initial velocity of the burst is shown in Lineweaver-Burk representation (scaling at the left). The enzyme (20 μM polypeptide chain) is mixed with the reaction partners and the effector. INC refers to the case of preincubation with 50 μM ADP. The dotted line shows the steady state activity (scaling at the right) under the same conditions (with 0.15 μM enzyme). Insert: Secondary plot of the K_M values. 0 are measurements with preincubation. [glu] = 8.5 mM, other conditions as in Fig. 1.

the model has to be further revised. Model calculations on the base of non-equilibrium kinetic competition between ADP and NAD^+ are under progress. Also for leucine the effect of preincubation is clearly visible. K_I of the first inhibition step without preincubation can be estimated to be around 25 mM.

It should be noted that the discontinuities that are observed in the steady-state reaction with the same concentrations of glutamate and NAD^+ (dotted line with changed y-scale) are completely absent in the burst. This shows that the self-activation by the coenzyme is rather due to accelerated dissociation of the product complexes ("flip-flop"-effect) than to anticooperative coenzyme binding as it was earlier proposed [5].

THE KINETICS OF COENZYME BINDING

The kinetics of NADPH binding to glutamate dehydrogenase have been studied in Tris buffer [19]. For comparison with the steady-state and equilibrium data we need, however, measurements in phosphate buffer and also with NADH. In the following we present some of those results which differ not only quantitatively from the known data but show qualitatively new effects.

It was known [14] that in addition to the active binding site for NADH there exists a second one. Kinetic data for this site were not available and it was thought to induce no fluorescence enhancement. On the contrary, the time course of the fluorescence upon mixing enzyme with NADH is clearly biphasic. A plot of the apparent time constants k_a versus [NADH] is shown in Fig. 5. K_D for the slow process coincides with the value determined by equilibrium measurements. NADPH shows, in agreement with the equilibrium data, only a very small amplitude of the slow phase. Dilution of a mixture of enzyme with NADH into buffer also causes a biphasic decay of fluorescence. In the presence of ADP (preincubated), which is known to suppress the second site, the slow phase disappears (whereas the other one only becomes faster) and the displacement of NADH by NAD^+ or ADP is also biphasic. These observations support the attribution of the slow phase to the second binding site. This finding allows differentiation of the effects of other ligands on either binding sites. In this way we studied the antagonistic

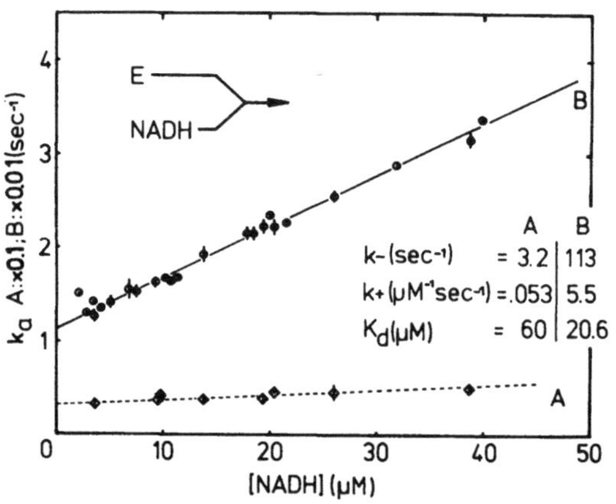

Fig. 5 The apparent time constants of the fast (B) and the slow (A) phase in the fluorescence enhancement upon mixing of NADH with GDH plotted versus [NADH]. Regard the different scales on the k-axis. M/15 phosphate buffer, pH 7.6, t = 20°C.

effects of leu cited above.

Nonequality of the six binding sites in the hexameric enzyme

Equilibrium studies have shown that the binding of NADPH is described by two sets of 3 sites with different K_D due to either intrinsic asymmetry or to negative interaction between the 6 active centers [4]. In contrast to earlier reports we find this effect also in the stopped-flow experiments. Fig. 6 shows a plot of k_a versus [NADPH] with significant nonlinearity. The data are well fitted by computer simulation of the binding process using a model with negative homotropic interaction between functional dimers. No additional step describing a conformational change is needed for the fit and we thus favor at the moment a model with direct interaction between the active sites instead of allosteric interaction.The same parameters introduced in a model with intrinsic asymmetry lead to large deviation from the experiment (dashed line) but a fit is also obtained by this model if the difference between both sites is assumed to be stronger. In order to see the reason why this effect has not been found earlier we repeated the ex-

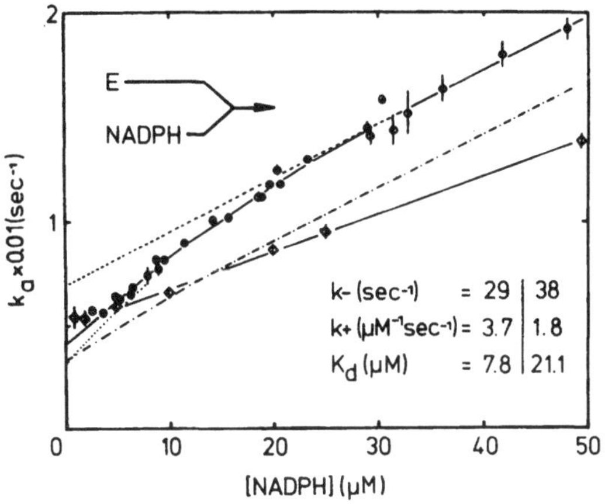

Fig. 6 The apparent time constants of fluorescence enhancement upon mixing of NADPH with GDH. O: measurements in phosphate buffer as in Fig. 5. —— theoretical fit using the parameters in the picture and a model of negative interaction with second order kinetics, ····· the same model calculated with first order kinetics. -·-·-·- the same parameters used in a model of a priori different sites. ◇ : measurements in M/10 Tris buffer, pH 7.6, + 1 mM phosphate.

periments in Tris buffer and, indeed, we obtained a linear plot. Cooperative effects in the enzyme thus depend strongly on the buffer conditions. K_D in Tris buffer corresponds to the higher value in phosphate so that some type of cooperativity between phosphate and coenzymes is suggested which has to be studied in more detail. (First steady-state-kinetic studies in Tris buffer showed also the absence of the discontinuities in the Lineweaver-Burk plots at least with $[NAD^+] > 100$ µM.) The "anticooperativity" is much less expressed for NADH [4] and our experiments with this coenzyme are not yet fully conclusive.

Direct determination of k^- for the ternary complexes with glutamate

From the published measurements with the ternary complexes [19] k^- could not be determined as a consequence of the relatively high coenzyme concentrations. The indirect method using K_D values obtained by other methods is less accurate and requires, in addition, the assumption that the observed stopped-flow amplitude describes exactly the equilibrium process. In order to determined k^- directly we reduced [NAD(P)H] drastically. Since for sensitivity reasons the enzyme concentration could not be diminished in the same ratio, the first order plots are no longer linear. In a first approximation we fitted them by least square parabola and plotted the initial values versus [NAD(P)H] as in Fig. 7.The data were fitted with theoretical second order initial rates for which first approximations are obtained from the slope of the high concentration part (k^+) and the plateau value p ($k^- = p - [GDH]\ k^+$). In the presence of 100 mM glutamate we found the values $k^- = 0.7 \cdot sec^{-1}$ (NADH) and $k^- = 0.18\ sec^{-1}$ (NADPH) showing a large difference between the two coenzymes which explains some differences in the steady-state kinetics. In the presence of ADP these values increase by a factor 36 (NADH) or even 50 (NADPH). In agreement with the preceding conclusions the limitation of the steady-state reaction by this step is thus fully removed by ADP. The acceleration factors induced by leucine (10 mM) are smaller (8 and 20) but obviously sufficient for removing the bottleneck for the total reaction.

The interaction of NAD(P)H with ADP and leucine

After preincubation of the enzyme with saturating amounts of ADP (100 μM) or

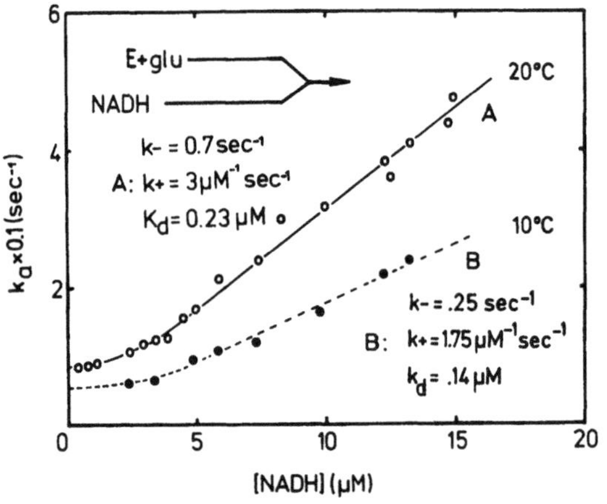

Fig. 7 The apparent time constants of NADH binding at GDH in the presence of 100 mM glutamate (in both syringes) versus [NADH] for two different temperatures. Buffer conditions as in Fig. 5.

leucine (10 mM) the k^- values of the coenzymes increase significantly whereas k^+ remains practically unchanged. Anticooperativity is no longer observed but it could become apparent at much higher coenzyme concentrations. If the enzyme is mixed with ADP and coenzyme simultaneously fast fluorescence enhancement is followed by a slower decay demonstrating slow binding of ADP.

The time constants of the fluorescence decay after adding ADP to a mixture of enzyme and NADH (keeping the total NADH concentration constant) plotted versus [ADP] show a curved graph (Fig. 8). At first glance this is in agreement with the two step binding model cited above [18], the plateau value, however, is much higher than the reported slow isomerisation step (50 sec^{-1}). Since the rate of this reaction is also limited by the dissociation rate of NADH in the presence of ADP which we found to be in the region of 140 sec^{-1}, we doubt if a two-step mechanism for the ADP binding exists at all. The binding process itself then is correspondingly slower. Out curve is fitted with $k^+ = 0.4\ \mu M^{-1} sec^{-1}$ but this value might be higher (up to 4-fold) in the absence of NADPH. k^- cannot be determined in this way, but is certainly below 5 sec^{-1}. The slower relaxation in the difference spectroscopy might describe a process which has no significant influence on the binding of ADP.

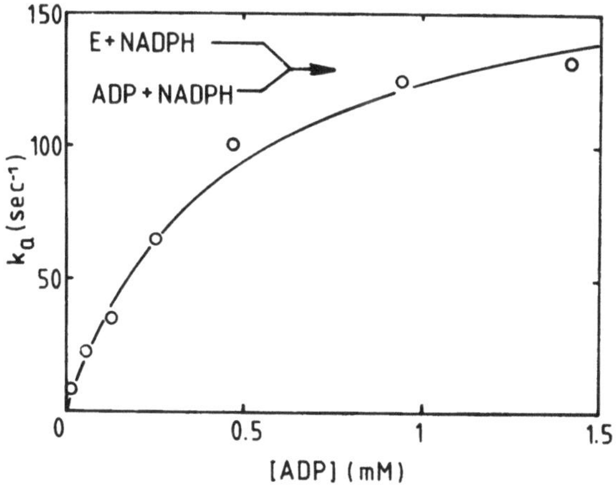

Fig. 8 The time constants of the fluorescence decay after addition of ADP to a mixture of enzyme with NADPH (same concentration in both syringes) versus [ADP]. Conditions as in Fig. 5.

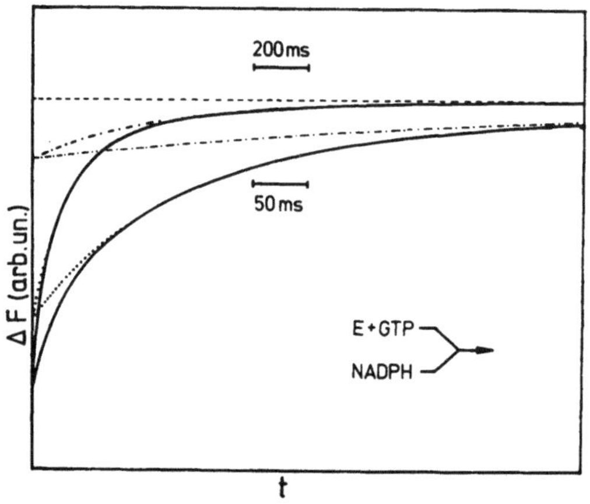

Fig. 9 The time course of the fluorescence upon addition of NADPH to a mixture of enzyme with 5 mM GTP (in both syringes) in two different time scales.

The ternary complexes with GTP

The fluorescence enhancement in the presence of GTP is very complex and comprises at least 3 phases (Fig. 9). At low [GTP] ($<$500 μM) the plots of k versus [NAD(P)H] are curved upwards and suggest positive cooperativity in agreement with the equilibrium data. At 5 mM GTP, however, cooperativity is no longer observed. The cooperative effect thus concerns obviously GTP and not NAD(P)H (possibly a strong negative effect in the binary GTP complex is removed in the ternary one with NAD(P)H). The fast phase has not yet been analysed even with NADPH since the experiments were performed either with very high NADPH [20], when the reaction is too fast, or by difference spectroscopy [21] when this step is not seen. We find for this step $K_D \approx 40$ μM. This shows that in the first step NADPH is not only not tightened by GTP but even loosened demonstrating probably an overlap of the binding sites. No indication was obtained for an isomerisation step following the binding which has been postulated [21]. The second phase is, in agreement with earlier suggestions, exactly correlated with the dissociation into hexamers (as we checked by measuring the Rayleigh scattering). A linear plot of k versus [NAD(P)H] is obtained also in this case. In agreement with difference spectroscopic data a third phase is also found. Its time constants are very near to those for the second binding site of NADH and we find a linear dependence of k on [NADH] with a reasonable K_D value. In the presence of GTP, however, a significant amplitude of this slow process is also found with NADPH. It thus must be assumed that GTP strengthens the binding of NADPH at the second site by a very large factor, a hypothesis which is not supported by the equilibrium studies [14]. The alternative interpretation of the slow phase as an isomerisation step, on the other hand, is excluded by the linearity of the plot k versus [NADPH]. The analysis of the whole process in a larger range of NAD(P)H concentrations and computer simulations are under progress.

We hope that the present results together with the replacement experiments by $NAD(P)^+$ which are under study will give us the parameters for the simulation of the whole enzymic reaction.

References

1. H. Sund, K. Markau and R. Koberstein: Glutamate Dehydrogenase in "Subunits in Biological Systems" Part C (Vol. 7 of Biological Macromolecules, a Series of Monographs, ed. by S.N. Timasheff and G.D. Fasman), Marcel Dekker, New York 1975, p. 225.

2. K. Markau, J. Schneider and H. Sund: Kinetic Studies on the Mechanism of the Action of ADP on the Glutamate Dehydrogenase reaction. FEBS Letters 24, 32-36 (1972).

3. K. Markau and I. Steinhübel: Kinetic Measurements with Monocarboxylic Acids as Substrates and Effectors of Glutamate Dehydrogenase. FEBS Letters 28, 115-120 (1972).

4. J. Krause, M. Bühner and H. Sund: Studies of Glutamate Dehydrogenase. The Binding of NADH and NADPH to Beef-Liver Glutamate Dehydrogenase. Eur. J. Biochem. 41, 593-602 (1974).

5. K. Dalziel and P.C. Engel: Antagonistic Homotropic Interactions as a Possible Explanation of Coenzyme Activation of Glutamate Dehydrogenase. FEBS Letters 1, 349-352 (1968).

6. R.A. Prough, J.M. Culver and H.F. Fisher: Spectrophotometric Evidence for a Glutamate Dehydrogenase-L-Leucine Complex. Arch.Biochem.Biophys. 149, 414-418 (1972).

7. R.A. Prough, J.M. Culver and H.F. Fisher: The Mechanism of Activation of Glutamate Dehydrogenase-catalyzed Reactions by Two Different, Cooperatively Bound Activators. J.Biol.Chem. 248, 8528-8533 (1973).

8. A. Hershko and S.H. Kindler: Mode of Interaction of Purine Nucleotides and Amino Acids with Glutamate Dehydrogenase. Biochem. J. 101, 661-664 (1966).

9. A. d'Albis and D. Pantaloni: Fast-Kinetic Studies of the Oxidative Deamination of Glutamate Catalyzed by Glutamate Dehydrogenase. Eur. J. Biochem. 30, 553-559 (1972).

10. P. Neumann, K. Markau and H. Sund: Yeast Glutamate Dehydrogenase from Candida utilis as a Model for the Mammalian Enzyme - the Interaction with Effectors at High Concentrations. 10th Int. Congr. Biochem. Hamburg 1976, Abstr. No. 07-5-160.

11. K. Markau: NMR Studies on the Mechanism of Glutamate Dehydrogenase. 10th Int. Congr. Biochem. Hamburg 1976, Abstr. No. 07-5-132.

12. R. Koberstein, H. Dieter, K. Markau and H. Sund: The Interaction of Glutamate Dehydrogenase with Ligands. This book.

13. K. Dalziel and R.R. Egan: The Binding of Oxidized Coenzymes by Glutamate Dehydrogenase and the Effects of Glutarate and Purine Nucleotides. Biochem. J. 126, 975-984 (1972).

14. R. Koberstein and H. Sund: Studies of Glutamate Dehydrogenase. The Influence of ADP, GTP, and L-Glutamate on the Binding of the Reduced Coenzyme to Beef-Liver Glutamate Dehydrogenase. Eur. J. Biochem. 36, 545-552 (1973).

15. K. Dalziel: Initial Steady-State Velocities in the Evaluation of Enzyme-Coenzyme-Substrate Reaction Mechanisms. Acta Chem. Scand. 11, 1706-1723 (1957).

16. A.L. Lehninger: Biochemistry, p. 566, Worth Publishers, New York 1975.

17. S. Subramanian, D.C. Stickel, and H.F. Fisher: Thermodynamics of Complex Formation between Bovine Liver Glutamate Dehydrogenase and Analogs of ADP. J. Biol. Chem. 250, 5885-5889 (1975).

18. A.H. Colen, D.G. Cross, and H.F. Fisher: Two-Step Binding of Adenosine Diphosphate to L-Glutamate Dehydrogenase. Effect on the Binding of Reduced Nicotinamide Adenine Diphosphate and on Enzymatic Catalysis. Biochemistry 13, 2341-2347 (1974).

19. A. di Franco and M. Iwatsubo: Reaction Mechanism of L-Glutamate Dehydrogenase. Characterization of Optical and Kinetic Properties of Various Enzyme Reduced-Coenzyme Complexes. Eur. J. Biochem. 30, 517-532 (1972).

20. J.M. Jallon, A. di Franco et M. Iwatsubo: Etude cinêtique de changements conformationnels de la L-glutamate dêshydrogênase provoquês par le couple effecteur GTP + NADH. Eur. J. Biochem. 13, 428-437 (1970).

21. Ch.Y. Huang and C. Frieden: The Mechanism of Ligand-induced Structural Changes in Glutamate Dehydrogenase. Studies of the Rate of Depolymerization and Isomerization Effected by Coenzyme and Guanine Nucleotides. J. Biol. Chem. 247, 3638-3646 (1972).

Received March 24, 1977

DISCUSSION

Werber: Could you comment on the difference between the results in phosphate and Tris buffers? Would it not be wiser to do the experiments in Tris?

Markau: We do not know the exact reasons for this difference. The weaker binding in the first step and the lacking anticooperativity can be due to the general destabilisation of the enzyme in Tris or to an accumulation of positive charges by preferential binding of the amine. Measurements in phosphate buffer are of special interest not only because we just want to study the anticooperative effects but also because Tris is a rather unphysiological system.

Sund: We have shown by treatment with iodo 2-^{14}C acetic acid as well as with Ellmann's reagent that the six -SH groups of the polypetide chain are much better accessible in Tris than in phosphate buffer (Eur. J. Biochem. 52, 221 1975) indicating a more flexible structure in Tris. Also the enzyme is apparently stable in phosphate buffer whereas in Tris it is very sensitive to denaturation and undefined aggregation. Therefore, Tris is in general not adequate for the investigation of glutamate dehydrogenase.

Kempfle: You said that you saw for the first time these two binding sites. Well, we saw it already in T-jump relaxation kinetics and reported it in 1974 at the 8th FEBS meeting in Budapest. We find the same values. Are your concentrations total concentrations or equilibrium concentrations? How high is your enzyme concentration?

Markau: As we already discussed you find two dissociation constants without reporting the rate constants in your abstract. No attribution of the constants to the two types of sites has been proposed and I could not see how you could exclude a sequential mechanism. Our experiments differ in the properties of either sites: in your experiments the high affinity site is removed by ADP whereas in ours it is the low affinity site. I do not know the reason for this difference but I have principal doubts about the T-jump method because of the electrical polarisation effects. The concentrations are total, but with [E] = 3 μM the corrections are very small.

Engel: You argue against a rapid-equilibrium random-order mechanism. Without doubt at low glutamate concentrations the reaction pathway through the enzyme-NAD^+ complex must predominate, but what do you think happens at higher glutamate concentrations? Since at the last pyridine nucleotide-symposion the the binding of glutamate in the absence of coenzyme has been directly demonstrated. Also our product inhibition results and Silverstein's isotope exchange results support a random-order mechanism.

Markau: I agree, and specified in the abstract, that the ordered mechanism is only induced by the low substrate concentrations which enable us to study the effectors under simplifying conditions. This degeneration of a random order mechanism is based, however, on the relatively slow equilibration of NAD^+. We confirmed by NMR that glutamate forms a binary complex with K_D = 50 mM and added the information that k^- is 4000 sec^{-1} (this symposion). Glutamate thus is always rapidly equilibrated. k^+ of NAD^+ to the enzyme-glu complex must be much higher than to the enzyme alone in order to explain the high burst velocity. Since k^- from the complex, on the other hand, is very low, the statement remains true for all values of [glu] that the coenzyme is not rapidly equilibrated. Due to the fast equilibration of glu things are more complicated under reversed conditions (very low NAD^+) the more as the determination of $\emptyset_S$ is difficult.

REGULATION OF ISOCITRATE OXIDATION BY TPN- AND DPN-ISOCITRATE DEHYDROGENASES

Gerhard W.E. Plaut and Colleen M. Smith
Temple University School of Medicine
Philadelphia, Pennsylvania 19140 USA

EXPERIMENTS WITH SOLUBLE ENZYMES

Isocitrate dehydrogenase. Studies on the reactivity of isocitrate dehydrogenases with several analogues of isocitrate (1) have shown that DL-threo-α-methylisocitrate (DL-threo-α-meIc) is a potent inhibitor, competitive with isocitrate, of the TPN-linked isocitrate dehydrogenase (TPN-IDH) from bovine and porcine heart and rat liver (2). The TPN-IDH activities from rat liver cytosol and mitochondrial extract were inhibited to the same extent by α-methylisocitrate (α-meIc) when Mg^{2+} was used as the activating cation. The DPN-specific enzyme (DPN-IDH) from extracts of rat liver mitochondria or purified from bovine heart was not inhibited by α-meIc (2).

A more detailed examination of the reactions catalyzed by purified pig heart TPN-IDH has shown that DL-threo-α-meIc is an inhibitor competitive with isocitrate, α-ketoglutarate, oxalosuccinate (decarboxylation), oxalosuccinate (reduction) and TPN^+ (3). The interaction coefficient ($\alpha = 0.11$) obtained in double inhibitor studies (α-meIc and TPN^+) of the reductive carboxylation of α-ketoglutarate to isocitrate suggests that α-meIc facilitates TPN^+ binding. With magnesium isocitrate varied at fixed levels of TPN^+, the line of the double reciprocal plot had zero slope at functionally infinite TPN^+ concentration (> 1 μM). These data seem consistent with an ordered rapid equilibrium reaction whose Km(app) of the substrate adding to the enzyme first (magnesium isocitrate) approaches zero with high levels of the second substrate (TPN^+).

This could imply that under physiological conditions Km(app) of magnesium isocitrate may vary considerably with relatively small changes in TPN^+

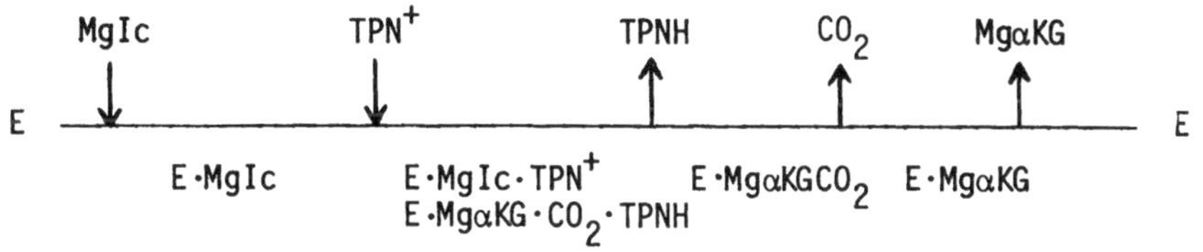

levels.

Identification of the active isomer of α-methylisocitrate. Examination of the stereospecificity of α-meIc revealed that: (a) DL-threo-α-meIc prepared by an unequivocal chemical synthesis is the inhibitor of TPN-IDH, whereas, the DL-erythro-diastereoisomer is inactive (2). (b) Gawron and Mahajan (4) showed previously that α-methyl-cis-aconitate is a substrate for horse heart aconitase (eq. 1).

(1) α-meIc $\leftrightharpoons$ α-methyl-cis-aconitate $\leftrightharpoons$ methylcitrate

DL-threo-α-MeIc is a substrate for bovine heart aconitase, whereas, DL-erythro-α-meIc is a weak uncompetitive inhibitor of the enzyme (5). (c) The product formed by aconitase from α-methyl-cis-aconitate inhibits TPN-IDH. A study of the stoichiometry of α-methyl-cis-aconitate disappearance and α-meIc formation (as measured by inhibition of TPN-IDH) shows that α-meIc appears to be the only product. In other experiments, formation or utilization of methylcitrate, the third expected component of the aconitase equilibrium between the methyltricarboxylate derivatives (eq. 1), could not be demonstrated. Furthermore, synthetic methylcitric acid, shown by NMR spectroscopy to contain both diastereoisomeric pairs, did not inhibit TPN-IDH (5). (d) Examination by optical rotation of α-meIc formed from α-methyl-cis-aconitate and of α-meIc remaining after incubation with aconitase has established that the D-threo-α-methylisocitrate isomer is the inhibitor of TPN-IDH and the substrate of aconitase (5).

EXPERIMENTS WITH INTACT RAT LIVER MITOCHONDRIA

The relative contributions of DPN-IDH and TPN-IDH to mitochondrial isocitrate oxidation has been the subject of conjecture and experimentation. Ernster and Navazio (6) found that the DPN-IDH activity was about one-fifth that of the TPN-IDH activity in rat liver mitochondria. The specific DPN-IDH and TPN-IDH from mitochondrial extracts of livers from several species (rat, rabbit, guinea pig) have been separated (7). The purified DPN-IDH from porcine liver exhibited properties similar to the enzyme from bovine heart, including activation by ADP and inhibition by DPNH, TPNH, ADPR, and ATP (8, 9). Stein et al. (10) in studies with pyridine nucleotide-depleted mitochondria from a number of tissues concluded that in the presence of ADP, the oxidation of isocitrate in rat liver mitochondria proceeded at about equal rates through DPN-IDH and the TPN-IDH-transhydrogenase pathway. Investigations by Ernster and Navazio (6, 11), Ernster and Glasky (12) and Nicholls and Garland (13) indicate that the oxidation of isocitrate by rat liver mitochondria occurs predominantly by the DPN-IDH step. However, Moyle and Mitchell (14) were unable to detect a DPN-IDH-dependent pathway in rat liver mitochondria and attributed isocitrate oxidation entirely to the TPN-IDH-transhydrogenase system. In contrast, Hoeck et al. (15), reported the occurrence of isocitrate oxidation in rat liver mitochondria when pyridine nucleotide transhydrogenase was inhibited by palmityl CoA.

In preliminary communications (16, 17) we have reported studies with α-meIc which indicate that oxidation of isocitrate in liver mitochondria occurs primarily via DPN-IDH. In the present report, the effect of DL-threo-α-meIc on isocitrate oxidation by intact rat liver mitochondria has been examined. This study includes the effect of the inhibitor of TPN-IDH on formation of intermediates of the citric acid cycle and accompanying changes in levels of intramitochondrial TPNH and DPNH, as well as the rates of formation of other products dependent on intramitochondrial generation of reduced pyridine nucleotides from isocitrate.

The uptake of α-meIc by liver mitochondria. Because the inner mitochondrial membrane poses a permeability barrier to the free transport of many anions (18), it was necessary to establish the time course and extent of uptake of α-meIc. The transport of α-meIc into mitochondrial matrix (sucrose-impermeable space) was demonstrated by measuring isotope exchange at equilibrium (19, 20) by a silicone oil centrifugation technique (21, 22). In Fig. 1, the α-meIc transported is plotted as a function of time and

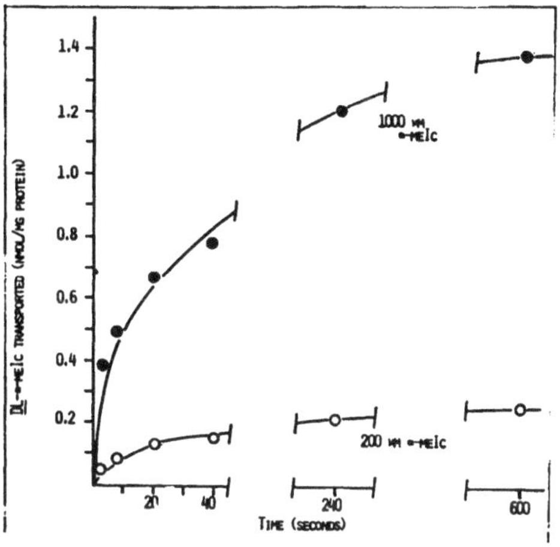

Fig. 1 Uptake of [^{14}C] α-meIc. Rat liver mitochondria (10-15 mg/ml) were pre-equilibrated for 10 minutes in iso-osmotic medium containing 0.08 M KCl, 0.032 M MOPS, .008 M $MgCl_2$, 0.0032 M EDTA, .080 M mannitol, 10 μg/ml rotenone, pH 7.2. At 0 time [^{14}C]-α-meIc was introduced into the mitochondrial suspension and, at the time designated on the abscissa, the exchange of isotope was terminated by centrifuging the mitochondria through a silicone oil layer into a layer of 14 % perchloric acid (w/v).

obeys the exponential form expected of isotope exchange. At two concentrations of [^{14}C]-*DL*-*threo*-α-meIc (200 μM and 1000 μM) over 50 % equilibration of isotope between medium and sucrose-impermeable space is reached in 30 sec. Using a value of 0.7 μl matrix water/mg mitochondrial protein, the equilibrium matrix concentration of α-meIc is 2000 μM at 1000 μM external α-meIc, and 390 μM at 200 μM external α-meIc. In experiments with mitochondria actively metabolizing pyruvate + malate or isocitrate + citrate, concentration gradients of matrix to external [^{14}C]-α-meIc were approximately 9:1.

Inhibition of isocitrate oxidation to α-ketoglutarate. A combination of 2 mM *DL*-isocitrate *plus* 20 mM citrate was used as the substrate in these experiments to provide a near equilibrium ratio of isocitrate and citrate for the aconitase reaction and, thereby, to maximize the steady state concentration of isocitrate during incubation. This substrate mixture was saturating for mitochondrial respiration, as expected from Km(app) values

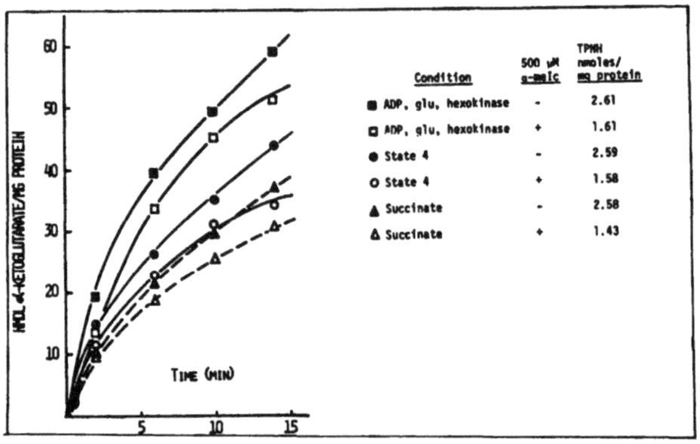

Fig. 2 Inhibition of isocitrate oxidation by α-meIc. Respiring mitochondria (4-8 mg/ml) were incubated in 130 mM KCl, 20 mM potassium phosphate, 10 mM MOPS, 25 mM $MgCl_2$, 40 mM glucose, 6 mM arsenite, 20 mM citrate, 2 mM DL-isocitrate pH 7.2, 28° C. (State 4, ■ and □). Where indicated, mitochondrial incubations also contained 1 mM succinate (▲ and △), or 1 mM ADP + 2-5 units/ml hexokinase (■ and □). Open symbols designate incubations which contained 500 μM α-meIc, added after 30 seconds of incubation.

for DL-isocitrate of 0.0159 mM (4.0 mM Mg^{2+}) for TPN-IDH from rat liver (2) and 0.5 mM (1.34 mM Mn^{2+} and 0.67 mM ADP) for the DPN-IDH from hog liver (8). With arsenite (6 mM) present to inhibit α-ketoglutarate dehydrogenase, the rate of isocitrate oxidation was determined by measuring the accumulation of α-ketoglutarate; this is shown for various conditions in Fig. 2. The production of α-ketoglutarate was nonlinear in State 4, but approached a steady-state in the presence of an ADP-generating system (State 3), which was used for possible activation of DPN-IDH by ADP. The addition of 500 μM α-meIc decreased isocitrate oxidation in these two states by 17 % and 25 %, respectively. A titration of isocitrate oxidation as a function of increasing concentrations of α-meIc was made for each of these conditions, and extrapolated to infinite inhibitor concentration. The maximal inhibition of isocitrate oxidation thus calculated was 31 % for State 4 and 34 % for State 3. Adding additional sources of reducing power, such as 1.0 mM succinate, 1.0 mM 3-hydroxybutyrate or 1.0 mM octanoate did not greatly decrease the rate of isocitrate oxidation or affect the inhibition resulting from the addition of α-meIc. (The effect of succinate is shown in Fig. 2.) Under each of these conditions, the addition of α-meIc resulted in a large decrease in the TPNH content of the mitochondria (Fig. 2). Under most conditions the decrease in TPNH was accompanied by a small increase in DPNH levels (for example in State 4 the average DPNH level increased progressively from 0.047 nmol/mg protein in the absence of α-meIc to 0.227 nmol/mg protein in the presence of 900 μM α-meIc). A similar effect of α-meIc on TPNH and DPNH levels was observed in subsequent experiments with pyruvate and malate as substrate.

Effect of α-meIc on citric acid cycle flux. Under the experimental conditions of Fig. 2, isocitrate concentration would be larger than the steady state levels expected during the catalytic operation of the citric acid cycle. To assess the relative contributions of DPN-IDH and TPN-IDH to isocitrate oxidation in liver under physiological conditions, mitochondria were incubated with 2 mM pyruvate and 1 mM malate in the presence and absence of 400 μM α-meIc. Isocitrate oxidation was estimated from the difference between pyruvate consumption and citrate accumulation under resting or State 4 conditions (Fig. 3A) and with an ADP generating system present to stimulate respiration (State 3, Fig. 3B). The steady state matrix concentrations of D-isocitrate without α-meIc present were approximately 40 μM and 60 μM in State 4 and State 3, respectively, and the matrix concentration of α-meIc calculated from the distribution of [^{14}C]-DL-threo-α-meIc was approximately 3400 μM. At these relative concentrations of isocitrate and α-meIc, less than 0.02 % of the TPN-IDH activity should have been retained. In State 4, citrate production was slightly larger with α-meIc present than in its absence and inhibition of isocitrate oxidation was approximately 16 % (Fig. 3A); no inhibition could be observed in State 3 (Fig. 3B). When dinitrophenol (10 μg/ml)was present in addition to the ADP generating system, the addition of α-meIc caused a 12 % decline in isocitrate oxidation. Thus, the inhibition of isocitrate oxidation in the operative citric acid cycle resulting from complete inhibition of TPN-IDH is hardly larger than the experimental error.

Although α-meIc had no significant effect on the rate of isocitrate oxidation, it affected the reduction state of pyridine nucleotides. The differences due to α-meIc addition were relatively small in State 4 (Fig. 3A) where both DPN and TPN are highly reduced (possibly because oxidation of

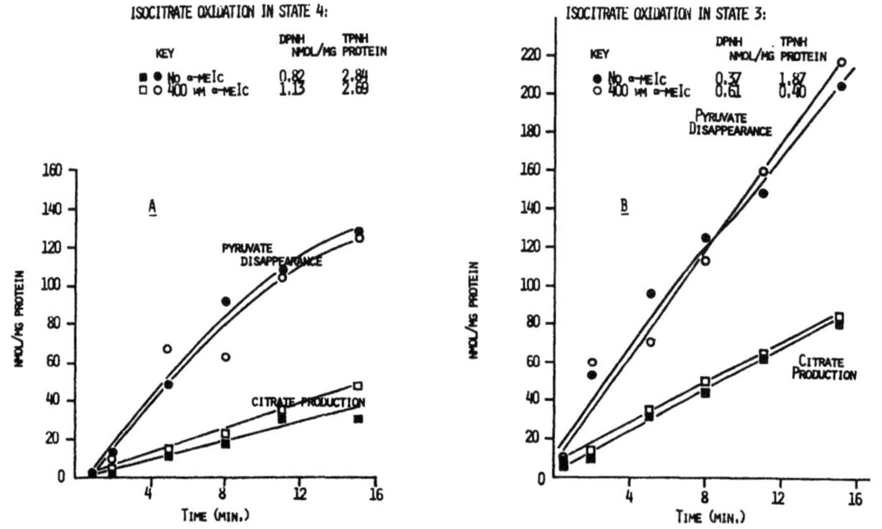

Fig. 3 Inhibition of the citric acid cycle by α-meIc. Rat liver mitochondria were incubated in medium containing 130 mm KCl, 20 mM potassium phosphate, 10 mM MOPS, 25 mM $MgCl_2$, 40 mM glucose, 2 mM pyruvate and 1 mM malate, pH 7.2, 28° C. Where indicated, 400 μM α-meIc was added to the incubation. Incubations shown in Fig. 3B also contain 0.10 mM ADP and 2-5 units hexokinase per ml. In these experiments both DPNH and TPNH reached steady state levels by 3 min, and the values given for each condition are averages from 3, 7 and 12 min of incubation.

DPNH is rate limiting). However, with respiration stimulated by the presence of an ADP generating system (State 3), α-meIc resulted in a 79 % decrease in TPNH (Fig. 3B); there was a small increase in DPNH in the presence of the inhibitor (Fig. 3A and 3B). The above experiments were performed a total of four times. In two of the experiments, as in Figs. 3A and 3B, there was no substantial inhibition of isocitrate oxidation by α-meIc. In the other two experiments, α-meIc caused an <u>increase</u> in isocitrate oxidation. The increase could have resulted from a stimulation of DPN-IDH activity by the decreased level of reduced pyridine nucleotides, especially TPNH. This possibility was supported by subsequent experiments in which substrates which oxidize TPNH were added.

<u>Isocitrate utilization with inhibited electron transport</u>. To test the

effect of α-meIc inhibition of TPN-IDH more directly, reduced pyridine nucleotides generated from isocitrate were utilized for reduction of substrates dependent on participation in the reactions of mitochondrial DPN- or TPN- specific dehydrogenases. In these experiments, electron transport was interrupted by specific inhibitors (cyanide, rotenone or antimycin A) or by the absence of oxygen. DPN or TPN was not added to the reaction mixtures, so that the coupled oxido-reductive disproportionation would depend on participation of the intra-mitochondrial cosubstrates. It has been reported that transport of isocitrate into liver mitochondria involves a dicarboxylate translocation system (23). Tartronate which is not metabolized by liver mitochondria has been found to increase transport of isocitrate into mitochondria (24). However, at the relatively high concentrations of DL-isocitrate (2 mM) used here, tartronate (2.5 mM) did not affect significantly the rate of product formation either in the absence or presence of α-meIc.

The reduction by isocitrate of acetoacetate, which requires participation of DPN-specific 3-hydroxybutyrate dehydrogenase

(2) Isocitrate + acetoacetate → α-ketoglut. + CO_2 + 3-hydroxybutyrate

was examined in a reaction mixture containing an ADP generating system to enhance the activity of DPN-IDH (for which ADP is a positive modifier) (9). The results of two experiments (Expts. A & B) are summarized in Table 1. The stoichiometry of acetoacetate consumed, and 3-hydroxybutyrate and α-ketoglutarate formed is in reasonable agreement with the stoichiometry of the reaction of eq. 2 (Expt. A). This reaction was not inhibited by 400 μM and 900 μM DL-threo-α-meIc in experiments A and B, respectively. When rotenone, the electron transport inhibitor used in Table 1, was replaced by antimycin A (0.02 mg/ml), 85 % of the original activity was retained in presence of 900 μM α-meIc.

In contrast to the experiments above α-meIc was an effective inhibitor of the intramitochondrial oxidation-reduction reaction

(3) Isocitrate + glutathione(ox) → α-ketoglut. + CO_2 + 2 glutathione(red)

in which utilization of reducing equivalents generated from isocitrate is coupled to TPN-specific glutathione reductase (25).

Table 1 EFFECT OF DL-α-METHYLISOCITRATE ON REACTION:

ACETOACETATE + ISOCITRATE → 3-HYDROXYBUTYRATE + α-KETOGLUTATE + CO_2

Metabolite Measured	Activity as nmol min^{-1} mg^{-1} protein			
	Concentrations of DL-threo-α-methylisocitrate			
	0 μM		400 μM	900 μM
	Expt. A	Expt. B	Expt. A	Expt. B
Acetoacetate (consumed)	6.3±1.4		6.4±1.0	
3-Hydroxybutyrate (formed)	7.4±1.6	10.0±1.3	7.5±1.5	9.6±1.5
α-Ketoglutarate (formed)	6.3±0.8		6.3±0.8	

Liver mitochondria (3 mg/ml) from rats starved for 18 hrs were incubated at 25° with 10 μg/ml rotenone, 2 mM DL-isocitrate, 0.5 mM acetoacetate, 226 mM sucrose, 20 mM Na-Hepes at pH 7.2, 20 mM KCl, 10 mM K-phosphate at pH 7.2, 15 mM Na_2AsO_3, 5 mM $MgCl_2$, 20 mM glucose, 0.1 mM ADP, 3 units of yeast hexokinase per ml, and DL-α-methylisocitrate where indicated. Samples were taken at 1 min intervals over a 5 min incubation period. The results represent the means of five 1-min incubation intervals.

The inhibition by α-meIc or reduction by isocitrate of oxidized glutathione to form reduced glutathione and α-ketoglutarate (eq. 3) was studied under several conditions. As shown in Fig. 4A, α-meIc inhibited formation of both products; in an experiment (not shown) under otherwise identical condition, but with levels of α-meIc varied (0, 100, 200, 300, 400 μM) the average ratio of reduced glutathione to α-ketoglutarate produced was 2.1±0.3, i.e., close to the stoichiometry expected (eq. 3). Increasing levels of α-meIc up to 500 μM led to decreasing formation of glutathione (Fig. 4B) and, under these conditions, 1000 μM α-meIc inhibited reduced glutathione production by 80-94 % (not shown). In the experiments shown in Fig. 4, cyanide was the electron transport inhibitor and a phosphate acceptor system was not added. Nevertheless, about the same extent of inhibition of glutathione production (45-60 %) was obtained with equivalent levels of α-meIc (333-500 μM) in the presence of an ADP generating system or when mitochondrial electron transport was inhibited by rotenone, antimycin A or by the absence of oxygen. If one makes the simplifying assumption that the matrix concentrations of

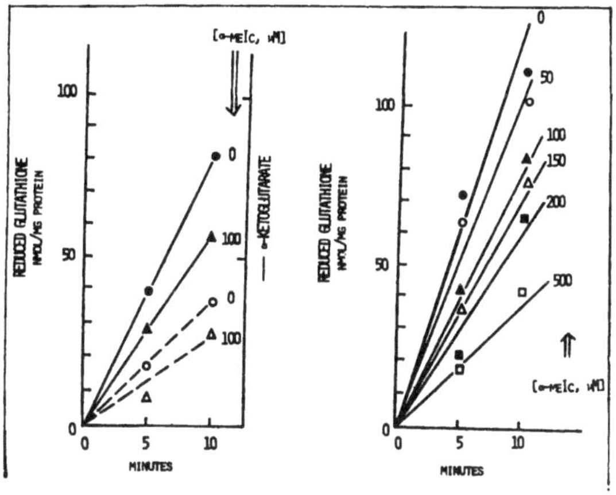

Fig. 4 Effect of α-meIc on reaction:
Glutathione (ox.) + isocitrate → α-ketoglutarate + CO_2 + 2 glutathione (red.). Rat liver mitochondria (0.6-2 mg/ml) incubated at 30° in 176 mM sucrose, 20 mM Na-Hepes, 20 mM KCl, 10 mM K-PO_4, 5 mM $MgCl_2$, 6-15 mM $NaAsO_3$, 1 mM KCN, 2 mM DL-isocitrate, 2 mM glutathione and varying α-meIc at pH 7.2.

DL-isocitrate and α-meIc are 2000 μM and 500 μM, respectively, one would expect (from the Km and Kis (2)) retention of 2-3 % of the activity if TPN-IDH were the rate determining step of the coupled reaction (eq. 3); this compares to an observed retention of about 40 % of the activity. However, glutathione reductase was found to be the rate determining step in extracts from sonicated mitochondria, with a ratio of TPN-IDH/glutathione reductase activities of about 5 at pH 7.2. Under these circumstances, the expected retention of activity of the coupled reaction (eq. 3) may be around 10-15 %.

Effects of TPNH utilizing systems on isocitrate oxidation. The addition of α-meIc caused a substantial decline in intramitochondrial levels of TPNH and a small (but reproducible) rise of DPNH in oxidation of isocitrate to α-ketoglutarate (Fig. 2). Similar changes in levels of reduced pyridine nucleotides upon addition of α-meIc were observed when a mixture of pyruvate + malate was oxidized (Fig. 3). These effects may be due to enhancement of DPN-IDH activity when α-meIc inhibits TPNH generation *via* TPN-IDH. This would be consistent with the potent inhibition by TPNH of soluble DPN-IDH from heart (9), liver (8) and ascites tumor (26). TPNH and DPNH are bound at separate sites on homogeneous heart DPN-IDH (27), and TPNH potentiates the inhibition by DPNH (8, 9), but TPN^+ has no effect on activity (28).

The stimulation of isocitrate oxidation by oxidized glutathione or NH_4^+ (Fig. 5A) and the marked decline in matrix TPNH levels compared to the control (Fig. 5B) would be consistent with the proposed role of TPNH as a negative modulator of DPN-IDH. A comparison of the increase in oxidation resulting from addition of TPNH oxidants in the presence and absence of α-meIc suggests that the decrease in TPNH increased the flux through both TPN-IDH and DPN-IDH. This is particularly apparent in the system oxidizing a combination of isocitrate and citrate where addition of 1 mM NH_4Cl caused a 3-fold increase in isocitrate oxidation (as measured by the sum of α-ketoglutarate and glutamate accumulated, Fig. 5A) and a 45 % decrease in TPNH level (Fig. 5B). The addition of 600 μM DL-α-meIc with NH_4Cl present inhibited isocitrate oxidation by about 40 % and caused a further decrease in TPNH (Fig. 5A and 5B). The stimulation of DPN-IDH resulting from the addition of NH_4Cl is shown by the differences in isocitrate oxidation ($\pm NH_4Cl$) with α-meIc present. The persistence of a substantial portion of the NH_4^+-enhanced isocitrate oxidation in the presence of α-meIc suggest that the decline of TPNH levels is also resulting in an activation of DPN-IDH.

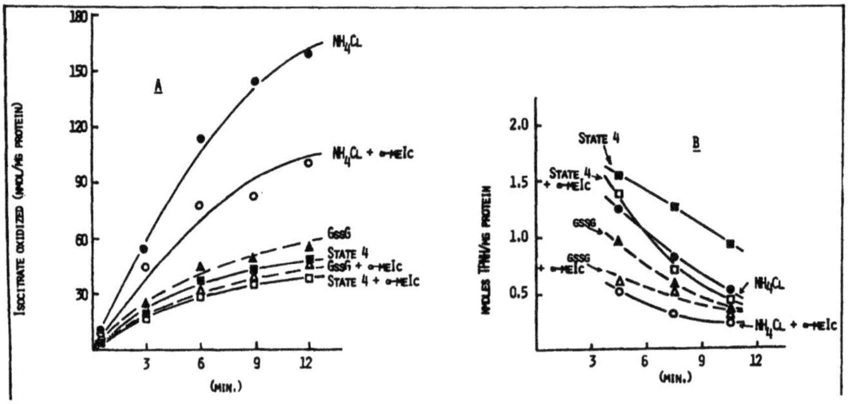

Fig. 5 Stimulation of isocitrate oxidation by TPNH oxidation. Rat liver mitochondria were incubated under the conditions described for Figure 2. Other additions included 600 μM α-meIc, 1 mM NH_4Cl, or 1 mM oxidized glutathione (GSSG), as indicated. The data are averages from two experiments in close agreement.

The decrease of TPNH on addition of NH_4Cl is consistent with the observation of Sies et al. (29) that lowering of TPNH levels by the transhydrogenase inhibitor rhein in rat hepatocytes and by t-butyl hydroperoxide in perfused liver caused a marked decline in urea formation from NH_4Cl. They attributed these effects to preferential utilization of TPNH by glutamate dehydrogenase for reductive amination of α-ketoglutarate. Although glutamate production and TPNH levels are decreased in the presence of α-meIc (Fig. 5A and 5B), the remaining reductive amination of α-ketoglutarate (Fig. 5A) may depend on the direct utilization of DPNH generated via DPN-IDH or on the subsequent transfer of the reducing equivalents of DPNH to TPN by way of the transhydrogenase reaction.

SUMMARY

Experiments with soluble enzymes show that D-threo-α-methylisocitrate is the substrate for aconitase and inhibitor of TPN-IDH. The regulation of isocitrate oxidation has been studied in intact rat liver mitochondria utilizing the TPN-IDH inhibitor α-meIc, with results as follows: (i) α-MeIc does not inhibit the oxidation of pyruvate + malate to α-ketoglutarate, though it decreases intramitochondrial TPNH by inhibition of TPN-IDH. This clearly demonstrates that citric acid cycle oxidation of isocitrate proceeds via DPN-IDH. (ii) Oxidized glutathione and acetoacetate are reduced by pyridine nucleotides generated within mitochondria from isocitrate. Under these conditions, the TPN-dependent formation of reduced glutathione is inhibited by α-meIc; the DPN-specific formation of 3-hydroxybutyrate is unaffected. (iii) The addition of TPNH oxidants, especially NH_4Cl, increases isocitrate oxidation by enhancing its flux through both TPN-IDH and DPN-IDH. This is partially reduced by α-meIc through inhibition of TPN-IDH; the remaining NH_4-stimulated isocitrate oxidation can be correlated with a further decrease of TPNH, which inhibits DPN-IDH.

ACKNOWLEDGMENT: Supported in part by grants AM 15404 and AM 18045 from the National Institutes of Arthritis, Metabolic and Digestive Diseases, National Institutes of Health.

REFERENCES

1. Plaut, G.W.E., Beach, R.L., and Aogaichi, T., Biochemistry 14, 2581-2588 (1975).

2. Plaut, G.W.E., Beach, R.L., and Aogaichi, T., J. Biol. Chem. 250, 6351-6354 (1975).

3. Kelly, J.H., and Plaut, G.W.E., Fed. Proceedings 36, 633 (Abs. #1914) (1977).

4. Gawron, O., and Mahajan, K.P., Biochemistry 5, 2343-2350 (1966).

5. Beach, R.L., Aogaichi, T., and Plaut, G.W.E., J. Biol. Chem. 252, 2702-2709 (1977).

6. Ernster, L., and Navazio, F., Exp. Cell. Res. 11, 483-486 (1956).

7. Plaut, G.W.E., and Aogaichi, T., Biochem. Biophys. Res. Commun. 28, 628-634 (1967).

8. Plaut, G.W.E., and Aogaichi, T., J. Biol. Chem. 243, 5572-5583 (1968).

9. Chen, R.F., and Plaut, G.W.E., Biochemistry 2, 1023-1032 (1963).

10. Stein, A.M., Stein, J.H., and Kirkman, S.K., Biochemistry 6, 1370-1379 (1967).

11. Ernster, L., and Navazio, F., Biochim. Biophys. Acta 26, 408-415 (1957).

12. Ernster, L., and Glasky, A.J., Biochim. Biophys. Acta 38, 168-170 (1960).

13. Nicholls, D.G., and Garland, P.B., Biochem. J. 114, 215-225 (1969).

14. Moyle, J., and Mitchell, P., Biochem. J. 132, 571-585 (1973).

15. Joek, J.B., Rydstrom, J., and Ernster, L., Biochim. Biophys. Acta 305, 669-674 (1973).

16. Smith, C.M., Beach, R.L., and Plaut, G.W.E., Abstr. Meet. Fed. Eur. Biochem. 10, (Abts.) 717 (1975).

17. Plaut, G.W.E., Beach, R.L., Mittnacht, S., Gabriel, A., and Aogaichi, T., Fed. Proceedings 35, 1436 (1976).

18. Meijer, A.J., and Van Dam, K., Biochim. Biophys. Acta 346, 213-244 (1974).

19. Lebing, W., and Smith, C.M., Biophys. J. 16, (Abts.) 133a (1976).

20. Coty, W.A., and Pedersen, P.L., J. Biol. Chem. 249, 2593-2598 (1974).

21. Werkheiser, W.C., and Bartley, W., Biochem. J. 66, 79-91 (1957).

22. Kraaijenhof, R., Tsou, C.S., and Van Dam, K., Biochim. Biophys. Acta 172, 580-581 (1972).

23. Chapell, J.B., Biochem. J. 90, 225-237 (1964).

24. Ferguson, S.M.F., and Williams, G.R., J. Biol. Chem. 241, 3696-3700 (1966).

25. Mize, C.E., and Langdon, R.G., J. Biol. Chem. 237, 1589-1595 (1962).

26. Stein, A.M., Kirkman, S.K., and Stein, J.H., Biochemistry 6, 3197-3203 (1967).

27. Harvey, R.A., Heron, J.I., and Plaut, G.W.E., J. Biol. Chem. 247, 1801-1808 (1972).

28. Lin, J.-P., and Plaut, G.W.E., unpublished observations.

29. Sies, H., Summer, K.-H., and Bücher, T., FEBS Lett. 54, 274-278 (1975).

Received February 18, 1977

CINNAMOYL-CoA:NADPH OXIDOREDUCTASE AND CINNAMYL ALCOHOL DEHYDROGENASE: TWO ENZYMES OF LIGNIN MONOMER BIOSYNTHESIS

H.Grisebach, H.Wengenmayer, D.Wyrambik
Lehrstuhl für Biochemie der Pflanzen, Biologisches Institut II der Universität, D-7800 Freiburg i.Br., Germany

INTRODUCTION

Lignin is an amorphous heteropolymer of substituted cinnamyl alcohols. It represents about 20 to 35% of the cell walls of leaf wood and conifers. The biosynthesis of cinnamyl alcohols starts with L-phenylalanine. This amino acid is deaminated to trans-cinnamic acid by action of the enzyme phenylalanine ammonia-lyase. In a hydroxylation and methylation sequence cinnamic acid is then converted to p-coumaric (4-hydroxycinnamic), ferulic (3-methoxy-4-hydroxycinnamic) and sinapic acid (3,5-dimethoxy-4-hydroxycinnamic acid). In the following step these acids are activated as the corresponding coenzyme A esters, which are the substrates for reduction to the respective cinnamyl alcohols, the primary building stones of lignin. This reduction proceeds in two steps via the free aldehyde according to the reaction sequence shown below

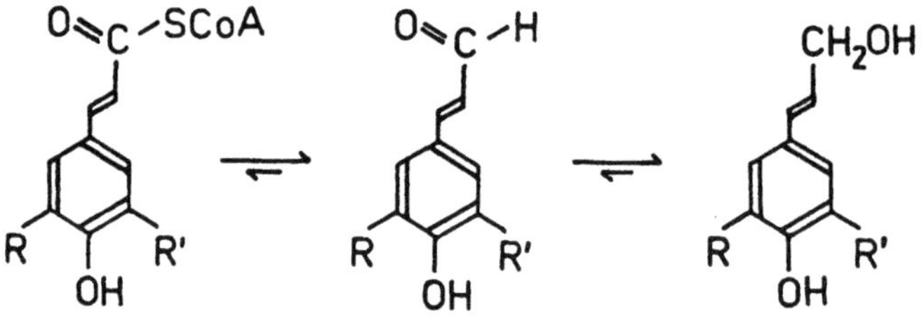

p-Coumaroyl-CoA (R=R'=H)
Feruloyl-CoA (R=H, R'= OMe)
Sinapoyl-CoA (R=R'=OMe)

These reactions are catalyzed by two NADPH-dependent dehydrogenases: cinnamoyl-CoA:NADPH oxidoreductase and cinnamyl alcohol dehydrogenase. These enzymes are the subject of the present article.

CINNAMOYL-CoA:NADPH OXIDOREDUCTASE

Enzymes catalyzing the reduction of cinnamoyl-CoA esters to the corresponding aldehydes have been reported to be present in cell suspension cultures of soybean (1), lignifying tissues of Forsythia (2) and aged swede root tissues (3). The enzyme from soybean cell cultures has been studied in more detail. A 1660-fold purification of the oxidoreductase with an overall yield of 39% was achieved by conventional purification procedures and affinity chromatography using the "general ligand" 5'-AMP attached to Sepharose. The enzyme could be specifically eluted with $NADP^+$ (1). On the basis of polyacrylamide disc electrophoresis the oxidoreductase was estimated to be approx. 90% pure and was free from cinnamyl alcohol and aliphatic alcohol dehydrogenase activity.

The apparent molecular weight of the enzyme obtained from the elution volume on Sephadex G-100 is 38 000. SDS-gel electrophoresis gave only one major band with a molecular weight of 38 000. These results indicate that the oxidoreductase consists of only one polypeptide chain.

The enzyme exhibits a high specificity for the reduction of differently substituted cinnamoyl-CoA esters in the presence of NADPH and belongs to the B-specific (H_S) dehydrogenases (2,4). The reaction rate with NADH is only about 5% of that found with NADPH. Double reciprocal plots for substrates and coenzyme were all linear, as were secondary plots of slopes and intercepts, from which the Km and V values listed in Table 1 were obtained. The enzyme has the highest affinity for feruloyl-CoA, 5-hydroxyferuloyl-CoA, and p-coumaroyl-CoA. A remarkable element is the strong dependence of the Michaelis constant for NADPH on the nature of the cinnamoyl-CoA substrate. A similar observation was, for example, made with glutamate dehydrogenase and NAD^+ as

Table 1 Substrate Specificity of the Cinnamoyl-CoA NADPH Oxidoreductase

Substrate	10^5 x Km	V	10^{-5} x V/Km
	M	nkat/mg	nkat x mg^{-1} x M^{-1}
NADPH	2.8	230	82
Feruloyl-CoA	7.3		31.5
NADPH	12	69	5.8
Sinapoyl-CoA	40		1.7
NADPH	29	30	1.0
p-Coumaroyl-CoA	10		3.0
5-Hydroxyferuloyl-CoA	5	58	11.6

coenzyme (5). Norvaline is a poorer substrate than glutamate not only because its Km value is much greater than that of glutamate but also because the Km for NAD^+ is much bigger (270 μM compared to 44 μM) and the maximum rate smaller with norvaline as substrate.

Because of the dependence of the Km-value for NADPH on the nature of the cinnamoyl-CoA substrate we have assumed a random mechanism and interaction between the binding sites for substrate and coenzyme (1). A detailed kinetic analysis of the enzyme mechanism has not yet been carried out because of the difficulty in obtaining product inhibition patterns with coniferaldehyde (see below). However, the product inhibition by $NADP^+$ and CoASH has been investigated. K_i-values were obtained from Dixon plots. The usefulness of this method is limited by the fact that it does not distinguish unambiguously between competitive and mixed inhibitors and that in the case of mixed or uncompetitive inhibitors it provides no measure of K_i', the dissociation constant of the EIS complex. The complementary plot of S/v against the inhibitor concentration (6) was therefore used to determine the inhibition type and K_i'. In all cases the intersection point was below the i axis a finding which corresponds to a mixed-noncompetitive inhibition type (6). The results, which were confirmed by Lineweaver-Burk plots, are summarized in Table 2.

Table 2 Inhibition of the Oxidoreductase by $NADP^+$ and Coenzyme A

Inhibitor	Variable Substrate	Inhibition Type	Ki	K_i'
			mM	mM
$NADP^+$	Feruloyl-CoA	mixed-non-competitive	0.6	2.0
$NADP^+$	NADPH	mixed-non-competitive	0.5	1.6
Coenzyme A	Feruloyl-CoA	mixed-non-competitive	1.0	1.5
Coenzyme A	NADPH	mixed-non-competitive	0.95	1.6

The third reaction product, coniferaldehyde, did not inhibit enzyme activity up to a concentration of 0.02 mM. Concentrations of the aldehyde exceeding 0.02 mM could not be tested because of interference with the spectrophotometric assay. Coniferylalcohol, the product of the following reduction step, was not inhibitory up to a concentration of 1 mM (1).

The double-reciprocal plots of initial velocities with cinnamoyl-CoA esters or NADPH as changing fixed substrates gave an intersecting pattern of straight lines, a result which proves a sequential mechanism. A random mechanism but not an Ordered one would be consistent with the mixed-non-competitive inhibition type found with NADPH versus $NADP^+$. As already mentioned a random mechanism could also explain the dependence of the Michaelis constant for NADPH on the type of cinnamoyl-CoA substrate.

The reversibility of the reduction is being investigated with $[^3H]$-coenzyme-A. Preliminary results have shown the formation of labelled feruloyl-CoA in an incubation of coniferaldehyde, $[^3H]$-CoASH, $NADP^+$ and enzyme, a finding which indicates that the reduction is reversible.

Some experiments on the enzyme mechanism have also been carried out. Since the enzyme is markedly inhibited by thiol reagents, transfer of the cinnamoyl-CoA to a sulfhydryl group on the enzyme before reduction was considered as one possible mechanism (1). The partial reaction has been in-

$$R\text{-}\underset{\underset{O}{\|}}{C}\text{-}SCoA + E\text{-}SH \rightleftharpoons R\text{-}\underset{\underset{O}{\|}}{C}\text{-}SE + CoASH$$

vestigated by measuring the formation of radioactive feruloyl-CoA in an incubation with feruloyl-CoA, ^3H CoASH and enzyme. So far, however, it has not been possible to find any radioactive feruloyl-CoA.

The change in enzyme activity during growth of the soybean cell cultures will be discussed together with the data on cinnamyl alcohol dehydrogenase.

CINNAMYL ALCOHOL DEHYDROGENASE

$NADP^+$ dependent aromatic alcohol dehydrogenases which are specific for cinnamyl alcohols have been isolated from Forsythia (7) and soybean cell cultures (8). It was shown that this type of dehydrogenase is present in a wide variety of taxonomically different plant groups (7). The enzyme from

Forsythia was purified about 600 fold. The apparent molecular weight (Sephadex G-200) is about 80 000 (7). In cell cultures of soybean two $NADP^+$ dependent isoenzymes are present which were separated from each other and from an NAD^+-dependent aliphatic alcohol dehydrogenase on DEAE-cellulose and hydroxyapatite (8). Whereas, isoenzyme 1 is specific for coniferyl alcohol, isoenzyme 2 can also oxidize cinnamyl alcohol and a number of substituted cinnamyl alcohols. Km values for substituted cinnamaldehydes (1.7-9.1 µM) are 3 - 11 times lower than for the corresponding alcohols (11-57 µM). For the equilibrium constant

$$K = \frac{[\text{coniferaldehyde}]\ [\text{NADPH}]\ [H^+]}{[\text{coniferyl alcohol}]\ [\text{NADP}^+]}$$

a mean value of $2.8\ (\pm 1.1) \times 10^{-9}$ was obtained. The fact that the Km values for cinnamaldehydes are lower than that for cinnamyl alcohols and the position of the equilibrium of the reaction indicate that under physiological conditions reduction of cinnamaldehydes is favoured.

Isoenzyme 2 was purified about 4000 fold (4). The last purification steps involved hydrophobic affinity chromatography on cinnamic acid-Sepharose, alkyl (C_2)-Sepharose and coniferylalcohol-Sepharose. The enzyme appeared homogeneous on polyacrylamide gels. SDS-gel electrophoresis gave only one band with a molecular weight of about 40 000. The apparent molecular weight estimated by the elution volume from a Sephadex G-100 column is about 69 000. Despite the discrepancy between the latter value and double the value obtained on SDS-gels we assume that the dehydrogenase is composed of two apparently identical subunits. A similar discrepancy between the molecular weight determined by gel filtration (65 000) and the subunit weight on SDS gels (37 000 - 41 000) has been reported for the ADH from rat liver (13).

With respect to NADPH the enzyme belongs to the A-specific (H_R) dehydrogenases (4,7) whereas the cinnamoyl-CoA oxidoreductase is B-specific. The different stereospecificity of the two dehydrogenases catalyzing consecutive reactions is an exception to the generalization expressed by Davies et al. (9,10) that NAD(P)-dependent dehydrogenases have the same stereospecificity when they catalyze consecutive steps in a metabolic sequence.

Cinnamyl alcohol dehydrogenase removes the pro-R hydrogen of coniferylalcohol in the oxidation to the aldehyde (4). This was proved in the following way: Coniferylalcohol labeled with ^3H in the methylene group was obtained by incubation of coniferaldehyde with the dehydrogenase and [A-^{3}H]-NADPH. The labeled coniferyl alcohol was then reoxidized with liver alcohol dehydrogenase, which has pro-R specificity (11). The resulting coniferaldehyde was inactive and the isolated NADH contained the calculated amount of tritium.

Enzyme activity is inhibited by various metal chelating agents and by thiol reagents. Table 3 gives the inhibitor concentrations for 50% inhibition. The inhibition by o-phenanthroline can be reversed to a certain extent by addition of Zn^{2+} (Fig. 1). These results indicate that like mammalian and yeast ADH this dehydrogenase is a Zn-enzyme, but a metal analysis has yet to be carried out.

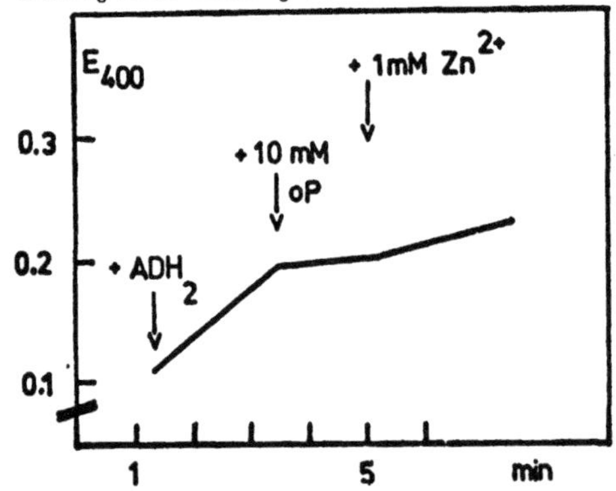

Fig. 1 Partial reversal of o-phenanthroline inhibition by Zn^{2+}. A higher concentration of Zn^{2+} interfered with the assay.

Table 3. Inhibition of Cinnamyl Alcohol Dehydrogenase with Metal Chelators and Thiol-Reagents

Compound	Concentration for 50% Inhibition
	mM
o-Phenanthroline [a]	0.38
4-Methylpyrazol [a]	4
α,α'-Dipyridyl [a]	2
8-Hydroxyquinoline [a]	0.15
N-Ethylmaleinimide [a]	10
p-Chloromercuribenzoate [b]	0.005

[a] without preincubation [b] 2 min preincubation

Steady state kinetics of the enzyme reaction were investigated. Lineweaver-Burk plots with substrates and coenzymes for oxidation and reduction gave an intersecting pattern of straight lines as expected for a sequential mechanism. Inhibition patterns were obtained with the combinations listed in Table 4. The inhibition type and Ki values were obtained from Dixon and Cornish-Bowden plots (6).

The inhibition pattern obtained is consistent with the ordered BiBi mechanism in which NADP(H) is the first substrate to bind and the last product released (12).

The cinnamyl alcohol dehydrogenase from soybean bears a close resemblance to alcohol dehydrogenase from yeast (14), horse liver (15), and rat liver (13) with respect to its subunit molecular weight, inhibition by metal chelators, steady state kinetics, and stereospecificity with regard to

Table 4 Inhibition Patterns of Cinnamyl Alcohol Dehydrogenase

Inhibitor	Variable Substrate	Inhibition Type	Ki	K_i'
			μM	μM
NADPH	NADPH	competitive	16	
$NADP^+$	NADPH	competitive	15	
NADPH	Cinnamyl alcohol	mixed-non-competitive	36	90
Cinnamaldehyde	$NADP^+$	mixed-non-competitive	8.6	15.6
Cinnamaldehyde	Cinnamyl alcohol	mixed-non-competitive	2	12.3

coenzyme and substrate. It would therefore be interesting to compare the primary structure of the plant enzyme with that of the other alcohol dehydrogenases.

CHANGES OF ENZYME ACTIVITIES DURING GROWTH OF CELL CULTURES

Concomitant large increases and subsequent decreases in the activities of phenylalanine ammonia-lyase (PAL), cinnamic acid 4-hydroxylase, and two isoenzymes of p-coumarate:CoA ligase occur prior to the stationary growth phase of fermenter grown cell suspension cultures of soybean. However, nosignificant change in the overall rate of reduction of feruloyl-CoA to coniferyl alcohol was observed (16). We have now determined the change in activity of the two enzymes involved in the reduction of feruloyl-CoA in fermenter-grown cultures. The cultures were kept either in a normal day-night rythm or in the dark. A significant increase and subsequent decrease in the activities of cinnamoyl-CoA:NADPH oxidoreductase and isoenzyme 2 of cinnamyl alcohol dehydrogenase were found (4). Both enzymes reached maximum specific activity slightly before the maximum of PAL (Fig. 2).

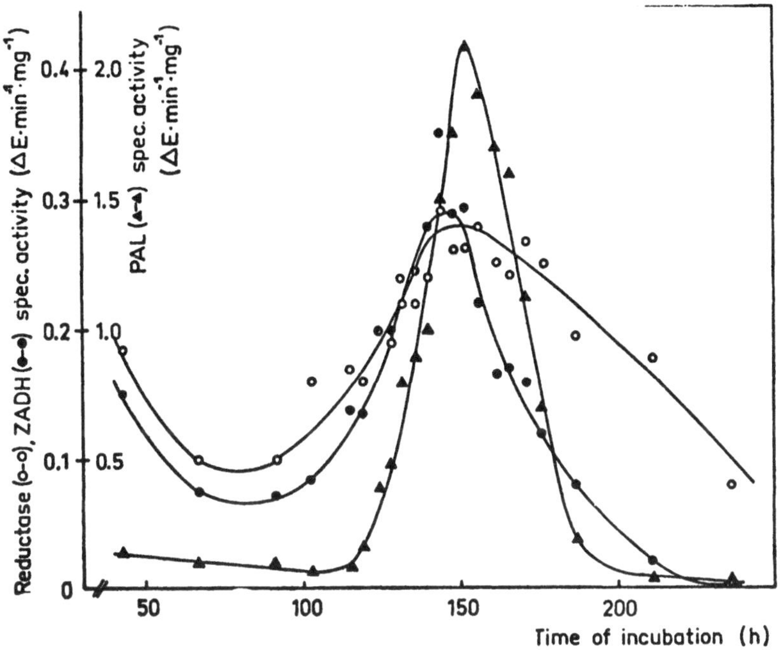

Fig. 2 Change in activity of cinnamoyl-CoA oxidoreductase (o—o), cinnamyl alcohol dehydrogenase (●—●), and phenylalanine ammonia-lyase (▲—▲) during growth of soybean cell cultures.

In contrast,no significant change in enzyme activity was observed with isoenzyme 1 of cinnamyl alcohol dehydrogenase. The aliphatic ADH showed a continuous rise in specific activity during the time of the experiment. No significant difference was found between the daylight and dark cultures. The earlier test system in which no change in the overall reduction was found (16) was quite complex, so that the findings cannot be correlated with the present results.

Our work described in this paper was supported by Deutsche Forschungsgemeinschaft (SFB 46) and by Fonds der Chemie.

REFERENCES

1) Wengenmayer, H., Ebel, J., Grisebach, H.: Enzymic Synthesis of lignin precursors. Purification and properties of a cinnamoyl-CoA:NADPH reductase from cell suspension cultures of soybean (Glycine max.). Eur. J. Biochem. 65, 529-536 (1976).

2) Gross, G.G., Kreiten, W.: Reduction of coenzyme A thioesters of cinnamic acids with an enzyme preparation from lignifying tissue of Forsythia. FEBS Letters 54, 259-262 (1975).

3) Rhodes, M.J.C., Wooltorton, L.S.C.: Enzymes involved in the reduction of ferulic acid to coniferyl alcohol during the aging of disks of swede root tissue. Phytochemistry 14, 1235-1240 (1975).

4) Wengenmayer, H., Wyrambik, D., Grisebach, H.: unpublished results.

5) Engel, P.C., Dalziel, K.: Kinetic studies of glutamate dehydrogenase with glutamate and norvaline as substrates. Biochem. J. 115, 621-631 (1969).

6) Cornish-Bowden, A.: A simple graphical method for determining the inhibition constants of mixed, uncompetitive and non-competitive inhibitors. Biochem. J. 137, 143-144 (1974).

7) Mansell, R.L., Gross, G.G., Stöckigt, J., Franke, H., Zenk, M.H.: Purification and properties of cinnamyl alcohol dehydrogenases from higher plants involved in lignin biosynthesis. Phytochemistry 13, 2427-2435 (1974).

8) Wyrambik, D., Grisebach, H.: Purification and properties of isoenzymes of cinnamyl-alcohol dehydrogenase from soybean-cell-suspension cultures. Eur. J. Biochem. 59, 9-15 (1975).

9) Davies, D.D., Teixeira, A., Kenworthy, P.: The stereospecificity of nicotinamide-adenine dinucleotide dependent oxidoreductases from plants. Biochem. J. 127, 335-343 (1972).

10) Do Nascimento, K.H., Davies, D.D.: The stereospecificity of sequential nicotinamide-adenine dinucleotide dependent oxidoreductases in relation to the evolution of metabolic sequences. Biochem. J. 149, 553-557 (1975).

11) Bentley, R.: Molecular assymetry in biology. Vol. II, p. 18. Academic Press, New York-London, 1970.

12) Cleland, L.W. in: The Enzymes, Boyer, P.D. editor, Vol. II, 3rd edition, p. 1-65, Academic Press, New York-London, 1970.

13) Arslanian, M.J., Pascoe, E.: Rat liver alcohol dehydroge-

nase. Purification and properties. Biochem. J. 125, 1039-1047 (1971).

14) Brändén, C.I., Jörnvall, H., Eklund, H., Furugren, B. in: The Enzymes, Boyer, P.D. editor, Vol. XI, 3rd edition, p. 171-173, Academic Press, New York-San Francisco-London, 1975.

15) Drum, D.E., Harrison, J.H., Li, T.K., Bethune, J.L., Vallee, B.L.: Structural and functional zinc in horse liver alcohol dehydrogenase. Proc. Natl. Acad. Sci. 57, 1434-1440 (1967).

16) Ebel, J., Schaller-Hekeler, B., Knobloch, K.H., Wellmann, E., Grisebach, H., Hahlbrock, K.: Coordinated changes in enzyme activities of phenylpropanoid metabolism during the growth of soybean-cell-suspension cultures. Biochim. Biophys. Acta 362, 417-424 (1974).

Received January 31, 1977

DISCUSSION

Sund: For the inhibition with methal-chelating agents you need rather high concentrations. The reactivation by addition of zinc ions could indicate that e.g. o-phenanthroline blocks a zinc ion which is necessary for activity but would also indicate that o-phenanthroline interacts with e.g. the substrate binding sites. In this connection it would be interesting to investigate m-phenanthroline. This compound has a similar configuration and surface as the ortho-compound but no metal-chelating properties. May I guess that you will find a competitive inhibition with respect to the substrate?

Grisebach: We have not carried out an inhibition experiment with m-phenanthroline but this is a good suggestion. In this connection I like to show the point that our results with metal chelators indicate a metal requirement but don't prove it.

Hollaway: Does the addition of Zn^{++} restore activity in the case of glutamate dehydrogenase inhibited by o-phenanthroline?

Sund: Yes, it does.

Hollaway: But presumably not at the concentration ratio of 10:1 o-phenanthroline:Zn^{++} where Professor Grisebach sees activity returning.

Dunn: As you know, we have found that 4-dimethylaminocinnamyl aldehyde reacts with the liver alcohol dehydrogenase-NADH complex to form an intermediate in which the carbonyl oxygen is directly coordinated to the active site zinc ion, resulting in a large spectral shift in the substrate spectrum. At high pH this intermediate is the most stable species on the reaction pathway, and therefore it can be observed without use of rapid kinetic techniques above pH 9 at high enzyme concentrations. If your alcohol dehydrogenase is a zinc enzyme, perhaps you may be able to detect a similar intermediate at high pH with your chemically similar substrates.

Grisebach: This is a very valuable suggestion. So far we have not carried out any spectroscopical investigation with the enzyme because of the small amount we had available.

Engel: I have two comments: Firstly it seems to me that your suggestion that cinnamyl alcohol dehydrogenase follows an ordered Bi-Bi mechanism is strongly supported by the constancy of K_m/V_{max} that you observe for several substrates. K_m/V_{max} is $\emptyset_1$ in the Dalziel nomenclature and should be constant if the coenzyme leads.

Secondly, with the cinnamoyl-CoA:NADPH oxidoreductase my feeling is that, from the nice steady-state data you showed us for the uninhibited reaction, you could obtain more information as to the mechanism. Your Table 1 gives V_{max} and the extrapolated K_m's for the two substrates. But the reaction rate is governed by four constants. From your data you could also determine the $\emptyset_{1,2}$ value (Dalziel nomenclature). The essential mechanistic test would be to see whether $\emptyset_1$ and $\emptyset_{1,2}/\emptyset_2$ are both constant. This would point to an ordered mechanism with coenzyme leading. If only $\emptyset_{1,2}/\emptyset_2$ is constant but not $\emptyset_1$, this would be consistent with a rapid-equilibrium random-order mechanism.

Grisebach: Thank you for your valuable suggestions. We will certainly do these calculations.

Südi: I'd be quite interested in calculating the third-order catalytic constants for the various substrates of your two interesting dehydrogenases with the procedure we have recently published with Bent Havsteen. Have you carried out a detailed kinetic analysis of any of your reactions (I'm meaning the complete set of kinetic coefficients according to Cleland) which would allow this?

Grisebach: No.

Heilmann: Referring to your K_m data for $NADP^+$ binding: Did you estimate the binding constant in absence of further substrates?

Grisebach: No.

Heilmann: To the -SH modification: There is a considerable disagreement between the effects of Hg^{++} and maleimide, respectively. (By the way: how long did you incubate the enzyme with the maleimide prior to stopping the reaction?). Can you reactivate after Hg^{++} inhibition?

Grisebach: In the experiment with maleimide no preincubation was carried out. In contrast, with p-chloromercuribenzoate two min preincubation was used. Reactivation after Hg^{++} inhibition was not tested.

Kempfle: In Table 1 you give K_m values and you give very different K_m values for NADP(H) with the different substrates. This may not be surprising since a K_m value is valid for the whole reaction. If you take dissociation constants for the coenzyme only it may be that you get always the same value.

Grisebach: I agree that the dissociation constant for NADP(H) could be the same for all substrates. Nevertheless, the finding that the K_m value for NADP is strongly dependent on the substrate is not so common. At least there are very few reports in the literature on this phenomenon. Most people publish only one K_m for the coenzyme for a range of substrates.

Sund: What was the substrate concentration in the determination of the K_m values for NADPH? This is important to know with respect to the result that the K_m value depends on the substrate.

Grisebach: The K_m values of NADPH for each substrate were determined from the slope and intercept replots of double reciprocal plots with NADPH as changing fixed substrate. These K_m values represent therefore values under saturating conditions of the substrates.

Luisi: How does the polymerization from cinnamyl alcohol to lignin proceed?

Grisebach: According to the present knowledge the polymerization is initiated by action of a cellwall-bound peroxidase. This forms a phenoxy radical which can have several mesomeric structures. These radicals then combine to form the network of lignin.

Kiltz: There is a very rough factor of two between the molecular weight of your two isozymes as determined by gelfiltration. Did you look into the possibility that isozyme 2 is simply the dimer of isozyme 1?

Grisebach: Yes, but isozyme 1 does not form a higher molecular weight species upon rechromatography. I therefore consider this possibility very unlikely.

OCTOPINE DEHYDROGENASE. SPECTROSCOPIC AND CONFORMATIONAL PROPERTIES OF BOUND COENZYME, AND A POSSIBLE TEMPERATURE-REGULATION FUNCTION.

P.L. Luisi, A. Olomucki, A. Baici, R. Joppich-Kuhn and F. Thomé-Beau
Technisch-Chemisches Laboratorium der ETH-Zürich, and Biochimie Cellulaire, Collège de France, Paris.

Octopine dehydrogenase (EC 1.5.1.11) is a single subunit enzyme (MW 38'000) catalyzing the following reaction:

$$\text{Pyruvate} + \text{L-Arginine} + \text{NADH} + H^+ \rightleftharpoons \text{NAD}^+ + H_2O + \text{D-Octopine}$$

The enzyme from the shell fish Pecten maximus L. was first isolated and characterized by the group of van Thoai (1), and later it was investigated in more detail by Olomucki and coworkers at the Collège de France. It was shown that the enzyme contains no metal atoms, has 7 SH groups, it has a B-stereospecificity for NADH, and it has a fairly restricted specificity for substrates and coenzymes (2-5). The spectroscopic properties of various complexes (6-8), the binding of coenzyme in the presence of various enzyme ligands (9-10), as well as the kinetic mechanism (11-12), have been the subject of detailed studies. Octopine dehydrogenase appears to be a lactate dehydrogenase substitute in some invertebrates, particularly in molluscs which contain phosphoarginine in their muscles.

In investigating NADH and NAD binding to ODH, it was found that the dissociation constants for the coenzymes, as directly determined by equilibrium measurements, are practically temperature independent in the range 5-40°C, and so are certain kinetically determined K_m-values (13). Similar findings have been reported for other enzymes extracted from poikilotherms, and have been ascribed to a temperature-regulation mechanism which would protect the particular enzymatic function from abrupt changes induced by fluctuations of the external temperature (see for instance ref. 13,14,15, and citations therein). (The body temperature of poikilotherms depends upon the environmental temperature).

An interesting question in this regard concerns the molecular mechanism which permits a temperature-independent binding process, i.e. a binding process in which the driving forces are solely entropic in nature (13). While the answer to this question is not easy, it is clear that the conformation of the bound coenzyme, as well as the interactions between the coenzyme and the aminoacids of the active site are important. We therefore determined to examine in more detail the spectroscopy of binary and ternary complexes of NAD and NADH bound to ODH to obtain information pertaining to conformation. In this paper we will also deal with the binding properties of thio-NADH, which is biologically inactive with ODH. The interest in this coenzyme analog lies in whether the anomalous temperature effect in the binding to ODH is specific only for the natural coenzymes or if it is exhibited by all coenzyme ligands.

EXPERIMENTAL

The enzyme ODH was prepared as previously described (2). Its binding capability was found to be 100% upon titration with NADH in the presence of D-Octopine. LADH, obtained from Böhringer (Mannheim), was prepared and assayed as already described (17). Thio-NAD was also a Böhringer product, and thio-NADH was prepared by enzymatic reduction of thio-NAD and purified as described earlier (19). UV-absorption measurements were carried out with a Cary 15 spectrophotometer, CD-measurements were carried out with a Jasco J-40 spectropolarimeter and fluorescence binding studies were performed with an Aminco-Bowman spectrophotofluorometer. Corrected fluorescence properties were measured with an Aminco-1000 ratio recording spectrophotofluorometer. All measurements were carried out using pH 7.0, 0.1 M phosphate buffer.

CD-PROPERTIES OF BOUND COENZYME

The circular dichroic properties of the ODH-NADH binary complex were investigated by Oriol and Olomucki some years ago (6). They found a considerable enhancement of the ellipticity of the bound coenzyme with respect to the free form. Fig.1 shows the CD properties of NADH bound to ODH in the presence of D-octopine. In this ternary complex, the dissociation constant of the coenzyme is only 0.9µM, i.e. ∿ 20 times smaller than in the binary complex (9), and it appeared therefore interesting to see whether this tightening was accompanied by spectral changes of the bound coenzyme. In fact, the ellipticity maximum is at 353 nm, namely there is an appreciable red-shift both with respect to the CD. λ_{max} of the free coenzyme (λ_{max}=340 nm) and to the binary complex (λ_{max}=348 nm). The molar ellipticity value in the ternary complex is about 9000 deg cm^2 $dmol^{-1}$. In the temperature range 5-40°C, the CD spectrum of the ternary complex is almost constant: this is remarkable, considering the high ellipticity value of the bound NADH, and considering that the temperature coefficient of optical activity is usually large in highly optically active compounds. In our opinion this indicates high conformational rigidity of the bound coenzyme.

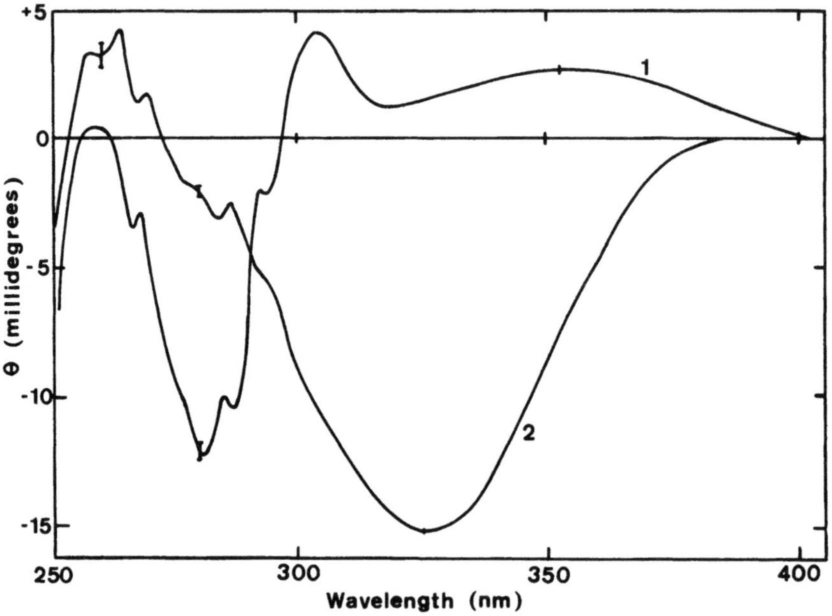

Fig. 1 CD-spectra of ternary complexes of ODH (1) and LADH (2) with NADH. (1) 30μM ODH, 30μM NADH, 0.02M D-Octopine. (2) 30μM LADH, 30μM NADH, 5mM Isobutyramide.

A significant difference between the CD spectra of binary and ternary complexes appears in the 300-310 nm region. In particular, there is an intense positive band ($[\theta]$=13'000 deg cm^2 $dmol^{-1}$) centered at 304 nm in the ternary, but not in the binary complex. This band seems to originate from the protein. In fact it is also present in the NAD complexes (see Fig. 2), and furthermore, it is known that CD spectra of protein can have bands around 300 nm (16). It is possible that this band corresponds to the 0-0 L_a transition of indole, which is being perturbed in the ODH ternary complexes due to the changes brought about by the ligand binding. It appears therefore possible to conclude that the binding of ligand to the binary complex, albeit not affecting the conformation of the bound coenzyme, does affect the structure of the protein. Furthermore, the fact that the same band is observed in the ternary complex with NAD, leads us to the conclusion that the ligand-induced perturbations in the protein structure are obtained with substrates as well as with products and are independent of the oxydoreduction state of the coenzyme (in this regard, see also Ref. 10).

Fig. 1 also shows the CD-properties of NADH bound to horse

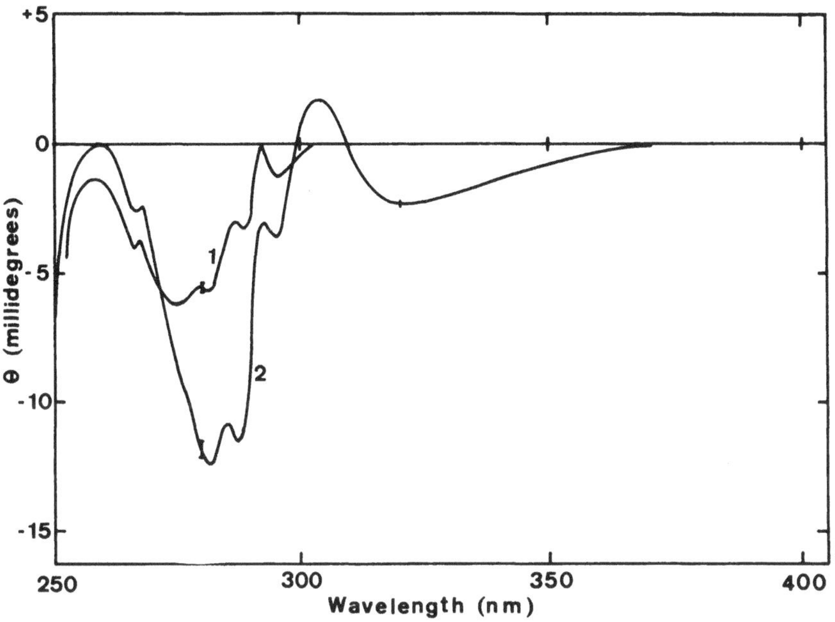

Fig. 2 CD-spectra of ODH (1) and ternary complex with NAD in the presence of Pyruvate and L-Arginine (2). 30μM ODH, 30μM NAD, 0.02M Pyruvate, 0.02M L-Arginine.

liver alcohol dehydrogenase (LADH) in the presence of isobutyramide. This comparison offers several points of interest. For example in LADH the molar ellipticity in the 310-360 nm region is the same in binary (not reported) as in ternary complexes, and the position of the ellipticity maximum is shifted by only 2 nm (towards the red in the ternary complex). It appears therefore, that there is no significant change of the conformation of NADH, at least around the nicotinamide chromophore, upon binding of substrate-like molecules to the binary complex. (The small changes observed in the 260 nm region might indicate some re-positioning of the adenine ring). As for ODH, the negligible temperature coefficient of the ellipticity (not shown) indicates a high degree of conformational rigidity of bound coenzyme. Note that in LADH the sign of the ellipticity band of bound NADH is opposite to that of ODH, and that the molar ellipticity value is much higher. These features certainly indicate different conformations of the bound coenzyme in the two enzyme systems. However, in both ODH and LADH the coenzyme conformation is with all likelihood an open one, as shown by our previous fluorescence studies (9,17). Furthermore, as shown by the negligible temperature coefficient of the molar ellipticity, the bound conformation is in both cases a rigid one. One possible explanation of the difference between the dichroic properties of NADH bound to ODH and to LADH is the opposite stereospecificity of the two enzymes. As it is well known, LADH has a A-stereospecificity towards NADH (18) and ODH has the opposite one (5). One could then assume that the conformation of bound NADH is different in the two enzymes, albeit being an open one in both cases. Studies are in progress in our laboratory with other dehydrogenases to ascertain the generality of the relationship between the CD signal of bound NADH and the enzyme's stereospecificity.

Let us now turn our attention to Fig. 2, which shows the CD spectrum of NAD bound to ODH in the presence of pyruvate plus L-arginine. The CD properties of NAD bound to various complexes, as well as the comparison with LADH, will be discussed elsewhere. Here it is important to point out, together with the band at 304 nm, previously discussed, the band in the 320 nm region which is with all likelihood due to pyruvate. This shows that this achiral substrate becomes highly dissymmetric ($[\theta]$=7500 deg cm^2 $dmol^{-1}$) when bound to the enzyme in the presence of L-arginine.

Another interesting CD property found in this investigation is the large ellipticity change of ODH, on going from free enzyme or binary to the ternary complexes (see Fig. 2). There is an almost 50% increase in the 250-300 nm region, without any significant shift of the position of the aromatic bands. This fact, together with the anomalous new protein band at 304 nm, confirms that important structural changes of the enzyme take place specifically upon binding of ligands to the binary complexes. Earlier UV-absorption studies by the group of Olomucki suggested that one (or more) tryptophyl residues are perturbed in the coenzyme complexes (6). At this stage of the investigation it is not yet clear whether the enzyme undergoes a conformational change, or whether there is some kind of direct interaction between the tryptophyl residue(s) and the coenzyme.

SPECTROSCOPIC INVESTIGATIONS WITH THIO-NADH

Among coenzyme analogs for dehydrogenases, thio-NAD(H) is one of the most interesting. In LADH, thio-NAD induces an enhancement of the turnover number with respect to the natural coenzyme, which has been ascribed to a change in the position of the rate limiting step (19). By contrast, thio-NAD has been shown to be practically inactive with ODH. Another interesting feature of thio-NAD lies in its bands in the 400-410 nm region, ascribed to the n-π^* transition of the sulfur, and which are very sensitive to environmental changes. In fact, thio-NAD and thio-NADH are being used as spectroscopic probes

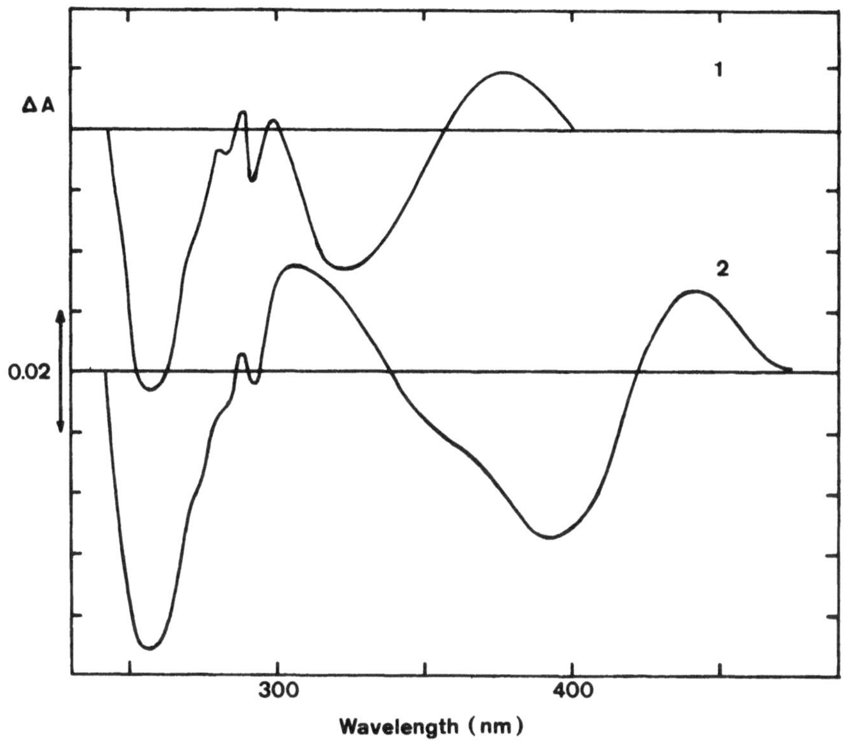

Fig. 3 Difference spectra of NADH (1) and thio-NADH (2) bound to ODH in the presence of octopine. 26.3μM ODH in 50mM phosphate buffer pH 7, 20mM octopine, 0.1mM EDTA, 0.1mM dithiothreitol, 66.2μM NADH, 108.9μM thio-NADH.

for the coenzyme conformational changes taking place in LADH when various ligands bind to the binary complexes (20).

In the present work we have investigated the binding of thio-NAD and thio-NADH to ODH by monitoring the quenching of the protein fluorescence, in analogy to our previous studies with the natural coenzymes (7,9). In agreement with the kinetic investigations by Olomucki et al (10), we could not find any measurable binding of thio-NAD, neither in the binary complex nor in the presence of pyruvate plus L-arginine.

Thio-NADH appears more interesting. Although it is also a very poor substrate for ODH, we did measure a binding with a K_d = 136μM. The binding can be studied also in the presence of D-octopine, which makes the binding tighter by a factor of

TABLE 1 Binding of thio-NADH to Octopine dehydrogenase as monitored by following the quenching of the protein fluorescence 0.1 M phosphate buffer, pH = 7.0.

System	t(°C)	K_{diss}(μM)	ΔF_{max}(%)
ODH·thio-NADH	23.0	136	100
ODH·D-octopine·thio-NADH[(a)]	5.5	77	94
	10.0	54	93
	20.0	44	88
	23.0	48	88
	30.0	47	88
	40.0	50	88

(a) the concentration of D-octopine was 0.02M; K_{diss} (measured at 23°C) does not change when D-octopine is varied in the concentration range 0.02 to 0.1M.

about 3. Kinetic studies are in progress to ascertain whether thio-NADH is a competitive inhibitor of NADH.

Fig. 3 shows the UV difference spectrum of the thio-NADH binding to ODH in presence of octopine. This spectrum is characterized by the perturbation of both the coenzyme and protein chromophores, similarly to that obtained for NADH binding (also shown in Fig. 3 for reference). The perturbation in the 280-310 nm region, which corresponds to a red shift of the aromatic amino acids absorption bands, could perhaps be attributed to a protein conformational change induced by the binding of the ADP-ribose part of both coenzymes (10).

The binding of thio-NADH to ODH is attended by a large quenching of the protein fluorescence, as shown in Table 1 (ΔF_{max} is the percent quenching at infinite coenzyme concentration, as obtained from double reciprocal plots). In particular, the fluorescence of the enzyme is ~90% (or more), quenched both in the binary and in the ternary complexes with thio-NADH, while with NADH the quenching is around 60% (9). This difference in quenching capability could be attributed to a different energy transfer efficiency, as caused by the different spectral properties of the two coenzymes and/or by a different orientation of the bound coenzyme relative to the tryptophyl residue(s) of the protein. However, it could also be due to different extents of protein structural change induced in the two cases by the coenzyme binding.

The poor reactivity of thio-NADH, despite its binding capability, indicates that this coenzyme binds in an unproductive way. Due to this basic difference in the binding properties between thio-NADH and NADH, it appeared interesting to study the temperature dependence of the thio-NADH dissociation constant. This study could solve the question whether the temperature-independence of K_d found for NADH was determined by the proper coenzyme binding or not. Results are shown in Table 1. Except for the data obtained at 5°C (which are experimentally the most critical), the K_d in the temperature range 10-40°C is temperature independent within the limit of experimental error (± 10%). It appears therefore that the peculiar thermodynamic features accompanying the coenzyme binding are also present for the coenzyme analog.

Two other interesting features could be noted in this investigation of the binding of thio-NADH to ODH by fluorescence. The first concerns the fact that thio-NADH acquires a measurable fluorescence when bound to ODH in the presence of D-octopine. Fig. 4 shows the corrected fluorescence emission spectrum of the ternary complex by excitation at 280nm. The peak at about 330nm belongs to the protein emission and that at 470 nm is the new one belonging to the ternary complex. The corresponding excitation spectrum (emission 470nm) shows two maxima at 286 and 360 nm. This fluorescence cannot be ascribed to impurities of NADH in thio-NADH samples as: i) the emission phenomenon is not observed in the binary or ternary complexes with LADH; ii) thio-NADH samples were found to be chromatographically and analytically pure.

The second effect concerns a time-dependent change in the fluorescence intensity of bound thio-NADH in the ternary ODH·D-octopine·thio-NADH complex upon irradiation. In particular the fluorescence intensity at 470 nm increases rapidly upon excitation at 280 nm. This can be ascribed to a photochemical reaction taking place in the bound thio-NADH, which induces a change of the electronic structure and/or of the conformation, with consequent changes in the fluorescence properties of the coenzyme.

Both fluorescence effects of the bound thio-NADH are rather interesting in view of the fact that they are unique to ODH ternary complexes and are therefore induced by the particular mode of binding of the coenzyme analog in the ternary complexes of this enzyme. Details of these fluorescence studies, as well as the comparison with LADH, will be given elsewhere.

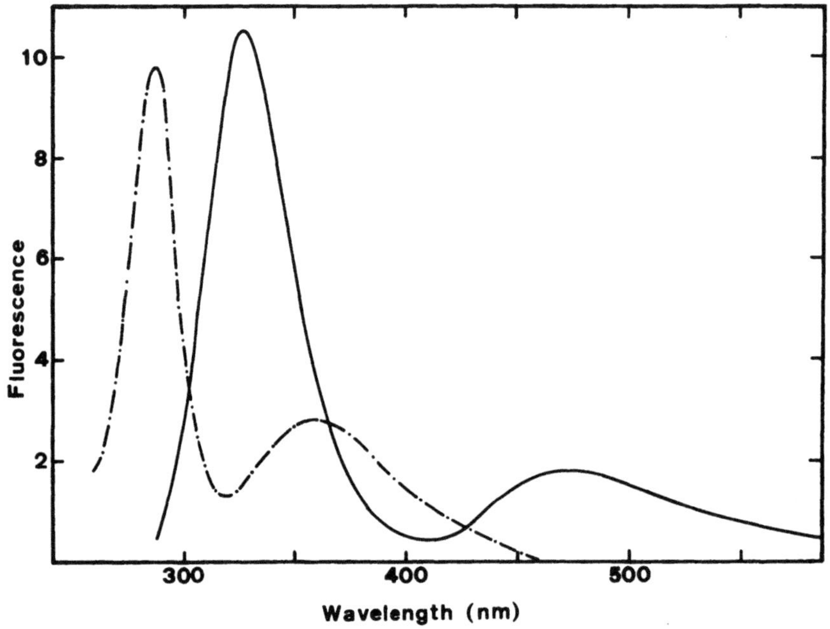

Fig. 4 Corrected fluorescence emission and excitation spectra of ODH·D-octopine·thio-NADH ternary complex. Concentrations: 13μM ODH, 140μM thio-NADH, 0.02M D-octopine. Emission spectrum (——), excitation 280nm. Excitation spectrum (—·—), emission 470nm.

CONCLUDING REMARKS

Circular dichroism appears to be a very useful method for a comparative study of the mode of binding of coenzyme to dehydrogenases. It is in fact possible to extract qualitative information concerning the conformation of the bound NADH using 1) the very large enhancement of molar ellipticity of the bound with respect to the free form, 2) the change in sign of the ellipticity in the different enzyme systems, 3) the temperature coefficient of the molar ellipticity, 4) and possible changes in the CD spectrum of enzyme or coenzyme on going from binary to ternary complexes. Particularly interesting is the possibility of correlating the sign of the dichroic band of the bound NADH with the enzyme's stereospecificity.

The coenzyme complexes of ODH and LADH appear to be provided with a large conformational rigidity, the coenzyme being bound in an open conformation. In addition to these similarities there are some significant differences between ODH and LADH coenzyme complexes. One is the total lack of activity of thio-NAD(H) in ODH, which is surprising in view of the binding capability of thio-NADH to ODH and in view of the good reactivity of thio-NAD(H) with other dehydrogenases. Clearly, the high coenzyme specificity of ODH obliges thio-NADH to bind in an unproductive way. This way of binding produces unique fluorescence properties in bound thio-NADH.

The other peculiar feature of ODH coenzyme complexes is correlated to the temperature independence of the dissociation constant in a temperature range where for example, the binding of coenzyme to yeast ADH changes by an order of magnitude (21). The enzyme therefore is able to accomodate the coenzyme with an athermic binding ($\Delta H^o=0$), regardless of the actual value of the total free energy of binding. The occurrence of this feature with quite different coenzyme complexes reinforces our hypothesis (13) that it is not due to a thermodynamic happenstance, but to a biologically relevant structure/function relationship of the enzyme active site.

Acknowledgments: This research has been partly supported by the Swiss National Science Foundation (Grant Nr. 3.3810.74), and by the CNRS (Grant Nr. 6)

REFERENCES

1. van Thoai, N. and Robin, Y.,: Métabolisme des derivés guanidylés. Biochim. Biophys. Acta 52, 221-233 and 233-240 (1961).
2. van Thoai, N., Huc, C., Pho, D.B. and Olomucki, A.: Octopine dehydrogenase: Purification et propriétés catalytiques. Biochim. Biophys. Acta 191, 46-57 (1969).
3. Olomucki, A., Huc, C., Lefébure, F. and van Thoai, N.: Octopine dehydrogenase - Evidence for a single-chain structure. Eur. J. Biochem. 28, 261-268 (1972).
4. Thomé-Beau, F. and Olomucki, A.: Presence of a single essential histidyl residue in Octopine dehydrogeanse as shown by photooxydation. Eur. J. Biochem. 39, 558-562 (1973).
5. Biellmann, J.-F., Branlant, G. and Olomucki, A.: Stereochemistry of the hydrogen transfer to the coenzyme by Octopine dehydrogenase. FEBS Letters 32, 254-256 (1973).
6. Oriol, C. and Olomucki, A.: Spectropolarimetric studies of binary and ternary complexes of Octopine dehydrogenase. Eur. J. Biochem. 29, 288-292 (1972).
7. Luisi, P.L., Olomucki, A., Baici, A. and Karlovic, D.: Fluorescence properties of Octopine dehydrogenase. Biochemistry 12, 4100-4106 (1973).
8. Pho, D.B., Olomucki, A., Huc, C. and van Thoai, N.: Spectrophotometric studies of binary and ternary complexes of Octopine dehydrogenase. Biochem. Biophys. Acta 206, 46-53 (1970).

9. Baici, A., Luisi, P.L., Olomucki, A., Doublet, M.-O. and Klincak, J.: Influence of ligands on the coenzyme dissociation constants in Octopine dehydrogenase. Eur. J. Biochem. 46, 59-66 (1974).
10. Olomucki, A., Thomé-Beau, F., Biellmann, J.-F. and Branlant, G.: Study on coenzyme binding site of Octopine dehydrogenase using analogues of NAD^+. Eur. J. Biochem. 56, 109-116 (1975).
11. Doublet, M.-O. and Olomucki, A.: Investigations on the kinetic mechanism of Octopine dehydrogenase. Eur. J. Biochem. 59, 175-183 (1975).
12. Doublet, M.-O., Olomucki, A., Baici, A. and Luisi, P.L.: Investigations on the kinetic mechanism of Octopine dehydrogenase. Eur. J. Biochem. 59, 185-191 (1975).
13. Luisi, P.L., Baici, A., Olomucki, A. and Doublet, M.-O.: Temperature-determined enzymatic functions in Octopine dehydrogenase. Eur. J. Biochem. 50, 511-516 (1975).
14. Hochachka, P.W. and Somero, G.N.: Biochemical adaption to the environment. In Fish Physiology VI, 99-156 (Hoar, W.S. and Randall, D.J. eds., 1971) Academic Press N.Y.
15. Baldwin, J. and Hochachka, P.W.: Functional significance of isoenzymes in thermal acclimatization. Biochem. J. 116, 883-887 (1970).
16. Strickland, E.H., Horwitz, J. and Billups, C.: Fine structure in the near-ultraviolet circular dichroism and absorption spectra of tryptophan derivatives and Chymotrypsinogen A at 77° K. Biochemistry 8, 3205-3213 (1969).
17. Luisi, P.L., Baici, A., Bonner, F.J. and Aboderin, A.A.: Relationships between fluorescence and conformation of εNAD^+ bound to dehydrogenases. Biochemistry 14, 362-368 (1975).
18. Levy, H.R. and Vennesland, B.: The stereospecificity of enzymatic hydrogen transfer from diphosphopyridine nucleotide. J. Biol. Chem. 228, 85-91 (1957).
19. Baici, A., Luisi, P.L. and Attanasi, O.: The kinetic mechanism of Horse liver alcohol dehydrogenase with thio-NAD as coenzyme. J. Mol. Catal. 1, 223-224 (1975/76).
20. Joppich-Kuhn, R. and Luisi, P.L., in preparation.
21. Karlovic, D., Amiguet, P., Bonner, F.J. and Luisi, P.L.: Spectroscopic investigations of binary and ternary coenzyme complexes of Yeast alcohol dehydrogenase. Eur. J. Biochem. 66, 277-284 (1976).

Received February 1, 1977

DISCUSSION

Werber: Are the rates of binding and dissociation ("on" and "off") of the coenzyme also temperature independent? Or maybe there is a compensation effect between the temperature dependence of these two processes?

Dunn: Have you been able to determine the specific "on" and "off" rate constants for coenzyme binding as a function of temperature?

Luisi: I assume there is a compensation. It should be pointed out that the determination of the rate-off of coenzyme (for example with replacement techniques using a fluorescence stopped-flow apparatus) is very difficult, if not impossible. In fact the rate off of coenzyme must be at least as large as the turnover number of the enzyme, i.e. over 1000 sec^{-1} which is too fast for our stopped-flow. T-jump techniques cannot be applied (as K_d is temperature independent). The only possibility then is to measure by fluorescence stopped-flow the bimolecular rate of complex formation - but I suspect that to determine the precise temperature dependence of this chemical step may prove rather difficult.

Koberstein: It is indeed tempting to look for a correlation between the stereospecificity of hydrogen transfer and the sign of the extrinsic Cotton effect of the dihydronicotinamide chromophore in enzyme-NADH complexes. It turns out, however, that there is no strict correlation, since e.g. LDH, known to be A-specific, induces the same type of extrinsic positive Cotton effect as octopinde dehydrogenase or glutamate dehydrogenase which are both B-specific. Thus, size and sign of the induced Cotton effect is not only determined by the orientation of the dihydronicotinamide ring on the enzyme surface, but also by the "frozen" angle of rotation around the 1'glycosidic bond.

Luisi: Thank you for the interesting information on LDH. This example alone would indeed show the inutility of any generalization attempt between stereospecificity and dichroic effect of bound NADH. I would like however to point out that when I mentioned the possibility that the different dichroic sign of bound NADH in liver ADH and octopine dehydrogenase could be ascribed to the different stereospecificity, I had only in mind the different angle of rotation around the 1'glycosidic bond. The X-ray data of Rossmann and collaborators have for example shown that the conformation of bound coenzyme to GAPDH and LDH are the same, with the exception of a 180^{0} rotation of the nicotinamide ring around the 1'glycosidic bond. It would be interesting to compare in detail the circular dichroism spectra of bound NADH in these two systems.

Olomucki: I would like to add some information concerning this monomeric dehydrogenase.

Comparing the amino acid composition of octopine dehydrogenase with some other dehydrogenases, an interesting feature is the relative constancy of the amount of sulfur and dicarboxylic amino acids, a relatively great number of aromatic amino acids (Tyr+Trp+Phe = 76 per 100000 g) and exceptionally high content of proline (55 residues per 100000 g).

The specificity of octopine dehydrogenase is very narrow: L-arginine can be replaced only by canavanine and homoarginine (weakly) and pyruvate by α-ketobutyrate. The specifiticy as coenzyme is also rather strict.

Results of the investigations by spectroscopic methods (ref. 10), indicate that the pyridine and adenine parts are both involved in the binding of the coenzyme. The nicotinamide binding site of octopine dehydrogenase seems to be the most specific and restricted one among the dehydrogenases so far described. The binding of ADP ribose is sufficient to induce local conformational changes necessary for the good positioning of substrates.

Investigation of the kinetic mechanism of octopine dehydrogenase by initial velocity and product inhibition studies (ref. 11) of the reaction are consistent with a sequential ordered bi-ter mechanism. NAD^+ binds first to the enzyme followed by D-octopine, and the products are released in the order L-arginine, pyruvate and NADH.

The use of stopped-flow and isotope replacement techniques (ref. 12) allowed us to locate the rate limiting steps for enzyme turnover.

Under conditions at large excess of substrate and coenzyme, the rate-limiting step is located after an intial rapid hydrogen transfer; when the substrate concentration is below its K_m, the rate-limiting step is largely governed by the hydrogen transfer step.

Woenckhaus: Bromopyruvate reacts with LDH modifying a cysteine residue but in presence of NAD^+ there is only a modification of His 195. The substrate analog is faster bound in presence of NAD^+. Is it possible to modify the pyruvate binding place with this analog?

Olomucki: In octopine dehydrogenase there are two reactive -SH groups, and only one - the essential one - can be protected by NAD(H). This enzyme has also two equally reactive histidine residues. A priori it seems difficult to obtain a specifically labeled His peptide, but we will attempt to do that in the near future.

Holbrook: May I ask Dr. Olomucki if she has followed up her early experiments (which showed that this was a histidine enzyme) by attempting to isolate a covalently modified histidine peptide, perhaps using bromopyruvate (as did Woenckhaus with another pyruvate requiring dehydrogenase).

Olomucki: We did not perform yet a covalent modification of the essential histidine, we only confirmed by Rose Bengal photo-oxydation experiments that octopine dehydrogenase contains only one essential histidine residue.

AN OIL-WATER-HISTIDINE MECHANISM FOR THE ACTIVATION OF CO-ENZYME IN THE α-HYDROXYACID DEHYDROGENASES

David M. Parker and J. John Holbrook
Department of Biochemistry, University of Bristol,
Medical School, Bristol BS8 1TD, U.K.

Catalysis by the simple NAD^+-dependent dehydrogenases has been usefully considered as two processes: firstly the binding of the reactants to the enzyme surface and secondly their activation and interconversion (Pfleiderer, 1963).

In lactate dehydrogenase the binding of NADH is rapid, bimolecular and without kinetically detectable rearrangements of the protein. NAD^+ also binds rapidly (k_{on} - 5.10^6 $M^{-1}.s^{-1}$; Parker, 1976) and this is consistent with the observation at 0.5 nm resolution that this coenzyme does not alter the conformation of the protein (Chandrasekhar _et al_., 1973). For other simple dehydrogenases the distinction between binding and activation is not always so clear cut. In both supernatant malate dehydrogenase (Lodola, 1976) and yeast alcohol dehydrogenase (Dickenson & Dickinson, 1977) there is a slight shift in the apparent pK_a of an essential histidine residue when the coenzyme NADH is bound. Horse liver alcohol dehydrogenase represents the extreme case where binding and activation of NAD^+ occur together. In this enzyme the binding of NAD^+ can release a proton from the enzyme in such a way that the oxidation of ethanol can then take place without further liberation of a proton (see the paper by Shore in this volume). The generation of the "hot potato" (Perham, 1975) before the alcohol binds could account for the relatively low substrate specificity of this enzyme.

The main theme of this paper is an attempt to understand the process of coenzyme and substrate activation in lactate and malate dehydrogenases. Ideally for lactate dehydrogenase this understanding should have been possible from a comparison of the sequence of amino acids, the three dimensional electron density map and the solution properties. An initial comparison (Holbrook et al., 1975) did not result in entirely satisfactory explanations of all the enzyme properties. One problem was that the structure, with the then known sequence, did not explain the very large difference in the affinities of the enzyme for NAD^+ and NADH (NADH binds to the enzyme 500-fold tighter than NAD^+ (Stinson & Holbrook, 1973)). The structure placed glutamate-140 close to the N^1-position of the bound pyridinium ring in NAD^+ and this interaction should have resulted in NAD^+ binding tighter than NADH; quite the reverse of what was actually observed in solution.

This anomaly prompted us to investigate in detail those ionic species which could be accommodated within stable ternary complexes of lactate dehydrogenase. The preliminary results (Shore et al., 1975) of these investigations showed that stable or transiently stable ternary complexes were formed when there was an overall zero nett charge (summed over the nicotinamide rind of the cofactor, the substrate or inhibitor and the imidazole of histidine-195). The nett zero charge was observed with complexes containing both NAD^+ and NADH and with complexes with singly or doubly negatively charged ligands. This observation caused us to suggest that the region of the active centre around the nicotinamide ring was non-aqueous in the complexes that we had examined; for had solvent water been available then it should have been possible to accommodate non-zero charges by solvation.

This conclusion raises apparent difficulties. The crystallographic study of the structures of the enzyme in ternary complexes with NADH and with NAD^+ (Hackert et al., 1973; Chandrasekhar et al., 1973) did not reveal any differences in the position of the nicotinamide ring in relation to the protein depending upon whether it was NAD^+ or NADH. Thus if glutamate-140 provided the negative counter-ion for the pyridinium cation in bound NAD^+ then how could that counter-ion be accommodated when the NAD^+ was replaced by the neutral NADH ring? An un-neutralised negative charge would be difficult to accommodate in the non-aqueous region we envisaged.

These difficulties were avoided if it was proposed, as did Shore _et_ _al_. (1975), that no counter-ion is present on the protein to neutralise the positive charge on N^1-position of the nicotiniumamide ring in NAD^+. Not only does this proposal avoid the difficulties to which we refer above, but it also provides an explanation for the well established observation that NAD^+ when bound to the enzyme is activated for attack by anions at its 4-position. The proposal is formalised in what we have called the "oil-water-histidine" mechanism for the enzyme (Fig. 1). This mechanism attempts to unify structural and mechanistic findings about the enzyme to provide an explanation of how the 4-position of the bound-NAD^+ pyridinium ring can abstract H^- from the α-carbon of the bound lactate.

The first step (Fig. 1) is the rapid and reversible binding of NAD^+ to the protein surface. After this step the active centre is still exposed to solvent water and the pyridinium cation is either solvated or is paired with a buffer anion. Collision of lactate with the active centre introduces an extra negative charge.

The next step in the mechanism is the movement of a loop of polypeptide chain to entrap the coenzyme and the substrate and to dehydrate the region around the B-face of the pyridinium ring by pushing it onto a hydrophobic surface. The loop of polypeptide chains (residues 98-110) can either extend into the solution (in the apoenzyme) or reside close to the main surface of the protein (in ternary complexes). The position of the loop at the protein surface is stabilised by favour-

FIGURE 1

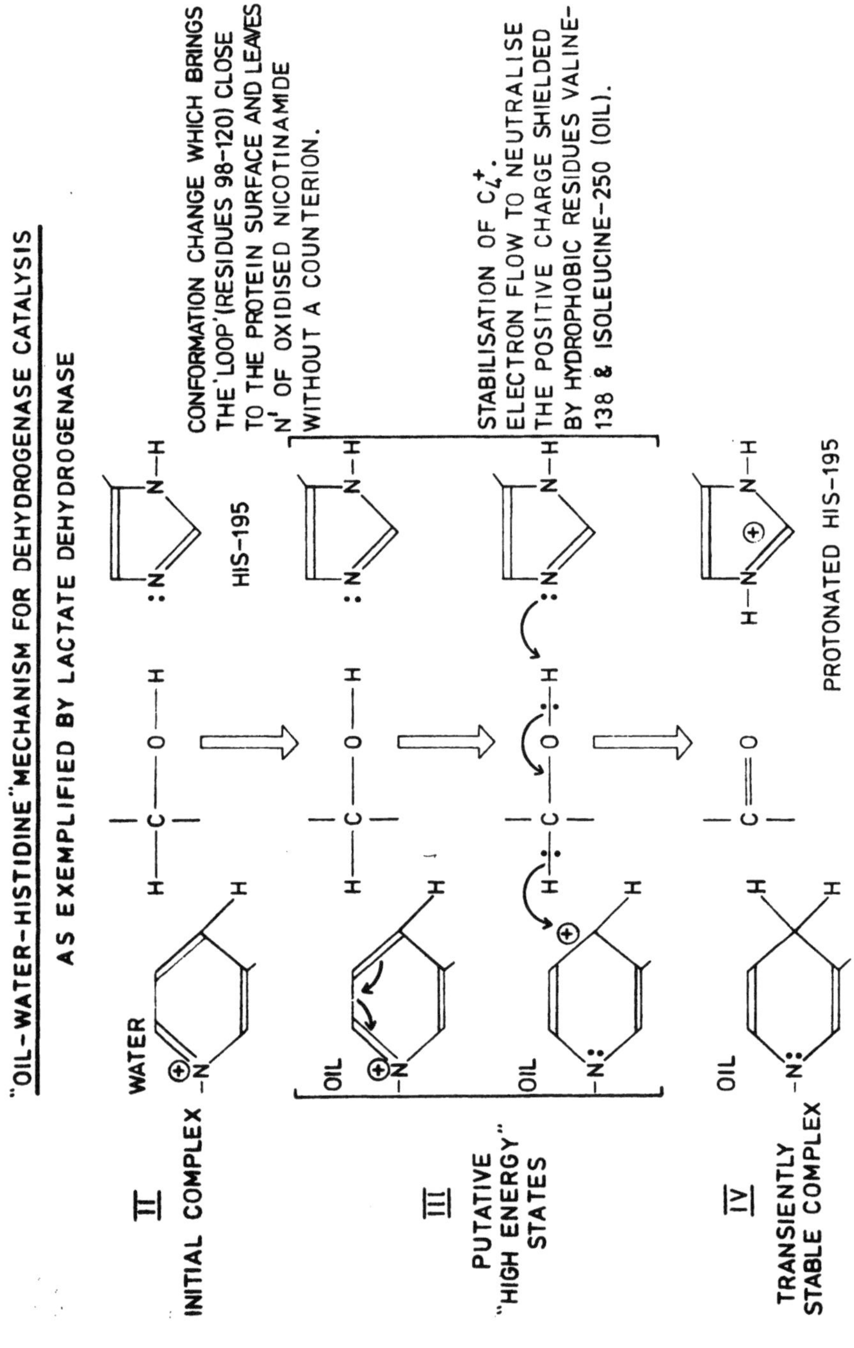
"OIL-WATER-HISTIDINE" MECHANISM FOR DEHYDROGENASE CATALYSIS
AS EXEMPLIFIED BY LACTATE DEHYDROGENASE
II
INITIAL COMPLEX
WATER
HIS-195
CONFORMATION CHANGE WHICH BRINGS THE 'LOOP' (RESIDUES 98-120) CLOSE TO THE PROTEIN SURFACE AND LEAVES N' OF OXIDISED NICOTINAMIDE WITHOUT A COUNTERION.
III
PUTATIVE "HIGH ENERGY" STATES
OIL
OIL
STABILISATION OF C4+. ELECTRON FLOW TO NEUTRALISE THE POSITIVE CHARGE SHIELDED BY HYDROPHOBIC RESIDUES VALINE-138 & ISOLEUCINE-250 (OIL).
IV
TRANSIENTLY STABLE COMPLEX
OIL
PROTONATED HIS-195

able ionic and hydrophobic interactions with the bound coenzyme and by ionic interactions with the negative charge of the substrate. It has been established crystallographically that in ternary complexes the pyridine ring is pushed further into the main domain of the protein (Chandrasekhar et al., 1973). This is supported by the enzyme in ternary complexes having a lower sedimentation coefficient.

The putative high energy state (Fig. 1), represented by the un-neutralised positive charge in an hydrophobic environment, is reduced by flow of electrons from the 4-position of the ring to leave the transient C_4^+-compound. This abstracts H^- from the bound lactate and induces a flow of electrons from the originally unprotonated histidine-195. This electron flow results in the formation of the transiently stable compound of the enzyme, NADH, keto acid and protonated imidazole. This compound has been observed with nitrophenylpyruvate (Holbrook & Stinson, 1973) and with pyruvate (Whittaker et al., 1975). The driving force for catalysis in this, the forward, direction is the very favourable interaction which can be made when the "loop" closes onto the protein surface. The favourable interactions are sufficient to balance the 10^5-fold unfavourable perturbation of the reaction equilibrium required to form NADH and pyruvate at neutral pH.

Quantitative estimation of the activation of enzyme-bound NAD^+.

It has been known for many years that NAD^+ and also pyridine model compounds will form adducts at the 4-position of the pyridine ring. NAD^+ will form an adduct with the sulphite dianion. This addition takes place at much lower sulphite concentrations when the NAD^+ is bound to the enzyme than when the NAD^+ is free in solution (Pfleiderer et al., 1957). Figures 2-5 show the method we have used to attempt to obtain a quantitative measure of the degree to which the sulphite adduct is stabilised on the surface of the enzyme. The concentration of the sulphite dianion needed to half saturate

FIGURE 2

THE PRINCIPLE OF MEASUREMENT OF THE ACTIVATION OF NAD^+ AT THE 4-POSITION

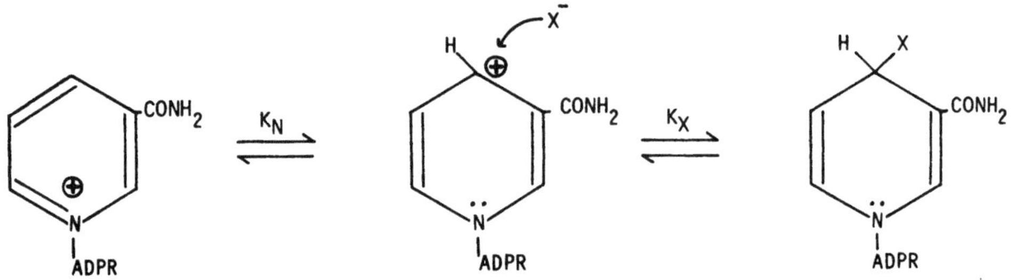

$$K_N = \frac{NAD^+}{\text{ACTIVATED } NAD^+} \qquad K_X = \frac{\text{ACTIVATED } NAD^+ . X^-}{\text{COMPLEX}}$$

K = APPARENT DISSOCIATION CONSTANT FOR X^- TO $NAD^+ = K_N . K_X$

IF K_X IS THE SAME ON AND OFF THE ENZYME, THEN

$$\text{ACTIVATION} = \frac{\text{K IN SOLUTION}}{\text{K MEASURED WITH ENZYME}} = \frac{K_N \text{ IN SOLUTION}}{K_N \text{ ON ENZYME}}$$

FIGURE 3

MEASUREMENT OF SULPHITE APPARENT DISSOCIATION CONSTANT IN SOLUTION.

$$HSO_3^- \rightleftharpoons SO_3^{2-} + H^+ + NAD^+ \overset{K_N}{\rightleftharpoons} NADSO_3^-$$

$$K_N = \frac{SO_3^{2-} . NAD^+}{\text{COMPLEX}} \qquad SO_3^{2-} = \frac{Ka}{[H] + Ka} . S_T$$

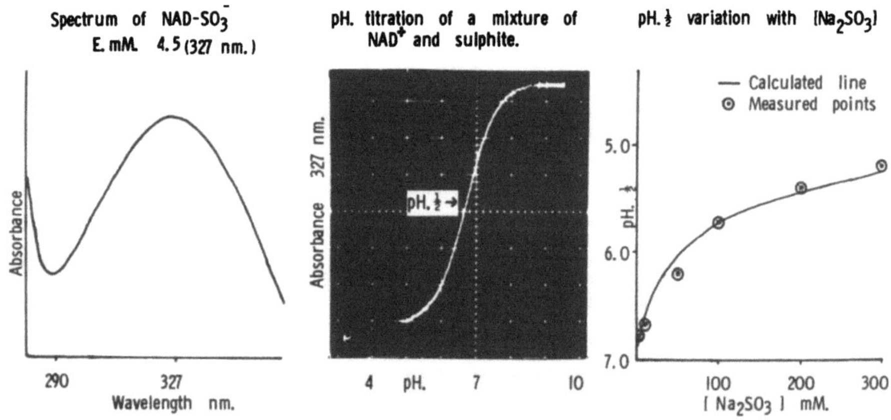

FIGURE 4(above) FIGURE 5(below)

MEASUREMENT OF SULPHITE APPARENT DISSOCIATION CONSTANT TO ENZYME-BOUND NAD^+.

$NAD\text{-}SO_3^-$ WHEN BOUND TO LDH QUENCHES TRYPTOPHAN FLUORESCENCE (F) BY ENERGY TRANSFER.
SATURATION OF PROTEIN WITH $NAD\text{-}SO_3^-$ $(\alpha) = (1 - F^{\frac{1}{4}})/(1 - 0.7)$

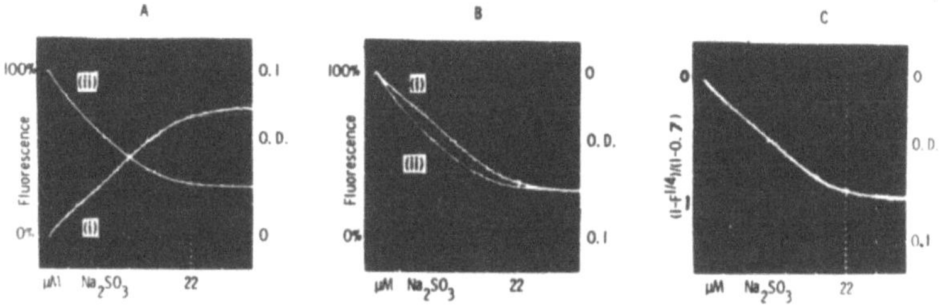

A. The continuous titration of 22μM PHLDH in 3mM NAD^+ at pH 6.3 with Na_2SO_3. The binding was monitored (i) by the linear extinction change at 327 nm and (ii) by the non-linear quenching of the fluorescence of the protein (ex. 297 nm, em. Wratten 18A).

B. Attempted superimposition of the two curves to emphasise the non-linear protein fluorescence quench due to energy transfer.

C. Successful superimposition of the two traces when fractional saturation (α) was calculated as $(1 - F^{\frac{1}{4}})/(1-x)$ using geometric quenching (Holbrook, 1972).

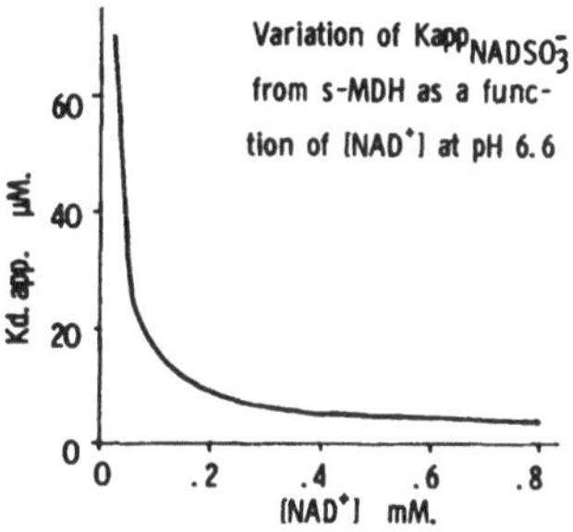

Variation of $Kapp_{NADSO_3^-}$ from s-MDH as a function of $[NAD^+]$ at pH 6.6

EQUILIBRIUM FORMATION OF ENZYME-NAD-SULPHITE

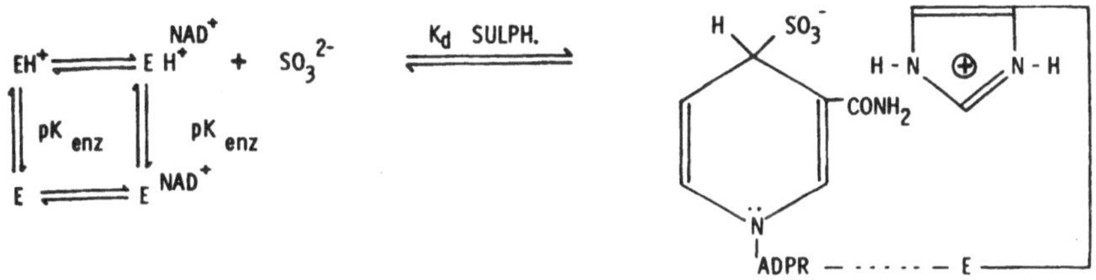

$$\alpha = \frac{[\,SULPH\,]\,/\,K_d\ SULPH}{(1 + pK\ enz\,/\,[H])\,(1 + K_d\ NAD\,/\,[NAD])\,(1 + [H]\,/\,pK\ sulph) + [\,SULPH\,]\,/\,K_d\ SULPH}$$

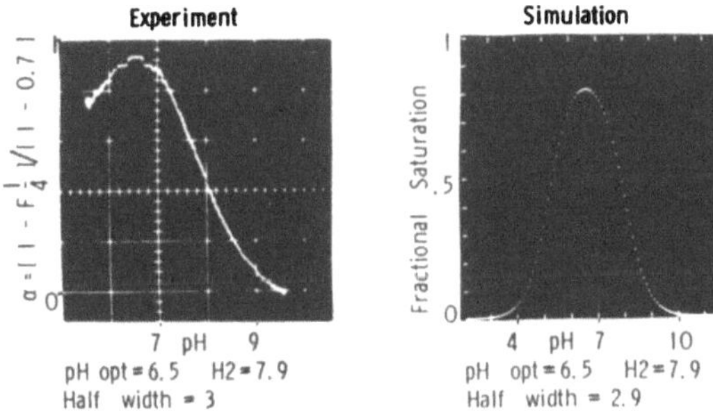

$NAD^+ = 10\mu M$
$Na_2SO_3 = 10\mu M$
Sulphite pK = 6.9
Kd. NAD^+ = 0.18 mM
Kd. NAD. SO_3^- = 80 nM
pK Enzyme = 6.1

NAD^+ in free solution is about 10 mM (Fig. 3). When NAD^+ is bound to the enzyme the measurement is more complex. The formation of the NAD^+-sulphite adduct on the enzyme can be detected from the decrease in the intrinsic fluorescence of the protein. After correction for energy transfer effects (Holbrook, 1972) the decreased tryptophan fluorescence is a quantitative measure of the degree of saturation (α) of the nucleotide binding sites of the protein with the NAD^+-sulphite adduct (Fig. 4). However the degree of saturation of the enzyme with the adduct does not follow a simple pH titration curve (as would be predicted if $HSO_3^- = SO_3^= + H^+$ was the only pH dependent step involved). In contrast a bell-shaped curve is observed and this indicates that, in addition to the sulphite equilibrium, a group on the protein with pK = 6.1 must be protonated for the stable complex to form (Fig. 5). The concentration of the sulphite danion calculated to half suturate NAD^+ bound to the enzyme in which histidine-195 is protonated is 80 nM. Thus the NAD^+ ring is activated 10 mM/ 80 nM = 125 000 times. It is suggested that this very large activation reflects the fact that when bound in a complex with the "loop" down the NAD^+ ring is held on a hydrophobic surface which transiently stabilises the positive charge on the C_4-atom (Shore et al., 1975). Stabilisation of the C_4^+-compound also accounts for the facilitated addition of other anions, such as enolpyruvate, cyanide, sulphide and hydroxylamine in the presence of enzyme.

No counter-ion for the positive charge in bound NAD^+.

The oil-water-histidine mechanism as originally proposed required that there was no counter-ion to pair with the positive charge on the N^--position of the pyridinium ring in enzyme-bound NAD^+, in spite of the apparent presence of glutamate-140 close to this position (Shore et al., 1975). However since that time Kiltz et al. (1977) have determined the primary structure of porcine lactate dehydrogenases H_4 and M_4. In these sequences there is no negative charge in a position equivalent to that occupied by glutamate-140 in the dogfish preliminary sequence. The residue which occupies this position is asparagine. The peptides in dogfish M_4 lactate - dehydrogenase can also be arranged to give a very strong homology to that in the porcine enzymes (Kiltz et al., 1977). Dr. S. Taylor (University of California at San Diego) has kindly communicated to us the revised sequence of the dogfish enzyme (Taylor S. (1977) J. Biol. Chem., in press). As predicted by the oil-water-histidine mechanism the residue which originally was glutamate-140 is now the neutral asparagine-138.

Other dehydrogenases.

Does the oil-water-histidine hypothesis apply to other dehydrogenases? Soluble malate dehydrogenase is closely related in its three dimensional structure to lactate dehydrogenase (Webb et al., 1973) especially in the conformation in which NAD^+ is bound. NAD^+ in this enzyme is also bound much less tightly than NADH (Holbrook & Wolfe, 1972). NAD^+ when bound to the enzyme can add sulphite 10^5-times more easily than free NAD^+ (Shore et al., 1975; Lodola, unpublished observations). The facilitated addition of sulphite depends upon the presence of a protonated enzyme group with apparent pK = 6.3 and to judge by chemical modification results this group may be a histidine. The preliminary sequence results from the Banaszak-Bradshaw groups in St. Louis suggest that supernatant malate dehydrogenase contains a histidine residue in a sequence of amino acid residues which is strongly homologous to the sequence around histidine-195 in lactate dehydrogenases. All these observations are just as would be expected if the oil-water-histidine mechanism applied to malate dehydrogenase. It will be informative to examine the sequence of amino acids around the bound nicotinamide ring in this enzyme in an attempt to find a counter-ion for the pyridinium cation.

In the case of alcohol dehydrogenases it is unlikely that coenzyme activation is brought about by the same mechanism as in the two α-hydroxyacid dehydrogenases. In these enzymes the activation appears to involve the zinc atom of the enzyme (see the papers by Shore and by Brändén in this volume). A mechanism similar to oil-water-histidine may apply to beef liver glutamate dehydrogenase. In this enzyme NAD^+ is bound less tightly than NADH although the situation is complicated by negative interactions (Dalziel, 1975) between the NAD^+ sites. However the bound NAD^+ is activated for the addition of sulphite (Wallis, 1972). There is no evidence that a histidine is required in this activation and indeed lysine-126 has often been implicated in the reactions of substrate binding to this enzyme.

The "oil-water-histidine" mechanism offers a structural explanation for many of the solution catalytic properties of lactate and malate dehydrogenases. The exploitation of a hydrophobic-hydrophilic boundary to generate an electron flow has formal similarity to the charge-relay mechanism advanced by Blow et al. (1969) to explain serine proteinase catalysis. The requirement for a stable hydrophobic-hydrophilic boundary could be a further rationalisation for the existence of proteins which are far larger than their active centres.

The work reported in this paper has been funded by generous grants from The Science Research Council (London). Special fluorescence instrumentation has been funded by The Royal Society (London) and by a NATO grant to travel to Detroit. We should like to thank Dr. A. Lodola for permission to quote some of his as yet unpublished results on malate dehydrogenase.

REFERENCES

Chandrasekhar, K., McPherson, A., Adams, M.J. & Rossmann, M. G. (1973) J. Molec. Biol. 76, 503

Dalziel, K. (1975) The Enzymes, 11a, 1

Dickenson, C.J., & Dickinson, F.M. (1977) Biochem. J. 161, 73-82

Hackert, M.L., Ford, G.C. & Rossmann, M.G. (1973) J. Molec. Biol. 78, 1665

Holbrook, J.J. (1972) Biochem. J. 128, 921

Holbrook, J.J., Liljas, A., Steindel, S.J. & Rossmann, M.G. (1975) The Enzymes, 11a, 191

Holbrook, J.J. & Wolfe, R.G. (1972) Biochemistry 11, 2499

Kiltz, H.-H., Keil, W., Griesbach, M., Petry, K. & Meyer, H. (1977) Hoppe-Seyler's Z. Physiol. Chem. 358, 123

Lodola, A. (1976) Ph.D. Thesis, University of Bristol

Parker, D.M. (1976) Unpublished results

Perham, R.N. (1975) Phil. Trans Roy. Soc. (London) Series B 272, 123

Pfleiderer, G. (1963) Colloq. Ges. Physiol. Chem. 14, 300

Pfleiderer, G., Jeckel, D. & Wieland, Th. (1957) Biochem. Z. 329, 104

Pfleiderer, G., Jeckel, D. & Wieland, Th. (1957) Biochem. Z. 329, 370

Shore, J.D., Weidig, C.F., Lodola, A., Parker, D.M. & Holbrook, J.J. (1975) in "Enzymes. Electron Transport System". Volume 40 Proc. FEBS Symp. XIIth, North Holland, Amsterdam, p. 63

Stinson, R.A. & Holbrook, J.J. (1973) Biochem. J. 131, 719

Wallis, R.B. (1972) Ph.D. Thesis, University of Bristol

Webb, L.E., Hill, E.J. & Banaszak, L. (1973) Biochemistry 12, 5101

Whittaker, J.R., Yates, D.W., Bennett, N.G., Holbrook, J.J. & Gutfreund, H. (1974) Biochem. J. 139, 677

Received February 7, 1977

DISCUSSION

Brändén: Your theory really fits very beautifully to what is known about the structure and solution properties of liver ADH. The B side of the nicotinamide ring is in a very hydrophobic region. There are no net charges in the active site region of the apoenzyme. Shore has shown that when NADH binds, a base is deprotonated,presumably the zinc bound water, so that the charges are balanced. Inhibitor studies have shown that the net negative charge of coenzyme and inhibitors is zero, for strong complex formation, e.g. NAD^+ and fatty acids form ternary complexes, as well as NADH and fatty acid amides, whereas the reverse combinations form much weaker complexes.

Thus with the exception of the specific role of histidine, your oil drop mechanism is in full agreement with available data on liver ADH.

Holbrook: I certainly agree that the mechanism of alcohol and lactate dehydrogenases are very similar if it is accepted that the zinc in ADH plays the same role as the histidine in LDH. It might even be that ADH is a histidine dehydrogenase in which a zinc (and serine) are interposed between a histidine (histidine-51) and the substrate.

Rossmann: John Holbrook has talked about an oily patch in LDH in the B side of the nicotinamide ring (this also exists in liver ADH). Carl Brändén has talked about a hydrophobic pocket for the substrate site in liver ADH (this does not exist in LDH). These two oily surfaces should not be compared.

Holbrook: I agree with Dr. Rossmann that there is danger of confusion. Perhaps the following sketch will help:

LDH Liver ADH

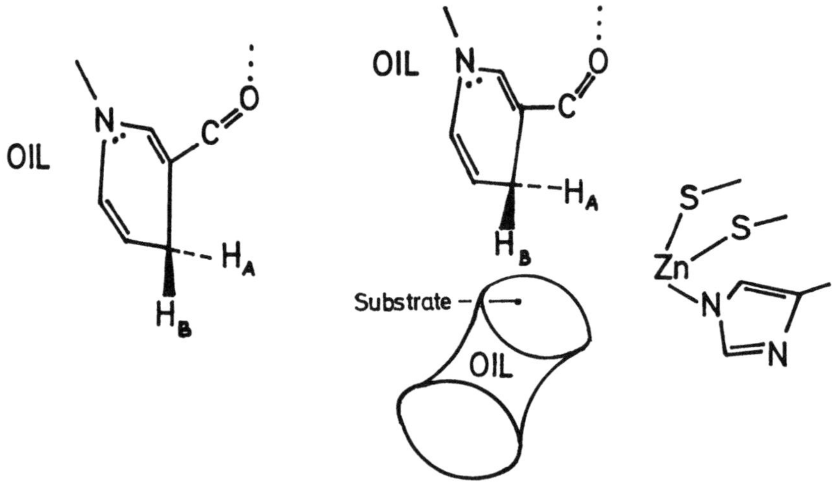

The mechanism about which I talked depends upon a hydrophobic area on the B face of the nicotinamide ring. This is present in LDH and, as I see from Dr. Brändén's comment, in liver ADH also. However in liver ADH there is a second important hydrophobic area: the "barrel" which selects for non-ionic substrates.

Hollaway: At first sight it might be considered that GAPDH is incomparable with your hypothesis as the affinities for NAD^+ and NADH are about the same with the free enzyme. However when you go to the thiohemiacetal and acyl enzymes the story is quite different and it could well be that these species bind NADH more tightly than NAD^+ and so conform to your oil-water model. I know that Trentham has shown that the affinity for NAD^+ is much lower in the acyl enzyme than in the free enzyme. Does anyone know the affinities of the acyl or thiohemiacetal species for NADH?

Seydoux: In contrast to NAD^+, NADH does interact strongly with acylated GAPDH. This fits nicely with Dr. Hollaway's suggestion. (Seydoux et al., Eur. J. Biochem. 64, 481-489 [1976]).

Holbrook: I did not include glyceraldehyde 3-phosphate dehydrogenase among those which are activated by hydrophobic contact. The number of groups present close to the nicotiniumamide ring in the active centre of this enzyme makes an unequivocable decision difficult. However, it is certainly true that the NAD^+/NADH affinities in the acyl enzyme are very different than in the binary complex. The presence of a histidine residue close to the 4-position of the nicotinamide ring has led several research groups to consider that this residue might initially carry the proton involved in the reaction.

Hinz: Would you agree with me that on the basis of a model which emphasises charge compensation inside a hydrophobic binding domain as an important prerequisite for ternary complex formation in lactate dehydrogenase, one should expect significant differences of that reaction in the case where there is no charge compensation, or a difference in charge? We have determined the thermodynamic parameters ΔG_B, ΔH_B, ΔS_B and ΔC_{P_B} for the following reactions involving pig heart muscle LDH:

a) the binding of oxalate in the presence of saturating concentrations of NAD^+ binding

b) the binding of oxalate in the presence of saturating concentrations of NADH.

The results of the binding studies and of the calorimetric measurements at 25° in 0.2 M potassium phosphate buffer pH 7.0 are:

	NAD^+, oxalate	NADH, oxalate
ΔG_B	-7.75	-4.59 kcal/mole ligand
ΔH_B	-7.6	-6.8 kcal/mole ligand
ΔS_B	+0.5	-7.4 cal/mole ligand·K
ΔC_{P_B}	339	-308 cal/mole ligand·K

(F. Schmid, Thesis, Regensburg 1977).

Quite clearly there are no large differences between the parameters recognisable , except for the free energy value. Is it possible to rationalise these experimental results with your model?

Holbrook: Yes, we certainly agree that there would be large differences in ternary complex formation if there was no charge compensation, and we also agree that there is a large difference in K_{eq} (i.e. ΔG^0) between oxalate binding to E-NAD^+ and to E-NADH. The apparent contradiction arises, however, from your assumption that oxalate is as the di-anion when it binds to E-NADH. We have attempted to measure the charge state of the ternary complex enzyme-NADH-oxalate and, although because of the high buffering of oxalate at pH 6 the results are not as precise as we have obtained with other ternary complexes, we conclude that the stable ternary complex has the structure

$$EBH^{+}\text{-}NADH\text{-}{COOH \atop COO^{-}} \quad .$$

The evidence for this conclusion is that we observe the binding of oxalate to the E-NADH complex continues to become tighter as the pH is decreased (K_{eq} = 2.1 mM at pH 8 and 0.15 mM at pH 6). Secondly, the experiment described below (cf. Fig.) qualitatively shows that the binding of oxalate to E-NADH is tightened not only with the protonation of a group of pK = 6.7 (His 195) but also as the pH is further decreased and approaches the pK_2 of oxalic acid (4.2). Unfortunately, it is hard to obtain results in the ideal region since the enzyme rapidly denatures at pH <5.3. Thus, in answer to your question, there is very little difference in the binding of the mono-anion of oxalic acid to E-NADH-BH^+ and the binding of the dianion of oxalic acid to E-NAD^+-BH^+. The observed difference comes from the proportion of oxalic acid which is in the correct ionisation state at pH 7. Your results are in excellent agreement with the oil-water-histidine mechanism. Oxalate will bind to the apoenzyme (D.A. Kolb & G. Weber (1975) Biochemistry 14, 4471). We have shown that at pH 8 oxalate (80 mM, 80 % saturation) causes the uptake of 0.74 mol H^+/subunit. These findings point to the danger in interpreting thermodynamic parameters in the absence of stoichiometry.

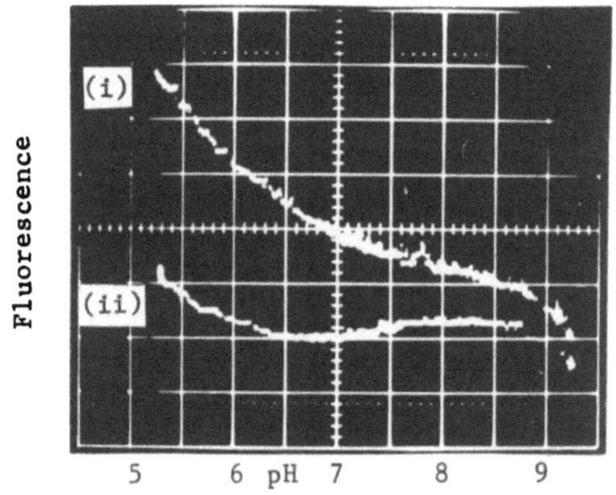

A titration of a mixture of 5 µM LDH and 10 µM NADH with pH observing enhanced nucleotide fluorescence (λex. = 366 nm, λem. = Wratten 98 filter) in the presence (i) and absence (ii) of 10 µM oxalate

Rossmann: Can you comment on the presence of Arg 109 in the active center of LDH (ii) the reason why the substrate cannot bind until after the finding of coenzyme?

Holbrook: Arg 109 is a residue on the "loop" and in the ternary complex is close to glutamate 107 on the same loop. One can only speculate, but it is possible that these two residues interact ionically. An even wilder speculation would be that the anionic substrate initially enters the active site as an ion pair with Arg 109 and that one of the "isomerisations" in the mechanism is the rearrangement of the Arg 109-substrate ion pair to give the substrate-Arg 171 ion pair and the Glu 107-Arg 109 ion pairs.

It was work from your laboratory which showed that there are two movements of the protein on going from the apo-structure to the ternary structure. One is the movement of the "loop". However thre is also a movement of the residues around His 195 and it may be that this rearrangement is the one which makes it possible for the substrate to bind after the coenzyme.

Blankenhorn: I am wondering about the possible function of Arg 109 in LDH. In free solution, the stability of SO_3^{--} or CN^- adducts to nicotinamide is proportional to the nicotinamide redox potential. In your enzyme the NAD^+-potential is increased by almost a factor of 10^3. The stability of the enzyme-sulfite complex, however, increases by a factor of 10^6. Wouldn't it be possible that this activation is due to Arg 109? To test this possibility one should look at the sulfite-enzyme complex which has the active center histidine alkylated. If NAD^+ is still bound you might be able to see whether it is still activated similarly as the native enzyme. If it is, activation by arginine would be a strong candidate in addition to the oil-water-histidine mechanism.

Holbrook: Perhaps Dr. Woenckhaus, who originally alkylated His 195 would like to answer this question.

Woenckhaus: LDH alkylated by bromopyruvate or 3-bromoacetylpyridine at the essential histidine residue binds further NAD^+ or NADH but we have not measured the K_D value. The abortive ternary complex from NAD^+, pyruvate and LDH is not formed after inactivation.

Holbrook: We can also confirm that enzyme in which His 195 is carbethoxylated will bind coenzymes quite normally. The activation of the NAD^+ ring has not however been examined because the acyl residue on the histidine occupies the position which would be used by the sulphite in a ternary complex.

Rossmann: We have studied apo LDH X (mouse) and apo LDH M_4 (dogfish). These structures are closely similar. We have also studied ternary LDH M_4 (dogfish) and ternary LDH H_4 (pig) complexes. These structures are also similar to each other. Thus, as far as LDH is concerned, the conformation is highly dependent on the state of the enzyme, not on species or isozymes. This is in surprising contrast to L. Banaszak's suggestion who proposes a difference in the behavior of soluble malate dehydrogenase and LDH. Perhaps it will be necessary to reexamine his proposed difference of soluble malate dehydrogenase to LDH.

Holbrook: I agree that the position of the loop in malate dehydrogenase and LDH is not the same in the apocrystals. However it is necessary that the loop move in malate dehydrogenase otherwise it is hard to see how the large coenzyme can enter the active site. Perhaps this emphasises that we badly need a physical probe which will allow us to interpret the position of the loop in the various binary and ternary complexes of these two kinds of dehydrogenases.

Jeck: The K_D of the LDH complex with dihydro-3-acetylpyridine butyl adenosine diphosphate (1) is 250 x larger as compared to the complex of AP-NADH (2). In contrast the K_D values of both LDH complexes with the oxidized analogs differ by a factor of 7. Can you explain this result without assuming a small movement of the functional riboside or of amino acid residues during the redox reaction?

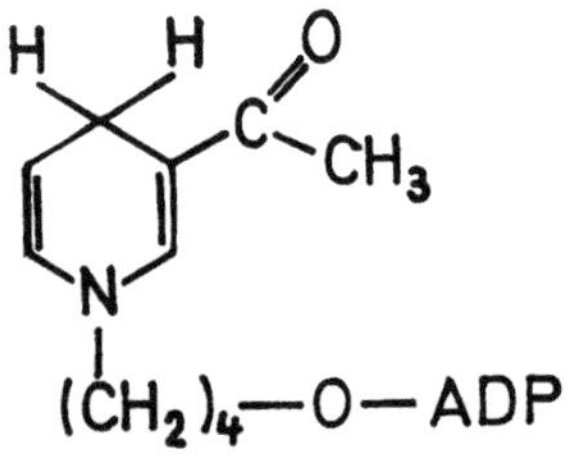

(1)

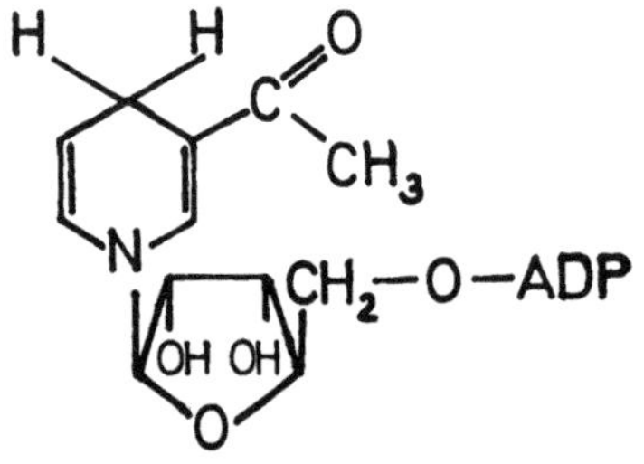

(2)

Holbrook: Compound 2 has five atoms between N^1 and the phosphate, whereas compound 1 has only four. Compound 1 is only bound to the enzyme poorly, with about the same affinity as AMP or ADP. This suggests that the $-(CH_2)_4$-nicotinamide moiety makes either no, or unfavorable, interactions with the protein and I see no reason to suppose that the oxidised nicotinamide ring in the analogue compound will be bound and orientated in the same position as in NAD^+.

Südi: It appears to me that your paper brings you in a rather awkward position. (1) You are trying to sell us the idea that available X-ray crystallographic data provide the final truth on fine details about the proton transfer involved in the LDH reaction. (2) The main point of your paper is, that in a critical region of the active site, X-ray crystallography has until very recently shown (among others) the presence of an extra methylene group (glutamate, instead of asparagine). Now, are you for or against the notion?

Holbrook: The main point of my talk was not that an extra methylene group was present in lactate dehydrogenase but rather that it was difficult to envisage an un-neutralised negative charge in position 140 in the ternary complex with NADH. It is not clear which notion you wish me to be for or against.

Engel: In relation to the comparison you make between LDH and glutamate dehydrogenase it may be of some interest to note that we find in a variety of stoichiometric substrate assays with NAD^+-dependent dehydrogenases: (1) That NAD^+ reacts with the hydrazine used to trap reaction products - presumably by reaction at the 4 position of the nicotinamide ring. (2) That this "blank" reaction is enhanced by certain dehydrogenases. Among these, lactate and glutamate dehydrogenase seem to give the most marked effect.

We have not, however, done a thorough systematic survey to establish which other enzymes have the same effect. I wonder whether you have such information for the sulphite addition reaction.

Holbrook: We have indeed shown, as would be expected if an oil-water-base mechanism were to apply to glutamate dehydrogenase, that in the bovine enzyme the NAD^+ when bound is activated for attack by SO_3^{--}. As yet we have not been able to measure the ionisations of the protein which contribute to the activation, nor to evaluate whether the base is Lys 126 (Holbrook, J.J. and Jeckel, R., Biochem. J. (1969) 111, 689-694), or some as yet undetected histidine (Wallis, R.B. and Holbrook, J.J. (1973) Biochem. J. 133, 183-187). There is of course the added complication that with this enzyme it is always hard to distinguish catalytic coenzyme binding from activator coenzyme binding.

Hemmerich: Dr. Dutler has put forward yesterday the interesting idea that an in-and-out of plane turn of the carbonamide group of nicotinamide might help to get the hydride on and off (cf. Table). Can this be correlated to the oil-water mechanism you propose?

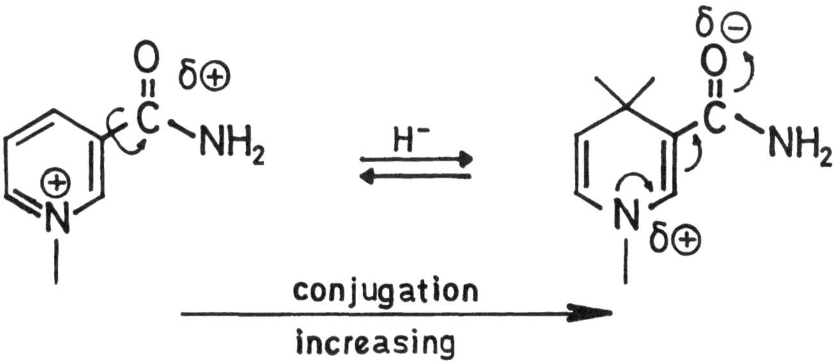

	Free State (Resting)	Enzyme Bound (Activated)	
NAD^+	less coplanar	more coplanar	$CONH_2$
NADH	more coplanar	less coplanar	

Holbrook: Interesting as I find the carboxyamide hypothesis I do not see how it accounts for the activation of the nicotiniumamide ring specifically in the 4-position rather than also in the 2- and 6-positions.

Concluding Remarks

Cees Veeger

Department of Biochemistry, Agricultural University,
Wageningen, The Netherlands

Looking back on the seven-and-a-half years between the first and the second pyridine nucleotide meeting here in Konstanz, very little seems to have changed. After almost every lecture hands were raised by people eagerly, either wanting to ask questions, or to present results supporting or contradicting the speaker. One major difference is however quite apparent: the much less aggressive way of the discussion during this meeting. It is my opinion, that the discussion is of more importance that the actual presentation of new data, since it develops new ideas for further research.

One can wonder about the need for a meeting of this kind and argue whether research in the field of pyridine nucleotides and pyridine nucleotide-dependent enzymes is of such relevance that this meeting was and a further meeting is needed. Expressing my own view, but it reflects the opinion of many participants, I believe there is a need for such meetings. Basic science has led in the past and will undoubtedly lead in the future to developments in the applied field. Just to mention an example: the rapidly-rising interest in the technological application of matrix-bound enzymes is in fact based on our knowledge of protein structure and on our insight in catalytic sites and mechanisms. Basic research in this field will be the source for chemical-technological applications of the future. Many of those applications even are not apparent to us today. Especially knowledge on biological redox-systems will lead to rapidly increasing insights in the prevention of pollution, and in the conversion of biological waste and probably of solar energy into useful products like CH_4, H_2 and NH_3.

It is, however, apparent from the data presented at this meeting that our insight in structure-function relations of enzymes has made enormous progress. Questions concerned with the conformation in which pyridine nucleotides are bound to dehydrogenases have been answered partly - at least for the diphosphopyridine nucleotide - by beautiful X-ray crystallographic studies. The conformation in which triphosphopyridine nucleotide is bound to enzymes will be known within a few years. Explanations for site- and stereo-specificity of reactions are available today. Detailed structural insights on active sites did lead to mechanistic proposals, interestingly enough by the X-ray crystallographers and not by kineticists or organic chemists.

Although some time was devoted to the chemistry of pyridine nucleotides and the hydride-ion transfer mechanism in catalysis brought forward as the most likely mechanism, in my opinion the problem has not been entirely solved. For instance in oxidative phosphorylation NADH is oxidized by an enzyme system containing flavin and non-heme iron protein. In this process radical-pair splitting takes place and the membrane is able to synthesize ATP. Whether this radical-pair splitting process occurs after hydride-ion transfer, from NADH to the flavin, or that it occurs by simultaneous attack of both the non-heme iron and the flavin on the NADH, like in succinate dehydrogenase, is not clear at all yet. Other evidence also points in that direction. For instance, we recently obtained evidence that the electron donor for nitrogenase in _A.vinelandii_ is flavodoxin hydroquinone. Electrons are transferred in one-electron transfer steps from flavodoxin hydroquinone - being itself oxidized to flavodoxin semiquinone - to the nitrogenase. Simultaneously we obtained evidence for a membrane-bound NADH-flavodoxin reductase. Thermodynamically the latter reaction is hardly feasible, because of the large difference in E_m of the redox couples ($NADH/NAD^+ \sim -320 mV$, flavodoxin hydroquinone/semiquinone ~ -490 mV at pH 7.0). It seems to us that such a reaction (which also involves radical-pair splitting) can only be achieved by a membrane-energization; both the membrane potential and pH-gradient could make circumstances proper for radical-pair splitting and a potential low enough to make formation of flavodoxin hydroquinone possible. In my opinion, in case a new conference on pyridine nucleotides will be organized in the future, some time must be devoted to such problems. Furthermore, as we have heard, data obtained from model redox-reactions point to the fact that in these pure systems, redox-reactions proceed via intermediates between the reactants, which means that some time must be devoted to pure organic chemistry.

Another interesting development which, as we have heard on this conference, might lead to insight in the dynamics involved in proteins, are pure calorimetric data. Although not-unequivocally the present evidence points to considerable contribution of vibronic modes in proteins, which has its implications for studies of the thermodynamics of protein unfolding and refolding and of subunit interaction. From this type of studies, in combination with the application of cross-linking techniques either to a matrix or to the interacting sites of subunits, one can expect very fundamental insights in structure/function relations in the near future. Especially in case these studies are carried out in combination with molecular physical techniques like NMR-, EPR- and fluorescence decay measurements. Although, as we heard, by observing with the latter technique more than one life-time in reduced pyridine nucleotides bound to enzymes, the problem of interpreting the results in terms of either non-equivalency of the sites or excited state phenomena has to be solved. But apart from the latter question, which can be answered, it is my opinion that we start gradually in the near future to understand dynamics in proteins and its relation to protein structure and biocatalysis.

Overlooking the data presented at this meeting, I must come to the conclusion that lactate dehydrogenase is a normal enzyme in terms of equivalency of subunits. Liver alcohol dehydrogenase could be a more complex enzyme, depending on the view of the speaker. At the moment I am inclined to believe one of the speakers who told me privately that abnormal kinetics might be observed as a stopped-flow apparatus becomes available with a dead-time of a few micro-seconds.

Glyceraldehyde 3-phosphate dehydrogenase is really a complex enzyme in all aspects. It is clear even after seven years, that the study of glyceraldehyde 3-phosphate dehydrogenase raises differences of opinion. Seven years ago we had exhausting discussions on either the Monod-Wyman-Changeux model or the Koshland-Nemethy-Filmer model as base for the regulatory properties, today the question is 'half-of-the-site to be or not to be'. I am used to work with coloured proteins, but I became worried after finding out here that this colourless enzyme contains subunits which have different colours! Having followed the interesting data presented by many speakers, I have come to the conclusion that when we succeed to isolate this enzyme from a lobster which grows in boiling Dead Sea water we will know all the answers and have solved the standing question: is the enzyme half sited or are we near-sighted?

Studies with glutamate dehydrogenase, at least the bovine enzyme, still leads to frustrations. I came to a similar conclusion about six years ago here in Konstanz, when I succeeded in denaturing this enzyme easily, but never renaturing it. Nobody succeeded since then, which is good for me, but not for understanding the enzyme. But does not frustration preceed understanding in science? I have the feeling that the more excellent and advanced studies one undertakes, the more complex this enzyme becomes. Questions I am left with after the meeting are: is the hexamer built up out of two layers of trimers or out of three times two dimers or out of a tetramer plus a dimer. The latter structure could account for the accumulating evidence for four sites saturating differently than the other two sites, as GTP-binding studies and spin-label studies seem to suggest.

In these past seven years the field of dehydrogenase has matured tremendously. We moved away from a rather descriptive phase, and made a start to understand the factors involved in structure-function relations. Nevertheless, we need the phenomenological description as a start for the more sophisticated approach. It is also clear to me that our present state of knowledge about the field of pyridine nucleotides and pyridine nucleotide-dependent enzymes was only made possible by the multi-disciplinary studies as presented here.

During this meeting it became apparent to me that there is strong support to have a third meeting to be held in Konstanz in about four years. Why in Konstanz? Everybody agrees that the excellent organization by Horst Sund is the major reason. The fact that only few of the participants left before the end of the meeting - even after four-and-a-half full days of hard working - points to the success of but also the need for such a meeting. An additional factor to the success of this meeting was the presence of Mrs. Silvia Lecointre, who got never tired from demanding participants, invading her secretarial desk during every break with questions and wishes. Thus I hope that we meet in good health and spirit in four years from now, to discuss during the 3rd Konstanzer Symposium on Pyridine Nucleotides and Pyridine Nucleotide-Dependent Enzymes, our latest results and newest ideas.

Index of Contributors

* Symposium paper

SUBJECT INDEX

Bei Fragen zur Produktsicherheit wenden Sie sich bitte an:
If you have any questions regarding product safety,
please contact:

Walter de Gruyter GmbH
Genthiner Straße 13
10785 Berlin
productsafety@degruyterbrill.com